VII International Congress of Engineering Physics 2014

Journal of Physics: Conference Series
Volume 582

Mexico City, Mexico
24 - 28 November 2014

Editors:

Ernesto Rodrigo Vazquez Ceron
Luis Enrique Norena Franco
Alberto Rubio Ponce
Oscar Olvera Neria
Jamie Granados Samaniego
Roberto Tito Hernandez Lopez
Anatolio Martinez Jimenez
Eusebio Guzman Serrano

ISBN: 978-1-63439-913-5
ISSN: 1742-6588

Printed from e-media with permission by:

Curran Associates, Inc.
57 Morehouse Lane
Red Hook, NY 12571

Printed by Curran Associates, Inc. (2015)

For permission requests, please contact the Institute of Physics
at the address below.

Institute of Physics
Dirac House, Temple Back
Bristol BS1 6BE UK

Phone: 44 1 17 929 7481
Fax: 44 1 17 920 0979

techtracking@iop.org

Additional copies of this publication are available from:

Curran Associates, Inc.
57 Morehouse Lane
Red Hook, NY 12571 USA
Phone: 845-758-0400
Fax: 845-758-2634
Email: curran@proceedings.com
Web: www.proceedings.com

Table of contents

Volume 582

VII International Congress of Engineering Physics
24–28 November 2014, Mexico City, Mexico

Accepted papers received: 18 December 2014
Published online: 14 January 2015

Introduction

In the frame of the fortieth anniversary celebration of the Universidad Autónoma Metropolitana and the Physics Engineering career, the Division of Basic Science and Engineering and its Departments organized the "**VII International Congress of Physics Engineering**". The Congress was held from 24 to 28 November 2014 in Mexico City, Mexico. This congress is the first of its type in Latin America, and because of its international character, it gathers experts on physics engineering from Mexico and all over the globe.

Since 1999, this event has shown research, articles, projects, technological developments and vanguard scientists. These activities aim to spread, promote, and share the knowledge of Physics Engineering. The topics of the Congress were:

- Renewable energies engineering
- Materials technology
- Nanotechnology
- Medical physics
- Educational physics engineering
- Nuclear engineering
- High precision instrumentation
- Atmospheric physics
- Optical engineering
- Physics history
- Acoustics

This event integrates lectures on top trending topics with pre-congress workshops, which are given by recognized scientists with an outstanding academic record. The lectures and workshops allow the exchange of experiences, and create and strengthen research networks. The Congress also encourages professional mobility among all universities and research centres from all countries.

CIIF2014 Organizing and Editorial Committee

Dr. Ernesto Rodrigo Vázquez Cerón
Universidad Autónoma Metropolitana - Azcapotzalco
ervc@correo.azc.uam.mx

Dr. Luis Enrique Noreña Franco
Universidad Autónoma Metropolitana - Azcapotzalco
lnf@correo.azc.uam.mx

Dr. Alberto Rubio Ponce
Universidad Autónoma Metropolitana - Azcapotzalco
arp@correo.azc.uam.mx

Dr. Óscar Olvera Neria
Universidad Autónoma Metropolitana - Azcapotzalco
oon@correo.azc.uam.mx

Professor Jaime Granados Samaniego
Universidad Autónoma Metropolitana - Azcapotzalco
jgs@correo.azc.uam.mx

Dr. Roberto Tito Hernández López
Universidad Autónoma Metropolitana - Azcapotzalco
hlrt@correo.azc.uam.mx

Dr. Anatolio Martínez Jiménez
Universidad Autónoma Metropolitana - Azcapotzalco
amartinez@correo.azc.uam.mx

Dr. Eusebio Guzmán Serrano
Universidad Autónoma Metropolitana - Azcapotzalco
gse@correo.azc.uam.mx

Peer review statement

All papers published in this volume of *Journal of Physics: Conference Series* have been peer reviewed through processes administered by the proceedings Editors. Reviews were conducted by expert referees to the professional and scientific standards expected of a proceedings journal published by IOP Publishing.

Journal of Physics: Conference Series **582** (2015) 012001 doi:10.1088/1742-6596/582/1/012001

Fabrication and optical testing of hybrid sio2:azo-polymer based planar waveguides for nlo/shg laser emission

V Torres-Zúñiga[1], O G Morales-Saavedra[2], A L Pérez-Martínez[3]

[1] Facultad de Ciencias, Universidad Nacional Autónoma de México, Circuito Interior, Ciudad Universitaria, C.P. 04360, México D.F., México.

[2] Centro de Ciencias Aplicadas y Desarrollo Tecnológico, Universidad Nacional Autónoma de México, Circuito Exterior S/N, Ciudad Universitaria, A.P. 70-86, C.P. 04510, México D.F., México.

[3] División de Ciencias Basicas, Departamento de Química, Facultad de Ingeniería, Universidad Nacional Autónoma de México, Circuito Exterior S/N, Ciudad Universitaria, México D.F., México.

E-mail: vicentz@gmail.com

Abstract. Predesigned push-pull azo-dye polymers were homogeneously dispersed within a SiO_2 sol-gel matrix synthesized via the sonogel (SG) route. High-quality spin-coated films were obtained with these hybrid structures in the liquid sol-phase. The spectroscopic UV-Vis analyses reveal the appropriate insertion of these organic compounds within the highly pure SG-environment whereas the thermal (DSC) analysis and photoacoustic measurements evidence the thermomechanical stability of the amorphous hybrid layers. As the optical attenuation, refractive index and film thickness values of the obtained films are adequate for opto-electronic applications; these hybrid films were implemented to fabricate optical waveguiding prototypes. In this sense, functional planar waveguides were fabricated for nonlinear optical (NLO) applications after performing a molecular ordering via a corona DC-poling procedure in order to achieve a macroscopic polar order (ferroelectric and noncentrosymmetric arrangement of the organic chromophores). The poled films were then able to exhibit stable NLO-waveguiding effects as excited with a Nd:YAG laser system in order to generate second harmonic waves travelling within the planar layer.

1. Introduction

Nowadays linear and NLO applications require the development of active organic compounds in order to improve the efficiency of several optical effects for practical technological applications. In this sense, the push-pull functionalized azo-benzene polymers have received increasing attention in the last decade [1]. Among their attractive features, these compounds exhibit a rod-like dipolar structure which is convenient to perform a collective molecular alignment, achievable by applying strong DC- fields on the material. This gives rise to a macroscopic ferroelectric polar order which is equivalent to a non-centrosymmetric crystalline arrangement. In this kind of polar structures quadratic $\chi^{(2)}$-NLO effects such as second harmonic generation (SHG) are allowable [2]. Nevertheless, poled organic compounds are susceptible to thermal relaxation which tends to annihilate the induced NLO response. Different efforts, including molecular cross-linking, preparation of Langmuir-Blodgett films and the use of high glass transition polymers have been attempted in order to produce stable NLO-polymeric materials [2]. So far, a conclusive and universal methodology to permanently preserve the poled phase of organic materials is still under research. Given that our research group has recently achieved some new sol–gel technology (via a physicochemical sonogel route: catalyst-free sonolysis approach) for the synthesis of high quality mesoporous SiO_2 glasses, we have implemented this approach for the development of optically active organic-inorganic hybrid glasses. In the current case, we report on the fabrication of optically active sonogel hybrid films and demonstrate their plausible applicability in NLO waveguiding photonic devices implementing promising polymeric molecules with current potential applications in opto-electronic technologies. The sonogel route is a low-cost technique which has been successfully implemented to increase the thermal and mechanical stability of organic compounds embedded within a vitreous sol-gel matrix [1, 3]. The SG-glasses are optically transparent and chemically inert allowing the observation of the photophysical properties of the guest organic compounds in a quasi-pristine form. In this way, we fabricated hybrid planar waveguides implementing two predesigned azo-polymers. These organic materials were dispersed in the liquid sol-phase of the sonogel precursor and spin-coated onto ITO-glass substrates in order to obtain hybrid film layers ready for optical inspections.

2. Materials and experimental methods

Chemical structure of the implemented NLO azo-benzene polymers

We implemented two photo-active push-pull polymers: Poly(hexa-2,4-diynilene-1,6-dioxydibenzoate) and Poly(hexa-2,4-diynilene-1,6-dioxydicinnamate) [4]; named here PB (para-benzoate) and PC (para-cinnamate), respectively. Both reddish polymers contain rod-like molecular units comprising electron-donor and electron-acceptor functional groups linked by an azobenzene molecule (see Fig. 1); this molecular architecture gives rise to an electronic dipolar configuration, adequate to perform electrical poling procedures for collective molecular alignment.

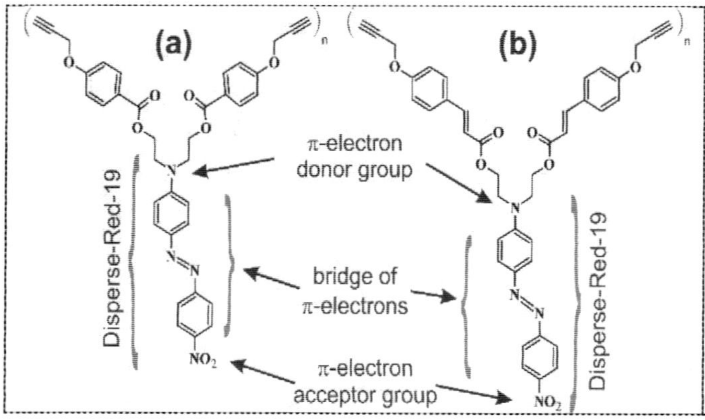

Figure 1. Molecular structure of: **a)** the PB and **b)** PC azo-polymers. These compounds contain delocalized electrons (π-electrons) and exhibit large dipolar moments.

2.2 Synthesis of the highly pure SiO_2–based sonogel

The sonogel synthesis is a catalyst-free sol-gel process [3]. Briefly, a stainless steel tip (Cole-Parmer- CPX) provides an ultrasonic irradiation (~3.2 W/cm^3 at 20◦kHz) on the interface of the two immiscible precursors: 25 mL of tetraethyl-orthosilicate (TEOS: Fluka 99% purity) and 25 mL of tri-distilled water, stabilized at 1°C. After 3 h of acoustical irradiation (with 5 s on/off intermittent sequences) and 24◦hrs of repose, two new immiscible phases emerge. Only the denser one is capable of producing (after drying/poly-condensation) a highly pure and transparent SiO_2 amorphous network.

2.3 Fabrication of organic-inorganic waveguiding hybrid structures

In order to fabricate organic-inorganic sonogel hybrid films, the chromophore doping must be performed in the liquid sol-phase. Hence, highly-loaded dopant dissolutions (DD) of the PB- and PC-polymers (in Tetrahydrofuran: THF) must be prepared (at 3.0 mg/mL). These DD must be filtered through 0.22 μm Teflon membrane, added and ultrasonically mixed to the liquid sonogel (sol-phase) in a 50-50% DD:sonogel-concentration. Undoped reference sonogel films (0.0% DD) were also prepared for reference and calibration spectroscopic purposes.

For the fabrication of planar waveguides we first deposited two thick layers of the pure sonogel (0.0% DD) onto Indium-Tin-Oxide-coated substrates (ITO: Aldrich 25 Ω/sq) via standard spin-coating procedures (each layer deposited at 100 rpm for 6 s; drying conditions: room temperature for 48 hrs). The obtained SiO_2 film showed an average thickness of ~3 μm (according to profilometry measurements), this layer plays the role of an optical buffer since its low refractive index (n_{SG}=1.31±0.05, measured by the Brewster's angle methodology) is essential in the cladding of waveguide. The second film (PB or PC active hybrid layer) was deposited with a (50.0-50.0% DD:sonogel, in sol-phase. Each layer deposited at 80 rpm for 4 s). The average film thickness of the hybrid layers was ~4 μm. The refractive index values of these layers were: n_{PB} = (1.48±0.06) and n_{PC} = (1.50±0.05). At this stage, the organic compounds are randomly oriented within the film geometry showing a quasi-isotropic macroscopic structure. Thus, in order to achieve a macroscopic noncentrosymmetric molecular alignment, as required for $\chi^{(2)}$-NLO effects, we implemented the corona-poling technique [2]. Here, the ITO-layer operates as the anode whereas a vertical non-oxidable iron needle placed above the active film layer of the sample (film-needle gap: 1 cm) works as electrode. The electrodes are then connected to a high voltage supply (at 5.2 KV) in order to apply a strong DC-electric field over the sample. During the whole electrical poling procedure the sample was mounted onto an electrical oven operated around the glass transition temperature (T_g) of the implemented azo-polymers (~180°C for ~1 hr) in order to allow highest molecular mobility for optimal collective alignment. Thereafter, the multilayered hybrid sample was cooled down to room conditions, keeping the DC-field on in order to freeze the collective molecular alignment within the sonogel network. In this way, the poled hybrid films were ready for morphological, photophysical and waveguiding characterizations.

2.4. Photophysical characterization methodology

Standard characterization methodologies were applied to undoped and hybrid sonogel film samples in order to determine their thermomechanical and optical properties: **i)** The absorbance UV-Vis analysis was performed using a double beam Shimadzu UV-Vis spectrophotometer (air as reference) within the 200–1100 nm spectral range. **ii)** The thermal and structural characterization to determine the T_g parameter and the degradation temperature of the organic compounds within the SG-environment were performed via the laser photoacoustic technique (LPA) [3] and Differential Scanning Calorimetry (DSC). **iii)** The NLO/SHG characterization was performed on the hybrid waveguides mounted onto a micrometric XYZ-stage; the active layers were directly excited with a pulsed laser beam (Continuum Nd:YAG: polarization state: P_{In}, at λ=1064 nm, pulse width: 15 ns). After overpass a linear polarizer (P_{Out}) and an interferometric filter, the SHG-output waves (at 532 nm) were collected by a photomultiplier tube connected to a digital oscilloscope. The planar waveguides were asymmetrically cut in order to perform SHG-intensity measurements as function of the optical path.

3. Results and discussion

UV-Vis spectroscopic measurements

The UV-vis spectra of the THF/polymer-based dissolutions present sharp and intense absorbance bands (see Fig 2) distinctive of organic compounds exhibiting efficient charge transfer properties [4]. In contrast, the pure reference sonogel film exhibits low absorbance at optical wavelengths and an intense UV-band which is typical of SiO_2 glassy materials [5]. Finally, as a result of the successful organic doping, the hybrid SG-samples exhibit the characteristic absorbance bands of their organic and inorganic constituents; with additional red-shifts of the main absorbance bands at optical wavelengths as a consequence of the extreme molecular packing (formation of head-to-tail aggregates) which is usual in the solid-state phase [3].

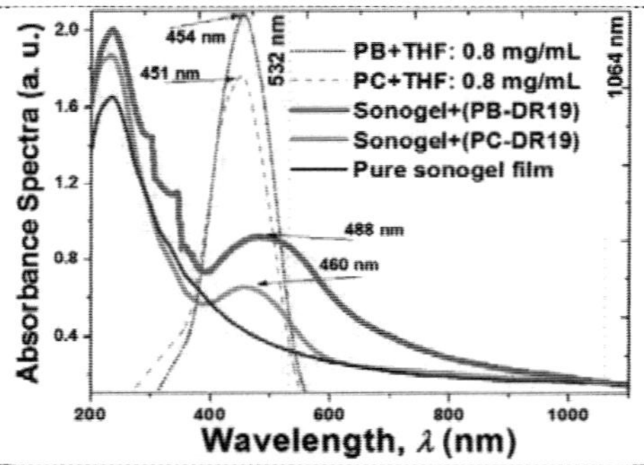

Figure 2 Comparative UV-Vis absorbance spectra of the hybrid film samples, a reference sonogel film, and the THF/polymer dissolutions.

3.2 Thermal (DSC) and (LPA) analyses

Fig. 3 illustrates the comparative DSC and LPA thermal analyses of the hybrid films. Before 100°C, the pure sonogel displays intense LPA signal-peaks associated to the evaporation of residual water and alcohols within the glassy matrix. Thereafter, the sonogel film is thermally stable. On the contrary, the pristine PB- and PC-powders show small DSC-variations until the corresponding T_g values (at 130 and 120°C, respectively) have been reached. However, these transitions are markedly delayed in the corresponding hybrid films, namely to 140 and 135oC, respectively. Besides, the pristine PB- and PC-compounds start their degradation process around 180°C whereas in the SG-environment this phenomenon occurs until 190°C. Thus, the thermal stability of the chromophores is guaranteed and can be enhanced within the glassy sonogel environment.

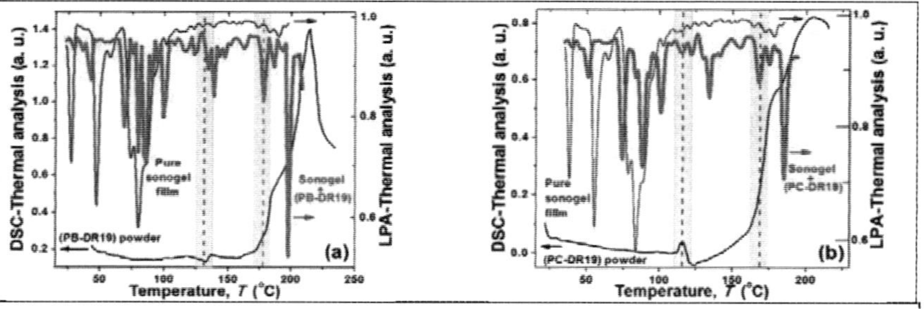

Figure 3 Comparative DSC and LPA thermal studies of the hybrid film samples, a reference sonogel film and the pristine PB- and PC-powders: **a)** PB-polymer and **b)** PC-polymer.

3.3 Planar waveguiding NLO/SHG-activity of the hybrid SG-layers

After laser excitation, both the PB- and PC-based hybrid waveguides were able to exhibit stable SHG-emission for more than 500 hrs (non-continuous working hours) at room temperature (see typical SHG-signals in Fig. 4). The SHG-conversion efficiency as function of the waveguide optical path is represented in Fig. 5 (normalized SHG-output signals). The PB-based sample show higher SHG-emissions than the PC-sample, and the optical attenuation of both PB- and PC-based hybrid waveguides is evidenced as the optical path (d) within the planar waveguide is increased.

However, under our experimental conditions and up to the 20 μm optical path ranges, we were not able to observe any significant oscillating behavior of the SHG-intensity signals (distinctive feature of phase-matching effects [1]). Thus, since the coherence-length cannot be measured in these samples, additional processes such as quasi-phase matching induced effects to improve the SHG-emissions were discarded for the current geometry [6]. Indeed, longer hybrid waveguides or planar samples at prepared different PB- or PC-molecular concentrations must be prepared. These experiments are currently underway in order to improve the SHG-conversion efficiency of our organic-inorganic hybrid waveguides.

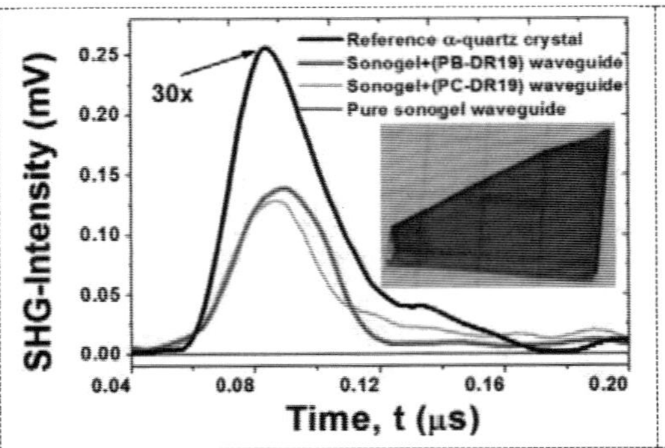

Figure 4. Comparative SHG-signals of the hybrid waveguides, the amorphous SG-glass and the reference α-quartz crystal. Inset figure: Picture of a typical hybrid waveguide with variable optical paths.

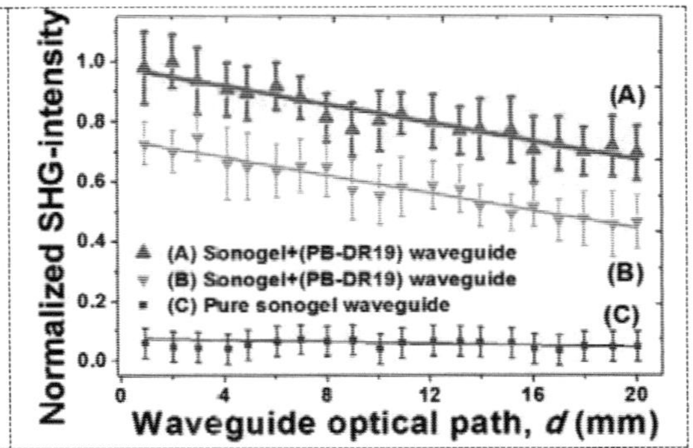

Figure 5. SHG-peak intensities recorded as function of the optical-path (d): (A)- and (B)-data correspond to the PB- and PC-based hybrid waveguides, respectively. (C)-data corresponds to the reference/pure SG-waveguide.

4. CONCLUSIONS

In this work, we briefly show some experimental details of the fabrication procedure and physical characteristics of planar waveguides comprising push-pull polymers embedded within a highly pure SiO_2-sonogel matrix. These low-cost hybrid materials could be important for the development of Si-based semiconductor technologies and evidenced excellent attributes such as: lower SG-refractive index (compared to the hybrid layers) or cladding effect, a stable mechanical structure (fracture-free hybrid samples), a homogeneous molecular dispersion within the host SG-network, an enhanced thermal protection of the guest organic compounds and a freezing effect in the collective molecular ordering induced by high DC/electric fields. Indeed, the corona-poling process allowed us to obtain a stable macroscopic noncentrosymmetric structuring of the hybrid SG-layers as required for quadratic SHG-effects. At this point however, it was not possible estimate the coherence-length of our samples in order to perform a periodic molecular poling along the waveguide layers, appropriate to obtain a quasi-phase-matched structure capable to enhance the SHG-activity as function of optical length of the waveguides. Future works must include several NLO-spectroscopic studies and geometry and molecular concentration variations of the samples in order to obtain the coherence-length and other optical characteristics of interest, such as the linear and nonlinear refractive index of the hybrid waveguides. Under this scope, the present work provides a starting reference in the study of the $\chi^{(2)}$/NLO properties of organic-inorganic SG-based photonic prototypes.

Acknowledgments

VTZ and OGMS acknowledge the financial support from DGAPA-UNAM (Postdoctoral Scholarship), SEP-CONACYT (Grant: 47421) and DGAPA/PAPIIT-UNAM (Grant: IG100713).

References

[1] Lebeau, B. and Innocenzi, P. 2011 *Chem. Soc. Rev.* **40** 886
[2] Bosshard Ch, Hulliger J and Florsheim M 1995 *Organic Nonlinear Optical Materials* vol 1 (Switzerland: CRC Press)
[3] Torres-Zúñiga V, Morales-Saavedra O G, Rivera E, Castañeda-Guzmán R, Bañuelos J G, and Ortega-Martínez R 2010 *J. sol-gel Sci. Techn.* **56** 7
[4] Pérez-Martínez A L, Gómez-Sosa G and Ogawa T 2005 J. *Macromol. Sci. Pure Appl. Chem.* **42** 1561
[5] Neuroth N 1998 *The properties of optical glass* ed. Bach H and Neuroth N (Berlin: Springer-Verlag) pp 90-93
[6] Warren N H and L. M Hayden 1995 *J. Opt. Soc. Am. B* **12** 416

Journal of Physics: Conference Series **582** (2015) 012002 doi:10.1088/1742-6596/582/1/012002

Rugosity and hardness determination in obsidianus lapis for the design of an Yb^{3+}-doped fiber laser

J A Alvarez-Chavez[1], A I Aguilar-Morales[1], G G Perez-Sanchez[1,2], A J Morales-Ramirez[1]

[1]Centro de Investigacion e Innovacion Tecnologica, Cerrada Cecati S/N, Col. Santa Catarina, D.F. C.P. 02250, Mexico D.F.

[2]Tecnologico de Estudios Superiores de Coacalco, Av. 16 de Septiembre 54, Col. Cabecera Municipal, 55700 Coacalco de Berriozabal, Estado de Mexico

E-mail: jose287@gmail.com

Abstract. Obsidianus lapis is a volcanic rock that has been worked into tools for cutting or weaponry by Teotihuacan people for hundreds of years. Currently, it is used in jewelry or for house decorative items such as elaborated sculptures. From the physico-chemical properties point of view, obsidianus lapis is considered a glass as its composition is 80% silicon dioxide. In México, there are different kinds of obsidianus lapis which are classified according to its colour: rainbow, black, brown, red, silver, golden and snowflake. The traditional grinding process for working with obsidianus lapis includes fixed grinders and sandpaper for the polishing process, where the craftsman grinds the rock manually for obtaining a variety of shapes. Laser processing of natural stones is a relatively new area. We propose the use of an Yb^{3+}-doped fibre laser for cutting and ablating obsidianus lapis into spherical, rectangular and oval shapes. By means of a theoretical analysis of roughness and hardness, which affect the different surfaces and final shapes, and by considering the changes in material temperature during laser interaction, this work will focus on parameter determination such as: laser fluence, incidence angle, laser average power and peak pulse energy, from the proposed Q-switched fibre laser design. Full optical, hardness and rugosity, initial and final characterization will be included in the presentation.

1. Introduction

Obsidianus lapis is classified as one of the natural volcanic glasses. It has been expelled mixed with basaltic strata and pumice rock, it is produced when molten igneous rock (magma) comes out to the earths surface as lava and cools down so quickly that their ions have not time to crystallize. The magma that forms the obsidianus lapis has its origin approximately 40 km below the earths surface and has a temperature of nearly 1200°C [1].

Obsidianus lapis is composed with 80% of silicon dioxide. The chemical composition of silicon dioxide is 46.7% silicon and 53.3% oxygen in the empirical formula SiO_2. The presence of hematite produces a variety of colors in obsidianus lapis. Hematite is a mineral composed of ferric oxide (Fe_2O_3). Obsidianus lapis could contain small bubbles of gas, which are aligned along the layers created in the molten rock, which flow just before cooling. These bubbles can produce interesting effects such as golden, green, gray, black, red-brown, silver, rainbow and blue shine [2].

Journal of Physics: Conference Series **582** (2015) 012002 doi:10.1088/1742-6596/582/1/012002

Characterization

For obsidianus lapis characterization, we performed Vickers hardness testing in a cube with 2cm of edge. Different results of Vickers hardness were obtained, according to the indentation, by using a force of 10 Kg. The results are in a range of 110HV10 - 150HV10.

The manufacturing of obsidianus lapis into shapes is performed with a process including the 7 following steps: 4 for grinding, 2 for polishing and 1 for shining. Using a Mitutoyo surftest 301 roughness tester, the data shown on Table 1, were obtained.

Table 1. Roughness average

Process	Roughness average (µm)
Diamond disc	3.4
Grinding wheel grain size 35	3.3
Grinding wheel grain size 90	2.83
120-grit sand paper	1.98
180-grit sand paper	1.23
Cerium oxide	1.02

The average roughness for artistic finishing (100nm) that we need to obtain from the proposed fibre laser crafmanship process is 1.02 µm since polishing the surface into its final values, is the most attractive feature for the osidianus lapis samples.

Absorptivity calculation

In material cutting, absorptivity calculation relies on vaporization, the focused beam heats up the surface up to the boiling point first and then generates a keyhole. The keyhole causes a sudden increase in the absorptivity due to multiple reflections, causing the hole to deepen quickly [4].

Absorptivity in obsidianus lapis can be calculated as a function of average reflectivity R_{ave}, as the ratio of the reflected power $P_{L,R}$ of the total laser power P_L. It depends on the polarization state of the incident laser radiation. Reflectivity values for R_p y R_s, for parallel and perpendicular polarizations, are established for the Fresnel equations 1 and 2 [3], as follows:

$$R_p = \frac{(n * \cos\varphi_{In} - 1)^2 + (k * \cos\varphi_{In})^2}{(n * \cos\varphi_{In} + 1)^2 + (k * \cos\varphi_{In})^2} \quad (1)$$

$$R_s = \frac{(n - \cos\varphi_{In})^2 + k^2}{(n + \cos\varphi_{In})^2 + k^2} \quad (2)$$

Where φ is incidence angle, n is the refractive index and k the extinction coefficient. In case of no polarized radiation and for circularly-polarized radiation, the reflectivity R_{ave} can be estimated as the average value of the two reflectivity components, according to equation 3:

$$R_{ave} = \frac{R_p + R_s}{2} \tag{3}$$

The optical parameters of refractive index "n" and the extinction coefficient are a function of permittivity, according to equations 4 and 5:

$$n = \left(\frac{\left((\varepsilon_1^2 + \varepsilon_2^2)^{\frac{1}{2}} + \varepsilon_1 \right)}{2} \right)^{\frac{1}{2}} \tag{4}$$

$$k = \left(\frac{\left((\varepsilon_1^2 + \varepsilon_2^2)^{\frac{1}{2}} - \varepsilon_1 \right)}{2} \right)^{\frac{1}{2}} \tag{5}$$

Where: ε_1 is the real and ε_2 the imaginary part of the materials permittivity. These parameters can be calculated with equations 6 and 7:

$$\varepsilon_1 = 1 + \frac{\omega_P^2}{\omega_L^2 + V_c^2} \tag{6}$$

$$\varepsilon_2 = \frac{V_c}{\omega_L} \cdot \frac{\omega_p^2}{\omega_L^2 + V_c^2} \tag{7}$$

The laser frequency ω_L can be calculated using equation 8 and it is inversely proportional to the wavelength of the emitted radiation.

$$\omega_L = \frac{2\pi \cdot C_o}{\lambda} \tag{8}$$

c_o is the speed of light in free space. The laser frequency of an Yb^{3+}-doped fibre laser corresponds to $1.762X1015s^{-1}$ for a lambda of 1.07 µm. The plasma frequency is proportional to the square root of the free electron density and can be obtained with equation 9 as:

$$\omega_p = \left(\frac{e^2 \cdot \eta_e(T)}{\varepsilon_0 \cdot m_e} \right)^{\frac{1}{2}} \tag{9}$$

Where the value of the electron charge is $e=1.602X10^{-19}C$, vacuum permittivity is $8.85419X10^{-12}$ F/m and electron mass is $0.91095X10^{-30}$ kg. Equation 10 below shows how to calculate free electron density:

$$\eta_e(T) = N_A \cdot \frac{Z \cdot \rho(T)}{A_A} \tag{10}$$

In this equation, N_A= 0.6022X1024 mol^{-1} is Avogadros number. Z being the number of valence for silicon, and since obsidianus lapis is composed with 80% of this material, we use Z=4, and the atomic mass A_A=28.086X10^{-3} Kg/mol. The density of obsidianus lapis in liquid form is a function of temperature and it is calculated with equation 11:

$$\rho = \rho_{m.p.} - \Lambda \cdot (T - T_{m.p.}) \tag{11}$$

Where $T_{m.p.}$ is the melting point temperature and $\rho_{m.p.}$ at the melting point, for obsidianus lapis $\rho_{m.p.}$=1492kg/m^3, Λ=0.88kg/m^3-K and $T_{m.p.}$ = 1383K, the boiling point $T_{b.p}$ = 2628K. For calculating $\eta_e(T)$, one can use values from 1.279X1029m^{-3} up to 3.399X1028 m^{-3}. So the plasma frequency at the melting point has a value of ω_p= 2.018X1016 S^{-1}and at the boiling point ω_p =1.0402X1016 S^{-1}. For obtaining the collision frequency V_c, we use equation 12:

$$V_c = \frac{\rho_e(T) \cdot \eta_e(T) \cdot e^2}{m_e} \tag{12}$$

Where $\rho_e(T)$ is the electric resistivity is a function of temperature and is calculated with equation 13.

$$\rho_e(T) = \rho_{e,0} + \Lambda_e \cdot (T - 273K) \tag{13}$$

Where $\rho_{e,0}$= 112.3X10-8Ωm and Λ_e= 0.0154X10-8Ωm/K. Therefore the electric resistance at melting point is $\rho_e(T)$= 1.294X10^{-6}Ω and at boiling point it is $\rho_e(T)$= 1.485X10^{-6} Ω. We obtained the values for refractive index n, and for extinction coefficient k at melting point, boiling point and an average, as shown in table 2.

Table 2. Refractive index, and extinction coefficient of obsidianus lapis, at the melting point, boiling point and at the average.

Temperature	**Wavelenght λ =1.07 µm**	
	N	k
$T_{m.p.}$=1383K	5.646	3.812

T_{ave}=2005K	5.526	3.164
$T_{b.p.}$=2628K	4.980	1.689

The average theoretical absorptivity of obsidianus lapis as a function of the angle φ_{In} using an Yb^{3+}-doped fibre laser is shown in figure 1.

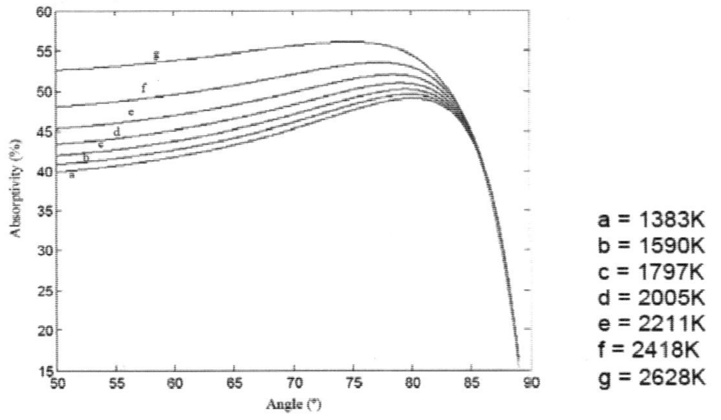

Figure 1. Behaviour of absorptivity of obsidianus lapis for an Yb^{3+}-doped fibre laser

The angle of incidence with better absorption is 79° with absorptivity higher than 50%. After reaching this angle the absorptivity decreases, as it is observed.

For the proposed laser design we define the laser beam power as P_{out} for the proposed obsidianus lapis cutting process. In the case of laser beam for cutting, we can calculate the power required for melting a volume of material per time unit, using equation 14 as:

$$P = \rho \cdot W_K \cdot T_C \cdot V_C \cdot \Delta h_m \tag{14}$$

Where ρ and Δh_m are the density and the necessary increase in the specific enthalpy to cause the melting of the material. For obsidianus lapis, we consider values of ρ=2360 Kg/m^3, W_K = 0.01m, T_C = 0.001m, V_C = 0.01m/s, Δh_m= 1650KJ/Kg, and we obtain a required average output power of P = 389.4 W.

Laser Design

Along with the increase in average power, the emergence of large mode area technology has made possible a number of achievements in the fundamental capabilities of fiber lasers, such as peak power and pulse energy, by means of an acousto-optic modulator. That is a valid reason for proposing a Q-switched fibre laser for obsidianus lapis processing [5].

Thus, a Q-switched fibre laser design with an Yb3+-doped for grinding and polishing obsidianus lapis is proposed. Fig. 2 shows the components for the system. It consists of a laser diode (A), a collimating lens (B), a dichroic mirror (C), a focusing lens (D), the Yb^{3+}-doped fibre piece (E), an acousto-optic modulator (F), and high-reflectivity, broad band mirror (G).

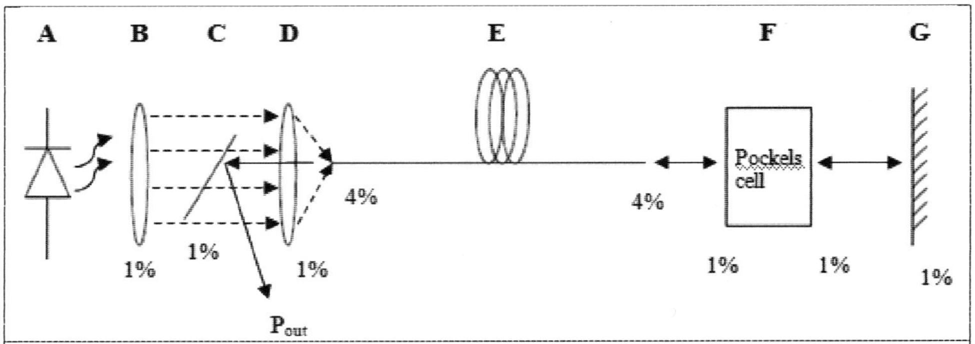

Figure 2. Yb^{3+}-doped fibre laser design and components

In the Q-switched fibre laser design, the pumping with a laser diode at 976 +/-10nm was made in a continuous wave output beam of 15W. After considering the losses in the system that we designated in figure 2, we calculate the average power P_{out} of the system. Background losses of 0.3dB/m is employed. A quantum efficiency of Ytterbium is considered 91%, absorption in the fibre is 3dB/m @976nm, and the average power has a value of P_{avg}=7.406W in continuous wave.

For obtaining theoretical parameters in the fibre laser design, we use a length of 3m for the Yb^{3+}-doped fiber, and from equation 15, the peak power of 1.33MW is obtained.

$$P_p = \frac{P_{avg}}{R_r \cdot \sigma_p} \quad (15)$$

Where P_{avg} =7.4 W, repetition rate is proposed at R_r = 500Hz and pulse duration σ_p = 11.11ns. Pulse energy is calculated via equation 16 as:

$$P_E = \frac{P_{avg}}{R_r} \quad (16)$$

A value of P_E = 0.0148J is then obtained. After varying the repetition rate, we obtain values shown in figure 3, for P_{avg} varying from 4.4 to 9.4 W.

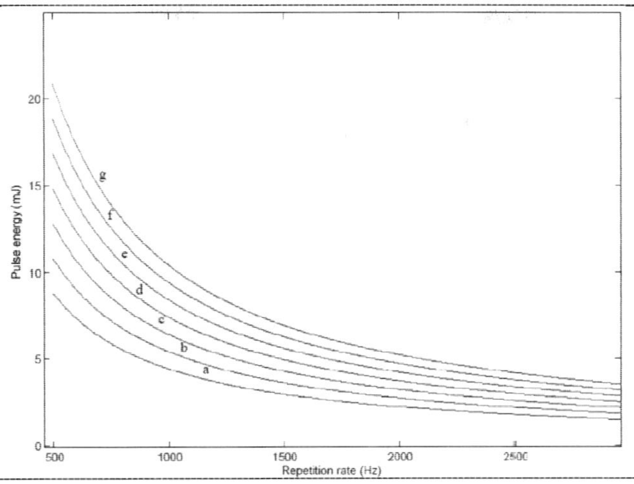

Figure 3. Repetition rate versus pulse energy in Q-switched Yb^{3+} doped fibre laser

Fluence is the ratio of the applied laser energy within the cross-sectional area of the incident laser beam. The fluence of an individual laser pulse is calculated by using equation 17[6].

$$F = \frac{8E_0}{\pi\omega_0^2} \tag{17}$$

Where ω_0 is the Gaussian beam diameter; ω_0 = 30+/-2μm, and E_0 is the energy of the incident beam; E_0 = 0.0148J. The final determination of fluence produces a value of F = 4187.54 J/cm^2.

Conclusion

The study of the main parameters and components in a laser design has been proposed for cutting obsidianus lapis. Cutting can be made in rock and in metal even there is less evaporation and considerable melting. The analysis of crucial design parameters involved in the interaction mechanisms between the fiber laser beam and rock, allows us the investigation of novel ways of rugosity and hardenes determination along with ablation in obsidianus lapis. The designed Q-switched Yb^{3+}-doped fibre laser, with short-pulse fibre technology, produces a diffraction-limited beam at the output of the system, with which different surfaces finishes can be processed. Theoretical results can be summarized as: i) Absorptivity in obsidianus lapis with Yb^{3+}-doped fibre laser is higher than 40% with incident angles from 50° to 79°, this range is important in beam efficiency. Calculated results show the effects on obsidianus lapis and different values for pulse energy with respect to repetition rate. Ii) Obsidianus lapis structure can be melt with a Q-switched Yb^{3+} doped fibre laser with a value of pulse energy of 0.0148J that can even be modulated.

Acknowledgements

The authors would like to express their gratitude to CONACYT, TESCO and CIITEC-IPN, for the support.

Journal of Physics: Conference Series **582** (2015) 012002 doi:10.1088/1742-6596/582/1/012002

References

[1] Bakken, B. *"Obsidian and its formation·,* Missuola, Montana, (1975).
[2] Villanueva, G. Thesis. *"Caracterización de obsidiana mexicana utilizando espectroscopia de resonancia pramagnetica, electrónica y medidas de magnetitaciòn"*, Mèxico D.F.,pp. 12, 15, 45-51. (200)
[3] Mahrle, A. Beyer, E. *"Theorical aspects of fibre laser cutting".* J. Phys. D: Appl. Phys. Dresden, Germany, pp 1-4 (2009)
[4] Steen, W. *"Laser material processing".* London: Springer, pp137 (2010).
[5] Galvanauskas, A. *"High power fiber lasers".* Optics and Photonics News, pp 43 (2004)
[6] Hooker, M. *"Laser-induced damage threshold and certification procedures for optical materials"*. NASA, 1395, (1997).

Current status and future trends of medical physics in Mexico

J Azorin Nieto
Universidad Autonoma Metropolitana-Iztapalpa, Av. San Rafael Atlixco 186 Col. Vicentina 09340 Mexico, D.F.

Email: azorin@xanum.uam.mx

Abstract. Medical Physics is an area that applies the principles of physics to medicine, particularly in the prevention, diagnosis and treatment of diseases using ionizing and non-ionizing radiation. The main attractive of medical physics is that it has a direct impact on the quality and safety of medical care in humans; this social component with direct implications for the population is of high value for Mexico. This paper describes the concepts of medical physics, trends and the current status of this discipline as a profession, which is directly related to the efforts of clinical research. It is also described what is, in my opinion, the future of medical physics in Mexico, emphasizing the fact that this field requires a substantial boost from universities and hospitals to recruit highly qualified young medical physicists and the support from government agencies such as Secretaria de Salud, Instituto Mexicano del Seguro Social and Instituto de Seguridad y Servicios Sociales para los Trabajadores del Estado through clinical research projects that allow the necessary evolution of medical physics into the hospital setting.

1. Introduction

Until the early twentieth century the attention of health problems was almost exclusive domain of the medical profession. However, based on the traditional concept of health as a state of bio - psycho- social well-being and not merely the absence of disease, from then and even more today, both nationally and internationally, the participation of professionals from other disciplines in medicine to solve health problems is becoming more common and more relevant.

In particular , applications of physics in medicine have been increasing in variety and quantity so that new diagnostic and treatment modalities have required that the number of physicists in hospitals be increased and their academic and clinical training suit to the demands of advanced technologies , giving rise to an interdisciplinary profession that allows addressing multiple therapeutic and diagnostic techniques, and provides the scientific basis for understanding, implementation and development of these technologies that are revolutionizing the practice of medicine, as well as the ability to establish criteria for the appropriate use of physical agents used in medicine ensuring the quality of the technical aspects involved in the processes, the effectiveness and quality of the data , thus reducing the likelihood of accidents in medical applications of radiation . This discipline is known as Medical Physics.

Journal of Physics: Conference Series **582** (2015) 012003 doi:10.1088/1742-6596/582/1/012003

2. History

During the decade of the 50s, In some European countries such as the UK and Sweden a new discipline, known as medical physics, was emerged. Medical physicists are scientists who apply their knowledge of physics to medicine, as made at their time, Roentgen, Becquerel and the Curies, especially in the field of diagnostic radiology and radiotherapy. Currently, the number of medical physicists in the world exceeds 20,000 and continues to increase with advancing technological development.

Cancer treatment in Mexico began around the twenties and consisted mainly of the use of Radium. This activity was performed by radiologists [1]. The arrival of the physicists to hospitals in Mexico was directly associated with the acquisition of the first radiotherapy equipment in the mid-twentieth century. In 1956 the first ^{60}Co irradiator was installed at the National Institute of Cancerology [2] where the first department of Medical Physics of the country was created

As pioneering examples of professional organization in medical physics, in May 1962 the Mexican Association of Physicists in Hospital (precursor of the current Mexican Association of Medical Physics) was founded. In those early days, and due to the lack of formal programs of specialization, most local physicists and engineers acquired their training with everyday experience. But fast forward in the complexity of the equipment made clear the need for a specific and rigorous training.

In the 1970s, with the invention of the first CT scanner, radiation physics changed again [3]. Today, much of the equipment used in this field is computerized, and is now being applied imaging techniques at the cellular level, so being able to identify the molecules that play a critical role in the development of certain diseases long before the clinical symptoms appear.

3. What is Medical Physics?

Medical Physics is an area that includes all applications of principles and techniques of physics to medicine, particularly in the prevention, diagnosis and treatment of diseases. The most attractive aspect of this branch of physics is that it has a direct impact on the quality and safety of medical care in humans; this social component with direct implications for the population is high.

Medical Physics is defined accurately by the International Organization for Medical Physics (IOMP) as follows [4]: *"Medical Physics is an interdisciplinary science that from knowledge, methods and techniques of physics, helps solve current problems of medicine, primarily in regard to medical imaging, radiotherapy, nuclear medicine and radiation protection"*. These application areas are not exclusive, but in practice determined the existence of the figure of "Medical Physicist" in hospitals, clinics, and other areas of the practice of medicine.

Therefore, medical physics ranging from the design of equipment and procedures for diagnosis and therapy to the development of computational algorithms to understand and explain the behavior of the human body, basing on the study, research and development of ionizing radiation, non-ionizing radiation, magnetic resonance, ultrasound, biophysics, bioengineering and imaging processing.

The possibility of researching for invent and optimize advanced techniques that use ionizing radiation, visible light, electromagnetic fields or biomaterials, in synergy with the extraordinary recent development of computing resources, makes of medical physicist a specialty of high scientific attraction for a current physisist, since it is the branch of physics of fastest growing nowadays.

4. Current Status of Medical Physics in Mexico

According to data provided by the National Health and Nutrition Survey 2012-Encuesta Nacional de Salud y Nutrición 2012- (ENSANUT 2012) [5] The population pyramid in our country, which currently has 117 million people whose life expectancy is 75.6 years, has been moving gradually showing a widening in the age groups older than 70. This situation has meant, in general, an increase in the demand for preventive and curative health services for those age groups in which the incidence of chronic degenerative diseases and cancer are higher.

In Mexico 128 000 new cancer cases per year are reported. This figure could rise to 150 thousand well considering that there are 25% of new cases not recorded, of which 78 000 die (60-61%) [6]. A shocking information that places Mexico in a difficult situation is that in the medium term is forecast that in the next two decades one of every two men and one in three women will have a diagnosis of cancer [7].

Until 2011, our country had 0.7 radiotherapy centers per million inhabitants; ie 83 radiotherapy centers with 223 irradiators (47 ^{60}Co, 65 LINACs, 10 IMRT, 7 RCS, 2 Gammaknifes, 47 Automatic brachytherapy and 45 Manual brachytherapy) which means 5.6 LINAC per million inhabitants [8]. In addition to these megavoltage radiotherapy systems (MV), Mexico has 157 nuclear medicine units where about 1.2 nuclear medicine studies are performed per million inhabitants (5000 SPECT studies per year), and about 7400 local radiology services with about 22 000 X-ray equipments (420 CT, 165 MRI and more than 600 mammography). In particular, the National Cancer Institute (INCan) published on its portal [8] that in 2011 the number of therapy sessions was about 63,000, while imaging studies totaling over 94,000.

In this context, the medical physicist, as part of a multidisciplinary health care team requires a formal university education. According to international recommendations, this education must be at least master's level [9]. In Mexico there are two graduate programs on Medical Physics which were created over 15 years ago. [10,11] (UNAM; Master, UAEMex, Master and Ph.D.) . To date these are the only graduate programs in the area, with more than 125 graduates (83 UNAM, 42 UAEMex). A high percentage of these professionals are working in radiotherapy departments (80%), and only a small fraction is associated with X ray services (15 %), nuclear medicine or magnetic resonance (5%). These medical physicists have acquired their clinical training through everyday experience due to the absence of formal clinical residency programs in the country. Their professional activities involve health care, teaching and research. It should be emphasized their participation as a liaison with universities and research centers for the development of projects involving the use of medical infrastructure and medical facilities. This is critical because of the technological sophistication of the equipment involved in diagnosis and treatment of diseases. The medical physicist must ensure the proper operation of the equipment and its use through standard protocols, so that their use in research does not interfere with patient care or result in deterioration of equipment. The reality is that today, half a century after the start of RT in Mexico, there are still medical physicists practicing in medical services without receiving the desirable university education and/or clinical training.

Research on medical physics in Mexico is scarce. At least in part, this is due to the lack of degree in the area that allow linking academic activities with professional practice in hospitals. In addition to the research conducted using the infrastructure in hospitals, there are laboratories in universities and research centers. But because of the high cost of equipment necessary to perform frontier research, their number is limited. However, the quality of research conducted in Mexico, in the area of medical physics, is at the level of international competition, but in terms of quantity is no doubt that the number of researchers is extremely small.

5. Future trends in Medical Physics in Mexico

Due to the rapid growth of radiotherapy and diagnostic radiology in our country, it takes on the order of 550 new medical physicists distributed as follows: 130 in radiotherapy, 400 in radio diagnostic and 20 in nuclear medicine. However, the future of medical physics in Mexico is linked not only to the modernization of existing health services , but also to the establishment of new hospitals, which must install high-tech equipment and incorporate medical physicists as part of multidisciplinary groups responsible for the diagnosis and treatment of patients . To strengthen research and development in the field of medical physics will be convenient to take the following actions:

- Encourage the development of academic programs (bachelor, specialist, master and doctorate) in medical physics at various universities and institutes of higher education in the country.
- Recruit young researchers with high-level preparation in medical physics in universities and in the health sector.
- Promote the creation of new medical physics laboratories in universities , research centers and hospitals with high technology having sufficient and appropriate government support.
- Increase collaboration agreements and the development of academic projects between universities and hospitals both domestic and foreign .
- Establish collaboration agreements with companies to have free access to new diagnostic and treatment equipment for conducting research,
- Increase communication links and cooperation with international organizations like the International Atomic Energy Agency (IAEA), the Pan American Health Organization (PAHO) and the World Health Organization (WHO).

These strategies, among others, will allow medical physicists have access to work with high-tech equipment, establish academic cooperation projects, making recommendations to improve the technology and conduct high-level research that results in the patient's welfare. For a long time, medical physicists have worked closely with physicians to exploit the diagnostic and therapeutic uses of radiation for the benefit of patients. Today, medical physicists play a crucial role in medical imaging in radiotherapy and nuclear medicine, to ensure the safety, convenience and efficiency of radiation -based clinical procedures

In the field of radiation therapy, the clinical function of the physicist was greatly enhanced by the arrival of LINACs in the latter part of the last century, which requires much more attention to the computing dosimetry and treatment planning. In the past 30 years, the advent of computed generated images (CGI) has the task of data acquisition separate from the image generation and display. This separation creates an essential role for medical physicists to ensure that any abnormalities in the images is due to expressions of pathological conditions in the patient and not artifacts introduced by the process of image formation.

The complexity of both radiology and radiation therapy has increased dramatically in the past three decades, increasing the importance and usefulness of clinical medical physicists. To meet this demand, the Universidad Autonoma Metropolitana-Iztapalapa (UAM-I) has decided to establish the program of "Specialization in Clinical Medical Physics" (Especializacion en Fisica Medica Clinica), which provides for the clinical training of the medical physicist in hospital by performing rotating and specialized residences.

6. Where is going the Medical Physics?

The Challenge of Medical Physics lies in a shift toward personalized therapies; ie, to accurately personalized treatment to those patients who need it, avoiding treating those who do not. Medical Physics should be able to predict how, when and in whom the disease develops. This will be implemented in the future with the advent of advanced techniques in the areas of molecular imaging and adaptive radiotherapy, which offers an unprecedented opportunity to characterize tumor biology. This would guide the treatment based on the biological properties of each patient's disease. This concept considers changes in the treatment plan during the course of it due to anatomical and biological changes in irradiated tissues.

The ultimate goal is to develop a technique called biology guided radiotherapy (dose painting) [12], to incorporate biological information of the tissues, so that based on this technique, treatment plans are designed to deliver high doses of radiation with accuracy and precision in space and time. This technique can be implemented in the coming years in radiotherapy services by developing high precision and accuracy irradiation procedures (as: IMRT, helical tomotherapy, image-guided radiotherapy, adaptive radiotherapy, robotic radiotherapy, hadrontherapy, etc.) combined whit sophisticated molecular imaging studies to identify the form, distribution, composition and biological response of different tissues to radiation.

Another great progress to develop personalized treatments is the use of intense beams of high energy antiparticles for irradiation of neoplastic tissues, where antiparticles annihilate locally releasing a lot of energy, greatly increasing the dose to the irradiated volume. Currently, the only successful application of antimatter in medicine is in nuclear medicine studies by positron emission tomography (PET), so the use of antiparticles is a challenge for the future, which due to the high cost and complexity of the involved technology could probably be a reality within decades.

7. Conclusion

As is evident, a number of new and emerging technologies appear regularly for possible medical applications that require skill and knowledge of the medical physicist to be developed and applied in clinical practice. The Medical Physicist not currently have the necessary knowledge in biology, genetics and molecular engineering to meet this complex area of development. Therefore, an important part of the development of medical physics is the establishment of programs for the training of medical physicists involving research and clinical practice.

In our country, to cope with this challenge, the Universidad Autónoma Metropolitana is making efforts to contribute to the training of specialists in Clinical Medical Physics by establishing the Specialty in Clinical Medical Physics which provides for the clinical training of the medical physicist in hospital by performing rotating and specialized residences.

In conclusion, the success of the Medical Physics in the immediate future lies in exploring the interface between diagnostic imaging and radiotherapy and exploiting the potential of expanding the horizons of the discipline oriented to personalized therapies. This will be possible only through medical physics training programs involving hospital practice in direct contact with the patients.

8. References

[1] SMeO Sociedad Mexicana de Oncología Historia Recuperado el 3 de septiembre de 2014,

de http://www.smeo.org.mx/quienessomos/historia.php.
[2] De la Garza J 1987 Editorial, *Revista del Instituto Nacional de Cancerología* **33**(1) 270.
[3] Beckmann E C 2005 Godfrey Newbold Housfield *Physics Today* **58**(3) 84-86.
[4] IOMP International Organization for Medical Physics. 3 de septiembre de 2012 http://www.iomp.org/?q=node/76.
[5] Encuesta Nacional de Salud y Nutrición 2012 (ENSANUT 2012) http://ensanut.insp.mx/
[6] Secretaría de Salud [SSA], Subsecretaría de Prevención y Promoción de la Salud [SPPS] 2013. Los 5 tipos de cáncer que más afectan a mexicanos. Recuperado el 3 de septiembre de 2014, http://www.spps.salud.gob.mx/noticias/1445-5-tipos-cancer-mas-afectan-mexicanos.html.
[7] INEGI Instituto Nacional de Estadística, Geografía e Informática 2014 Estadísticas a propósito del día mundial contra el cáncer (4 de febrero). Datos nacionales.
Recuperado 3 septiembre de 2014
http://www.inegi.org.mx/inegi/contenidos/español/prensa/aPropositom.asp?s=inegi&c=2009&ep=149
[8] Instituto Nacional de Cancerología www.incan.edu.mx
[9] OIEA Organismo Internacional de Energía Atómica 2010 El físico médico: Criterios y recomendaciones para su formación académica, entrenamiento clínico y certificación en América Latina IAEA *Human Health Reports* No. 1, Vienna 2010.
[10] Posgrado en Física Médica de la UNAM, www.fisica.unam.mx/fismed/
Maestría y Doctorado en Ciencias con Especialidad en Física Médica http://www.uaemex.mx/pestud/mae_doc/maestrias%20y%20doctorados/Medicina/Mtria%20y%20Doc.%20Ciencias%20Fisica%20Medica.pdf

Van der Heide U A. Houweling A C, Groenendaal G, Beets-Tan R G and Lambin P 2012 Functional MRI for radiotherapy dose painting *Magn. Reson. Imagimg* **30**(9) 1216-23

Numerical simulation of the shot peening process under previous loading conditions

B Romero-Ángeles[1], G Urriolagoitia-Sosa[2], C R Torres-San Miguel[2], A Molina-Ballinas[2], H A Benítez-García[2], J A Vargas-Bustos[2], G Urriolagoitia-Calderón[2]

[1]Instituto Politécnico Nacional, Escuela Superior de Ingeniería Mecánica y Eléctrica. Unidad Profesional Azcapotzalco, Av. de las Granjas 682, Col. Santa Catarina, Del. Azcapotzalco, CP 02250, México D. F., México

[2]Instituto Politécnico Nacional, Escuela Superior de Ingeniería Mecánica y Eléctrica, Sección de Estudios de Posgrado e Investigación, Unidad Profesional Adolfo López Mateos, Zacatenco, Edificio 5, 2do. piso, Col. Lindavista, C.P. 07738, México D. F., México

E-mail: romerobeatriz97@hotmail.com

Abstract.
This research presents a numerical simulation of the shot peening process and determines the residual stress field induced into a component with a previous loading history. The importance of this analysis is based on the fact that mechanical elements under shot peening are also subjected to manufacturing processes, which convert raw material into finished product. However, material is not provided in a *virgin* state, it has a previous loading history caused by the manner it is fabricated. This condition could alter some beneficial aspects of the residual stress induced by shot peening and could accelerate the crack nucleation and propagation progression. Studies were performed in beams subjected to strain hardening in tension ($5\varepsilon_y$) before shot peening was applied. Latter results were then compared in a numerical assessment of an induced residual stress field by shot peening carried out in a component (beam) without any previous loading history. In this paper, it is clearly shown the detrimental or beneficial effect that previous loading history can bring to the mechanical component and how it can be controlled to improve the mechanical behavior of the material.

1. Introduction

Residual stresses are defined as stress acting into a material that is free of the action of external agents. The effects of this stresses can be beneficial or detrimental depending on their magnitude and distribution. Residual stresses are auto-equilibrate, but in particular residual stresses are beneficial when they are compressive (they can increase service life, arrest crack growth and increase fatigue strength). It is important to mention that the residual stresses aren't the only kind of prior history found in the mechanical components, which are produce by non-homogeneous loading. In this sense, loading a material in a homogeneous manner beyond the yield stress, induce an increment in the elastic zone (yield stress) in the direction of load implicated, known as strain hardening. Consequently a decrease in the elastic zone (yield stress) at the opposite direction of the first load, the phenomenon is known as *Bauschinger* effect. This phenomenon not only affects the yield strength of the material, but also affects the residual stress field distribution when prior loading is present.

On the other hand, shot peening is one of the most used techniques to improve the performance of the components and has proven to be a very effective method that increase fatigue strength. During the shot peening process, each blast that strikes the material acts as a tiny peening hammer, introducing into the surface a small indentation or dimple. To dimple developed in the surface of the material must yield in tension. Below the surface, the material tries to restore its original shape. Nearly all fatigue and stress corrosion failures are originated at the surface of the component, but cracks will not initiate or propagate in a compressively stressed zone. Because the overlapping dimples from shot peening are developed, a uniform layer of compressive stress at metal surfaces is induce, the shot peening provides considerable increases in service life.

In this work it is shown a numerical evaluation on the effect of strain hardening and residual stress induces by shot peening. The shot peening effect is analyzed by considering a metallic beam with previous homogenous loading (tension or compressive loading respectively). Additionally, a numerical simulation of the shot peening process in a specimen free of previous loading history is performed. Results are compared, observing changes in the residual stress field distribution and magnitude, which depend on the magnitude and direction of the previous loading history.

2. Numerical analysis development.

The numerical simulation was performed by the application of commercial computational software that uses a Finite Element Method algorithm (ANSYS). The analysis was carried out under structural considerations, on a 2D configuration, applying plane stress theory and using a solid element (*Plane183*) with higher order (8 nodes). The geometry proposed was a rectangular metallic beam, with 100 mm long and 10 mm high. The mechanical properties simulated were related of SAE/AISI 1045 steel and considering the behavior as bilinear; Elastic Modulus (E) = 200 GPa, Poisson's Ratio (υ) = 0.28, Yield Strength (σ_y) = 622 MPa, Yield Strain (ε_y) = 0.00311, Ultimate Strength (σ_u) = 944.7 MPa, Ultimate Strain (ε_y) = 0.0964 (Figure 1). Additionally, kinematic hardening ruling option was considered to simulate the mechanical properties with the elastic and plastic modulus varying as the load is increased. A general material model consisting of a non-linear kinematic and isotropic hardening component was given by Equations 1 and 2 [11] and no contact consideration were taking into account.

$$d \propto = C\frac{1}{\sigma}(\sigma - \alpha)d\varepsilon^{-pl} - \gamma\alpha d\varepsilon^{-pl} \tag{1}$$

$$\sigma_0 = \sigma|_0 + Q_\infty(1 - e^{-\varepsilon^{-pl}b}) \tag{2}$$

Where ε^{-pl} is the equivalent plastic strain, α is the back-stress, C is the initial kinematic hardening modulus, γ determines the rate at which kinematic modulus decreases with plastic deformation, σ_o is the current yield stress, $\sigma|_o$ is the initial yield stress, Q_∞ is the maximum change in the size of the yield surface and b defines the rate at which the size of the yield surface changes as well as plastic straining develops. Equation 1 describes the translation of the yield surface in the stress space due to the back-stress, α, while Equation 2 describes the change of the equivalent stress defining the size of the yield surface, σ_o, as a function of plastic deformation [10 to 12]. For the numerical simulation of the shot peening process it was necessary to add into the numerical analysis a penetrate object, to simulate the shot peening process [13]. The shot effect cause by the ball in this manufacturing technique was simulated by rigid component which is harder than the beam (Figure 2)

Journal of Physics: Conference Series **582** (2015) 012004 doi:10.1088/1742-6596/582/1/012004

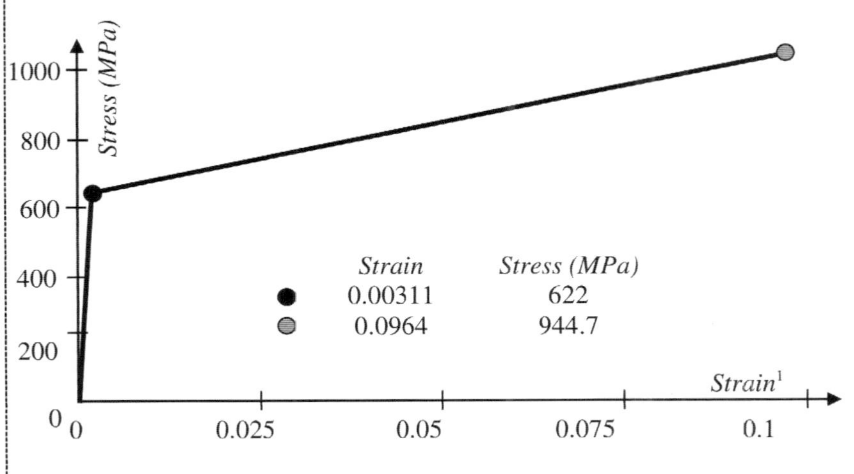

Figure 1. Mechanical properties applied to the numerical simulation.

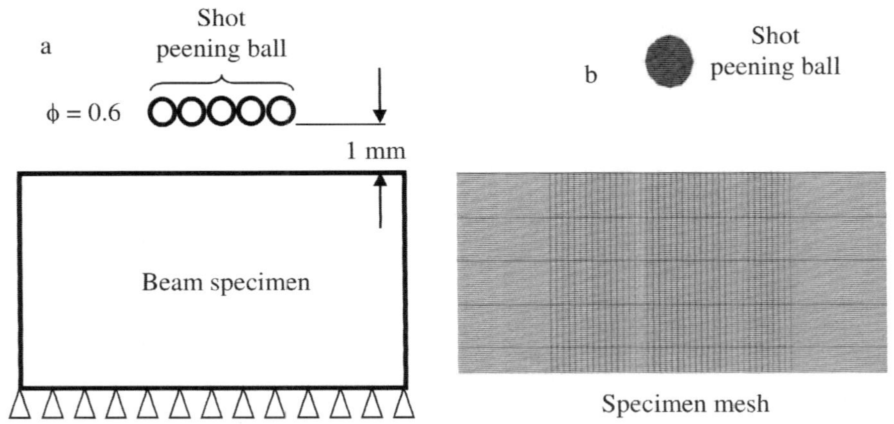

Figure 2. Shot peening process simulation. a) Geometries. b) FEM simulation.

Journal of Physics: Conference Series **582** (2015) 012004 doi:10.1088/1742-6596/582/1/012004

SAE J-441 standardizes the ball applied to the shot peening processes [14]. For this research it was designated *S-230* that has a radius of 0.3 mm. In previous studies *Urriolagoitia-Sosa* and coauthors have found good numerical relation against experimental shot peening data [6]. The discretization of the beam was developed in a controlled manner, making it finer at the center of the element and harsh at both ends of the specimen [13]. The solution of the system is a multi-step one, where the shot is consecutively loaded and unloaded to generate the residual stress field. A shot is produced followed by a second shot in a distance of 0.4 mm; this procedure was carried out until 7 shots were performed. The separation between the specimen and the ball was decided to be 1 mm and the penetration of each ball into the specimen was selected to be 0.2 mm (to simulate the experimental effect in the shot peening process). The specimen is restricted to move in the *y* direction at the bottom surface.

2.1 Numerical simulation of the shot peening process with a tensile previous loading in a beam.
For the case of study with previous loading history, the beam was subjected to an axial tensile homogeneous plastic deformation before the shot peening effect was induced. The beam was axially tensile pulled (by pressure to avoid stress concentrations) until a strain equal to $5\varepsilon_Y$ was reached (Figure 3). By this manner it was possible to assess the change in the isotropic condition of the material and transformed it to anisotropic behavior, which could be corroborated by a bending procedure [15].

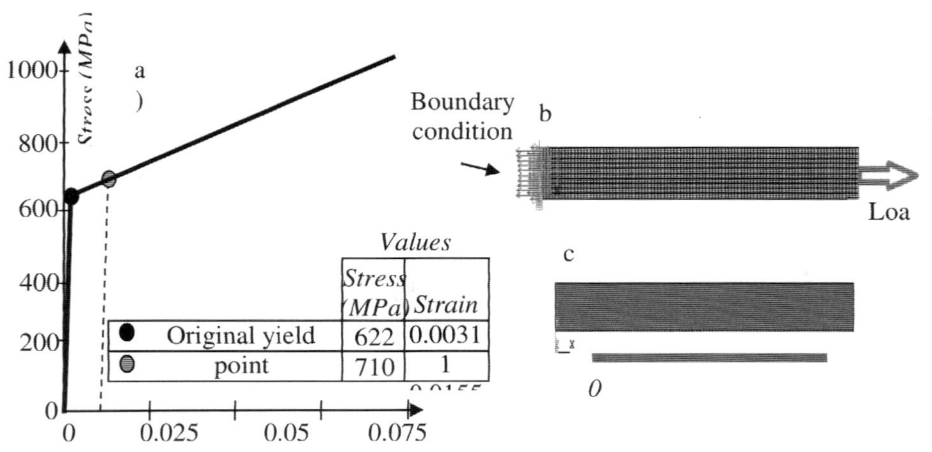

Figure 3. Strain hardening process (simulation of previous loading history). a) Material stress-strain curve. b) Numerical axial tensile homogeneous loading. c) Stress results by axial tensile procedure.

Journal of Physics: Conference Series **582** (2015) 012004 doi:10.1088/1742-6596/582/1/012004

After the axial tensile load was applied and the system was unloaded, the specimen was modified in its boundary conditions to apply the shot peening process. The boundary conditions were applied at the rear base of the beam by constraining displacements at *y* direction (UY) that will prevent movement of the element and will permit both elements to compress between them (Figure 4). The rigid element is move into the beam until a penetration of 0.2 mm is reach (Figure 4a), which is sufficient to generate a non-homogeneous plastic deformation into the material surface. The system is unloaded by retiring the rigid component until the original position and the residual stress field has been applied (Figure 4b) [16].

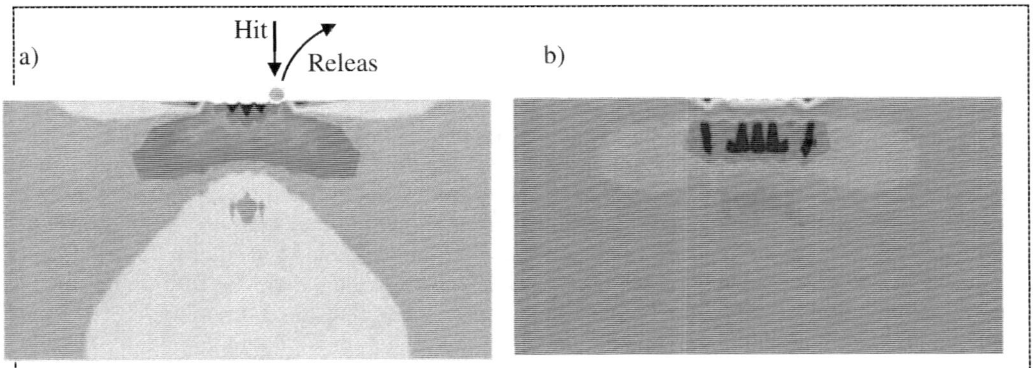

Figure 4. Shot peening simulation. a) Load step. b) Unload step.

2.2 Numerical simulation of the shot peening process without a tensile previous loading in a beam. This case of study was performed to evaluate the effect of prior loading history. The beam was subjected to the same shot peening process numerically simulated as in the section before. All the same conditions were applied.

3. Numerical results

The numerical results obtained by this research are presented in Figure 5.

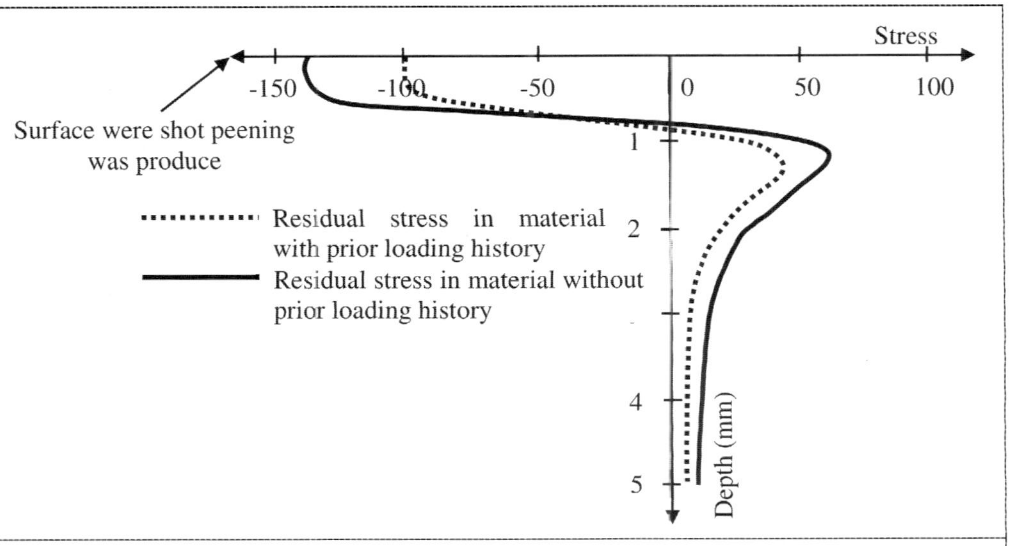

Figure 5. Comparison of residual stress field induced by shot peening.

In this figure it can be observed the comparison between the result on the shot peening numerical simulation process; with and without a prior loading history. Prior loading history has an effect on the resulting induced residual stress field, it could be seen a reduction on the magnitude of the compressive residual stress at the surface of the component. This residual stress field reduction could decrease the resistance to fatigue that the residual stress field introduces into the material without a prior loading history. In both cases were the residual stress field was generated by shot peening process in a beam, the stress distribution began in compression and end with small tensile magnitude.

4. Conclusions

In this paper it is presented a numerical study on the effect that a prior loading history has on the introduction of residuals stresses by shot peening procedure. Two different configurations are used; the first one is a metallic beam without a previous loading history which is subjected to shot peening and the induced residual stress field is determined. The second case is related to a metallic beam with a previous axial homogenous tensile loading on which it is then applied a shot peening process. Results in both cases show a compressive residual stress at the peened surface of the component. The depth of penetration in both cases is almost the same (0.9 mm approximately). Approximately at 1 mm depth (in both cases), appear tensile residual stress fields, which will promote crack nucleation and propagation.

In the Figure 4a, it can be observed that the strain hardening effect modifies the elastic range of the material by modifying the yield strength of the component. The homogeneous tensile loading has to be applied further than the original yield stress and by a pressure manner; otherwise (punctual load) stress concentrations can appear. If stress concentrations are present in the system, no homogenous loading will be produce causing the induction of residual stress and not a strain hardening effect (Figure 4b and 4c).

Compressive residual stress field could be enhanced by increasing the depth of penetration of the shot peening process, not by increasing axial tensile pulling. In fact, if homogeneous tensile pulling is increase, the beneficial effect of the compressive residual stress field can disappear. It is recommended, to increase the beneficial compressive residual stress field to strain hardening the material in opposite direction (the material axially compressed by homogeneous loading).

It is very complex to determine the best shot peening condition to increase the fatigue strength, because it depends on many variables. Nevertheless, in this paper it was demonstrated that shot peening effects depends on the prior condition of the material. In fact, higher compressive residual stress field at the surface of the material will produce the best results against fatigue and will increase service life.

Acknowledgements
The authors gratefully acknowledge the financial support from the Mexican government by the Instituto Politécnico Nacional and the Consejo Nacional de Ciencia y Tecnología. To Ing. Juan Ramón Benítez-Cabral, Mr. Luis Miguel López-Hernández and Mr. José Alfredo Rangel-Hernández for their technical guidance.

Journal of Physics: Conference Series **582** (2015) 012004 doi:10.1088/1742-6596/582/1/012004

References

[1] . K. Masubuchi, *Analysis of Welded Structures*, 1st edition, 1980.

[2] N. E. Dowling, *Mechanical Behavior of Materials*, 2nd Edition, 2007.

[3] A. Molina-Ballinas, *Evaluación y Determinación Experimental-Numérica del Endurecimiento por Deformación y el Efecto Bauschinger en las Propiedades Mecánicas de un Acero Inoxidable*, M. Eng. Thesis, SEPI ESIME Azcapotzalco, Instituto Politécnico Nacional, México, 2010.

[4] M. A. S. Torres and H. J. C. Voorwald, An evaluation of shot peening, residual stress and stress relaxation on the fatigue life of AISI 4340 steel, *International Journal of Fatigue*, Vol. 24, No. 8, pp 877-886, 2002.

[5] D. Natkaniec-Kocanda, S. Kocanda and K. L. Miller, Influence of shot peening on short crack behaviour in a médium carbón steel, *Fatigue and Fracture of Engineering Materials and Structures*, Vol. 19, pp 911-917, 1996.

[6] G. Urriolagoitia-Sosa, E. Zaldivar-González, J. M. Sandoval-Pineda and J. García-Lira, Assessment of the crack compliance method and the introduction of residual stresses by shot peening using the finite element method, *Applied Mechanics and Materials*, Vol. 15, pp 109-114, 2009.

[7] G. Urriolagoitia-Sosa, J. F. Durodola and N. A. Fellows, Determination of residual stress in beams under Bauschinger effect using surface strain measurement, *Strain*, Vol. 39, pp 177-185, 2003.

[8] B. Romero-Ángeles, *Aplicación de Multicargas para el Arresto de Grietas*, M. Sc., Thesis SEPI ESIME Zacatenco, Instituto Politécnico Nacional, México, 2009.

[9] R. M. Pelloux, Case studies of fatigue failures in aeronautical structures, *Proceeding of Conference Fatigue*, Vol. 93, pp 1727-1737, 1993.

[10] G. Urriolagoitia-Sosa, A. Molina-Ballinas, G. Urriolagoitia-Calderón, L. H. Hernández-Gómez, B. Romero-Ángeles and A. Michtchenko, Numerical and experimental analysis in the manipulation of the mechanical properties for enhancing the mechanical resistance of material, *Journal of Applied Research and Technology*, Vol. 3, pp 156-172, 2011.

[11] G. Urriolagoitia-Sosa, *Analysis of Prior History Effect on Mechanical Properties and Residual Stress in Beams*, PhD Thesis, Oxford Brookes University, 2005.

[12] G. Urriolagoitia-Sosa, J. F. Durodola and N. A. Fellows, Effect of strain hardening on residual stress distribution in beams determined using the crack compliance method, *Journal of Strain Analysis for Engineering Design*, Vol. 42, pp 115-121, 2007.

[13] E. Zaldívar-González, *Efecto del Tamaño de Bola en el Proceso de Granallado*, M. Sc. Thesis, SEPI ESIME Zacatenco, Instituto Politécnico Nacional, México, 2008.

[14] J. Champaigne, *Shot Peening; Theory and Application*, Ed. IITT International, 1991.
[15] G. Urriolagoitia-Sosa, J. F. Durodola and N. A. Fellows, Determination of tensile and compressive stress-strain curves from bend test, *Applied Mechanics and Materials*, Vol. 1-2, pp 133-138, 2004.

G. Urriolagoitia-Sosa, A. Molina-Ballinas, G. Urriolagoitia-Calderón, L. H. Hernández-Gómez and J. M. Sandoval-Pineda, Characterization of strain hardening behaviour and residual stress induction used for crack arrest in a biocompatible material, *Material Research Society Symposium Proceeding*, Vol. 1242, pp 233-238, 2009

Journal of Physics: Conference Series **582** (2015) 012005 doi:10.1088/1742-6596/582/1/012005

Counterion release from a discretely charged surface in an electrolyte: Monte Carlo simulation study

M Hernández-Contreras

Departamento de Física, Centro de Investigación y Estudios Avanzados del Instituto Politécnico Nacional, Apartado Postal 14-740, México Distrito Federal, México

E-mail: marther@fis.cinvestav.mx

Abstract. Monte Carlo simulations allowed us to determine the amount of released electric charges from a discretely charged surface in 1:1 aqueous electrolyte solution as a function of surface charge density. Within the restricted primitive model and for a fixed concentration of 0.1 M bulk electrolyte in solution, there is an increase in the number of released counterions per unit surface area as the strength of the surface charge is enhanced. A similar behaviour of the number of released counterions was also found through the use of mean field and liquid theory methods.

1. Introduction

Neutron reflectometry measurements indicate that there occurs adsorption of protein bovine albumin onto a planar polyacrylic acid brush [1]. It has been argued that one of the most important forces that drives this adsorption process is due to the counterion release rather than the direct electrostatic or Van der Waals forces. The counterion release entropic interaction was first reported in Refs. [2,3] to explain the salt dependence of the binding of proteins to DNA macromolecules [4]. The counterion entropy effect was also needed to understand the self-assembly of DNA and cationic liposomes in studies performed with conductivity measurements [5]. More recently, spectroscopy experiments on the protein adsorption into inorganic mesopores have revealed that the joined effects of both electrostatic, Van der Waals particle's interactions and the counterion entropy, can successfully reproduce the observed concentration adsorption of proteins [6]. Most of the theoretical modeling on these systems use an uncovered flat surface whereas those employed in the experiments are covered with a branched polymer. In order to incorporate such characteristic of the real system, in the present manuscript we report our preliminary calculations of the counterion entropy release from the surface with a discrete surface distribution of negative charges of 3 Å diameter. Such adsorbed charges are randomly distributed on the accessible surface area of the wall. Such a model system is embedded in an aqueous pure salt represented by the primitive model of a 1:1 electrolyte. We determined the overall amount of released counterions per unit area as a function of surface charge density. This structural property is found to enhance as the magnitude of the surface charge is increased.

1. Model system and theory

We performed canonical Monte Carlo simulation [7] for a system made of a flat surface with adsorbed negatively spherical charged ions of 3 Å diameter leading to a discrete surface charge $\sigma = \Delta ne / W2$ where ΔN is the number of adsorbed ions, $w = 104.125$ Å is the wall's lateral size, and e is the electronic charge [8,9]. The discrete charges were randomly distributed on the surface. We assumed that the discrete charges are not capable of being detached from the surface and therefore their chemical potential is the same as the ions in the bulk. A similar system to ours was studied with molecular dynamic simulation in Ref. [10]. The simulation cubic cell has volume $W3$ There is an aqueous 1:1 electrolyte at $c_0 = 0.1\,\mathrm{M}$ bulk concentration. The salt ions are spherical particles which have an effective hydrated diameter of $d = 4.25$ Å, and are dispersed in a continuous solvent represented by the dielectric constant of water $\in = 78.5$ at room temperature $T = 298K$. We varied the surface charge density in the interval , where $\sigma^* = \sigma d^{2/} e$ is the reduced density, and calculated the amount of released ions N per unit surface area A through the use of the equation

$$N / A = \int_0^W dz(c + (z) + c - (z) - 2c_0) \quad (1)$$

An overall charge neutrality was imposed with the expression $\int_0^w dz(ec + (z) + ec - (z) - 2ec_0 + \sigma = 0$ Here $c \pm (z)$ are the local concentrations of counterions (+) and co-ions (-) at a given distance from the charged wall. These thermodynamic properties were obtained from the simulations at increasing surface densities.

Results

Figure 1 depicts the logarithm of the reduced local density profiles of the two species of ions at the fixed surface charge $\sigma^* = -0.3$ and electrolyte concentration $c_0 = 0.1M$.

Journal of Physics: Conference Series **582** (2015) 012005 doi:10.1088/1742-6596/582/1/012005

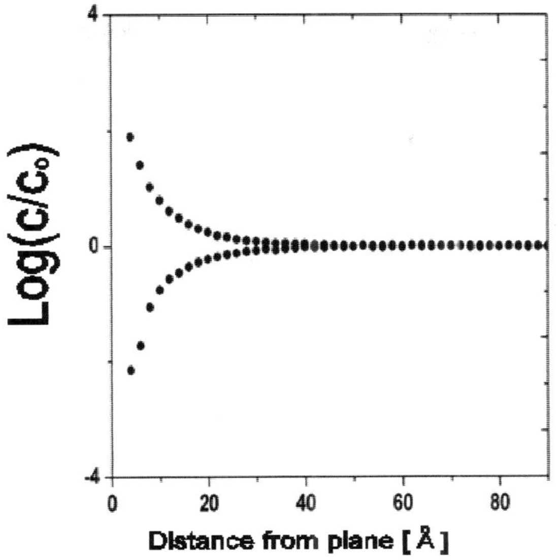

Figure 1. Monte Carlo calculation of the profile concentrations $c+(z)$ of counterions (upper curve) and $c-(z)$ f co-ions (lower curve) versus distance from the charged wall. Both structural properties are normalized with their bulk value c_0

The upper curve represents the double layer formed by counterions close to the charged wall, and it has approximately a thickness of 20 Å. Even though, the surface is highly charged it does not present layering of ions close to the wall.

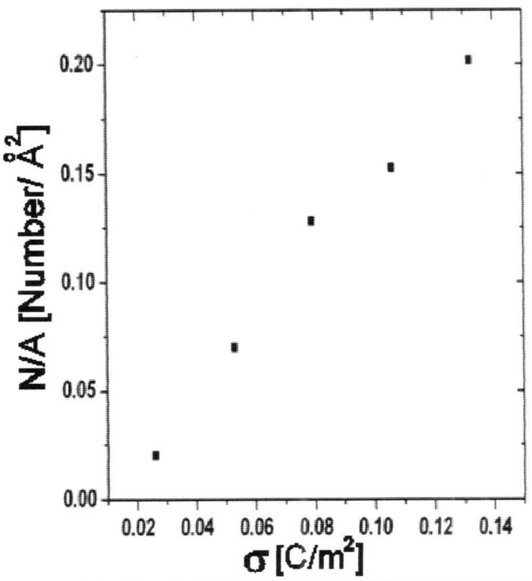

Figure 2. Monte Carlo calculation of the amount of counterions released N/A as a function of surface charge density σ at $0.1M$ of added 1:1 electrolyte in solution. Case of only one charged wall where its surface charge is provided by a random distribution of adsorbed negatively charged spheres of radius 3 Å. Use was made of the restricted primitive model

In Figure 2, we provide the plot of equation (1). It gives the amount of released ions from the double layer which, therefore, are driven into the bulk of the electrolyte. For an increase of the surface charge, there is a corresponding increase in the number of released ions. By electroneutrality, this result is expected since then there is a larger number of counterions that consequently are needed in order to compensate for the increase in surface charge density at the same bulk electrolyte concentration.

Figure 3 yields *N/A* as a function of σ for a system formed by two charged surfaces separated the mean distance of 9 Å in a 1:1 electrolyte solution at $0.1M$ concentration. Upper triangle curve is the prediction of the Poisson-Boltzmann (PB) theory for a smooth surface charge density σ. Whereas, the anisotropic hypernetted chain liquid (AHNC) theory was used for the curves with symbol ○ for discrete, and • smooth σ on surfaces, respectively. For the AHNC calculations the adsorbed charges on the surfaces are constituted by spheres of 4.25 Å diameter. They are in chemical equilibrium with those ions in the bulk and in the slit, thus, at equilibrium all charges have the same chemical potential [11,12]. In this case both surfaces are of equal strength σ but of opposite sign [11,12]. Other parameters are as in Figure 2. Liquid theory is able to incorporate correlations of charge density fluctuations which are not taken into account by PB theory.

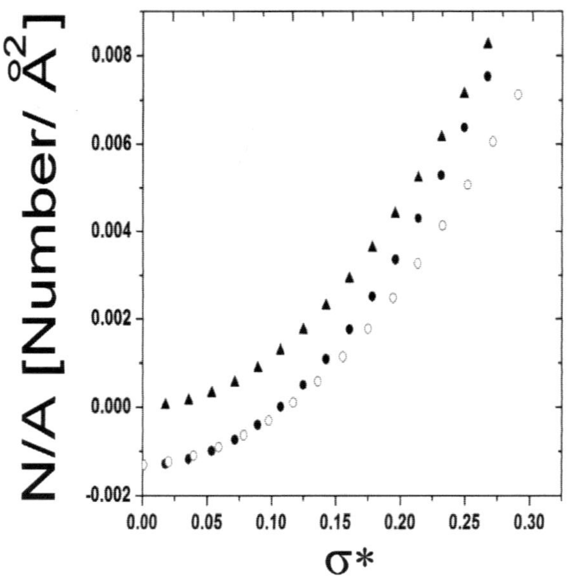

Figure 3. Plot of N/A as a function of the reduced surface charge density $\sigma^* = \sigma d^{2/}e$ at $0.1M$ of added 1:1 electrolyte in solution. In this case, there are two discretely charged walls which surface charge is provided by a liquid like distribution of charged spheres of 4.25 Å diameter and which are adsorbed onto each wall. Both walls have equal strength of σ and they have opposite sign. Their separation distance is 9 Å. Upper triangle curve is the mean field PB theory result. The two bottom curves are liquid theory predictions corresponding to symbols ○ for discrete, and • for smooth surface charge, respectively, after using the restricted primitive model. Other parameters used are as in Figure 2.

2. Conclusion

We determined with Monte Carlo simulations in the canonical ensemble the amount of released counterions as a function of the surface charge for the case of one discrete surface density on the plane in an 1:1 electrolyte solution. As σ is increased, we observed there is a corresponding increase in N/A. For two discretely charged surfaces with the same strength of σ and of opposite sign, in the presence of bulk electrolyte, mean field predicts larger values for N/A than liquid theory does. It can be observed that for a single plane the magnitude of N/A is of larger magnitude than for the two surfaces case with asymmetric surface charge density.

Acknowledgments

The author acknowledges the General Coordination of Information and Communications Technologies (CGSTIC) at CINVESTAV for providing HPC resources on the Hybrid Supercomputer "Xiuhcoatl," which has contributed to the results reported within this paper. This work was supported by CONACYT Grant No. 48794-F, Mexico.

References

[1] Czslik C, Jackler G, Steitz R and von Grünberg H H 2004 *J. Phys. Chem. B* **108** 13395
[2] Record M T, Lohman T M and de Haseth P 1976 *J. Mol. Biol.* **107** 145
[3] de Haseth P L, Lohman T M and Record M T 1977 *Biochemistry* **16** 4783
[4] Meier-Koll A A, Fleck C C and von Grünberg H H 2004 *J. Phys.: Condens. Matter* **16** 6041
[5] Wagner K, Harries D, May S, Kahl V, Rädler J O and Ben-Shaul A 2000 *Langmuir* **16** 303
[6] Moerz S T and Huber P 2014 *Langmuir* **30** 2729
[7] Allen M P and Tildesley J D 1989 *Computer Simulation of Liquids* (Oxford University Press)
[8] Torrie G M and Valleau J P 1980 *J. Chem. Phys.* **73** 5807
[9] Boda D, Chan K Y and Henderson D 1998 *J. Chem. Phys.* **109** 7362
[10] Bengaveti S and Travesset A 2012 *J. Chem. Phys.* **137** 064708
[11] Flores-Amado A and Hernández-Contreras M 2011 *Rev. Mex. Phys.* **57** 356
[12] Flores-Amado A and Hernández-Contreras M 2007 *Phy. Rev. E* **75(3)** 031404

Journal of Physics: Conference Series **582** (2015) 012006 doi:10.1088/1742-6596/582/1/012006

Dose measurements in intraoral radiography using thermoluminescent dosimeters

C Azorín[1], J Azorín[2], F Aguirre[2] and T Rivera[1]
[1] Centro de Investigación en Ciencia Aplicada y Tecnología Avanzada-Legaria Instituto Politécnico Nacional, México, D.F., México
[2]Universidad Autónoma Metropolitana-Iztapalapa, México, D.F., México

E-mail: cazorin@ipn.mx

Abstract. The use of X-ray in medicine demands to expose the patient and the professional to the lowest radiation doses available in agreement with ALARA philosophy. The reference level for intraoral dental radiography is 7 mGy and, in Mexico, a number of examinations of this type are performed annually. It is considered that approximately 25% of all the X-rays examinations carried out in our country correspond to intraoral radiographies. In other hand, most of the intraoral X-ray equipment correspond to conventional radiological systems using film, which are developed as much manual as automatically.
In this work the results of determining the doses received by the patients in intraoral radiological examinations made with different radiological systems using LiF:Mg,Cu,P+PTFE thermoluminescent dosimeters are presented.
In some conventional radiological systems using film, when films are developed manual or automatically, incident kerma up to 10.61 ± 0.74 mGv were determined. These values exceed that reference level suggested by the IAEA and in the Mexican standards for intraoral examinations.

1. Introduction

Medical applications represent the most important source of population exposure to radiation; among these, dental radiology is the most common radiological examination type used in industrialized countries1, being intraoral radiography one of the X ray examinations most used in human beens2. One of the techniques used in intraoral radiography is periapical radiography.

The intraoral radiography is an exploratory technique that consists in placing within the mouth of the patient radiography films of different sizes that are exposed from the outside by a X-ray machine. Intraoral periapical radiography serves to explore two to four teeth, from the crown to the apex, the periodontal space and the surrounding bone tissue.

The measurement of the dose received by patients is considered as an important factor in the quality control in medical radiology, so it is important to expose the patient to radiation doses as low as possible, according to the ALARA principle. One of the most important parameters in diagnostic radiology is the surface entrance dose3, so it is important to determine the entrance dose received by patients in intraoral radiographic examinations performed with different imaging systems.

Journal of Physics: Conference Series **582** (2015) 012006 doi:10.1088/1742-6596/582/1/012006

The reference level for a periapical dental radiography proposed by the International Atomic Energy Agency (IAEA) and adopted by Mexico in the Mexican Official Standard NOM -229-SSA1 -2002 , is 7 mGy4 5. In Mexico a lot of annual examinations of this type is performed, it is estimated that approximately 25 % of all radiographs made in our country are for intraoral radiographs. In addition to most dental radiography systems correspond to conventional periapical radiographic film systems whose development is done either manually or automatically.

The entrance surface dose (ESD) is a measure of the dose absorbed by the skin at the point of entrance of the X ray beam. The ESD may be determined directly or indirectly due that is proportional to factors such as the current, the exposure time the kilovoltage , the filtration and the beam collimation ; This determination could be made by direct measurement using thermoluminescent dosimeters (TLD) or ionization chambers or calculated indirectly. The method recommended for measuring entrance dose is thermoluminescent dosimetry due to the specific characteristics of this type of dosímeters6.

1. Materials and methods

The study was performed on 60 patients in several dental radiology installations of the city of Leon, Guanajuato, Mexico. Twelve X-ray dental systems of different brands were considered, setting the voltage and current in the range of 60 to 70 kVp and 5 to 10 mA respectively. The exposure time was varied between 0.2 and 0.5 s for lower right first premolar and left first molar respectively with a total filtration of 2 mm Al. All patients were studied only during their prescribed examinations, so they were not submitted to any extra examination or any unnecessary exposure to radiation.

The dosimetric control was performed on dental patients previously having their informed consent. Thermoluminescent dosimeters (TLD) of LiF:Mg,Cu,P + PTFE of national manufacturing were used7. These TLDs were wrapped and sealed in black plastic previously cleansed with alcohol. The DTLs were placed in the center of the radiation beam on the skin of the patient without interfering with the image For each measurements were used three DTLs by taking the average of the dose measured by each of these three dosimeters. TLDs were read at the end of the workday using a Harshaw 3500 TL reader; subsequently they were thermally erased to be to be reused subsequently.

The preparation method of the LiF:Mg,Cu,P + PTFE has been described in the previous work8. The characterization of the TLDs was performed as follows: from a batch of 500 TLDs, those showing linearity in the dose range 1 to 10 mGy and repeatability of ± 5% were selected. For calibration of the dosimeters a 55 keV X - ray beam was used. Average absorbed dose was checked with an ionization chamber within a standard deviation of 2%. The thermoluminescent response of the dosimeters was converted to air kerma using the calibration equation as in (1) and subsequently to dose equivalent by applying the appropriate weighting factor which in this case is equal to the unit due that radiation are X - rays. Calibration curve obtained was the following:

$$D(mGy) = 49.15L - 29.231 \quad (1)$$

Where L is the readout in nC.

2. Results

This study included a total of 60 adult patients (33 women and 27 men). Table 1 summarizes the basic information about patients and exposure parameters. The average age of the patients was 36.9 ± 0.5 years being statistically significant the difference between the ages of women and men. The purpose of the examinations was: periapical diagnosis 45 %, treatment 28 %, surgery 15 % and other reasons 12 %. The average exposure time was 0.303 s without statistical significance between exposure times for men and women.

Table 1. Basic information about patient age, purpose of the examination and exposure time.

Basic information	Women (N = 33)	Men (N = 27)	Total (N = 60)
Age (years)	36.7 ± 0.5	37.2 ± 0.5	36.9 ± 0.5
purpose of the examination			
a) Diagnosis	25%	20%	45%
b) Treatment	15%	13%	28%
c) Surgery	8%	7%	15%
d) Other	7%	5%	12%
Exposure time (s)	**0.293 ± 0.118**	**0.314 ± 0.116**	**0.303 ± 0.121**

Table 2 shows the values obtained for the incident air kerma measured at the center of the beam in periapical radiography examinations. The average kerma for radiography of molar teeth was 6.1 ± 1.3 mGy . The highest kerma values obtained were 10.6 ± 0.8 mGy and 9.2 ± 1.4 mGy for first molar teeth right and left respectively. While the lowest kerma value (3.2 ± 0.9 mGy) was measured for the upper right first premolar.

Table 2. Incident kerma values measured at the center of the beam in periapical radiography examinations.

	Women (N = 33)	Men (N = 27)	Average
Average value of the surface incident kerma (mGy)	6.0 ±1.1	6.1 ± 1.2	6.0 ±1.2
a) Pre-molar	5.7 ±1.0	5.7 ±1.1	5.7 ±1.1
b) Molar	6.3 ±1.4	6.5 ±1.5	6.4 ±1.5
c) Incisor	6.0 ±1.3	6.1 ±1.9	6.1 ±1.3

3. Discussion

Cases where kerma values higher than the reference level (7 mGy) were obtained, may be due that these X rays systems are short cone in which the focus-skin distance is 5.0 cm, obtaining dispersion higher than that obtained in the long cone X ray equipment and also due that some dental instalations does not have a Quality Assurance Program.

TLD-100 were used for dental dosimetry because they fulfill with the characteristics of a good dosimeter for this purpose: small size, tissue equivalence, independence of radiation energy.

It is recommended to make known these results to dentists for creating awareness about the fact that they there should not be indiscriminate use of X-rays and they must have a quality assurance program.

For further studies, we suggest taking into account the physical characteristics of the patient, such as thickness of the cheek, and molar size, in order to relate with the factor causing of the increase the dose in molar radiography.

4. Conclussions

The results of this study show that exposure of patients undergoing periapical examinations , on average, do not exceed the dose reference levels specified in the standard (7mGy); however, in some cases if these levels (10.6 ± 0.7 mGy and 9.2 ± 1.4 mGy) for right and left first molar teeth respectively these levels were exceeded.

It was learned that it is necesary to do quality control to those X-ray systems that exceeded the limit.

Results obtained are useful and important to propose modifications to the standard (NOM-229-SSA1-2002) to include periapical dental X-ray equipment, due that this standard currently only takes into account to the dental panoramic systems.

5. References

[1] Horner K 1994 Radiation protection in dental radiology. *British Journal of Radiology* **67** 1041-1049

[2] Kalinowski P, Kalinowski P, Rozylo-Kalinowska I, Rozylo T K 2001 Demographic structure of patients taking dental X-rays in the Lublin region. Annals of University Mariae Curie Sklodowska [Med.] **56** 431-435

[3] Williams J R and Montgomery A 2000 Measurement of dose in panoramic dental radiodiology. *British Journal of Radiology*; **73** 1002-1006

[4] González L, Vano E, Fernández R 2001 Reference doses in dental radiodiodiagnostic facilities. *British Journal of Radiology* **74** 153-156

[5] NORMA Oficial Mexicana NOM-229-SSA1-2002 Salud ambiental. Requisitos técnicos para las instalaciones, responsabilidades sanitarias, especificaciones técnicas para los equipos y protección radiológica en establecimientos. Diario Oficial de la Federación, septiembre de 2006

[6] Azorín J. 1990 Luminescence Dosimetry. Theory and Applications. Ediciones Técnico-Científicas. México, D.F.

[7] González P R, Quiroz M C, Azorín J, Furetta C and Ávila O 2005 Improvement in the preparation method of LiF:Mg,Cu,P thermoluminescent phosphor. *Journal of Applied Science* **5(8)** 1408-1411

[8] Azorín J, Furetta C and Scacco A 1993 Preparation and properties of thermoluminescent materials. *Physica Status Solidi A*. **138(1)** 9-47

Journal of Physics: Conference Series **582** (2015) 012007 doi:10.1088/1742-6596/582/1/012007

Development of an exploration land robot using low-cost and *Open Source* platforms for educational purposes.

D Hernández, H Trejo and E Ordoñez.
Instituto de Ingeniería y Tecnología. Departamento de Física y Matemáticas.
Universidad Autónoma de Ciudad Juárez.
Av. del Charro no. 450 Nte. Col. Partido Romero, CP 32310, Tel. +52 656 688-4801

E-mail: hdz.david@me.com

Abstract. In this paper we present the didactic experience of building a low-cost robot composed of sensors, actuators, general electronics and already available frameworks. The control of the robot is through the usage of commercial *Open Source* platforms as ***Arduino***; and the ***Raspberry Pi***. The experience ranges from general conceptualization, mechanical, electric and electronic design, microcontroller programming and communications.

1. Introduction

Engineering physics can be defined as the direct application of mathematics and physics in the resolution of engineering problems. A direct example of application is robotics. Robots are devices composed of actuators, sensors and electronic circuits, which find their foundation in electrodynamics, classical mechanics and solid state physics. Another important subject around robots that has a direct relationship with physics engineering is control theory, where physics and advanced mathematics are used to solve robot control problems. This two approaches make robots a good supporting tool for teaching and demonstrating physics in class. It motivates and assist students to develop problem-solving skills, as well as help them understand the real life applications of the concepts they are learning.

A strong drawback of the usage of robots as a tool to teach physics is their high cost, which range from **$350 USD** (*Lego Mindstorms* [1]) up to **$10,100 USD** (*Robotino* [2]). This can make them inaccessible to low-budget schools and universities. Many authors have tackled this problem in several ways, for example, with the use of hobbyist equipment for their projects [3-6].

In this paper we illustrate the experience of building an inexpensive robot with the main objective of explaining the details and shortcuts for the general idea to be implemented in any school. Even if the cost of this robot is not as low as other options (**$14 USD**, *KiloBot* [7]), it provides the student the experience of building a robot, experience that he will never get only by experimenting with a pre-assembled robotic kit.

2. Development platforms.

In order to create the systems that monitor and control the robot, we need to use microcontrollers and computers. For this purpose, we used already available development platforms. They were

Journal of Physics: Conference Series **582** (2015) 012007 doi:10.1088/1742-6596/582/1/012007

chosen by their low cost, availability, ease of use, large user base, support, example code, but mostly by the usage of free and/or open source hardware of software [8].

2.1. Raspberry Pi.

The ***Raspberry Pi*** is a **$35 USD**, credit-card sized computer that uses the Broadcom *BCM2835 System-On-a-Chip*. With a power consumption of 5.0W and the equivalent computing power of the original *X-Box*, it supports video outputs in *RCA* and *HDMI* formats, has an Ethernet port, two *USB* ports, up to 512MB of *RAM*, a *GPIO* port and solid state disk drive [9]. It runs on *Linux* operating systems, being *Raspbian* (based on *Debian*) the default option [9].

2.2. Arduino.

Arduino is a family of boards oriented to prototyping based on Open Source hardware and software [8]. It consist of a board with two microcontrollers. The principal microcontroller is where user's programs are loaded and executed. Carries a set of digital and *ADC* inputs besides digital and PWM outputs. The secondary microcontroller acts as a *Serial to USB converter* that interacts with the computer. The *IDE* includes the means for writing programs (refereed as sketches), compile and upload them to the board [10].

In this project we used an ***Arduino Uno*** board and four extra ***ATMEGA328P*** microcontrollers preloaded with the ***Arduino*** bootloader. Many authors have used ***Arduino*** to operate their robots, usually employing the whole board in their project [3-4, 6], having the inconveniences of false connections, inefficient use of the ports, space and cost increment [I]. The present robot design requires four microcontrollers, making the use of four ***Arduino*** boards (**$25 USD** each) unsuitable. Our solution was the deployment of the ***ATMEGA328P*** microcontrollers (**$1.00 USD** each) to a standalone *PCB*s using resonators [II], surrounding electronics and power supply.

3. Architecture and modules

The robot developed in this project is shown in Figure 1, where the main components are highlighted. Figure 2 illustrates the main blocks of architecture of the robot. It shows the major components, modules and the means of communication between them. An ***ATMEGA328P*** microcontroller is used in every module except the *Control* module.

3.1. Frame and support

The robot is manufactured over a recovered electric scooter, where the front wheel was eliminated in favour of a welded Ackermann steering geometry [11], in which the two wheels came from an old bicycle. The main stand was removed and four pillars added to mount acrylic platforms to accommodate modules, components and cables.

3.2. Steering and traction.

The displacement of the robot is achieved through the management of a traction motor that powers a rear wheel. The original electric motor, gears, chain and rear wheel were kept, but a rotary encoder for motion measurement, adapted from a *PS/2 Mouse* board and acceded with a microcontroller using a library [12], was attached to the wheel. With the help of this encoder, the microcontroller monitors the number of turns that the traction wheel makes, becoming an odometer for the robot.

The motor is driven by an *H-bridge* using a *PWM* signal that is generated by means of a *PID* controller implemented from a library [13] also running in the microcontroller. Finally, direction control is fulfilled by governing the steering mechanism with a digital servomotor. Figure 3 illustrates this module.

The most expensive component in this module is the *L298 H-bridge* costing **$2.45 USD**. The price of manufacturing the *PCB* was **$6 USD**. The total amount was approximately **$10 USD**.

3.3. Power.
The main source of energy is a 4Ah, 12V, lead-acid battery. A microcontroller monitors the currents that are flowing through the modules, and control the charge rate of the battery with a *PWM* signal that manages a switching-mode power supply. The module also contains a *Step-Down Power*

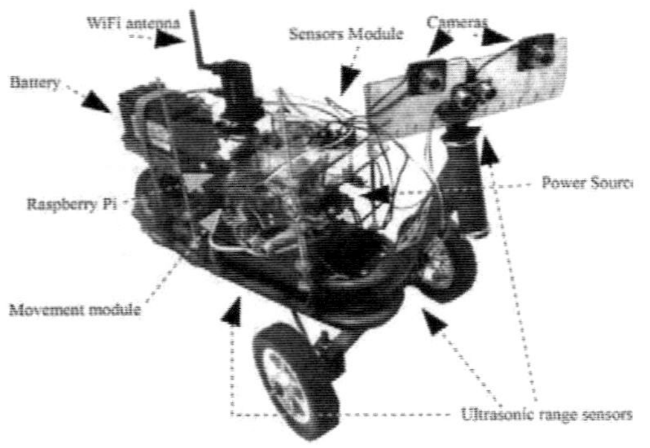

Figure 1. Robot.

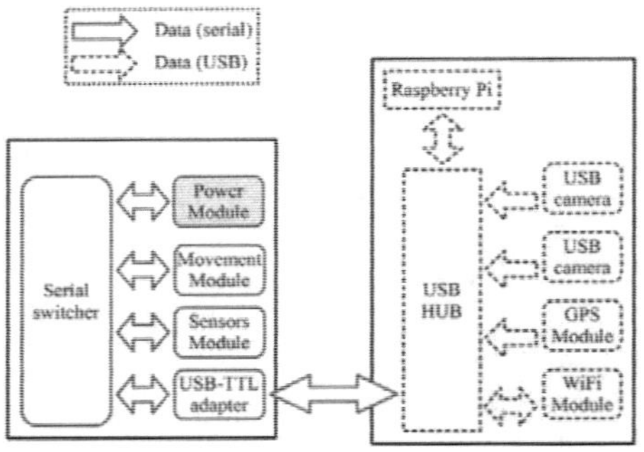

Figure 2. Architecture of the robot.

Converter that feeds all the electronic circuits in the robot, and a *LM7805* circuit that powers servomotors. Every component is protected by a fuse to avoid over currents or short-circuit damage. Figure 4 shows the steps required to regulate and condition the power from the battery in order to be used by the different components in the robot.

The most expensive components in this module were two capacitors and a power transistor (**$2.00 USD** each, used to build the switching-mode power supply), six fuse mounts (**$0.50 USD** each) and the Step-Down Power Converter (**$1.55 USD**). The price of the board was approximately **$14 USD** for a total cost of about **$27 USD**.

3.4. Sensors.
In order of the robot to navigate and monitor its internal conditions, it needs to obtain physical information of its environment via sensors. Some sensors need to be implemented in a board with a microcontroller in order for their output, which can be analog (as the temperature sensor [III]) or in the form of an **I²C** bus [IV] (like the *IMU* and the *Magnetometer*), to be read and adapted to a format the control module can understand.

Sensors implemented in a board with an ***ATMEGA328P*** microcontroller are: four *Ultrasonic Range sensors*, a *Temperature sensor*, an *Inertial Measurement Unit* [V] (*IMU*; combines *Accelerometer* and *Gyroscope*), and a *Magnetometer*. Figure 5 illustrates the sensors module, its components and interfaces. The total cost of the sensors is about **$11 USD**, plus **$1 USD** for the microcontroller and **$3 USD** for the board. Total is about **$15 USD**.
Other sensors are connected directly to the ***Raspberry Pi*** via an *USB* cable. Sensors connected directly to the ***Raspberry Pi*** are: two video cameras (**$5 USD** each) and a *GPS* module (**$20 USD**).

3.5. Communication.
A microcontroller is used to interact with the *Movement*, *Power Source* and *Sensors* modules. This microcontroller acts as a *"network switch"*, receiving packages of data and retransmitting them to.

Journal of Physics: Conference Series **582** (2015) 012007 doi:10.1088/1742-6596/582/1/012007

the adequate port, from the modules to the computer and vice versa. The price of this module is very low (approx. **$3 USD**), as only a microcontroller, terminals and a very small *PCB* are needed.

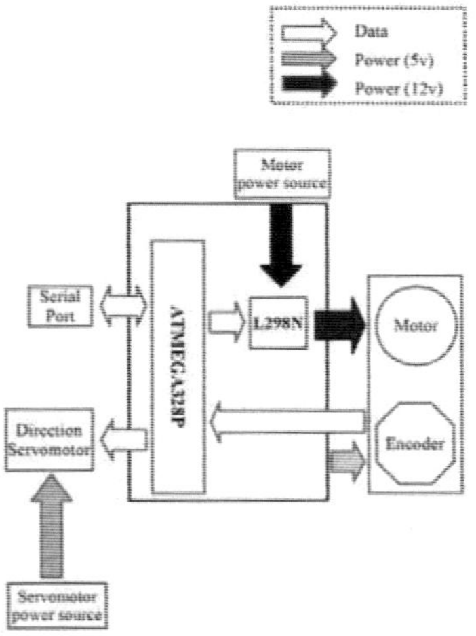

Figure 3. Movement module.

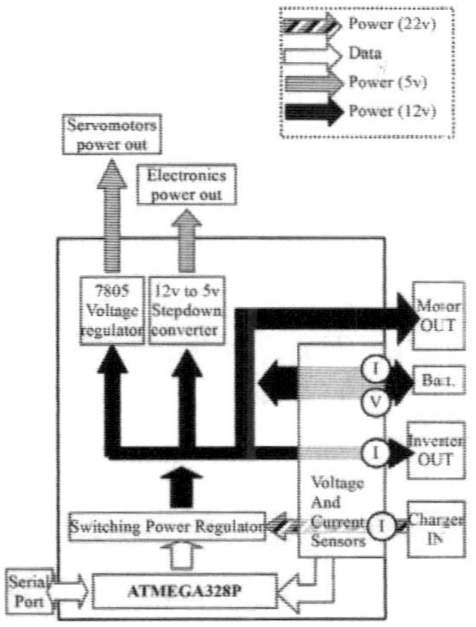

Figure 4. Power module.

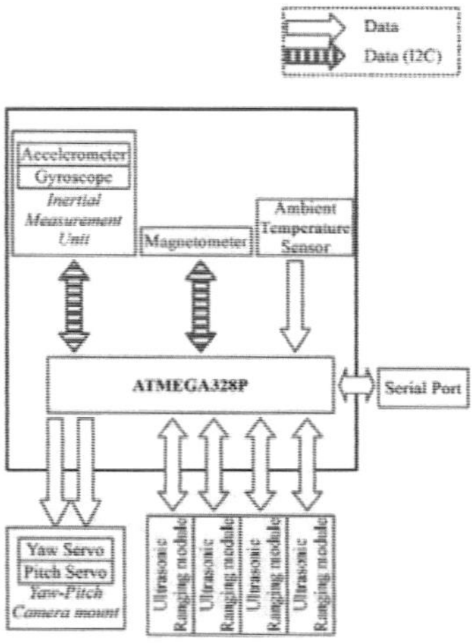

Figure 5. Sensors module.

3.6. Control.

The control module is composed of the ***Raspberry Pi*** and the devices directly connected to it [VI]. The computer seizes data from the modules, the *GPS* and the cameras; processes and stores information, send orders to the microcontrollers and receive instructions over the internet through the *Wi-Fi* module. The serial port information is analysed and catalogued by *Python* scripts. Data from the cameras is processed with the *OpenCV* library and the *GPS* data is shown in an interactive map. The robot can stream live video, and can be controlled with an *X-Box* controller connected to a remote computer.

The most expensive components are: the ***Raspberry Pi*** (**$35 USD**) and the *Wi-Fi* module (**$14 USD**). Total price is around **$50 USD**.

4. Results and future work

The robot has been completely assembled and the different modules were independently tested with satisfactory results. All the sensors and actuators are working properly within their intrinsic error and the computer communicates effectively with the operator. What follows is the theoretical mode of operation for the simple case of sending the robot to a specific point on the planet.

An automatic-travel algorithm is yet to be implemented. It would use information obtained from *GPS* to know a very good approximate of its position at a fixed moment, and estimate from there by using the odometer, accelerometer, gyroscope and magnetometer.

Obstacle avoidance and path finder algorithms will obtain their data from *GPS*, ultrasonic range sensors and stereo reconstruction from the cameras. Obstacle avoidance will have precedence over path finder, and top speed will be limited in unknown environments. Stereo reconstruction will focus in mapping the floor in front of the robot in furtherance of evading holes and other obstacles.

5. Conclusion

Using recycled parts from an electric scooter and a bicycle, microcontrollers and a computer, the full system for an exploration terrestrial robot for educational purposes was constructed. The total cost of the bought components is less than **$200 USD**, whereas many parts were already available (scooter, battery, mouse, cameras, ***Arduino***) so the total price starting from zero could raise up to **$300 USD**. Even if this is the case, the experience of:

- gathering information from various sources to implement different components and sensors;
- application of theoretical physics, mathematics and statistics for movement control, sensor output interpretation and calculations;
- development of microcontroller and computer software in several programming languages;
- design and implementation of communications schemes, mechanical parts, electric interfaces, electronic circuits and *PCB*s;
- putting it all together to accomplish the building of a robot; leaves more to the student than just experimenting with a pre-assembled kit.

Notes

I Nevertheless, one can state that ***Arduino*** is a single electronic board which provides convenient communication, ways of programming, surrounding electronics and electrical terminals for inputs and outputs, that make it ideal as a development platform.

II The ***ATMEGA328P*** can work at several frequencies, but 16MHz is what timing functions are optimized for in the ***Arduino Uno***. This means that if other frequency is used, functions as serial communication will not work unless the code for them is compensated for the frequency change.

III This sensor outputs 10mV/K and can be read with the *ADC* in the ***ATMEGA328P*** microcontroller with a precision of ±0.45K.

IV The ***ATMEGA328P*** microcontroller can connect to an **I^2C** bus with the use of the wire library.

V The *Digital Motion Processor* in the *IMU* is not very easy to use, and is still not completely open. The maker of the sensor is not clear on how to program it, and its use is almost not documented at all. Up to now, the only available source is a compiled *Demo Code* that **Jeff Rowberg** managed to reverse-engineer to create a very extensive library for the use of this sensor [14].

VI The ***Raspberry Pi*** computer has only two *USB* ports. In order to accommodate all this devices, an externally-powered *USB HUB* had to be used. It is recommended to be externally-powered for the current of the connected devices don't go through the ***Raspberry Pi.***

References

[1] The LEGO Group. *"LEGO MINDSTORMS EV3 LEGO Shop"*. Accessed 07-Jun-2014. 2013. URL: "http://shop.lego.com/en-US/LEGO-MINDSTORMS-EV3-31313".

[2] RobotShop. *"Festo Robotino Mobile Robotic Development Platform"*. Accessed 07-Jun-2014. 2013. URL: "http://www.robotshop.com/ca/en/festo-robotino-mobile-robotic.html".

[3] Luiz A. Junior et al. *"A Low-Cost and Simple Arduino-Based Educational Robotics Kit"*. In: Cyber Journals: Multidisciplinary Journals in Science and Technology, Journal of Selected Areas in Robotics and Control (JSRC), December edition, Volume 3, Issue 12 (2013).

[4] Wan Muhamad Hanif Wan Kadir, Reza Ezuan Samin, and Babul Salam Kader Ibrahim. *"Internet Controlled Robotic Arm"*. In: International Symposium on Robotics and Intelligent Sensors (IRIS 2012) (2012), pp. 1061-1071.

[5] Ricardo Neves and Anibal C. Matos. *"Raspberry Pi based stereo vision for small size ASVs"*. In: Oceans – San Diego. (2013), pp. 1-6.

[6] Wojciech Maślak and Bohdan S. Butkiewicz. *"Autonomous vehicle with fuzzy control"*. In:

Signal Processing Symposium (SPS), (2013). pp 1-6.

[7] Kteam. *"Kilobot"*. Accessed 11-Jun-2014. 2014. URL: "http://www.k-team.com/mobile-robotics-products/kilobot".

[8] Joshua M. Pearce. *"Building Research Equipment with Free, Open-Source Hardware"*. In: Science, 14 September 2012: Vol. 337 no. 6100 pp. 1303-1304.

[9] The Raspberry Pi Foundation. *"FAQs – Raspberry Pi"*. Accessed 27-May-2014. 2014. URL: "http://www.raspberrypi.org/help/faqs/".

[10] Arduino SA. *"Arduino"*. Accessed 16-May-2014. 2014. URL: "http://arduino.cc/".

[11] Wikipedia contributors. *"Ackermann steering geometry, Wikipedia, The Free Encyclopedia"*. Accessed 07-Jun-2014. 2014. URL:"http://en.wikipedia.org/wiki/Ackermann_steering_geometry".

[12] Chris J. Kiick. *"PS2 mouse interface for Arduino"*. Accessed 02-May-2014. 2008. URL: "http://playground.arduino.cc/ComponentLib/Ps2mouse".

[13] Brett Beauregard. *"Arduino PID Library"*. Accessed 28-May-2014. 2008. URL: "http://playground.arduino.cc/Code/PIDLibrary".

[14] Jeff Rowberg. *"I2Cdevlib. MPU-6050 6-axis accelerometer/gyroscope"*. Accessed 28-May-2014. 2013. URL: "http://www.i2cdevlib.com/devices/mpu6050#source".

Journal of Physics: Conference Series **582** (2015) 012008 doi:10.1088/1742-6596/582/1/012008

Development of Al-killed/Ti stabilized steels

A L Ramirez-Ledesma[1], M A Aguilar-Mendez[1], R A Rodriguez-Diaz[1], G Aramburo[2], J A Juarez-Islas[1]
[1]Instituto de Investigaciones en Materiales[1], Facultad de Quimica[2], Universidad Nacional Autonoma de Mexico, Circuito Exterior S/N, Ciudad Universitaria, C.P. 04510, Del. Coyoacan, Mexico, D.F., Mexico.

E-mail: julioalb@unam.mx

Abstract. Several Al-killed/Ti-stabilized low carbon steels were developed in a Mexican steel industry with the aim of obtaining an interstitial free steel for automotive applications. The steelmaking route involved the use of 100% sponge iron which was feed into an electric arc furnace, vacuum degassed, ladle treated and continuously casted. The resulting slabs were then hot rolled at 1100 °C and coiled at 650 °C. Then, the steel plates were cold rolled at room temperature and sheets annealed at 700 °C. As-cast microstructure showed the presence of α-ferrite with titanium nitrides in matrix and grain boundaries while in the as-hot rolled condition, elongated grains showed the presence of titanium nitrides, titanium sulfides and titanium carbosulfides. The annealed sheets showed, additionally to the other precipitates, the presence of titanium carbides. Microstructure, texture, the Lankford ratio and mechanical properties of fully recrystallized coils fulfilled the target properties established by the automobile industry.

1. Introduction

Al-killed/Ti-stabilized interstitial low carbon steel sheets have been developed for applications in the automotive industry because they fulfill the requirements for excellent drawability in order to produce panels with complex shapes [1]. High formability has been achieved as a result of lowering the interstitial content of carbon and nitrogen during the steelmaking practice and by the addition of stabilizing elements such as titanium, niobium and/or a combination of titanium plus niobium [2]. As a result, the ferrite matrix of heavily cold rolled interstitial low carbon and nitrogen sheet steels, recrystallized during annealing to a polycrystalline microstructure with a strong γ-fiber (<111>//ND) texture [3]. Attention has also been paid to the improvement of the final product by controlling hot rolling parameters [4]. For instance, finishing the hot rolling schedule in the ferrite region followed by annealing recrystallization resulted in stronger γ-fiber [5], and increasing the strain rate during ferrite rolling stage enhances the development of this texture [6]. Furthermore, lowering the rolling temperature from 850 to 600 °C also resulted in higher γ-fiber intensities [4] which have an impact on the drawability of annealed sheets. The recrystallization texture in metals has been discussed by means of the oriented nucleation [3] or the selective growth [7, 8] mechanisms; which for high drawability steels, implies that the first mechanism will be dominant [9, 10]. Additionally, it has been shown [11, 12] that γ-fiber texture is nucleated at the grain boundaries and proceeds to grow consuming the deformed neighboring grains. The present study shows the results of an industrial trial in order to obtain a low carbon Al-killed/Ti-stabilized steel which is intended to be used as an annealed sheet for deep drawing applications in the automotive industry.

2. Experimental procedure

The experimental steels were produced in a Mexican steel industry in order to fulfill a chemistry as that designated for interstitial free steels. The as-cast slab was thermomechanically treated, with sample dimensions of 0.03X0.1X0.2 m^3, at a temperature of 1100 °C and finishing at 840 °C, achieving 80% of accumulative hot reduction. The steel plate $5X10^{-3}$ m in thickness was air cooled in a run out table and coiled at 650 °C. Room temperature cold rolled plate resulted in a sheet with $3X10^{-4}$ m in thickness and with an accumulated reduction of 85%. Then, sheets were isothermally annealed under an argon atmosphere at 700 °C. Specimens were characterized by using a 440 Stereoscan scanning electron microscope and a 2100 Jeol scanning transmission electron microscope. The progression of recrystallization was followed by a point counting technique [13] while flat tensile room temperature test on fully recristallyzed samples were conducted on an Instron 1125 (10 t) test machine at a strain rate of 5×10^{-3} s^{-1}. From the tensile test experiments in the annealed condition, it was calculated the normal plastic anisotropy ratio value at 0°, 45° and 90° with respect to the rolling direction; from these values, the average plastic anisotropy ratio value (r) was obtained. Texture measurements on fully recristallyzed samples were carried out on a Siemens D5000 texture goniometer using cobalt radiation; the pole figures were measured up to a tilt angle χ = 80°; the steps were $\Delta\chi = 2.5°$ in the tilt direction and $\Delta\varphi = 4°$ in the azimuth direction.

3. Results and Discussion

Table 1 shows the chemical composition of experimental steels slabs where the steel AM3 was chose for performing the thermomechanical, mechanical and annealing treatment because it was the steel which showed a C + N_2 content <25 ppm falling in the designation of interstitial free steels.

Table 1. Chemical composition of experimental steels (in wt. %).

Steel designation	C	Si	Mn	P	S	Al	Ti	N_2
AM1	0.0050	0.328	0.534	0.0423	0.0089	0.0797	0.0260	0.0015
AM2	0.0034	0.0015	0.0741	0.0055	0.0110	0.0312	0.0602	0.0027
AM3	0.0023	0.0006	0.1050	0.0057	0.0089	0.0428	0.0692	0.0016

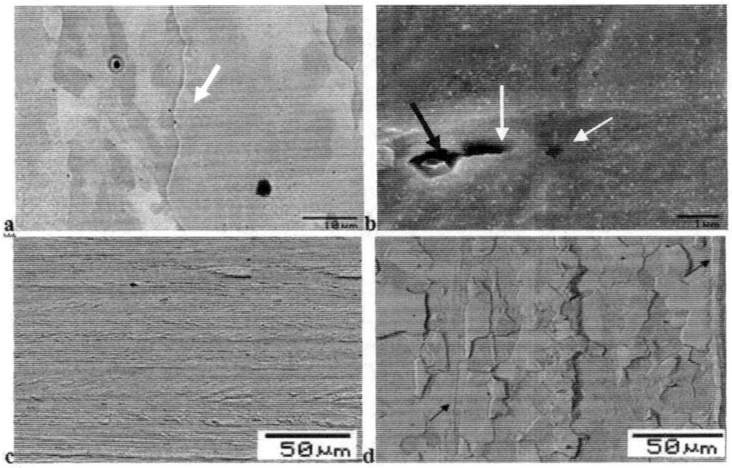

Fig. 1 Microstructure (a) As-cast, (b) hot rolled and coiled, (c) cold rolled, (d) annealed.

Journal of Physics: Conference Series **582** (2015) 012008 doi:10.1088/1742-6596/582/1/012008

In Fig. 1 are shown the microstructure obtained in as-cast ingots, hot rolled, cold rolled and annealed samples, for instance, Fig. 1a shows α-Fe grains with the presence of rhombohedral precipitates, which were identified by SEM-microanalysis as TiN (white arrow). The microstructure obtained in hot rolled plates, from 1100 to 840°C is shown in Fig. 1b where it was observed a microstructure of α-Fe grains elongated in the rolling direction with the presence of TiN, TiS (white arrow) and TiC_4S_2 (black arrow) precipitates and with an grain size of 20±0.9μm. Fig. 1c shows the microstructure of as-cold rolled specimens. As may be observed, the ferrite grains are flattened, and inside the grains, some shear bands can be observed, which corresponded to the narrow regions of intense shear that carry large strains during deformation and appear to become the major deformation mode [14]. For instance, in some interstitial free-steels, it was observed that in-grain shear bands were inclined principally at angles of 30° to 35° to the rolling plane [15]. In the steel sheet under study, the shear bands were inclined ~32° with respect to the rolling plane and were relatively planar and parallel to each other within the individual grains. The grain boundary, after the cold rolling operation, remaining closely parallel to the sheet plane and when the steel sheet was annealed, recrystallization initiated preferentially in the region of the original grain boundary (pointed by a black arrow in Fig. 1d); spreading from there into the adjacent grains. It appears that the coiling temperature of 650 °C in combination with the 85% per cent of accumulative deformation was sufficient to induce recrystallization at 700 °C in 180s. Fig. 1d shows nonequiaxed grain shapes during recristallization of sheets..

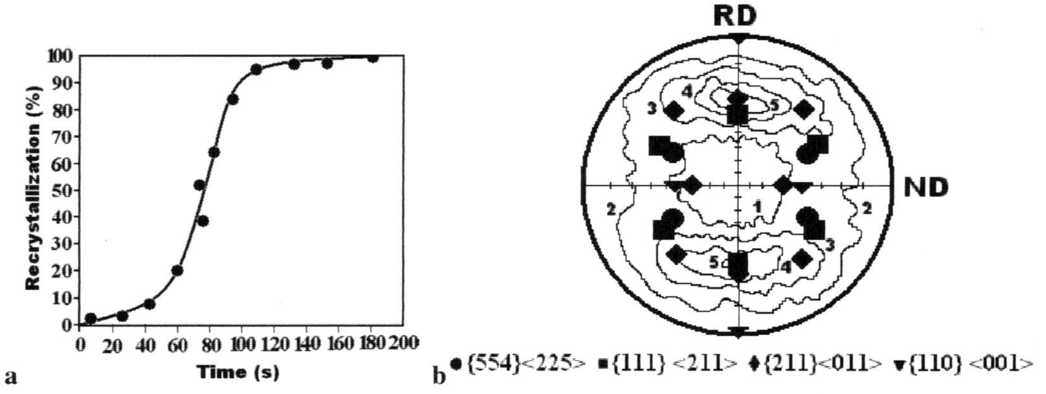

Fig. 2 Recrystallization curve and {100} pole figure for the steel under study in the annealed condition.

The recristallization rate of this IF-steel (Fig. 3a), is significantly lower than the rate of recrystallization of Al-killed steels (i.e. 100% of recrystallization in 20 s [2]). This delay in the recrystallization time is attributed to the contribution of Ti-precipitates, which retarded grain boundary migration by a pinning effect [16]. The ferrite grain size achieved in samples with 100% of recrystallization was 12±0.6 μm and the microstructure observed in this condition, consisted of a nonequiaxed grain shape. Fig. 3b presented a {100} pole figure for the steel under study in the annealed condition showing an appreciably strong texture than in the as cold rolled condition, from {554}<225> component near {111}<112> to {211}>011>. As reported in ref [6], during recrystallization, the pinning force ($3.7X10^5$ to $20X105N/m^2$) exerted by precipitates on the grain boundaries is one of the most important factors controlling the recrystallization texture. For instance, Barnett and Kestens [15] reported that for observations carried out in low carbon, ultra low carbon and interstitial free steels, regarding the presence of in-grain shear bands; the increasing density and severity of these in-grain shear bands lead to a bulk recrystallization texture dominated by {111}<112> near the normal direction–rolling direction (ND–RD), especially when titanium was added to the low carbon steel. On the other hand, industrial considerations in automotive technology are to improve safety and fuel efficiency by reducing vehicle weight and to shorten the manufacturing process of body panels. To achieve those goal s, interstitial free steels have been hot rolled in the austenite region. Reports show that when the finish discharge temperature was lower than the α/γ transformation temperature, the appearance of a coarse grain structure in hot bands and a decrease in the intensity of {111} texture in cold rolled and annealed sheets was observed, resulting in a deterioration of its drawability [16]. Further research [17] showed that hot rolling temperatures of interstitial free steels at 1050 °C resulted in similar values of the average plastic anisotropy ratio as compared to those IF steels hot rolled at 1200 °C. However, the resulting mechanical properties of annealed sheets may depend on the diversity of the steel composition, hot rolling process; finish discharge temperature, coiling, cold rolling process and annealing temperature. For instance, the most commonly produced steel corresponds to the interstitial free steel with carbon contents up to 50 ppm and nitrogen contents from 20 to 310 ppm, which have been stabilized with Ti and/or Nb additions. In spite of having similar compositions fulfilled target properties required in the automobile industry (see Table 3).

Table 2. Target and mechanical properties of interstitial free steel.

0.2 Y.S	T.S	El	r	n
138 to 173*	242 to 345*	46*	>2*	>0.21*
140	297	53.1	2.3	0.24

*Target properties

4. Conclusions

The low carbon Al-killed/Ti-stabilized steel shows outstanding behavior in terms of mechanical properties suggesting its use as an annealed sheet in the automobile industry. The automobile target properties were achieved in annealed sheets after employing low hot rolling temperatures, 1100 °C, instead of the typical 1250–1200 °C and annealing temperatures of 700°C, instead of the high recrystallization temperatures of 850 °C, normally needed in interstitial free steel processes.

The use of low rolling and annealing temperatures may represent industrial advantages from the point of view of reducing energy consumption of at least 40 MCal/t per 100°C. The use of titanium element to stabilize the low carbon steel was effective in retarding its recrystallization rate, enhancing formation of {111}<112> texture and Lankford values >2.

Acknowledgements

The authors are grateful for the technical assistance of Eng. Antonio Sanchez during the mechanical tests.

References

[1] C.M. Cady, S.R. Chen, G.T. Gray, D.A. Korezekwa, J.F. Bingert, 2000 Metall. Mater. Trans. A **31A** 2439.
[2] D.O. Wilshynsky-Dresler, D.K. Matlock, G. Krauss, 1994 International Forum for Physical Metallurgy of IF Steels-94, The Iron and Steel Institute of Japan, **May** 13.
[3] Y.Y. Hayakawa, J.A. Szpunar, 1997 Acta Mater. **45 (6)** 2425.
[4] M.R. Barrett, J.J. Jonas, 1997 ISIJ Int. **37** 706.
[5] T. Senuma, H. Yada, Y. Matsumura, K. Yamada, 1987 Tetsu-to-Hagané **73** 1598.
[6] S. Matsuoka, K. Sakata, S. Satoh, T. Kato, 1994 ISIJ Int. **34** 77.
[7] T. Urabe, J.J. Jonas, 1994 ISIJ Int. **34** 435.
[8] L. Kestens, J.J. Jonas, 1996 Metall. Mater. Trans. **A27** 155.
[9] W.B. Hutchinson, 1989 Acta Metall. **37** 1047.

[10] E. Lindh, B. Hutchinson, P. Bate, 1994 Mater. Sci. Forum **157–162** 997.
[11] M. Abe, Y. Yokobu, Y. Hayashi, S. Hayami, 1982 Trans. Jpn. Int. Metals **23** 718.
[12] H. Inagaki, 1984 Trans. Iron Steel Inst. Jpn. **24** 266.
[13] ASTM E112-82, 1991, ASTM, Philadelphia, Pennsylvania, p. 280.
[14] Y.B. Park, L. Kestens, J.J. Jonas, 2000, ISIJ Int. **40** 393.
[15] M.R. Barrett, J.J. Jonas, 1997 ISIJ Int. **37** 706.
[16] I. Gupta, T. Parayil, L.-T. Shiang, in: R. Pradhan, G. Ludkovsky (Eds.), Proceedings of Symposium On Hot and Cold Rolled Sheet Steels, TMS-AIME, Warrendale, 1988, p 139.
[17] O. Kwon, K.Z. Min, International Forum for Physical Metallurgy of IF Steels-94, The Iron and Steel Institute of Japan, May 1994, p. 9.

Journal of Physics: Conference Series **582** (2015) 012009 doi:10.1088/1742-6596/582/1/012009

Effect of rapid solidification and heat treatment on Co-20 wt. %Cr alloy for biomedical applications

A L Ramirez-Ledesma[1], M A Aguilar-Mendez[1], R A Rodriguez-Diaz[1], J A Juarez-Islas[1]
[1] Instituto de Investigaciones en Materiales, Universidad Nacional Autonoma de Mexico, Circuito Exterior S/N, Ciudad Universitaria, C.P. 04510, Del. Coyoacan, Mexico, D.F., Mexico.

E-mail: alramirez1303@hotmail.com

Abstract. A series of cobalt-base alloys with different chromium contents (20, 25, 30, 35, 40 and 44 wt. %) were melted into an induction furnace with argon atmosphere and casted into a chill cooper mold. The characterization of samples was carried out with a scanning electron microscope in order to evaluate the effect of chromium additions on microstructure. The resulting microstructure consisted mainly of columnar dendrites with randomly distributed precipitates in primary and secondary dendrite arms. X-ray diffraction patterns in as-cast samples identified the presence of both ε-hcp and the metastable α-fcc cobalt solid solution. As the Cr-content increased, the amount of both interdendrite segregation and precipitates increased too. From the Co-Cr alloys under study, the Co-20 wt. % Cr alloy showed a microstructure nearly free of interdendrite segregation and precipitation therefore was subject to an additional heat treatment to improve elongation from 2.6%, in the as-cast condition to 25.5% in the as-heat treated condition.

1. Introduction

Is well known that cobalt-base alloys are widely used in biomedical applications due to their excellent resistance to corrosion, mechanical properties and biocompatibility. A dental implant and/or prosthetic material suitable must function mechanically without permanent deformation or failure while leaving the biological tissues completely unaltered. Therefore, the materials used for surgical implants must meet principal requirements to perform in the human body: resistance to corrosion by physiological fluids; strong enough to avoid fracture because the normal physiological forces; and finally they must have a high degree of biocompatibility to ensure correct healing after insertion [1, 2]. These features can be achieved with a controlled solidification process and subsequent heat treatments to improve microstructure and mechanical properties of biomedical materials. ASTM F-75 is the most study cobalt-base alloy for biomedical applications and generally is fabricated with the investment casting process. Its standard specification establishes the mechanical properties as follows: yield strength, 450MPa (min); UTS, 650MPA (min); and a minimum ductility of 8% of elongation to fracture [3]. Many investigations has reported that the main defects in the as-cast cobalt alloys are: large grain size, porosity, chemical inhomogeneity and a microstructure with thicker interdendritic zones that may lead to low of mechanical properties of this materials [4,5]. Also, these alloys present the formation of various microstructural features including stacking faults, dislocations and twins which can affect the final properties of the alloys [6, 7]. Finally, recent studies report that precipitates presents in cobalt- base alloys affect the mechanical properties, wear and corrosion resistances [8]. In this work, we present results of solidified Co-20, 25, 30, 35, 40 and 44 wt. %Cr, in order to obtain an optimum as-cast microstructure free of interdendritic zones and a minor amount of precipitates. Also, were proposed a heat treatment to improve mechanical properties of cobalt-chromium alloy for biomedical applications.

2. Experimental procedure

Cobalt-base alloys with different chromium contents (20, 25, 30, 35, 40 and 44 wt. %Cr) were melted in a vacuum induction furnace under an argon atmosphere. Materials with high purity grade were placed into an alumina crucible and the liquid melt was casted in situ by gravity into a chill cooper molds with two isolated faces in order to keep the heat flux in the x-direction. The cooling rate was measured by placing a Pt/Pt-18%Rd thermocouple into the mold reaching values of ~$5X10^3$K/s. The resulting ingots were sectioned longitudinally, mechanically polished and electrolitically etched with 60% HNO_3 and 40% H_2O solution. Microstructure characterization was carried out in a scanning electron microscopy Stereoscan 440 coupled with a wavelength dispersed x-ray microanalysis. X-ray diffraction was performed in a Siemens D-5000 by using a $K_{\alpha\text{-}Co}$ radiation, a scanning angle (2θ) from 0 to 120° and a scanning speed of 5min/degree. Hardness HRc measurements were performed in a Matzusawa Hardness tester and mechanical properties were performed in a tensile test Instron machine 1210, using a travel speed of 0.5mm/min. From the experimental results in terms of microstructure, we selected the Co-20 wt. %Cr alloy in order to perform a heat treatment at 750°C during 60 minutes and air cooled.

3. Results and Discussion

A rapid solidification casting technique was performed to obtain ingots from chill-cast cobalt-chromium alloys because it has been reported that casting parameters have an impact on the mechanical properties, resistance corrosion and biocompatibility on these alloys [9, 10]. The microstructures obtained from chill cast solidification of alloys casted into cooper molds at $5X10^3$K/s cooling rates are shown in Figure 1.

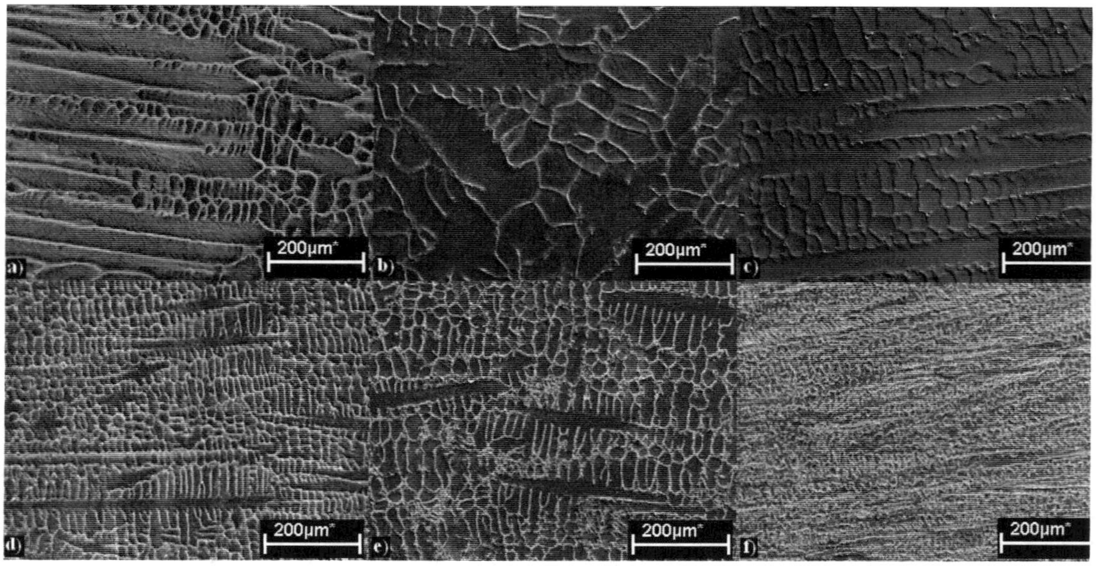

Figure 1. Microstructures observed in chill cast directional solidification of Co-Cr alloys; a) Co-20 wt. %Cr, b) Co-25 wt. %Cr, c) Co-30 wt. %Cr, d) Co-35 wt. %Cr, e) Co-44 wt. %Cr.

The resulting microstructures in Co-20, 25, 30, 35, 40 and 44 wt. %Cr alloys showed a columnar and/or equiaxed dendrites and, the effect of rapid solidification in cobalt-chromium alloys was observed regarding dendrite arm refinement as the chromium content increases and the extension of solid solubility in the eutectic Co-44 wt. %Cr alloy. From these alloys, we choose the Co-20 wt. %Cr alloy for apply in a heat treatment because its microstructure was nearly free of interdendritic segregation and a minimum of precipitates, see Figure 1a). On the other hand, Co-25 wt. %Cr and Co-30 wt. %Cr alloys have a major amount of precipitates respect to the Co-20 wt. %Cr alloy, Co_3Cr and CoCr precipitates were identified with wavelength dispersed x-ray microanalysis, showing a volume fraction of ~3X10-3, randomly distributed in the cobalt-chromium matrix, see Figure 1b) and Figure 1c). The interdendritic regions thickness was measured in Figure 1a) based on the scale of the micrograph taken by scanning electron microscopy and, the value was of the order of 5-6X10^{-3}mm in both Co-25 wt. %Cr and Co-30 wt. %Cr alloys. As the chromium content increased we can observed the presence of a major quantity of precipitates that corresponded to σ phase. Furthermore, there is an increment of interdenditic segregation and reduction in magnitude of primary dendrite arm spacing. All these features are in agreement with other research [11, 12]. The results obtained in this work showed a reduction or absence of interdendritic segregation and small volume fraction of precipitates in the Co-20 wt. %Cr alloy which is a very important result, consequence of control of microstructure during rapid solidification. Based on present results were obtained various ingots of as-cast Co-20 wt. %Cr alloy and were thermally treated at 750°C during 1 hour and air cooled.

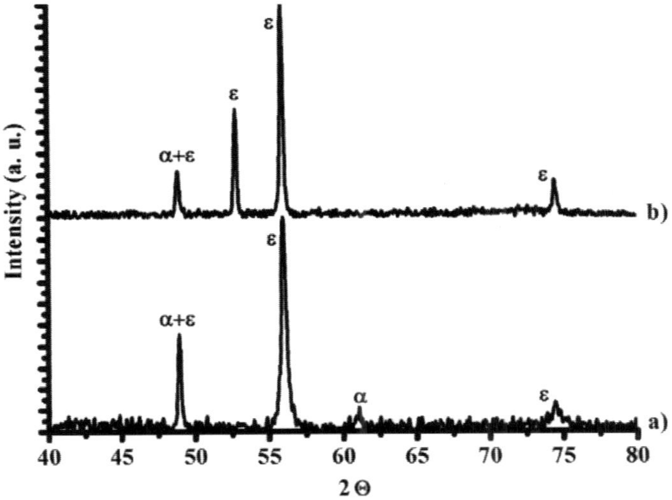

Figure 2. X-ray diffraction pattern: a) as-cast Co-20 wt. % Cr alloy, b) heat treated Co-20 wt. % Cr alloy.

The presence of both α-Co, fcc and ε-Co, hcp phases (metastable alpha -cobalt and epsilon-cobalt, respectively) are confirmed by x-ray diffractometry and shown in Figure 2. It's important to remark that generally cobalt-base alloys consists by a mixture of α-Co/ε-Co phases and, depending of solidification process and/or heat treatments applied to alloy, the α-Co or ε-Co phases are in minor or major percentage (independently of chromium content). Some studies reported an increase of ε-Co, hcp phase from heat treatments applied to cobalt-base alloys due to the promotion of fcc → hcp transformation and, as consequence an increase of mechanical properties in these materials. In agreement with those studies, we confirm a mayor presence of ε-Co phase as is seen in Figure 2b), that correspond to the heat treated Co-20 wt. %Cr alloy, in comparison with the diffraction pattern of as-cast Co-20 wt. %Cr alloy shown in Figure 2a). Also, C. Montero and co-workers indicated that the mechanical behavior of these two phase Co-27Cr-5Mo-0.05C alloys have shown that the hardness and yield strength of a 50% hcp alloy can be increase by at least 30% without ductility looses [13]. Those results are in agreement with this study where it was observed an improvement on mechanical properties from heat treated Co-20 wt. %Cr alloy, regarding ductility, we observed an important increase in ductility from 2.8% in the as-cast condition to 25.5% in the heat treated condition as is shown in Figure 3. This is a very important result since these values of elongation only are reported for cobalt-base alloys in as-cast condition (superalloys) with certain nickel percentage or in cobalt-base alloys in forged condition (that contain determined percentage of Ni too), because an amount of nickel is added to these alloys to improve their ductility by stabilizing the α-Co, fcc phase, in spite of the fact that nickel causes skin allergies or cancer in living organisms [14, 15].

Table 1. Mechanical properties of as-cast and heat treated Co-20 wt. %Cr alloy.

Co-20%Cr alloy	UTS (MPa)	σ_Y (MPa)	% Elongation	Hardness (Rc)
As-cast	357.09	240	2.6	18
Heat treated	596.243	325	25.5	24

Furthermore, various authors reported that nickel produce allergic reactions in human bodies, women mainly [16,17]. Table 1, show the results of mechanical testing made to as-cast and heat treated Co-20 wt. %Cr alloys. Hardness measurements are reported in Table 1 and we can note that as-cast alloy exhibit hardness value of 18HRc which is inferior to 24HRc value measured in heat treated sample.

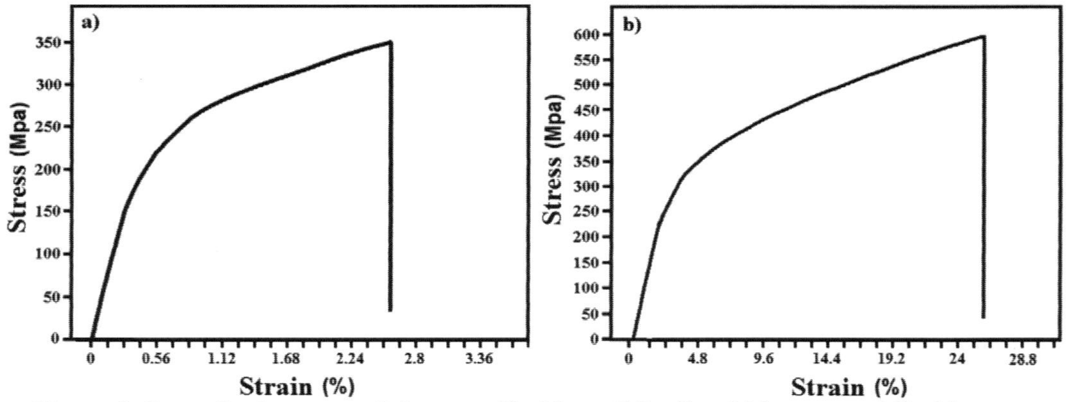

Figure 3. Stress-Strain curves of a) as-cast Co-20 wt. %Cr alloy, b) heat treated Co-20 wt. %Cr alloy.

Furthermore, ultimate tensile strength values exhibited by as-cast and heat treated alloys were slightly below to the lower limit for cobalt-base alloys. But is important to remark that the ASTM-F75 standard specification establishes the mechanical properties as follows: yield strength, 450MPa (min), UTS, 650MPA (min) and minimum ductility of 8% elongation to fracture and some researchers are in agreement with variation in mechanical properties data, particularity in ductility [3]. Beyond that, today there are greatest efforts to understand the complexity of cobalt-based alloys, for example, martensitic transformation mentioned above that been closely related with stacking faults and twins, which have a direct impact on mechanical properties as ductility and strengthening of these materials [18,19]. For this reason, is necessary to elaborate a controlled solidification process that permit to obtain a better characteristics in both microstructural (minimum precipitates and reduction or absence interdendritic segregation) and mechanical features in cobalt alloys. Since, it has been reported that even the materials employed in cast process and temperature of mold during casting are determinant in improving the mechanical properties of these materials [5,20,21]. The materials used in this investigation were selected carefully.

4. Conclusions

The results in this work showed that microstructures obtained from casting of Co-20, 25, 30, 35, 40 and 44 wt. %Cr alloys presented columnar and/or equiaxed dendrites of α-Co, fcc metastable phase and ε-Co, hcp phase. The effect of rapid solidification is present with the reduction of interdendritic segregation, minimum amount of precipitates and a notorious extension of solid solution that is evident in Co-44 wt. %Cr alloy. This is an encouraging result because we can obtained near free intedendritic segregation microstructure and a few precipitates in Co-20 wt. %Cr alloy at a cooling rate of $5X10^3$K/s. Therefore, an improvement in ductility was obtain in the Co-20 wt. %Cr alloy increasing from 2.6% in the as-cast condition to 25.5% after heat treatment. This response in the mechanical properties of alloy was consequence of fcc $\rightarrow$ hcp transformation that promotes an increase of ε-Co, hcp phase which was confirmed by X-ray diffraction patterns.

Acknowledgements
The authors are grateful for the technical assistance of Eng. Antonio Sanchez during the mechanical tests and M. C. Adriana Tejeda Cruz.

References

[1] Lee S H, Nomura N and Chiba A 2008 *Mater. Trans. JIM*. **49**, 260.
[2] Granchi D, Ciapetti G, Stea S, Savarino L, Filippini F, Sudanese A, Zinghi G and Montanaro L 1999 *Biomaterials*. **20**, 1079.
[3] Gomez M, Mancha H, Salinas A, Rodriguez J L, Escobedo J, Castro M and Mendez M 1997 *J. Biomed. Mater. Res.* **34**, 157.
[4] Giacchi J V, Morando C N, Fornaro O and Palacio H A 2011 *Mater. Charact.* **62**, 53.
[5] Kaiser R, Williamson K, O'brien C and Browne D J 2013 *Metall. Mater. Trans. A* **44**, 5333.
[6] Saldivar-Garcia A J and Lopez H F 2004 *Metall. Mater. Trans. A* **35**, 2517.
[7] Lashgari H R, Zangeneh Sh, Hasanabadi F and Saghafi M 2010 *Mater. Sci. Eng.* **527**, 4082.
[8] Narushima T, Mineta S, Kurihara Y and Ueda K 2013 *JOM* **65**, 489.
[9] Ramírez-Vidaurri L E, Castro-Román M, Herrera-Trejo M, García-López C V and Almanza-Casas E 2009 *J. Mater. Process. Technol.* **209**, 1681.
[10] Kaiser R, Browne D and Williamson K 2011 *Mater Sci. Eng.* **27**, 1.
[11] Rosenthal R, Cardoso B R, Bott I S, Paranhos R P R and Carvalho E A 2010 *J. Mater. Sci.* **45**, 4021.
[12] Mineta S, Namba S, Yoneda T, Ueda K and Narushima T 2010 *Metall. Mater. Trans. A.* **41** 494.
[13] Montero-Ocampo C, Juarez R and Salinas A *2002 Metall. Mater. Trans. A.* **33**, 22329.
[14] Jiang W H, Guan H R and Hu Z Q 1999 *Metall. Mater. Trans. A.* **30**, 2251.
[15] Yamanaka K, Mori M and Chiba A 2012 *Metall. Mater. Trans. A.* **43**, 5243.
[16] Roash M 2007 *Dent. Clin. N. Am.* **51**, 603.
[17] Niinomi M 2002 *Metall. Mater. Trans. A.* **33**, 477.
[18] Khlebnikova Y V, Rodionov D P, Sazonova V A, Tabatchikova T I, Antonova O V, Turkhan Y E and Kazantsev V A 2007 *Phys. MET. Metallography (USSR)*. **103**, 609.
[19] López H F and Saldívar-García A J 2008 *Metall. Mater. Trans. A.* **39**, 8.
[20] Montero C, Talavera M and Lopez H 1999 *Metall. Mater. Trans. A.* **30**, 611.
[21] Ghazvinizadeh H, Meratian M, Kermanpur A, Fathi M H and Minouei H 2010 *IJE Trans. A.* **24**, 49.

Journal of Physics: Conference Series **582** (2015) 012010 doi:10.1088/1742-6596/582/1/012010

Prediction of compression strength of high performance concrete using artificial neural networks.

A Torre[1], F Garcia[2], I Moromi[1], P Espinoza[3], L Acuña[3]
[1] Facultad de Ingeniería Civil, Universidad Nacional de Ingeniería. Av. Túpac Amaru, 210. Lima 25 – Perú
[2] Dpto. Ingeniería Forestal. Universidad Politécnica de Madrid. Ciudad Universitaria S/N, 28040 Madrid – España
[3]Facultad de Ingeniería Industrial y de Sistemas. Universidad Nacional de Ingeniería. Av. Túpac Amaru, 210. Lima 25 – Perú

E-mail: anatorre@uni.edu.pe

Abstract. High-strength concrete is undoubtedly one of the most innovative materials in construction. Its manufacture is simple and is carried out starting from essential components (water, cement, fine and aggregates) and a number of additives. Their proportions have a high influence on the final strength of the product. This relations do not seem to follow a mathematical formula and yet their knowledge is crucial to optimize the quantities of raw materials used in the manufacture of concrete. Of all mechanical properties, concrete compressive strength at 28 days is most often used for quality control. Therefore, it would be important to have a tool to numerically model such relationships, even before processing. In this aspect, artificial neural networks have proven to be a powerful modeling tool especially when obtaining a result with higher reliability than knowledge of the relationships between the variables involved in the process. This research has designed an artificial neural network to model the compressive strength of concrete based on their manufacturing parameters, obtaining correlations of the order of 0.94.

1. Introduction

High-strength concrete (CAR) is, nowadays, one of the most used construction material among those used worldwide. Manufacture from a number of basic components water, cement, aggregates (fine and coarse) and adding a number of additives, whose proportions largely determine their future mechanical properties. Of all compressive strength at 28 days is the most commonly used for quality control [1].

Since early twentieth century empirical model have been developed by various different authors that relate those parameters with final strength [1-5]. However, correlations obtained generally are rather low.

Artificial neural networks (ANNs) are known as a powerful modeling tools specially when is more important to obtain a highly reliable result from initial data than to know the relationship between the variables involved [6]. ANNs are a mathematical structure that attempts to imitate the functioning of a biological brain. Multilayer perceptron is the most used type of ANN. Its nature as universal aproximator [7] makes ita very useful tool to model and control industrial processes where traditional modeling techniques are not reliable [8]. Multilayer perceptron is made up of several nodes or neurons grouped in three layers named input layer, hidden layer and output layer (Figure 1). The main problem is that there is no set rule as to how many neurons the hidden layer should include or whether it should include a single or more than one sublayer. The input and output layers have as many neurons as the dimension of the input and output vectors. However, the only way to configure the hidden layer is by trial and error [9].

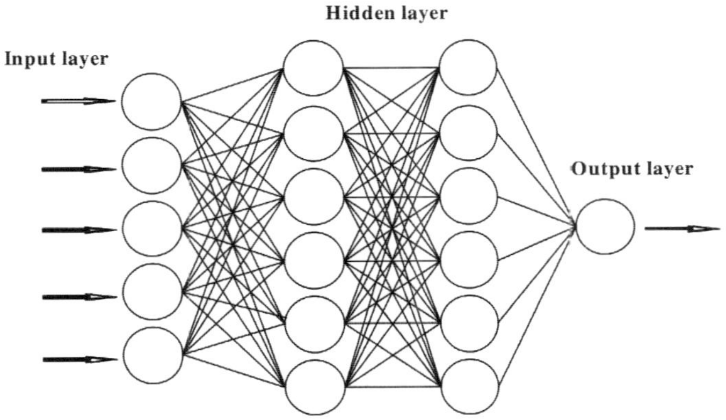

Figure 1. Scheme of a multilayer perceptron.

These structures have been widely used in fields ranging from structural wood based panels [10] to alloys [11] and concrete [3,4]

The aim of this study was to develop a multilayer perceptron artificial neural network in order to obtain the compression resistance of concrete by means of its manufacturing parameters.

2. Material and methods

This study used 296 manufactured concrete specimens with different types and quantities of cement, sand, and additives. Test pieces for compression test were made according to ASTM C 192 / C 192M [12] standards. Compression tests were performed according to ASTM C39 / C 39M [13] standards after different periods of curing.

Axial compression tests were performed on a machine: TECHNIK TONI with a 3.000KN cell and TINIUS OLSEN with 1.500KN cell. Of each specimen data input variables were the quantity of cement, amount of sand, cementitious microsilica and curing period. As output variable, resistance to compression was taken. Amount of water (186 l/m^3), stone (1066.2 kgf/m^3), fineness modules of sand (2.88) and stone(6.5), nominal maximum size of the stone (1) and specific gravity of stone (2.7 t/m^3) and sand (2.7 t/m^3) were maintained constant.

In order to obtain an optimal curing process, specimens were fully submerged in water pools throughout the process.

A principal components analysis was carried out before developing the multilayer perceptron in order to detect any possible correlations among the input variables. Variables that contribute more than 2% to the variability of the sample were selected [14].

Hyperbolic tangent sigmoid function (Eq. 1) was used as transfer function. This is equivalent to the hyperbolic tangent function and also improves network performance by producing a faster output

$$f(\theta) = \frac{2}{1+e^{(-2\theta)}} - 1 \quad (1)$$

f(θ): Output value of the neuron; θ: Input value of the neuron.

[14]

To improve the generalizing ability of the ANNs, input data were normalized [14] in accordance with Eq. 2.

$$\theta' = \frac{\theta - \theta_{min}}{\theta_{max} - \theta_{min}} \quad (2)$$

θ': Value after normalization of vector X; θ_{max} and θ_{min}: maximum and minimum values of vector X.

To avoid overfitting, which decreases ANN modeling capacity, the early-stopping method was used. For this propose the global set of data was divided in three auxiliary sets: training set with 164 data, validation set with 46 data and testing setwith 57 data. The first two sets were used while the development of the ANN and to avoid over-fitting. The test set was used to evaluate the degree of performance of the ANN avoiding any bias introduced by the groups of training and validation [15].

The network creation process was developed through a specific MATLAB® programme which applies automatically a principal component analysis to improve the learning process even when there are few input variables and successively increases the number of neurons in the inner sublayers and compares the results obtained in the training and validation sets every 100 training epochs. Around 20.000 neural nets were evaluated.

To evaluate the outcome a regression coefficient (*r*), the coefficient of determination (r^2), root mean square error (RMSE) were used.

3. Results and discucion.

Table 1 shows input variables and test results obtained in this study

Table 1. Test results. E_{ens}: time of curing; R: compression resistance.

	Variable						
	Water (l/m^3)	Mf_a	Mf_p	TNM	Stone (kgf/m^3)	PE_a (t/m^3)	PE_p (t/m^3)
Value	186.4	2.87	6.5	1	1066.2	2.7	2.7

	Property					
	Cement (kg/m^3)	Sand (kg/m^3)	% cementitious	% microsilica	E_{ens} (days)	R (kgf/cm^2)
Mean	632.5	376.2	2.1	7.2	-	752.3
Std. Dev.	63.3	82.3	0.3	2.1	-	95.8
Min.	506.0	282.5	1.6	5	7	565
Max.	708.3	546.9	2.5	10	56	1022

The principal components analysis showed a contribution level to the variability of the samples superior to 2% for each variable.

The best ANN obtained was a multilayer perceptron of three hidden sublayers of 7, 7 and 5 neurons each.

The results of the process of training, validation and testing are given in the following table (Table 2).

Table 2. Results obtained by the ANN development.

Net	Set	r	r^2	Equation	RMSE
Perceptron [5 7 7 5 1]	Training	0.92	0.84	y=1.12·x-78.1	36.6
	Validation	0.94	0.89	y=0.86·x+106.1	32.5
	Testing	0.91	0.82	y=0.71·x+198.2	34.6

Figure 2 shows the correlation between observed values and results obtained by the ANN for the testing set.

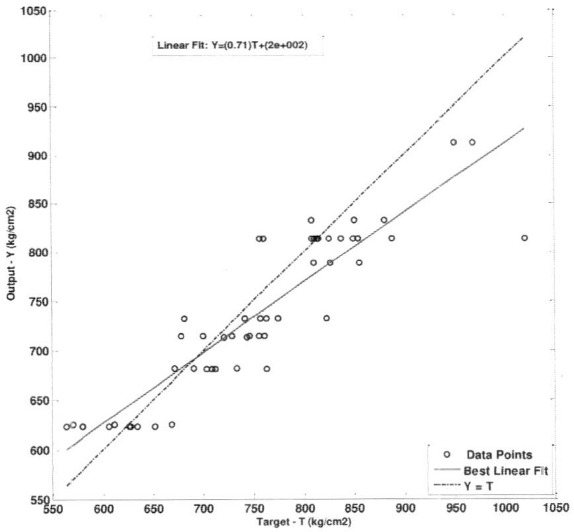

Figure 2. Correlation in the test set.

The coefficients of determination obtained in the test set indicate that the developed model fails to explain at least 82% of the variability of the samples.

There is not much works about the use of ANN in concrete properties modelling, and even less in CAR properties. Results obtained during the development of the network, with correlation coefficients between 0.90 and 0.94, are consistent with those obtained by other authors consulted, which obtained correlation coefficients between 0.81 and 0.98 [16-21].

Similarly, coefficients of determination obtained ($r^2 = 0.82$, $r^2 = 0.84$ and $r^2 = 0.88$) are higher than those obtained by Yeh [3] and similar to those obtained by Ozturan et al. [4] which were maximum of 0.78.

Yeh [3] and Ozturan et al. [4] also modelled concrete strength by means of regression models. However lower values were obtained in those studies. Thus, for these models, Yeh [3] obtained coefficients of determination between 0.46 and 0.89 and Ozturan et al. [4] obtained coefficients of determination between 0.55 and 0.79.

4. Conclusions

These results make ANN capable of being used to improve in-factory control quality as they predict the final resistance with high accurany and opens up a new area of application for ANNs in the field of concrete by applying neural networks to CARs for the first time.

5. Recognitions / Acknowledments

This work is part developed Project PIPA-1-P-676-13 FINCyT Research Project funded within the Programme of Applied Research Projects 2013 of Peruvian government.

6. References

[1] Kosmatka S H, Kerkhoff B and Panarese W C 2003 *Design and Control of Concrete Mixtures* (Illinois: Portland Cement Association)
[2] Abrams D A 1919 *Design of concrete mixtures* (Chicago: Structural Materials Research Laboratory, Lewis Institute)
[3] Yeh I C 1998 Modeling of strength of high-performance concrete using artificial neural networks *Cem. Concr. Res.* **28** 1797-808
[4] Ozturan M, Kutlu B and Ozturan T 2008 Comparison of concrete strength prediction
[5] Caldarone M A 2009 *High-Strength Concrete: A Practical Guide* (New York: Taylor and Francis).
[6] De Veaux R D and Ungar LG 1996 *A brief introduction to neural networks* http://www.cis.upenn.edu/~ungar/Datamining/Publications/nnet-intro.pdf. Last accessed: 2014/05/26
[7] Hornik K 1989 Multilayer Feedforward Networks are Universal Approximators, *Neural Netw.* **2** 359-66
[8] Sukthomya W and Tannock J 2005 The training of neural networks to model manufacturing processes *J. Intell. Manuf.* **16** 39-51
[9] Lin T Y and Tseng C H 2000 Optimum design for artificial networks: an example in a bicycle derailleur system *Eng. Appl. Artif. Intell.* **13** 3-14
[10] García Fernández F, de Palacios P, García Esteban L, García-Iruela A, González Rodrigo B and Menasalvas E 2012 Prediction of MOR and MOE of structural plywood board using an artificial neural network and comparison with a multivariate regression model, *Compos. Pt. B-Eng.* **43** 3528-33
[11] Reddy N S, Krishnaiah J, Hong S G and Lee J S 2009 Modeling medium carbon steels by using artificial neural networks Mater. *Sci. Eng. A-Struct. Mater. Prop. Microstruct. Process.* **508** 93-105
[12] ASTM C 192/C 192M 2000 *Standard Practice for Making and Curing Concrete Test Specimens in the Laboratory* (West Conshohocken, PA: The American Society for Testing Materials).

Journal of Physics: Conference Series **582** (2015) 012010 doi:10.1088/1742-6596/582/1/012010

[13] ASTM C 39/C 39M 2001 *Standard Test Method for Compressive Strength of Cylindrical Concrete Specimens* (West Conshohocken, PA: The American Society for Testing Materials)
[14] Demuth H, Beale M and Hagan M 2002 *Neural Network Toolbox User's Guide, Version 4* (Natick: The MathWorks Inc)
[15] Bishop C M 1995 *Neural networks for pattern recognition* (Oxford: Oxford University Press)
[16] Ozerdem M S and Kolukisa S 2009 Artificial neural network approach to predict the mechanical properties of Cu-Sn-Pb-Zn-Ni cast alloys *Mater. Des.* **30** 764-9
[17] Lee S C 2003 Prediction of concrete strength using artificial neural networks *Eng. Struct.* **25** 849-57
[18] Oztas A, Pala M, Ozbay E, Kanka E, Caglar A and Bhatti M A 2006 Predicting the compressive strength and slum of high strength concrete using neural network *Constr. Build. Mater.* **20** 769-75
[19] Ukrainczyk N and Ukrainczyk V 2008 A neural network method for analysing concrete durability *Mag. Concr. Res.* **60** 475-86
[20] Yaprak H, Karaci A and Demir I 2013 Prediction of the effect of varying cure conditions and w/c ratio on the compressive strength of concrete using artificial neural networks *Neural Comput. Appl.* ***22*** *133-41*
[21] Prasad B K R , Eskandari H and Reddy B V V 2009 Prediction of compressive strength of SCC and HPC with hay volumen fly ash using ANN *Constr. Build. Mater.* **23** 117-28

Journal of Physics: Conference Series **582** (2015) 012011 doi:10.1088/1742-6596/582/1/012011

Modeling of the Inductance of a Blumlein Circuit Spark Gap

V Aboites, L Rendón, A I Hernández, E Valdés
Centro de investigaciones en óptica, Loma del bosque 115, León, Guanajuato, México.

E-mail: aboites@cio.mx

Abstract. In this paper we present an analysis of the time-varying inductance in the spark gap of a Blumlein circuit. We assume several mathematical expressions to describe the inductance and compare theoretical and computational calculations with experimental results. The time-varying inductance is approximated by a constant, a straight line and two parables which differ in their concavity. This is the first time to our knowledge, in which the time-varying ignition inductance of a nitrogen laser is modeled.

1. Introduction

It is known that spark gaps are mainly used as fast switches in many circuits and in gas lasers such as CO_2 and N_2. The Blumlein nitrogen laser system is an example of a simple working tool used in many different branches of science [1-5].

Within a Blumlein gas pulsed laser there are two electrical discharge elements that play an important role on the laser efficiency, these are: the spark gap and the laser spine. Electrically a gas discharge represents a time-dependent resistance and inductance in series. However all electric analyses of these systems assume constant values for the inductance which we know is inaccurate. Experimentally it has been shown [6] that the inductance varies in a complicated manner in time. No doubt it is a difficult task the modeling of this element but an accurate electric analysis implies that the ignition system must be replaced by elements with a time variation.

In the literature there are few reports of inductance modeling in the spark gap. An exception is provided in [7-8] where it is show that assuming an arc-discharge with an electron Gaussian distribution, the resistance and inductance can be modeled as:

$$R = \frac{l^2}{\mu e N_t} \tag{1}$$

$$L = \frac{\mu_o}{4\pi} \int_{\varphi=0}^{\infty} \frac{\left(1 - e^{\lambda\varphi^2}\right)^2}{\varphi} d\varphi \int_{x=0}^{\frac{l}{2}} \left(\frac{\alpha}{\sqrt{\alpha^2 + \varphi^2}} + \frac{\beta}{\sqrt{\beta^2 + \varphi^2}}\right)^2 dx + \frac{\mu_o}{4\pi} \int_{\varphi=0}^{\infty} \frac{\left(1 - e^{\lambda\varphi^2}\right)^2}{\varphi} d\varphi$$

$$\int_{x=\frac{l}{2}}^{\infty} \left(\frac{\alpha}{\sqrt{\alpha^2+\varphi^2}} + \frac{\beta}{\sqrt{\beta^2+\varphi^2}}\right)^2 dx \tag{2}$$

Where R is the resistance, l is the inter-electrode distance, e is the electron charge, μ the electron mobility, N_t the total charge in the discharge channel, L the inductance, μ_o the magnetic permeability, φ the radial distance in cylindrical coordinates (perpendicular to l), $\alpha = (1 + 2x)/2$, $\beta = (1 - 2x)/2$ $\lambda = \ln(2/r_o2)$ and r_o is the radius of the cylindrical channel where the charge distribution has a value half its maximum value on the cylinder axis and is considered the radius of the discharge channel. Using these expressions it is possible to obtain a (constant) value for the resistance and inductance which is only an approximation.

The purpose of this article is to discuss possible analytic approximations for the time-dependent inductance. The objective is to find mathematically tractable expressions which may be used with an accuracy of ninety percent in the theoretical analysis of circuits involving spark gaps such as the Blumline gas laser. The temporal functions of inductance here discussed are approximations of the experimental time-dependent inductance observed.

This paper is divided into five sections. Section two describes the Blumlein circuit and explains the principle of operation of the ignition system. Section three presents the experimental and computational results of the voltage across one of the capacitors of the Blumlein circuit. This is done for each one of the inductance approximations. In section four, we comment on the agreement between the experimental data and the approximations done. Finally in section five we present our conclusions about which time dependent inductance approximation is more advisable to use both in relation to its mathematical simplicity and its accuracy.

1. Development

The laser Blumlein circuit is shown in figure 1. Here L.H. is the Laser tube Head, S.G is the Spark Gap or ignition system. This element consists of two electrodes separated by some distance. High voltage (H.V.) is fed to the system through R and L applying the same voltage across S.G. loading capacitors C_2 and C_1. When this voltage is sufficiently high to form an arc in the S.G. capacitor C_1 is discharged across S.G. and therefore is grounded.

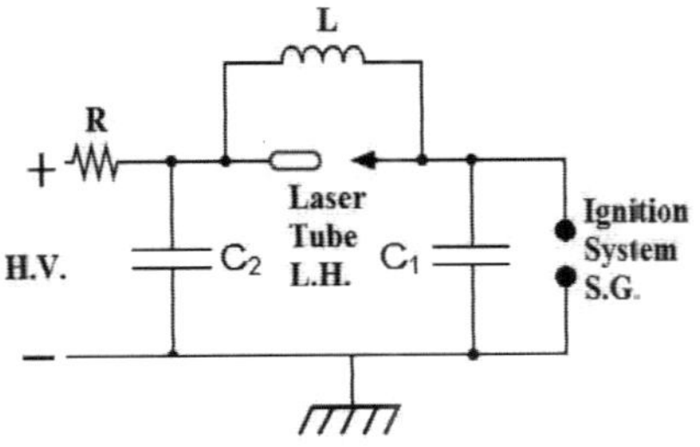

Figure 1. Electrical circuit Blumlein-line N_2 laser.

Inductor L avoids the discharge of C_2 through it, therefore instantaneously this capacitor will be at a higher voltage than C_1 (which is grounded), therefore the potential difference across this last element causes a discharge arc across the laser cavity head L.H. The process keeps repeating itself with a frequency that depends on the separation between the electrodes of the S.G.

The electric Blumlein circuit during the discharge of the spark gap and the laser head is shown in figure 2. Here it is noted that the L.H. discharge was replaced by a resistor R_2 and an inductor L_2. In a similar way the spark gap is replaced by a resistor R_1 and an inductor L_1. Typical values for U_o, R_1, R_2, C_1, C_2 and L_2 are: 10kV, 0.05 Ω, 0.06 Ω, 8 nF, 24.4 nF and 5 nH.

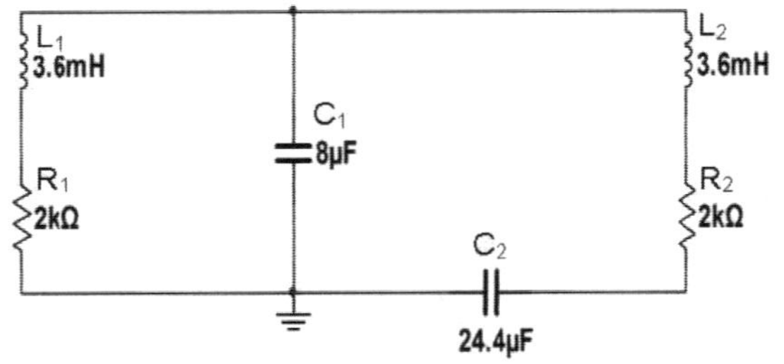

Figure 2. Equivalent electrical circuit Blumlein-line during the discharge of the spark gap and the laser head in a N_2 laser. L_1 and R_1 represent the spark gap discharge and L_2 and R_2 the laser head.

Before discharge occurs in the L.H. only the left hand loop of the circuit (L_1, R_1 and C_1) begins to oscillate due to the discharge of the spark gap and the dynamic of the net is governed by the following equation:

$$L_1 C_1 \frac{d^2 U_1}{dt^2} + R_1 C_1 \frac{dU_1}{dt} + U_1 = 0 \tag{3}$$

Where U_1 is the voltage across C_1. The particular solution of this equation making $U_1(0) = U_0$ and $U'_1(0) = 0$ is:

$$U_1(t) = U_0 e^{-at} \cos \omega t \tag{4}$$

Where:

$$a = \frac{R_1}{2L_1} \qquad \omega = \left(\frac{1}{L_1 C_1} - \frac{R_1^2}{4L_1^2}\right)^{1/2}$$

Once the discharge in the laser cavity head L.H. occurs, both nets of the above circuit are activated and the voltage starts to oscillate. Now the system dynamics is governed by the following differential equation [8]:

$$aU_i^{(4)} + bU_i^{(3)} + cU_i^{(2)} + dU_i^{(1)} + U_i = 0 \tag{5}$$

Where:

a=$L_1 L_2 C_1 C_2$,

b=$L_1 C_1 R_2 C_2 + L_2 C_2 R_1 C_1$,

c=$L_1 C_1 + (L_1 + L_2) C_2 + R_1 C_1 R_2 C_2$,

d=$R_1 C_1 + (R_1 + R_2) C_2$.

The subscript of U can be 1 ó 2 because the same differential equation applies for C_1 and C_2. Here attention is focused on the voltage across C_2.
The general solution of equation (5) has the form [8]:

$$U(t) = Ae^{-a_1 t} cos\omega_1 t + Be^{-a_2 t} cos\omega_2 t \tag{6}$$

Where A + B = U_o. As we may see solution (6) is the superposition of two oscillations.

2. Results

Figure 3 show an example of the experimental inductance of the ignition system [6]. The dashed lines shows the square of the flowing current. Which is important for our analysis. This experimental graph was also reproduced in Matlab with 2000 sampled points. This is shown in figure 4.

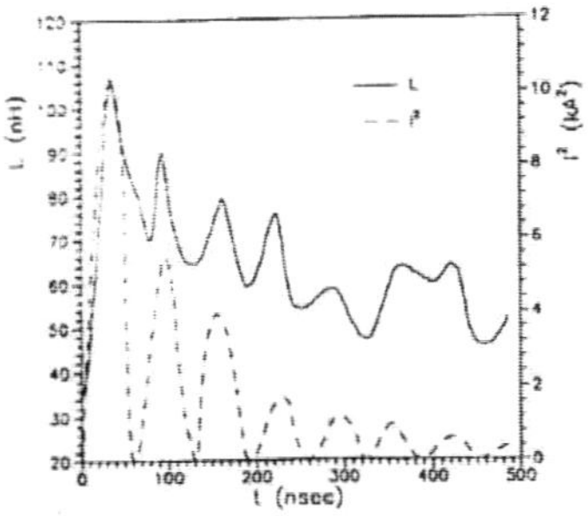

Figure 3. Temporal evolution of the inductance of the Spark Gap S.G.

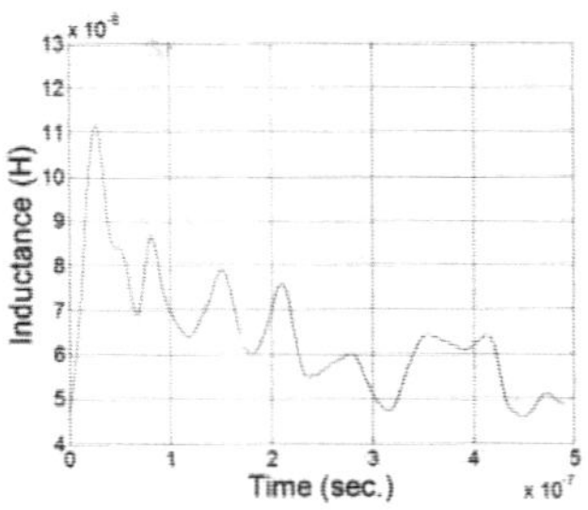

Figure 4. Sampling of the temporal evolution of theinductance of S.G.

As we may see from figure 4 the inductance at the spark gap is a very complex time varying function, which is not constant at all. Figure 5 shows the numeric solution of equation (6) taking the instantaneous value of inductance as is given in figure 5. This result should in principle be very close or identical to the one shown in figure 6.

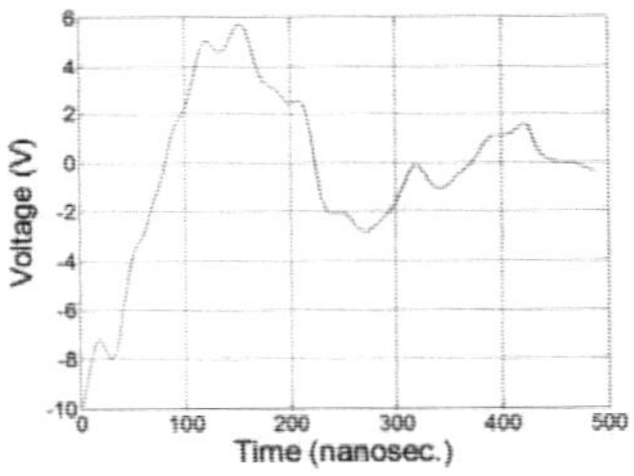

Figure 5. Numeric solution of the voltage at C_2 using the experimental inductance values shown in figure 3.

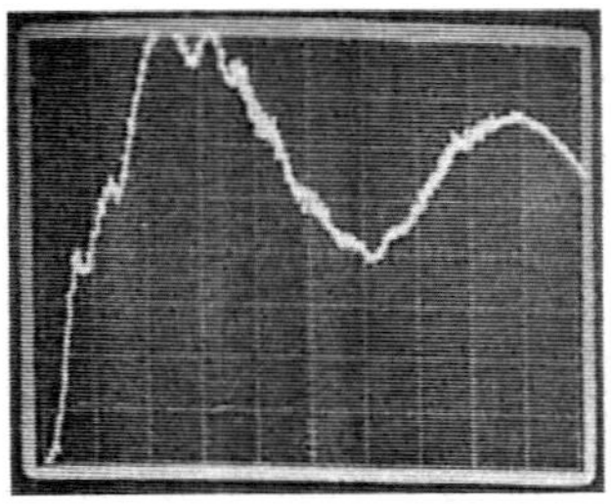

Figure 6. Voltage swing in C_2 when the laser discharge occurs.

Our task is to reproduce, as close as possible, the experimentally observed voltage variation at C_2 solving equation (5) using different approximations for the inductance, namely: i) constant, ii) straight line, iii) upward concave parable and iv) downward concave parable. Our objective is to find which time dependent inductance approximation is more advisable to use both in relation to its mathematical simplicity and its accuracy.

i) Approximation of inductance by a constant

The simplest approach is to take the inductance as a constant value. This constant is given by expression (7), which as we may see in figure 7 is just the average value of the experimentally observed inductance:

$$L = 6.4408x10^{-8}\text{H} \tag{7}$$

Figure 8 shows the voltage obtained when solving equation (5) using this approximation. The discussion of how good is this approximation is presented in the Discussion section.

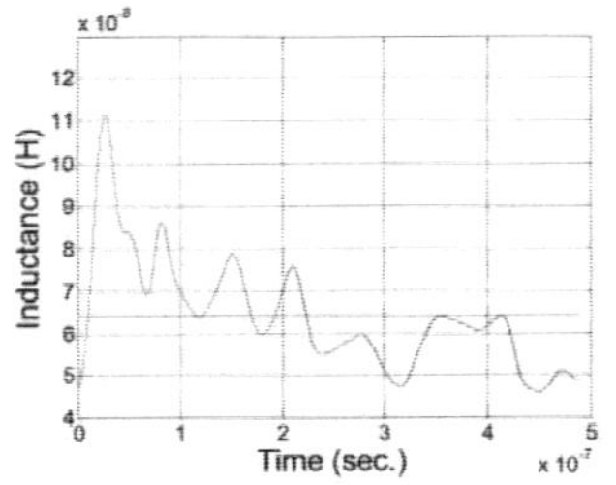

Figure 7. A constant approximation for the inductance.

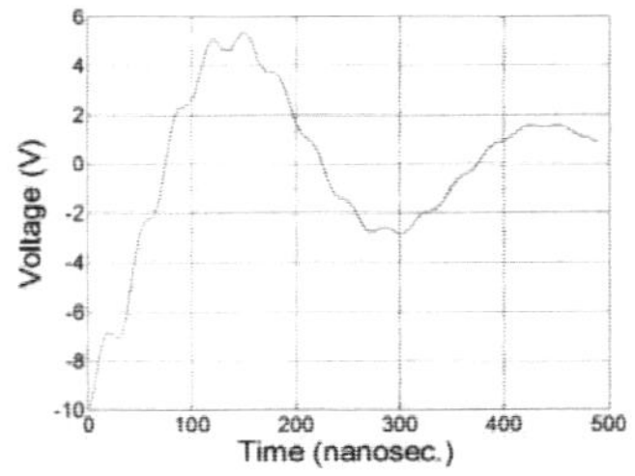

Figure 8. Voltage swing in C_2 using the constant approximation of the inductance

ii) Approximation of the inductance through a straight line y=ax+b.
The second approximation for the inductance is given by a straight line. In this case the temporal varying inductance is given by the next equation:

$$L = -6.3x10^{-11}t + 7.97x10^{-8}[\mathrm{H}] \quad (8)$$

This is shown in figure 9.
Numerically solving equation (5) using for the inductance the function given in expression (8) we obtain the voltage in C_2, which is shown in figure 10.

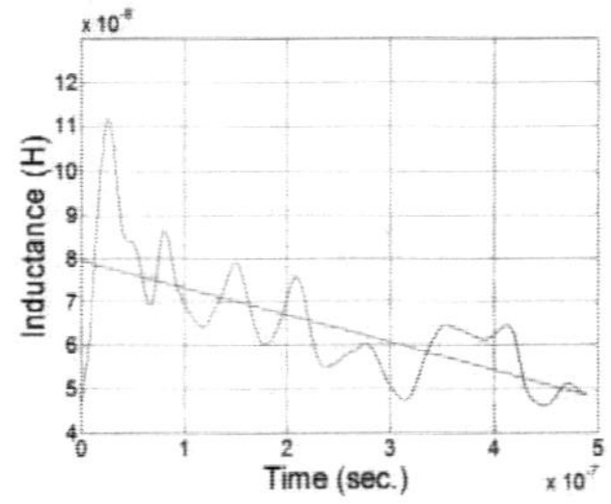

Figure 9. Linear approximation for the inductance.

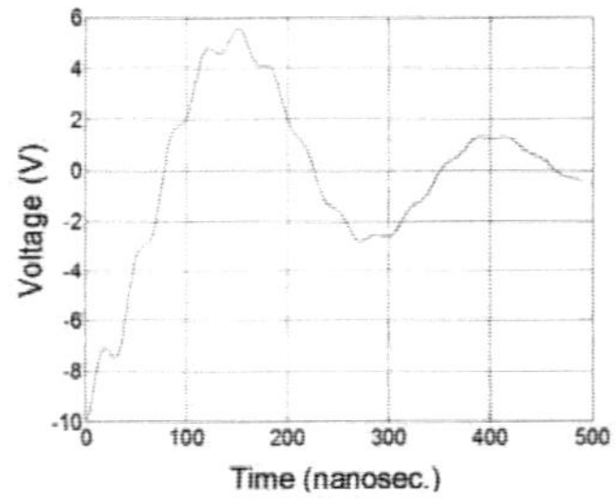

Figure 10. Voltage swing in C_2 using the linear approximation of inductance.

iii) Approximation of the inductance through an upward concave parable y= $\boldsymbol{ax^2 + bx + c}$.
The best fit of a upward concave parable for the time dependent measured impedance is show in figure 11. The parable equation is given by expression (9):

$$L = 7.7x10^{-14}t^2 - 1x10^{-10}t + 8.26x10^{-7}[\mathrm{H}] \quad (9)$$

The voltage at C_2 obtained using expression (9) to solve equation (5) is shown in figure 12.

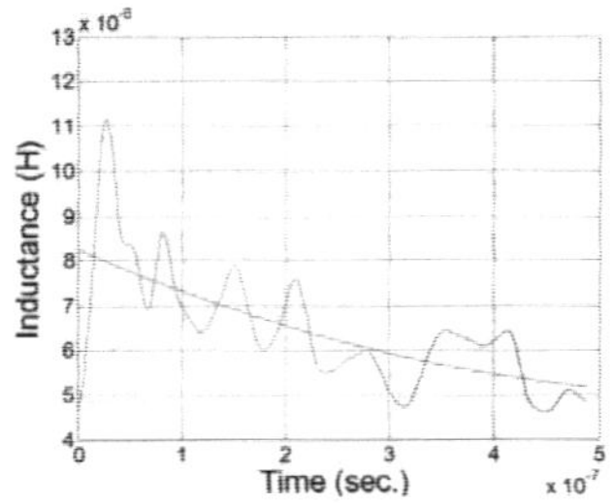

Figure 11. Parabolic approximation for the inductance.

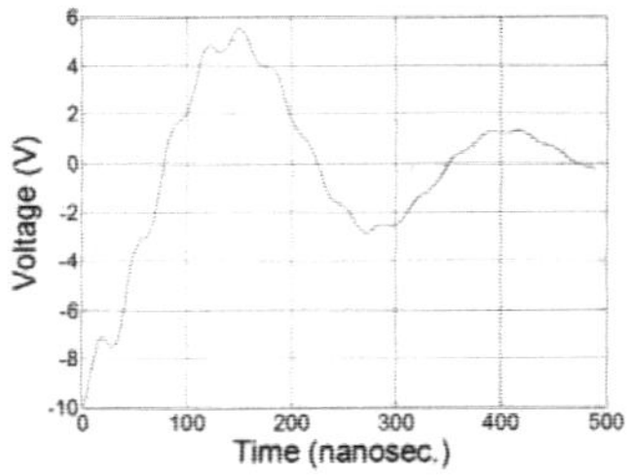

Figure 12. Voltage swing in C_2 using a parable approximation.

iv) Approximation of the inductance through an downward concave parable y=$-ax^2 + bx + c$.
The best fit of a downward concave parable for the time dependent measured impedance is show in figure 13. The parable equation is given by expression (10):

$$L = -3.81x10^5 t^2 + 0.19t + 4.6x10^{-8} \quad [\text{H}] \tag{10}$$

The voltage at C_2 obtained using expression (10) to solve equation (5) is shown in figure 14.

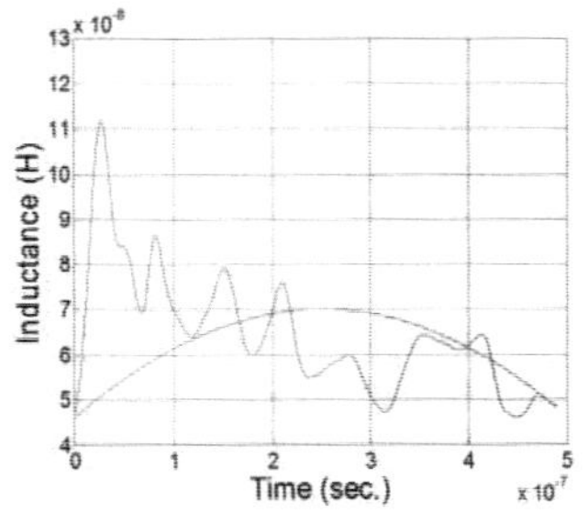

Figure 13. Second parabolic approximation for the inductance.

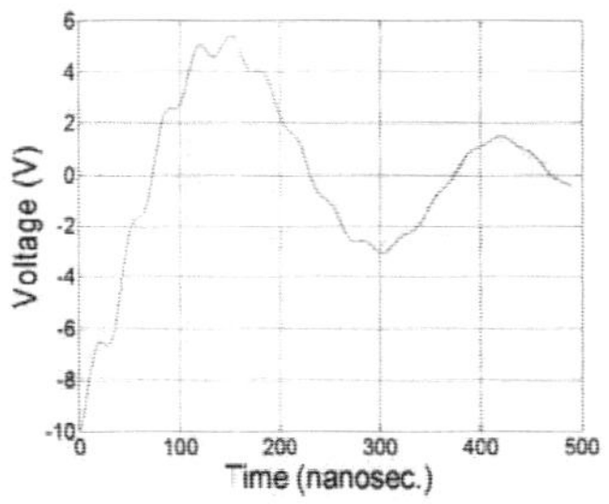

Figure 14. Voltage swing in C_2 using the second parable approximation of inductance.

3. Discussion

We deal with the problem of modeling the inductance of the spark gap S.G. within a pulsed Blumein circuit laser. Four different curves were proposed to simulate the time history of the actual inductance.

The four approximation used were taken due to their mathematical simplicity, since the aim of this work is to provide researchers with a good starting point for the modeling of the spark gap inductance in future designs of lasers with Blumlein circuits. It is therefore important to make a quantitative comparison of each result obtained with different time varying inductance approximation. With this purpose we define the r factor called determining factor [8]. This factor measure how closely two graphs resemble one to the other. When r = 1 both graphs are identical and when r=0 both graphs have not relation to each other. This factor is mathematically defined as:

$$r^2 = 1 - \frac{\Sigma(y_i - y_i')^2}{\Sigma(y_i - \mu'_{y_i'})^2} \tag{11}$$

Where y_i are the experimental results, in this case data of figure 5, y'_i are the approximation values and $\mu'_{y_i'}$ are the mean value of the approximation results.

The r values obtained for the graphs shown in figure 8,10, 12 and 14 taking as reference figure 5 are: for the constant approximation 0.9383, for the straight line approximation 0.9729, for the upward concave parable approximation 0.97404 and finally for the downward concave parable approximation 0.9249.

We may note that all the above approximations have r parameters that differ in a relatively small amount. In principle all of them are rather good approximations. We may summarize our results as follows table:

Journal of Physics: Conference Series **582** (2015) 012011 doi:10.1088/1742-6596/582/1/012011

Table 1. Results of values of r

Approximation	**Kind of approximation**	**Value of r**
Best	Upward parable	r = 0.97404
Excellent	Straight line approximation	r = 0.9729
Good	Constant approximation	r = 0.938
Fair	Downward parable	r = 0.9249

It is remarkable that the mathematical simplicity of a straight line is enough to guarantee very good results indeed. In fact the mathematical complexity of an upward parable does not justify - in our opinion- the very small gain in accuracy. On the other hand it is outstanding to realize that a constant approximation to the inductance gap can provide a very reasonable agreement with experimental results.

If computer power is not a limitation the greatest approximation may be obtained fitting the time varying inductance to a high degree polynomial for better accuracy

4. Conclusions

In this work we tested several approximations for the time-dependent inductance of the spark gap in a Blumlein laser. First we substituted an average constant value of inductance giving a value of 0.9383 to parameter r. It continued with the approximation of a straight line giving a value of 0.9729 to parameter r. The upward concave parabolic approximations gave an r parameter value of 0.9740 and the downward concave parable of 0.9249.The straight linear or constant approximation can be a good choice when modeling the time-dependent inductance of spark gap in a laser Blumlein. The result for the constant approximation is in agreement by more than 90 %. This result is in fact in agreement with common practice. This is the simplest approximation, which offers the best ratio of mathematical simplicity to accuracy. Therefore it is not a bad idea to use a constant value of inductance on a first calculation.

References

[1] Aboites V 2001 Handbook of Optical Engineering, Ed. Malacara-Thompson, Chapter 20, "Lasers", Marcel Dekker, New York

[2] Vazquez Martinez A and Aboites V 1993 High-efficiency low-pressure Blumlein Nitrogen laser *IEEE J. Quantum Electronics*, Vol. **29(8)** pp. 2364-2370

[3] Pinto V J and Aboites V 1991 High EfficientNitrogen laser *Revista Mexicana de Física* Vol. **37(3)** pp. 391-395

[4] Pinto V J, Aboites V and de la Rosa J 1991 N2 MOPA systemwith coaxial laser amplifier *Revista Mexicana de Fisica* Vol. **37(4)** pp. 665-670

[5] Kwek K and Chi O 1992 A Four Channel TEA N2 laser controlled by a multistage spark gap *Institute for Advanced Studies, University of Malaya,* 59100 *Kuala Lumpur, Malaysia*

[6] Persephonis P, Vlachos K, Georgiades C, and Parthenios J 1992 The inductance of the discharge in a spark gap *J. Appl Phys.* Vol. **71(10)** pp. 4755-4761.

[7] Persephonis P, Giannetas V, Georgiades C, Parhenios J and Ioannou A 1995 The influence of the external circuit on Arc-Discharge of a Spark-Gap: Its Application to a Pulsed Gas Laser *IEEE J.Q.E.* Vol. **31(3)** pp. 567-572

[8] Persephonis P 1987 Electrical behavior of a Blumlein-line N2 laser *J. Appl. Phys.* Vol. **62(71)** pp. 2651-2655.

Chapra S Aplied Numerical Methods With Matlab Third Edition pp 361

Journal of Physics: Conference Series **582** (2015) 012012 doi:10.1088/1742-6596/582/1/012012

Design and Construction of a pulsed Nd:YAG laser for LIBS applications

A Garcia-Villarreal[1] H Sobral[2]
[1] Departamento de Ciencias Básicas, Universidad Autónoma Metropolitana, Av. San Pablo180, Azcapotzalco, 02200 México D.F., México
[2] Centro de Ciencias Aplicadas y Desarrollo Tecnológico, Universidad Nacional Autónoma de México (CCADET-UNAM), Apartado Postal 70-186, México DF, 04510, México.

E-mail: lgvangel@gmail.com

Abstract. A multi-pulse Nd:YAG laser was designed and built, that can be used as an excitation source for Laser Induced Breakdown Spectroscopy (LIBS) experiments. A trigger and a power supply for the flash lamp has been successfully developed. A cooling system that uses a distilled water flow for the active medium was implemented. The laser has an output energy of 306 ± 7 mJ for 1 Hz repetition rate and its temporal profile is multi-pulse with 1 μs of average separation between them. To validate the system, the output was used as an ablation source of an aluminum target and the emission was investigated by spectroscopy analysis.

1. Introduction

In 1916 Albert Einstein established the theoretical basis for the laser and the maser, using Planck's black body radiation law and introducing the concept of stimulated emission of radiation [1]. The first laser was built in 1960 by T. H. Maiman in Hughes Research Laboratories. For this purpose, Cr^{3+} impurity ions present in Al_2O_3 crystal, known as "Ruby", served as an active medium in this laser. Later, Sorokin, Stevenson, Kaiser and Garret created the first 4 energy level, solid state laser. CaF_2 crystals activated by ions of a rare earth element provided the active medium. The next important step was taken by E. Snitzer in 1961 [2] who built the first solid laser doped with Nd^{3+} ions. The Nd^{3+}: YAG laser has been widely positioned, not only for its extensive use in scientific applications, but also for its increasing implementation in the medical field, the industry and even in cosmetology.

On the other hand, Laser Induced Breakdown Spectroscopy is an atomic emission spectroscopy (AES) analytical technique where a pulsed laser serves as an excitation, vaporization and atomization source of a sample. For this aim, a pulsed laser produces a plasma inside or onto the material, and the light is collected and spectrally analyzed [3]. Since the plume is produced by focusing of optical radiation, this method presents many advantages over other AES techniques that use alternate devices (such as electrodes and coils). LIBS allow *in situ* and remotely sample analysis without any target pretreatment. Quantitative and qualitative analysis is accomplished by following the intensities of the spectral lines. Although LIBS has more than 50 years of existence, many works still focus on the fundamentals of plasma physics. A review of LIBS capabilities shows that this method has a detection sensitivity comparable or greater than the one reached with other extended-field techniques.

1. Design and construction of the laser

The Nd:YAG laser basic scheme consists namely of an optical pumping system, an opto-mechanical system (to form the cavity) and a cooling system. To pump the active medium, the electrical conversion efficiency into optical energy is about 3% [1]. Since a significant fraction of this power is converted into heat, a cooling system must be designed, to avoid the warming of the optical pumping elements. Usually, the optical pumping system consists of a xenon or krypton flash lamp. The first one is preferred since it has emission peaks near the Nd:YAG crystal absorption bands. To couple the light from the flash lamp to the active medium a reflective elliptical cavity was implemented [4]. Typically, both, the flash lamp and the active medium have a thin cylindrical shape. Hence, the best coupling between them is achieved by placing both at the focal points of an ellipse.

The power supply for the flash lamp was designed taking in consideration that the laser frequency has to be variable up to 3 Hz and that the minimum energy output should be 100 mJ per pulse. Subsequently, a capacitor charging circuit with a current limiting resistor [1] and an external trigger was incorporated. Figure 1 shows the electrical circuit constructed for excitation of the xenon flash lamp.

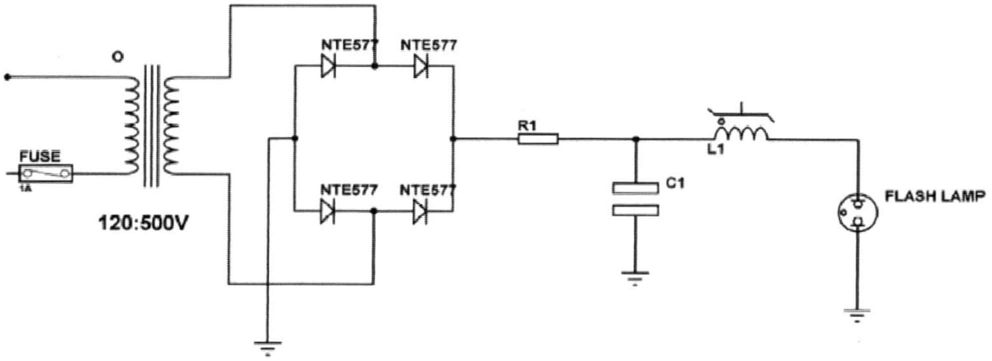

Figure 1. Flash lamp electrical circuit.

The capacitor C1 is charged through a 120:500V AC/200 mA mono-phasic transformer, a full-wave bridge rectifier and a 860 Ω/300 W current limiting resistor R1. With this circuit a 750 VDC output was obtained.

The Pulse Forming Network PFN circuit allows obtaining a critically damped current pulse according to the physical properties or the flash lamp [6]. The right size of figure 1 shows the corresponding circuit of the PFN. The flash lamp has an impedance 23 $\Omega \cdot A^{1/2}$ and the pulse width was set to 400 μs for an input energy of 31 J. With these values, a critically damped current pulse of 200 A peak is obtained when C1 = 110 μF and L1= 161 μH. C1 consists of three 330 μF electrolytic capacitors connected in series. L1 was built with an open ferrite core 2.6 cm in length and 1.0 cm in diameter, with a magneto wire of 0.62 mm in diameter turned 62 times around the core.

The trigger circuit is responsible for high voltage pulse generation, required to ionize the flash lamp gas, allowing the subsequent capacitor discharge through this. For low shooting frequency laser operation, a parallel or series external trigger configuration can be used [5]. In this work we implemented an external trigger where the high voltage pulse is directly applied to a thin copper wire wrapped around the flash lamp. Figure 2 shows the external trigger circuit designed for this purpose.

Here, the microcontroller PIC16F84A /20 MHz, was used as a pulse generator. A LCD screen was connected to this device to show the shooting frequency. Afterward, the generated pulses were used to control the SCRs IRF830 and 50RIA120. The BC547_1 and BC547_2 transistors amplify the signal from the microcontroller, and the SCRs are enabled or disabled by the signal. Moreover the SCRs in turn, allow charging and discharging the 1 μF capacitors, C3 and C4. The charging circuit consists of a 400 VDC /100 mA power supply and a 1.5 kΩ /100

W current limiting resistor. The discharge process is carried out through the pulse transformer which generates 12 kV peak per pulse in the secondary winding. A pulse transformer is usually constructed with a ferrite core, due its compact size. According to the ratio of the desired voltage, the windings on the secondary coil are usually between tens and hundreds of turns, whereas the primary coil has just a few turns [6]. The pulse transformer designed for this laser was assembled using a square ferrite core, with 2 turns of 1.12 mm in diameter of magneto wire for primary winding and 60 turns of 1.60 mm for secondary.

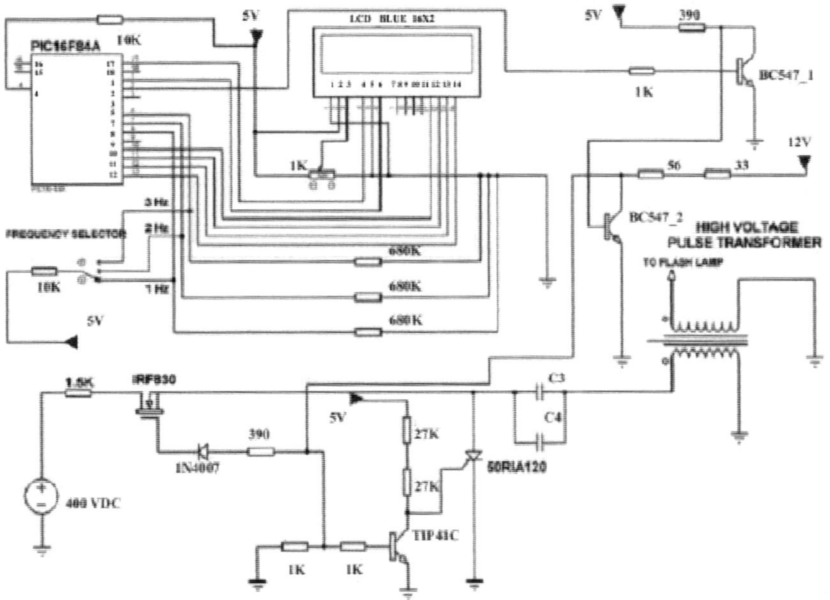

Figure 2. External trigger circuit.

The xenon flash lamp employed has an internal pressure of 470 torr, an arc length of 9.2 cm (between electrodes), and an internal diameter of 0.5 cm. The active medium is a Nd:YAG cylindrical rod of 11.1 cm in length and 0.8 cm in diameter. In addition, one end of the rod has a 99% reflective coating for 1064 nm; thus, only one mirror is needed to form the resonant cavity [7]. This is similar to the setup used in the first ruby laser, where the rod ends were polished to form the resonant cavity [1]. The optical coupling between the flash lamp and the active medium was performed through a ceramic single elliptical cavity. Finally, to avoid the flash lamp heating, a 60 L/hr flow of distilled water (conductivity of $2x10^{-6}$ $\Omega^{-1}\cdot m^{-1}$) was used. Water thermal dissipation mechanism consists of a radiator coupled to a fan.

2. Laser emission characterization and LIBS application

Laser emission from the built system (see Fig. 3a), was measured using a Power Meter (1918-C, from Newport) with a high power head (818P). The measured mean energy output was 306 ± 7 mJ at 1 Hz. Besides, the temporal output profile was obtained using a fast photodiode with a rise time of 1 ns (DET210, from Thorlabs), connected to a 1 GHZ bandwidth oscilloscope (DPO 4104B, from Tektronix). The laser output consists of a train of pulses of about 1 μs with a total duration of 250 μs as shown in Figure 3b.

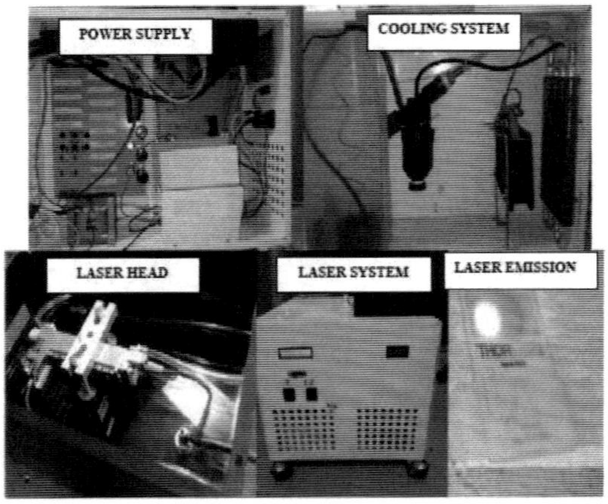

Figure 3a. Photograph of the built system.

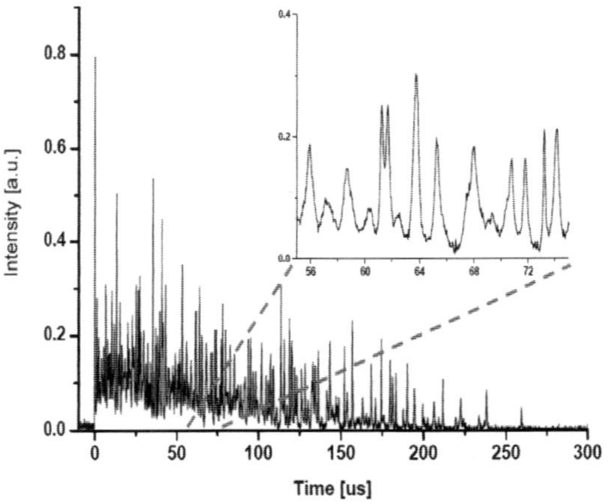

Figure 3b. Temporal profile of the laser emission.

To investigate the laser output as an excitation source for LIBS experiments, laser emission was focused onto a solid target, using a 100 mm plane-convex lens L1 (see Figure 4). The sample used was a 6463 commercial aluminum plate (Al 98%, Mg 0.8%, Si 0.5%, and other minor components), which had already been tested with a commercial Nd: YAG laser [8]. Ablated craters were found to be about 250 μm in diameter.

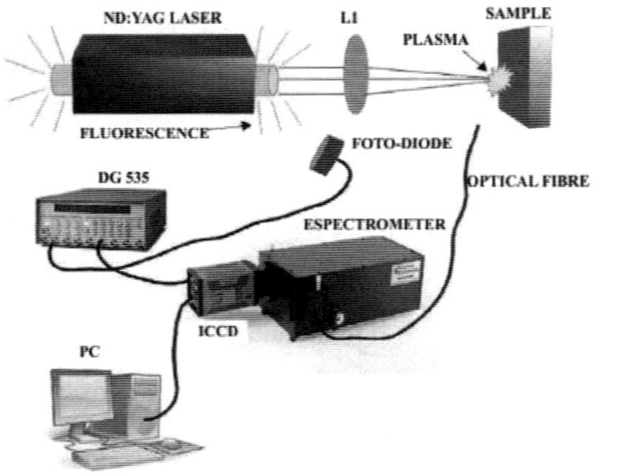

Figure 4. LIBS experimental setup used to analyze an aluminum sample.

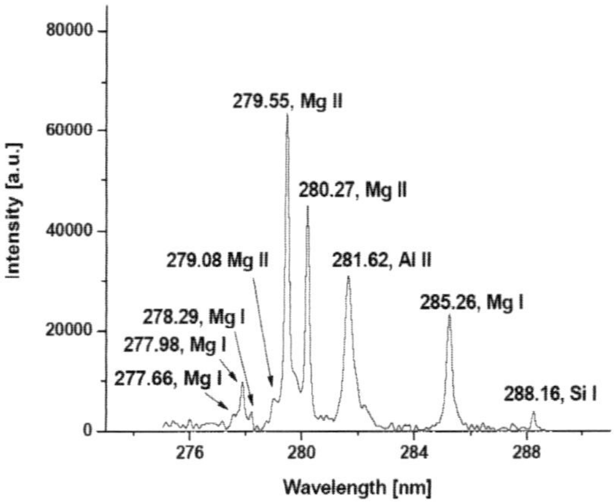

Figure 5. Typical LIBS spectrum of an aluminum sample.

The light emitted by the plasma is collected by a quartz optical fiber bundle and directed to a 500 mm spectrometer with 1800 lines/mm diffraction grating (Spectra Pro 2500, from Acton Research). The dispersed light was detected by an intensified CCD (PiMAX 1024x1024, from Princeton Instruments).

To reduce Bremsstrahlung continuum emission and radiative recombination processes [3], light collection was performed 500 ns after laser onset. Camera exposure time was set to 300 μs to acquire the whole light emission. The laser pulses and the diagnostics system were synchronized using a pulse/delay generator (DG535, from Stanford Research).

The collected emission for the spectral window centered at 282 nm is shown in Figure 5. This region has been chosen since it has several transitions corresponding to the main components of the sample. Emission spectrum is equivalent to the one obtained with a commercial nanosecond Q-switch laser [8], although the laser employed has a pulse duration two orders of magnitude larger. This could be due to its multi-pulse features, where successive pulses enhance material removal and simultaneously re-excitation of the produced plasma. Subsequently, it is possible to observe ionic and neutral species of aluminum, magnesium and silicon. Furthermore, the ions detected in this spectrum indicate that the generated plasma can reach a temperature of about 10,000 K.

3. Conclusions

A fully operational Nd:YAG laser, with a multi-pulse output was designed, constructed and characterized. The system has a high energy output, enough to induce breakdown in solid materials, and robust enough for portable LIBS applications. Re-excitation phenomena produced by this multi-pulse laser enhance ions and neutral emission, with figures equivalent to the obtained with other high-price commercial Q-switched lasers. The electrical circuits and all the parts used in this work are affordable and inexpensive in most countries. Besides, since this laser can be adapted according to a user requirement, it could be applied in different fields.

4. References

[1] Koechner W and Bass M 2003 *Solid-State Laser: A Graduate Text* (Springer-Verlag)
[2] Snitzer E 1961 *Phys. Rev.* **7** 444
[3] Miziolek A, Palleschi V, Schechter I 2006 *Laser Induced Breakdown Spectroscopy, fundamentals an aplications* (Cambridge University Press)
[4] Ciftan M, Luck F,Shafer G and Statz H 1961 *Proc, IRE,* 49:960
[5] McLeod A 2003 Desing Considerations for Triggering of Flashlamps *APP_Perkin Helmer*
[6] Markiewicz J and Emmett J 2003 Desing of Flashlamps Driving Circuits *IEEE J Quantum Electronics* **2** 11
[7] Marshall L 1972 *Laser Tecnology and Aplications* (McGraw-Hill Book Company)
[8] Sanginés R, Sobral M and Zauco E 2012 The Efect of Sample Temperature on the Emisión Lines Intensification Mechanism in Orthogonal Double-Pulse Laser Induced Breakdown Spectroscopy *Spectrochimica Acta* B **68** 40-45

5. Acknowledgments

This work was supported by the National Autonomous University of Mexico (DGAPA-UNAM: IN101212) and CONACyT-Mexico (221506). We thank E. Martinez Loran for the critical reading of this manuscript.

Preparation and characterization of bismuth nanostructures deposited by pulsed laser ablation

L Escobar-Alarcón[1], J G Morales-Mendez[2], D A Solís-Casados[3], S Romero[1], M Fernández[1], E Haro-Poniatowski[2]

[1] Departamento de Física, Instituto Nacional de Investigaciones Nucleares, Apdo. Postal 18-1027, México DF 11801, México.
[2] Departamento de Física, Universidad Autónoma Metropolitana Iztapalapa, Apdo. Postal 55-534, México, D.F., México.
[3] Centro Conjunto de Investigación en Química Sustentable UAEM-UNAM, Carretera Toluca-Atlacomulco Km 14.5, Unidad San Cayetano, Toluca, Estado de México, 50200, México.

E-mail: luis.escobar@inin.gob.mx

Abstract. Bismuth nanostructures, from nanoparticles to quasi-percolated films, were deposited by pulsed laser ablation (PLA) on different substrates using the 355 nm line of a Nd:YAG laser. The morphology and size distribution of the obtained nanostructures were investigated, as a function of the number of ablation pulses, by high resolution electron microscopy (HRTEM) and atomic force microscopy (AFM). Deposits with a small number of pulses, 50, are formed of separated isolated particles with diameters in the range from 5 to 20 nm. Further increase in the number of pulses (>100) results in coalescence of individual particles with the formation of dendritic structures and finally, for 500 pulses, quasi-percolated Bi films are obtained. Additionally, the nanostructures formed were characterized by XPS, and Raman spectroscopy in order to determine the physical and chemical properties of the deposited material.

1. Introduction

Nanostructured materials have attracted great interest because their physical properties are different from those of the same bulk material. The differences arise owing to the nanoscale dimensions of nanostructures that favour quantum effects due to spatial confinement of the carrier charges and corresponding modification of their density of states (DOS) [1]. It is worth mentioning that not only the size but also the shape of the nanostructures determine some important materials properties. Therefore investigating preparation techniques that allow the control of size and shape of nanostructures is a key issue in this field. It has been shown that pulsed laser ablation at low pressure can produce bismuth nanostructures in an easy way. In particular, the solid–liquid-solid transitions of nanocrystalline bismuth, embedded in amorphous germanium, prepared by PLA in vacuum have been studied [2]. Furthermore, the optical properties of nanocomposite films consisting of Bi nanostructures embedded in Al_2O_3 prepared by the same technique have been investigated [3] in detail. Bismuth is a semimetal which has unusual electronic properties resulting from its highly anisotropic Fermi surface, low carrier concentration, small effective mass and long mean free path of the charge carriers [3]. As a consequence, Bi nanostructures have been considered good candidates for nano-sized materials with novel physical, chemical and thermal properties. Therefore there are a number of research efforts looking to synthesize bismuth and bismuth-based compounds at the nanoscale controlling the properties of the deposited material.

2. Experimental Procedure

The nanostructures were produced by laser ablation in a vacuum chamber (Intercovamex) at low pressure, close to 2 x 10^{-5} Torr, ablating a high purity bismuth target. A Nd:YAG laser (Spectra Physics) emitting at the third harmonic (355 nm) with 10 ns pulse duration was used as energy source. The laser energy per pulse and the spot size were 5 mJ and 1.5 mm respectively, which correspond to a laser fluence of 300 mJ/cm^2. The deposits were performed at room temperature on different substrates: glass, TEM carbon-coated copper grids and pieces of silicon (100) wafers. The morphology and size distribution of the nanostructured films was investigated as a function of the number of pulses used to ablate the target. The substrates were placed 5 cm away and in front of the Bi target.

The as-prepared bismuth nanostructures onto the TEM carbon-coated grids were observed in a JEOL 2100 transmission electron microscope in order to analyze their shape, size and structure. Determination of the elements presents in the deposits as well as their bonding features were done by X-ray Photoelectron Spectroscopy (XPS) using a Jeol JPS 9200 XPS. Raman spectroscopy (RS) was used to study the vibrational features of the deposits using an HR LabRam 800 system equipped with an Olympus BX40 confocal microscope and a Nd:YAG laser (532 nm). The amount of Bi deposited on silicon substrates was estimated from Non-Rutherford Backscattering Spectroscopy (NRBS). Measurements were performed in a Tandem Van de Graff accelerator using a proton beam of 2.8 MeV and a detection angle of 165°. AFM images were acquired with an Asylum Research microscope operating in the non-contact mode.

3. Results and Discussion

3.1 Transmission Electron Microscopy (TEM)

In order to investigate, and eventually control, the morphology and size of the Bi nanostructures and therefore the properties of the obtained materials, the number of pulses used to ablate the Bi target was varied from 50 to 500. Representative TEM images of Bi nanostructures as well as their corresponding size distributions at different number of pulses (50, 100, 200, 300 and 500) are shown in figure 1. As it is observed in figure 1a, the sample prepared using 50 pulses consists of nearly spherical nanoparticles with a mean diameter close to 6 nm and a size distribution from 5 to 16 nm. The sample deposited using 100 pulses (figure 1b) is formed by well-separated nanoparticles (characteristic of a nucleation stage of growth) with bigger sizes, around 12 nm, and with wider size distribution. In this case, it is clearly observed that nanoparticles exhibit faceted irregular shapes and the covered area was increased. When 200 pulses are used, the mean diameter increases to 25 nm with a size distribution from 5 to 70 nm (figure 1c). This broad and asymmetric distribution is indicative of coalescence processes between the individual particles as a result of a larger covered area that favors their interaction. The deposit obtained with a number of laser shots greater than 300 shows a quasi-percolated state in which nanostructures with mean diameter of 33 nm are observed (figure 1d). At the highest number of pulses used in this work (500) a quasi-percolated film is observed leaving channels of uncovered substrate (figure 1e), with Feret´s diameters (defined as the measured distance between parallel lines that are tangent to the particle`s profile and perpendicular to the ocular scale) close to 60 nm. The HRTEM image of this sample shows the presence of nanocrystals with sizes ranging from approximately 10 to 20 nm. From this image the measured interplanar distances are 3.3 and 2.2 Å that agree well with the values of Bi (012) plane (3.3 Å) and (110) plane (2.2 Å) according to the data reported in the JCPDS database (00-044-1246). This result reveals the polycrystalline nature of the deposited Bi nanostructures. Figure 1 f) shows the plot of the Feret's diameter of the Bi nanostructures in the deposits, as a function of the number of laser pulses. As is clearly observed, the mean Feret's diameter increases with the number of laser shots following a linear relation. This is an interesting result because it allows the control of the mean size of the nanostructures in a straightforward way.

Journal of Physics: Conference Series **582** (2015) 012013 doi:10.1088/1742-6596/582/1/012013

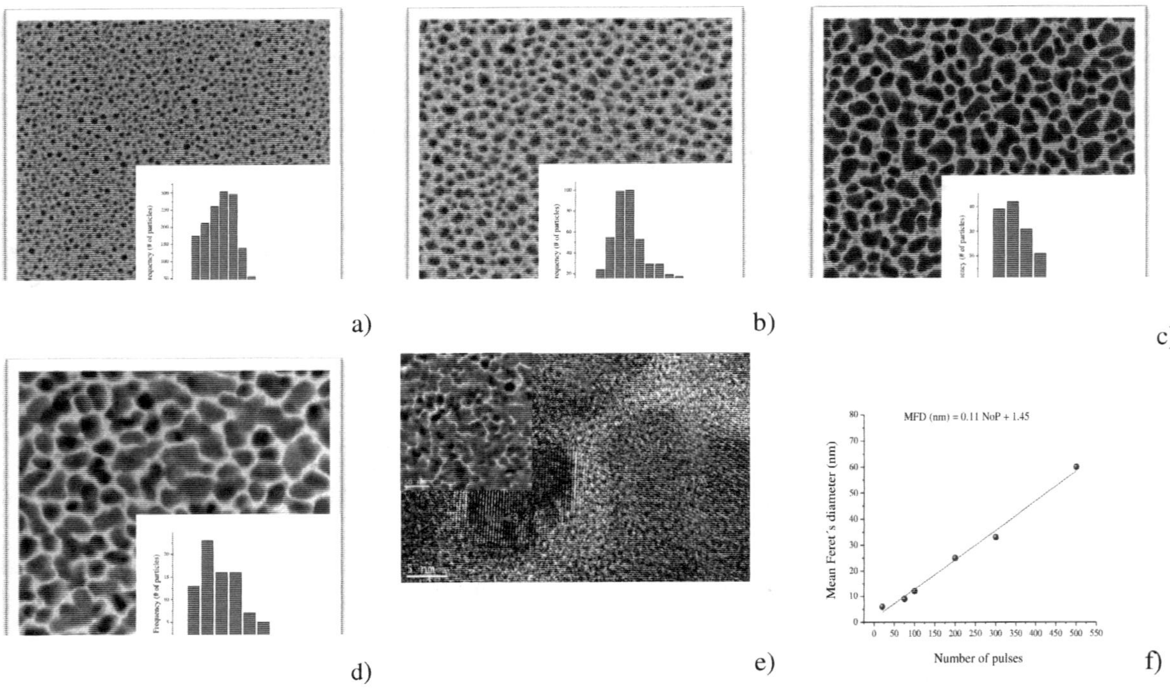

Figure 1. TEM images of Bi nanostructures and their corresponding size distributions at different number of pulses, a) 50, b) 100, c) 200, d) 300, e) 500 and the Feret's diameter as a function of the number of laser pulses.

3.2 Atomic Force Microscopy

Figure 2 shows, as an example, the AFM images obtained for the nanostructures deposited using 50 pulses (figure 2a) and 200 pulses (figure 2b). The images show that the nanostructures shape and size change with the number of laser pulses. Individual nanoparticles with a quasi-spherical shape and diameters of about 5 to 35 nm are observed for the sample deposited using 50 pulses. In contrast, nanostructures with sizes ranging from 15 to 45 nm, arranged in a rod-like shape, are observed in the case of the sample deposited with 200 pulses. These results agree reasonably with the TEM measurements. In general terms, in the case of a lower number of pulses the nanostructures preferentially grow with a quasi-spherical shape; whilst a higher number of pulses leads to larger nanostructures with more complex shapes covering completely the surface. It is worth noting that these nanostructures result from aggregation processes and seem to be formed actually by individual smaller nanoparticles. In some sense, these results reveal the early mechanisms that control the deposition of Bi thin films; starting with the arrive of incident atoms on surface defects that act as nucleation centres for nanoparticle growth followed by their agglomeration resulting in islands and finally promoting the formation of a continuous film.

3.3 Non-Rutherford Backscattering spectroscopy (NRBS)

The NRBS spectra of the same samples (not shown) reveal an increment of the signal associated to bismuth as the number of pulses increases. From these spectra, the number of atoms per unit area was estimated. It was found that the amount of Bi atoms varies monotonically, from about 3.8×10^{15} atoms/cm^2 to 1.5×10^{16} atoms/cm^2 as the number of pulses is increased from 50 to 500 respectively. It is clearly that the deposition rate of Bi depends on the number of pulses, decreasing as the number of pulses increases. This could be attributed to a reduction in the sticking coefficient and/or to sputtering processes.

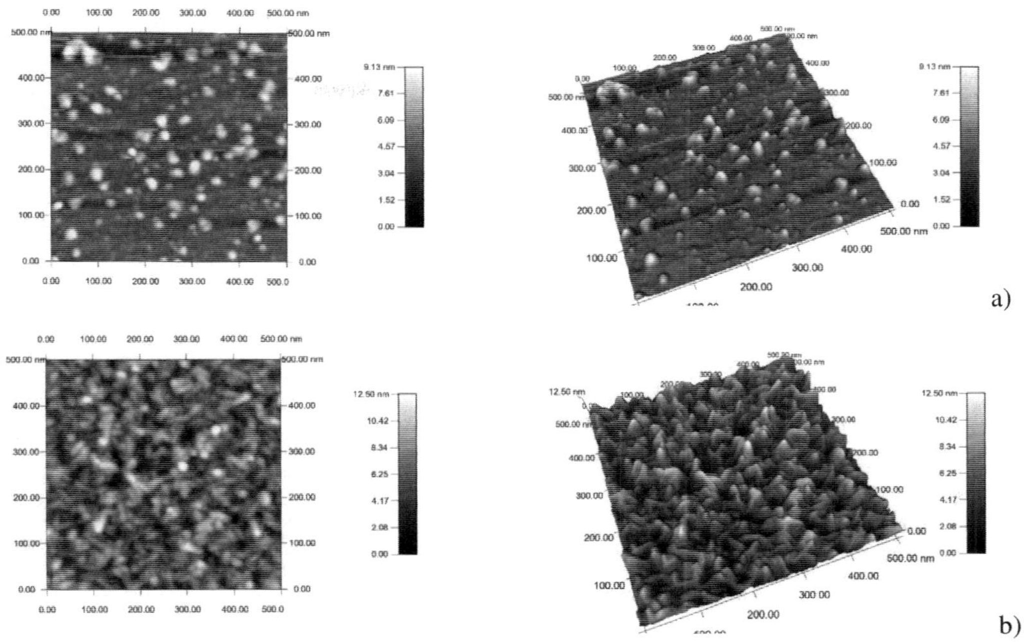

Figure 2 The AFM images of Bi nanostructures deposited using top) 50 pulses and bottom) 200 pulses.

3.4 Raman characterization

The Raman spectra of all samples showed peaks at 71 and 97 cm^{-1} as is presented in figure 3. These signals correspond to the E_g and A_{1g} first order Raman modes characteristic of rhombohedral bismuth [5] and reveal the crystalline state of the formed nanostructures. It is worth mentioning that the signal around 125 cm^{-1}, that could be attributed to the Bi—O stretches of β-Bi_2O_3 [6], was observed only in the sample prepared using 500 pulses, suggesting a partial oxidation of the Bi nanostructures due to exposure to the environmental moisture and/or laser induced oxidation.

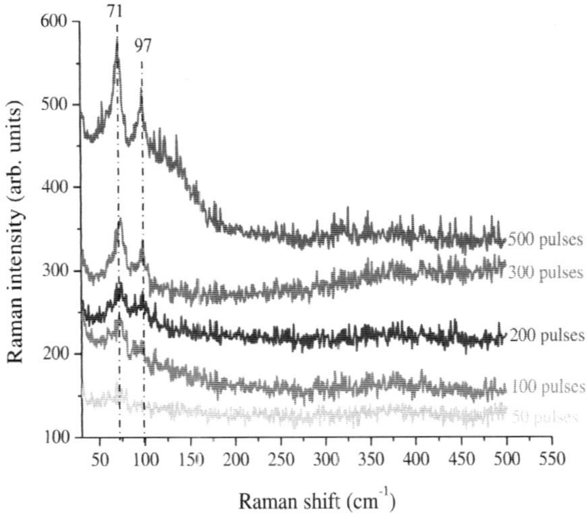

Figure 3. Raman spectra of the samples prepared using different number of pulses.

3.5 X-ray photoelectron spectroscopy (XPS)

Figure 4a shows the high-resolution XPS spectrum of the Bi 4f region of the sample prepared with 500 pulses. It is worth mentioning that the XPS spectra of all samples were very similar. The two peaks at 162.4 eV and 157.1 eV can be assigned to Bi 4f5/2 and Bi 4f7/2, respectively revealing the metallic state of the deposited nanostructures. The shoulders around 164.3 eV and 159.0 eV can be ascribed to the binding energies of Bi 4f5/2 and Bi 4f7/2 corresponding to Bi_2O_3 [7]. Therefore, these results reveal that the Bi nanostructures were deposited in the form of metallic bismuth with a slight amount of Bi_2O_3. Figure 4b shows the spectrum corresponding to the O 1s region. A low intensity signal suggesting a low amount of oxygen present in this sample is observed. The presence of oxygen in the samples could be due to adsorption from the environmental moisture.

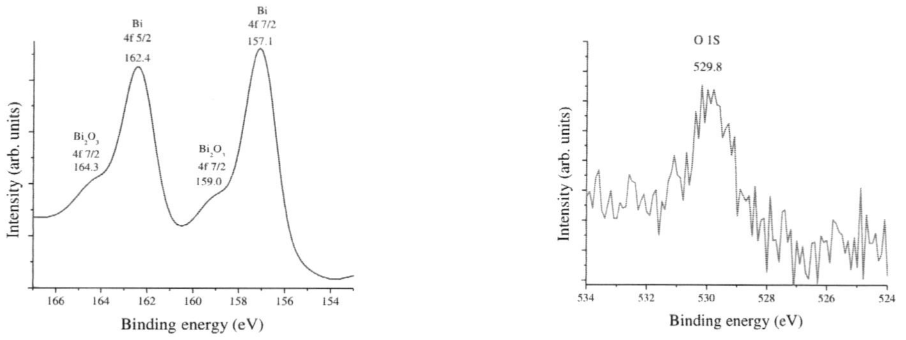

a) b)

Figure 4. High-resolution XPS spectra of the Bi 4f region (a) and the O 1s region (b) of the sample prepared with 300 pulses.

4. Conclusions

Bismuth nanostructures were obtained by pulsed laser ablation under vacuum conditions. Their size and size distribution were controlled by varying the number of pulses used to ablate the Bi target following an approximately linear relation. Bi NPs with Feret´s diameter as small as 7 nm were obtained using only 50 pulses owing to the high ablation rate of Bi. Compositional and Raman analysis showed the metallic nature of the deposited material.

6. References

[1] Choi D S, Balandin A A, Leung M S, Stupian G W, Presser N, Chung S W, J. R. Heath J R, Khitun A and Wang K L 2006 Applied Physics Letters **89** 141503

[2] Haro-Poniatowski1 E, Serna R, Suárez-García A, Afonso C N, Jouanne M, Morhange J F 2004 *Appl. Phys.* A **79** 1299–1302

[3] Serna R, Jiménez de Castro M, Toudert J, Haro-Poniatowski E, García López J 2013 *Appl. Phys.* A **110** 863-867

[4] Carotenuto G, Hison C L, Capezzuto F, Palomba M, Perlo P, Conte P 2009 J. Nanopart. Res. **11** 1729

[5] Trentelman K, Raman J 2009 Spectrosc. **40** 585

[6] Latha Kumari, Jin-Han Lin, Yuan-Ron Ma 2008 J. Phys. D: Appl. Phys. **41** 025405

[7] Dharmadhikari V S 1982 J. Electron Spec., **25** 181

5. Acknowledgements

This research project was partially supported by the EU under the project FP7-NMP-2010-EU-Mexico (Grant no.263878): Functionalities of Bismuth Based Nanostructures (BisNano).

Drop deformation in two-roll mills considering wall effects

C A Escalante-Velázquez[1], M A Huesca-Reyes[2], I Yescas Rosas[1] and E Geffroy[1]

[1] Instituto de Investigaciones en Materiales, Universidad Nacional Autónoma de México, México.

[2] Departamento de Termofluidos, Facultad de Ingeniería, Universidad Nacional Autónoma de México, México.

E-mail: cescalante@ciencias.unam.mx

Abstract. Experimental, theoretical and numerical results of dynamics of drop deformation in strong flows generated by a co-rotating two-roll mill and considering the influence of near rigid walls are presented. The drop dynamics is altered, with respect to a drop free of wall effects, by the proximity of the rigid boundaries as well as caused by a non-linear and non-uniform flow due to gradients of flow-type parameter and shear rate. Simulations were carried out using the Boundary Element Method (BEM). Since the inclusion of the whole boundaries (drop and rollers surfaces) is not an easy and trivial task, bi-dimensional numerical simulations was performed as a first approach. The experimental and numerical results were obtained for a flow type of $\alpha = 0.03$ and two values of viscosity ratio $\lambda = 0.012$ and 16. In general, numerical results for the stationary deformation parameters, up to intermediate confinements, are in agreement with the experiments, with and without wall effects. Since the case of drops with a high viscosity ratio did not match existing theoretical models, the wall-effect theory of Shapira and Haber was modified, considering Cox´s second-order theory as the converging theory without wall effects. From low to intermediate confinements, the new Cox-Shapira-Haber model fitted the observed experimental deformations.

1. Introduction

Deformation and breakup of a droplet in a continuous immiscible liquid phase under confined flow is of interest in several applications, such as microfluidics technologies and emulsion processing [1], among many other where the flow dynamics can be significantly altered by the presence of channel walls.

Most of previous studies have investigated the deformation of drops in confined simple shear flows (which corresponds to $\alpha = 0$) by the relative motion of parallel plates [2-4]. The experimental [2, 3] and numerical [4] results remit to Shapira and Haber theory (SH model) [5] that is valid for small deformations. This theory predicts the *total* deformation of a drop under simple shear flow given by:

$$_w = {}_0\left(1 + \left(\frac{R}{W}\right)^3 \frac{1+2.5\lambda}{1+\lambda} C_S\right), \quad (1)$$

where D_0 correspond to deformation parameter of Taylor [6] —being a function of the capillary number *Ca*—, R is the initial radius of drop, W is the separation between the walls, λ is the viscosity ratio, and C_S is the so-called shape factor, which depends on the position of the drop. Recently, Maffetone and Minale [7] adapted this model (we will call it SH-MM Model), where D_0 corresponds to the deformation of ellipsoidal drops proposed previously by them [8].

Experimental and numerical studies [2-4] have compared the SH theory and found it valid for small capillary numbers ($Ca < 0.3$) and intermediate ratios of confinement ($2R/W < 0.35$). Beyond these limits the shapes of stationary drops shift from ellipsoidal to sigmoidal.

In this work, the flows are non-linear, two-dimensional and capable of generating a wide spectrum of values of its kinematic parameters: in one extreme the flow is very close to simple shear, sweeping smoothly towards the other end where significant elongation effects can be present. The flow generated by two-roll mills (2RM´s), has been studied analytically and numerically in the past, with general solutions being available [9, 10, 11] and more recently experimentally [12] as well. The size of the cylinders and their separation determine the value of the flow-type parameter α at the stagnation point, while the shear rate γ also at the stagnation point depends on the angular velocities of the cylinders. For 2RM´s, the shear rate has a minimum value at the stagnation point and a maximum value on the surface of the cylinders while the flow-type has a maximum value at the stagnation point and a minimum value on the cylinders surface [9, 10].

In order to study the effect of walls on drops, the size of the drop (diameter) must be comparable to the gap between the rolls. Under those conditions, the drop deforms in flow fields with gradients on the flow-type parameter α and the shear rate γ around the stagnation point. This combination of features makes the dynamics of droplets influenced by wall effects a non-idealized system not fully analysed up to date.

Since the experimental and numerical results of this work did not match the space of parameters of existing studies, a model that preserves the dependence of the confinement factor on deformation (in the spirit of the initial model) is developed, assuming as well drops with high viscosity ratios. Thus, D_0 corresponds now to the steady free deformation of Cox theory [13].

1. Experimental

An extensive description of the 2RM flow can be found in [9-12]. The radius of the cylinders was 49 mm, with a separation between their centers of *52* mm giving a gap (g) of *3* mm. The experiments were carried out for two viscosity ratio ($\lambda = 0.012$ and *16*). In both cases the suspending fluid was PolyDiMethylSiloxane (PDMS), DMS T35 from Gelest Inc., with viscosity $\mu_m = 51$ Poise and relative density $\rho_m = 0.973$ at *25* °C. For the low viscosity ratio system the drop fluid consisted of vegetable canola oil with a viscosity $\mu_{d1}=0.6$ Poise and a relative density $\rho_{d1} = 0.917$ at 25 °C. The interfacial tension for this system was 2.7 mN/m. For the high viscosity ratio system the drop fluid consisted in PolyIsobutilene by Polysciences Inc., with a viscosity $\mu_{d2} = 816$ Poise and a relative density $\rho_{d2}=0.92$ at 25° C. The corresponding interfacial tension was 3.5 mN/m. All experiments were performed at 25°C ± 0.1°C.

2. Numerical method

The boundary integral equations take into account explicitly the presence of rigid surfaces and deformable fluid interfaces, simulating the deformation of a drop between two cylinders. The integral equation for the flow field within the drop, $\mathbf{u}_2(\mathbf{x})$, or exterior to the drop in the matrix fluid, $\mathbf{u}_1(\mathbf{x})$, will be dependent upon boundary integrals taken on the drop interface S_{drop} as well as on the cylinder surfaces $S_{rigid,1}$ and $S_{rigid,2}$. These integral equations can be found in reference [14].

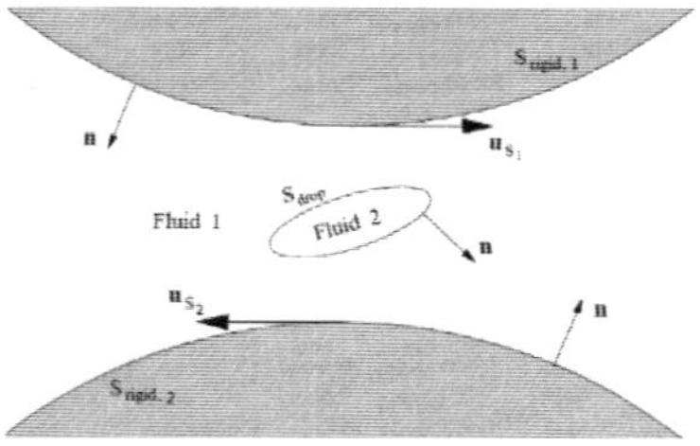

Figure 1. Liquid-drop interface deformed by the flow field generated by solid boundaries. In the middle of the rollers there is a stagnation point where the drop is initially placed.

In Figure 1, a schematic representation is depicted. We assume a Newtonian, inertial-less drop with initial radius R and viscosity μ_{drop}—located between the two cylinders— and immersed in a Newtonian matrix fluid of viscosity μ_m. The viscosity ratio was $\lambda = \mu_{drop}/\mu_m$. The position of the stagnation point is located exactly halfway between the cylinders when the angular velocities are same.

3. Results

The numerical and experimental results obtained in a two-roll mill device for stationary drop deformation values are presented as a function of capillary number and also compared with theoretical models of: (a) small deformation Cox theory [13], (b) Chaffey and Brenner model [15], for angle orientation in the free case, and (c) SH-Cox model, adapted here in the case of confined drop deformation.

3.1. Stationary Drop deformation. The case of Low viscosity ratios

The lowest viscosity ratio ($\lambda = 0.012$) was studied experimental and numerically for different capillary numbers (between *0.05-0.8*). Figure 2 shown the deformation and orientation for the stationary drop deformation as a function of capillary number for two confinement factors $d/g = 2R/g$.

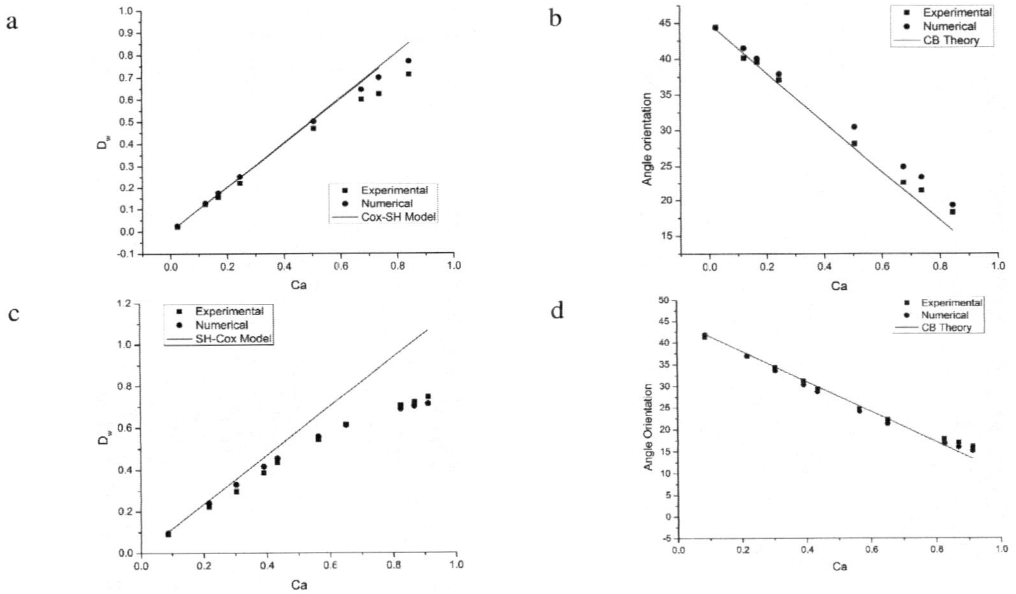

Figure 2. Deformation and orientation angle for the stationary drop deformation as a function of capillary number. (a) And (b) corresponds to $d/g = 0.08$, (c) and (d) to $d/g = 0.3$: ■, experimental results; ●, numerical results; —, SH-Cox model (a) and (c), Chaffey and Brenner theory (b) and (d).

For weak confinement a good agreement was found between the experimental, numerical, and theoretical models. In the numerical results, when $Ca > 0.8$, no stationary deformation was observed (drop break up occurs), whereas in the experimental results the critical capillary was found around $Ca = 0.95$. In Figures 2c and 2d a good agreement between numerical and experimental results can be observed in the case of confined drops, whereas the SH-Cox model exhibited a higher deviation with respect to the experimental results for $Ca > 0.4$ whereas no change was observed in the final angle orientation due to confinement. It was corroborated that there is not difference in the use of the Taylor, SH-MM or Cox theories for these viscosity ratios. Such theories coincided and predicted a linear dependence between capillary number and the deformation parameter.

3.2. Stationary Drop deformation. High viscosity ratio case

For weaker confinements, *d/g = 0.1*, the numerical results agreed with the values obtained using SH-Cox model, as it can be seen in Figure 3a; however, numerical and theoretical results gave larger deformation values. This discrepancy might be the result of the experimental control system that tends to enhance the flow vorticity, thus inhibiting slightly the experimental drop deformation. On the other hand, good agreement for *d/g = 0.205* can be observed in Figure 3b. Numerical results and the SH-Cox model were very close to experimental results. Under this condition, the non-linear effects of the walls may mitigate the non-idealities of the control scheme of the flow

For a higher confinement, *d/g = 0.337*, a good agreement was also found, as it is shown in Figure 3c. However, the theoretical models predict larger deformation than experimental or numerical data. For more pronounced confinement effects, *d/g = 0.433*, Figure 3d is shown. In this last case the numerical simulations and theory gave deformations larger than those occurring in experimental results. Both, SH and SH-MM models fail to predict this asymptotic behavior observed for high viscosities, even in the no-wall effects case, that's because these model preserves in some way, the linear dependence on Ca. Alternatively it is possible use the Taylor asymptotic model [6] but this not describe the initial dependence on Ca.

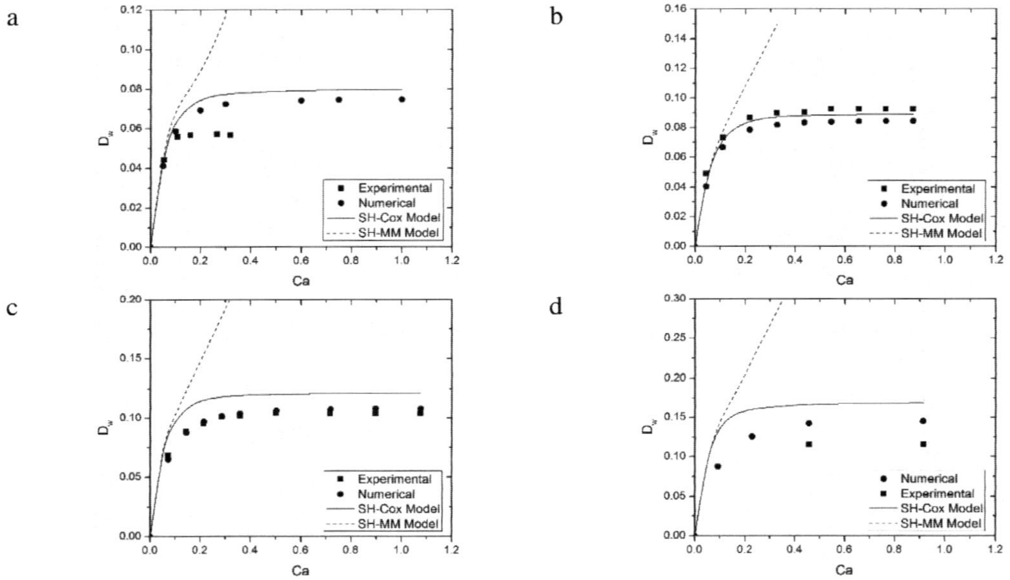

Figure 3. Stationary deformation versus capillary number. Comparison of experimental (■), numerical (●) and theoretical (—, SH-Cox model), (- -, SH-MM model) for four confinements factors, *d/g = 0.1* (a), *d/g = 0205* (b), *d/g = 0.337* (c) and *d/g = 0.433* (d).

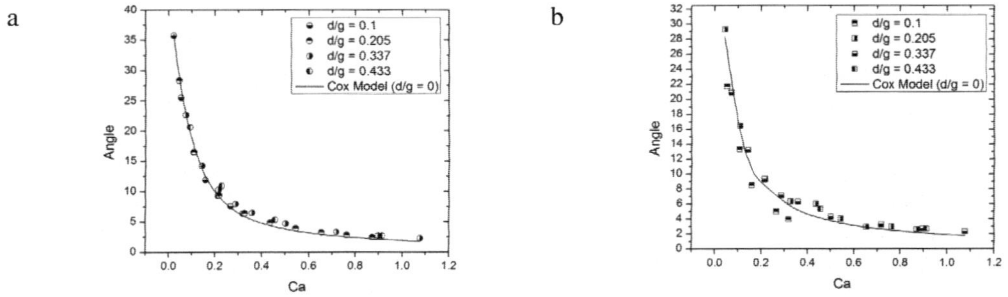

Figure 4. Comparison of theoretical model (Cox theory) with numerical (a) and experimental (b) results obtained for the orientation angle of the stationary drop deformation in a flow field generated by the two-roll mill device for different confinement factors.

The steady state orientation angle of the deformed drops did not manifest wall effects, for neither the experimental or numerical results. In fact, both methods agreed with Cox Theory [13] as it can be seen in Figure 4. A comparison of numerical, experimental and theoretical results for stationary drop deformation values, as a function of confinement for high viscosity ratio, is shown in Figure 5. It can be observed that the agreement with predictions of the SH-Cox model were mostly qualitative. This difference may be attributed to the flow dynamics of the non-linear flow of the 2RM´s.

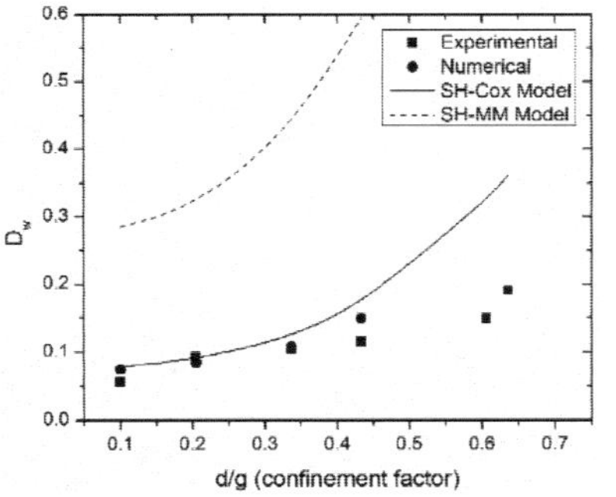

Figure 5. Stationary drop deformation vs. the confinement factor. SH-Cox model predicts a larger deformation than numerical and experimental data. It can be also observed that the SH-MM model fails to predict at high viscosity cases.

4. Conclusions

Numerical simulations using BEM in the study of stationary drop deformation with confinement effects showed a noticeable agreement with experimental results obtained in a device able to generate non-idealized flows (non-linear flow fields with gradients in shear rate and flow type, and curved walls). All phenomena observed in the experiment were reproduced numerically. Despite the bi-dimensional simulations nature, the numerical model captured the basic hydrodynamics, which confirm its predictive nature, mainly due to the fact that drop deformation remain small for most experimental conditions, and the drop curvature is quasi-homogenous. This work is the first effort to encompass the full kinematics existing in the two roll mills in the study of confined drop deformations. The adapted model SH-Cox model predicts the drop deformation on intermediate confinements and small deformations, covering the full regime of high viscosities drop deformations under wall effects.

6. Acknowledgments

The first author (Escalante-Velázquez) is thankful to National Council of Science and Technology (CONACYT) and PCeIM-UNAM for providing the financial support for his PhD studies at UNAM.

5. References

[1] Stone H A, Stroock A D and Ajdari A 2004 *Ann. Rev. Fluid Mech.* **36** 81—411
[2] Sibillo V, Pasquariello G, Simeone M Cristini V and Guido S 2006 *Phys. Rev. Lett.* **97** 054502
[3] Vananroye A, Van Puyvelde P and Moldenaers P 2007 *J. Rheol.* **51** 139-53
[4] Janssen P J A and Anderson P D 2007. *Phys. Fluids*, **19** 043602
[5] Shapira M and Haber S 1990 *Int. J. Multiphase Flow* **16** 305-21
[6] Taylor G I 1932 *Proc. Roy. Soc. A* **138**, 41.
[7] Minale M 2008 *J Rheologica Acta* **47** 667-75
[8] Maffettone P and Minale M 1998 *J. Non-Newtonian Fluid Mech.***78** 227-41
[9] Reyes M A H and Geffroy E 2000 *Phys. Fluids*, **12** 2372-75
[10] Reyes M A H and Geffroy E 2000 *Rev. Mex. Fís.* **46** 135-47
[11] Reyes M A H 2005 PhD thesis, UNAM, México
[12] Rosas Y I 2013 PhD thesis, UNAM, México
[13] Cox R G 1969 *Journal of Fluid Mechanics* **37** 601-23
[14] Pozrikidis C 1992. *Boundary Integral and Singularity Methods for Linearized Viscous Flow* (Cambridge University Press)
[15] Chaffey C E and Brenner H 1967 *J. Colloid Interface Sci.***24** 258-69

Journal of Physics: Conference Series **582** (2015) 012015 doi:10.1088/1742-6596/582/1/012015

Structural instability of atmospheric flows under perturbations of the mass balance and effect in transport calculations

M A Núñez, R Mendoza
Departamento de Física, Universidad Autónoma Metropolitana Iztapalapa, A. P. 55-534, C. P. 09340, D. F. México

E-mail: manp@xanum.uam.mx

Abstract. Several methods to estimate the velocity field of atmospheric flows, have been proposed to the date for applications such as emergency response systems, transport calculations and for budget studies of all kinds. These applications require a wind field that satisfies the conservation of mass but, in general, estimated wind fields do not satisfy exactly the continuity equation. An approach to reduce the effect of using a divergent wind field as input in the transport-diffusion equations, was proposed in the literature. In this work, a linear local analysis of a wind field, is used to show analytically that the perturbation of a large-scale nondivergent flow can yield a divergent flow with a substantially different structure. The effects of these structural changes in transport calculations are illustrated by means of analytic solutions of the transport equation.

1. Introduction

Several methods to estimate trajectory models for long-range atmospheric transport calculations, have been proposed [1]. However, conservation of mass is not always exactly satisfied [2,3]. Although several authors have pointed out that the use of a wind field that does not satisfy exactly the continuity equation, can introduce significant errors in transport calculations of chemical species [3,4], the accuracy with which the continuity equation is satisfied, is not always reported in works that propose methods for estimating wind fields for transport calculations. Since large scale atmospheric flows are nearly nondivergent, it is generally accepted that a mass-balanced wind field has to satisfy the continuity equation in its simplest form $\nabla \cdot \mathbf{v} = 0$. Simple methods applied to wind field analysis can yield a divergence of order $10^{-5}s^{-1}$. In this work we show that a divergence of this order can change the structure of a large-scale nondivergent flow.

A proposed method to reduce the effect of using a divergent field in transport calculations, consists in subtracting the value of $\nabla \cdot \mathbf{v}$ at each grid point where the transport equations are solved with standard spatial-discretization methods [4]. The results of this work show that this procedure can yield results with a significant error because the structure of a nondivergent wind field can be different from that of a divergent one with a divergence of order $10^{-5}s^{-1}$.

2. Types of mass-balanced flows and their perturbations

The structure of an atmospheric flow $\mathbf{v}=u\mathbf{i}+v\mathbf{j}$ in the vicinity of an arbitrary point $(x_{0c,}y_{0c})$ can be studied by means of the linear terms of its Taylor series. Let us consider a Cartesian system with its origin at $(x_{0c,}y_{0c})$, then we have

$$u = \mathrm{M}_{11}\, x + \mathrm{M}_{12}\, y + u_{00}\,, \qquad v = \mathrm{M}_{21}\, x + \mathrm{M}_{22}\, y + v \tag{1}$$

where the matrix $M=\{\mathrm{M}_{jk}\}$ is the velocity gradient at the origin. In terms of the parameters

$$\delta = \nabla\cdot\mathbf{v}/2, \quad \omega = \mathbf{k}\cdot\nabla\mathrm{x}\,\mathbf{v}/2\,, \quad a = (\mathrm{M}_{11} - \mathrm{M}_{22})/2\,, \quad b = (\mathrm{M}_{12} + \mathrm{M}_{21})/2 \tag{2}$$

we have

$$u = (\delta + a)x\, + (b - \omega)\, y + u_{00}\,, \qquad v = (b + \omega)x\, + (\delta - a)\, y + v_{00}\,. \tag{3}$$

Hence the velocity field $\mathbf{v}$ can be written as $\mathbf{v}^{\delta} = \mathbf{v}^{0}+\mathbf{u}^{\delta}$ where $\mathbf{v}^{0}$ and $\mathbf{u}^{\delta}= \delta\ (x\mathbf{i}+y\mathbf{j})$ are the nondivergent and divergent components of $\mathbf{v}^{\delta}$. Since large scale atmospheric flows are nearly nondivergent, $\mathbf{v}^{0}$ will be considered an ideal mass-balanced flow, and $\mathbf{u}^{\delta}$ will quantify the perturbation of the mass balance of $\mathbf{v}^{0}$. Accordingly, the mass-unbalanced flow $\mathbf{v}^{\delta}$ with $\delta \neq 0$ and the mass balanced one $\mathbf{v}^{0}$ will be referred to as a divergent flow (DF) and a nondivergent flow (NDF), respectively.

The discriminant of equation $\det(M - \lambda I) = 0$ and the eigenvalues of M are given by

$$\Delta^{\delta}= a^2 + b^2 - \omega^2, \quad \lambda_1^{\delta} = \delta + \sqrt{\Delta}, \quad \lambda_2^{\delta} = \delta - \sqrt{\Delta} \tag{4}$$

For a NDF $\mathbf{v}^{0}$ the eigenvalues $\lambda_1 = \sqrt{\Delta}$, $\lambda_2 = -\sqrt{\Delta}$, imply that $\mathbf{v}^{0}$ has one of the following structures. *Case (NDF-a)*: For $\Delta > 0$ the eigenvalues are real and have different signs, $\mathbf{v}^{0}$ has a hyperbolic structure whose principal directions are given by the eigenvectors of M. *Case (NDF-b)*: For $\Delta < 0$ the eigenvalues are imaginary and $\mathbf{v}^{0}$ has an elliptic structure. *Case (NDF-c)*: For $\Delta = 0$ there is one (real) eigenvalue, $\mathbf{v}^{0}$ has a linear structure since the trajectory of each fluid particle is a straight-line and the eigenvector of M yields the unique principal direction. A DF $\mathbf{v}^{\delta}$ admits the same structures but the eigenvalue expression (4) implies that $\mathbf{v}^{\delta}$ can have additional structures. *Case (DF-a)*: For $\delta \neq 0$ and $\Delta^{\delta}> 0$ there are three types of flows: (*i*) For $|\delta| < \Delta^{\delta}$ the flow $\mathbf{v}^{\delta}$ is hyperbolic. (*ii*) For $\delta = \Delta^{\delta}$ one eigenvalue λ_j is zero and another λ_k has the sign of δ. The principal axis associated to λ_j consists of equilibrium points which are sources for $\delta > 0$ and sinks for $\delta < 0$. (*iii*) For $|\delta| > \Delta^{\delta}$ the eigenvalues have the sign of δ and $\mathbf{v}^{\delta}$ has a parabolic structure, the trajectories behave like parabolas that converge toward the origin for $\delta > 0$ and diverge for $\delta < 0$. *Case (DF-b)*. For $\delta \neq 0$ and $\Delta^{\delta}< 0$ the DF $\mathbf{v}^{\delta}$ is similar to the corresponding elliptic NDF, the trajectories are spirals that converge toward to the origin for $\delta > 0$ and diverge for $\delta < 0$.

Cases *(DF-a)-(i)*, *(DF-a)-(ii)*, show that the perturbation $\mathbf{u}^{\delta}$ of a NDF $\mathbf{v}^{0}$ can yield a DF $\mathbf{v}^{\delta}$ with a different structure. To determine the structural differences between $\mathbf{v}^{0}$ and $\mathbf{v}^{\delta}$ with characteristic values of large scale atmospheric flows, we can define the characteristic value of a variable $f(x,y)$ in a region $R=\{0\leq x\leq L,\ 0\leq y\leq L\}$ with length scale L, by its the mean value (denoted by $\overline{f}$) on R. According to equation (3) the relation between mean values $\bar{u}$, $\bar{v}$, $\bar{x}$, $\bar{y}$ and the parameters (2) is

$$\overline{u} = (\delta + a)\overline{x}\, + (b - \omega)\,\overline{y} + u_{00}, \qquad \overline{v} = (b + \omega)\overline{x}\, + (\delta - a)\overline{y} + v \quad , \tag{5}$$

with $\overline{x} = \overline{y} = L/2$, By simplicity we set $u \quad = v \quad = 0$. These expressions yield a, b, in terms of δ, ω, and the mean values, namely,

$$a = (\alpha + 2\omega\overline{x})/2\overline{x}\,, \qquad b = (\beta - 2\delta\overline{x})/2\overline{x} \tag{6}$$

where we set $\alpha = \bar{u} - \bar{v}$, $\beta = \bar{u} + \bar{v}$. Since the last relations imply that there is a one-to-one relation between the pairs $\{\bar{u},\bar{v}\}$, $\{\alpha, \beta\}$, the latter will used in our study. The discriminants of $\mathbf{v}^{0}$, $\mathbf{v}^{\delta}$, are

$$\Delta = (\alpha^2 + \beta^2 + 4\alpha\, \bar{x}\, \omega)/4\, \bar{x}^2, \quad \Delta^{\delta}= \Delta\ + q(\delta,\beta,\bar{x}) \tag{7}$$

where we set $q(\delta,\beta,\bar{x}) = \delta(\delta - \beta/\bar{x})$. As is shown below, the characteristic values for large scale atmospheric flows $\delta, \omega{\sim}10^{-5}\ \mathrm{s}^{-1}$, α, $\beta{\sim}\ 10\ \mathrm{ms}^{-1}$, $L=10^3$ km ($\overline{x}=5{\times}10^5$m), of large scale atmospheric flows can yield fields $\mathbf{v}^{0}$, $\mathbf{v}^{\delta}$ with different structure.

To analyze the effect of a mass imbalance in transport calculations we consider the transport equation

$$\partial_t c^\delta + \nabla\cdot\left(c\,\mathbf{v}^\delta\right) = -k\,c^\delta. \qquad (8)$$

Superscript δ indicates that c^δ is solution of equation (8) with a field $\mathbf{v}^\delta$ having divergence 2δ. A scale analysis shows that the diffusion term can be neglected for large scale flows. Replacing $\nabla\cdot\mathbf{v}^\delta = 2\delta$ into equation (8) it takes the form

$$\partial_t c^\delta + \mathbf{v}^\delta\cdot\nabla c^\delta = -kc^\delta - 2\delta c^\delta\,. \qquad (9)$$

We see that the perturbation term $\mathbf{u}^\delta$ in a DF $\mathbf{v}^\delta = \mathbf{v}^0 + \mathbf{u}^\delta$ generates a fictitious reaction term $-2\delta c^\delta$. Since characteristic values of k are of order $10^{-6}s^{-1}$, a divergence $|\delta|$ of order $10^{-5}s^{-1}$ yields unrealistic solutions c^δ. The solution proposed in the literature [3,4] to reduce the effect of using a DF $\mathbf{v}^\delta$ is to substract the term $c^\delta\nabla\cdot\mathbf{v}^\delta$ on the left hand side of equation (8). This is equivalent to solving the equation

$$\partial_t c^\delta + \mathbf{v}^\delta\cdot\nabla c^\delta = -kc^\delta \qquad (10)$$

Transport calculations reported below were made by solving equation (10) analytically with the characteristics method and $c = \exp[\{-(x-x_c)^2 - (y-y_c)^2\}/\sigma^2]$, $\sigma = 0.5$km at $t = 0$. For brevity, we shall consider some cases where the structure of a DF $\mathbf{v}^\delta$ differs from that of the NDF $\mathbf{v}^0$.

Let us consider cases where an hyperbolic NDF $\mathbf{v}$ becomes an elliptic DF $\mathbf{v}^\delta$. A condition to go from $\Delta^0>0$ to $\Delta^\delta<0$, is $q = \delta(\delta - \beta/\bar{x})<0$. Since the minimum value of q is $-\beta^2/4\bar{x}^2$ at $\delta_m = \beta/2\bar{x}$, condition $\beta \neq 0$ is sufficient to guarantee the existence of an interval of δ values with which $q<0$. We have the following cases:

(*a*) Let ω be fixed and α variable. For $\delta = \delta_m$, $\Delta^\delta = \alpha\,(\alpha + 4\bar{x}\omega)$ has the minimum value $-\omega^2$ at $\alpha = -2\bar{x}\omega$. Therefore, for ω, $\beta \neq 0$, δ, such that $\Delta^0 > 0$, there is a set of values of δ, α, for which $\Delta^\delta < 0$, and the hyperbolic NDF $\mathbf{v}$ becomes an elliptic DF $\mathbf{v}^\delta$. For instance, let $\omega=10^{-5}s^{-1}$, $\beta = \pm 11ms^{-1}$, $L=10^3$km, $\alpha = -2\bar{x}\omega = -10ms^{-1}$, $\delta = \delta_m$, then $\Delta^0 > 0$, $\Delta^\delta = -\omega^2$ and, by continuity, $\Delta^\delta < 0$ holds for δ near δ_m.

(*b*) Consider fixed $\alpha, \beta \neq 0$, and variable δ, ω. According to equation (7) the critical values

$$\omega_c = -(\alpha^2 + \beta^2)/4\,\bar{x}\alpha\,, \qquad \omega_c^\delta = \omega_c - q\,\bar{x}/\alpha \qquad (11)$$

are the unique ones where Δ and Δ^δ change their sign. The sets $E_{\alpha\beta} = \{\,\omega : \Delta < 0\,\}$, $H_{\alpha\beta} = \{\,\omega : \Delta > 0\,\}$, are ω values where a NDF $\mathbf{v}$ is elliptic and hyperbolic, respectively. Let $E_{\alpha\beta}^\delta$ and $H_{\alpha\beta}^\delta$ be the corresponding sets of a DF $\mathbf{v}^\delta$ with the same α, β, and $\delta \neq 0$. Consider δ, β, such that $q<0$ and $\alpha>0$. This yields $\omega_c < \omega_c^\delta$, $H_{\alpha\beta}^0 = (\omega_c^0, \infty)$, $E_{\alpha\beta}^\delta = (-\infty, \omega_c^\delta)$, and the intersection $I_{\alpha\beta}^\delta = (\omega_c^0, \omega_c^\delta)$ are ω values with which an hyperbolic NDF $\mathbf{v}$ becomes and elliptic DF $\mathbf{v}^\delta$. Example 1: Consider $\delta=10^{-5}s^{-1}$, $\beta=10$ ms^{-1}, $L=10^3$km, which yield $q=-10^{-10}s^{-2}$, and let $\alpha=1$ ms^{-1}, then $\omega_c^\delta = -0.05\times10^{-5}$ s^{-1}, $\omega_c^0 = -5.05\times10^{-5}$ s^{-1}, and any ω in $I_{\alpha\beta}^\delta$ yields an hyperbolic NDF $\mathbf{v}$ that becomes an elliptic DF $\mathbf{v}^\delta$. Figures 1, 2, show $\mathbf{v}^0$, $\mathbf{v}^\delta$, with $\omega = -10^{-5}s^{-1}$ and the corresponding solutions of equation (10) by means of the movement of the iso-line with c=0.75 at $t=0$. For $\alpha<0$ and δ, β, such that $q<0$, we get $\omega_c^\delta < \omega_c$ and the sets $H_{\alpha\beta} = (-\infty, \omega_c)$, $E_{\alpha\beta}^\delta = (\omega_c^\delta, \infty)$, whose intersection $I_{\alpha\beta}^\delta = (\omega_c, \omega_c^\delta)$ are ω values with which an hyperbolic NDF $\mathbf{v}$ becomes an elliptic DF $\mathbf{v}^\delta$.

Let us consider cases where an elliptic NDF $\mathbf{v}$ becomes an hyperbolic DF $\mathbf{v}^\delta$. A condition to go from $\Delta^0 < 0$ to $\Delta^\delta > 0$ is $q>0$, this occurs with $|\delta| > |\beta|/\bar{x}$. Let $\alpha>0$, then $\omega_c^\delta < \omega_c^0$, $E_{\alpha\beta} = (-\infty, \omega_c)$, $H_{\alpha\beta}^\delta = (\omega_c^\delta, \infty)$, and the set $I_{\alpha\beta}^\delta = (\omega_c^\delta, \omega_c^0)$ are ω values with which an elliptic NDF $\mathbf{v}$ becomes and hyperbolic DF $\mathbf{v}^\delta$. Example 2: Consider $\alpha=\beta=10ms^{-1}$, $L=10^3$km, then $\omega_c=-10^{-5}s^{-1}$, for $\delta=-10^{-5}s^{-1}$ we have $q = 3\times10^{-10}s^{-2}$, $\omega_c^\delta = -2.5\times10^{-5}s^{-1}$, and any ω in $I_{\alpha\beta}^\delta = (\omega_c^\delta, \omega_c^0)$ yield a NDF $\mathbf{v}$ and a hyperbolic DF $\mathbf{v}^\delta$. Figures 3 and 4 show the fields $\mathbf{v}^0$, $\mathbf{v}^\delta$, with $\omega=-1.5\times10^{-5}s^{-1}$ and the effect in the solution of equation (10). In a similar manner, for $\alpha<0$ and β, δ, such that $q<0$, we have $\omega_c^0 < \omega_c^\delta$, $E_{\alpha\beta}^0 = (\omega_c^0, \infty)$, $H_{\alpha\beta}^\delta = (-\infty, \omega_c^\delta)$, and the ω values in the interval $(\omega_c^0, \omega_c^\delta)$ yield an elliptic NDF $\mathbf{v}^0$ whose perturbation $\mathbf{v}^\delta$ is hyperbolic.

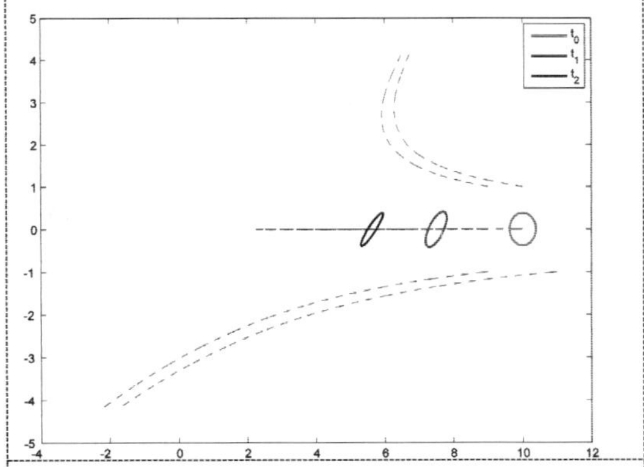

Figure 1. Hyperbolic NDF $\mathbf{v}^0$ and iso-lines of concentration for example 1, at $t_0=0$, $t_1=8.88h$, $t_2=17.75h$. Unit of length in km.

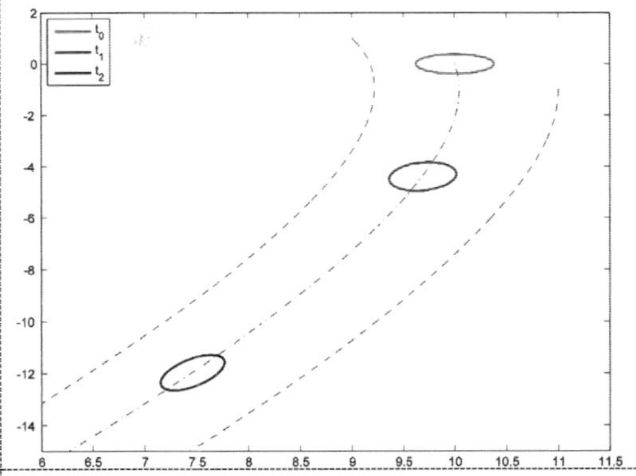

Figure 2. Elliptic DF $\mathbf{v}^\delta$ and iso-lines of concentration obtained by perturbation of the NDF of figure 1 with $\mathbf{u}^\delta$, $\delta=10^{-5}s^{-1}$.

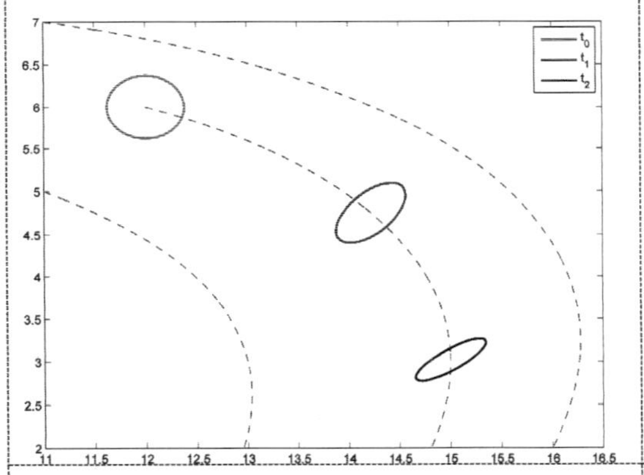

Figure 3. Elliptic NDF $\mathbf{v}^0$ and iso-lines of concentration for example 2.

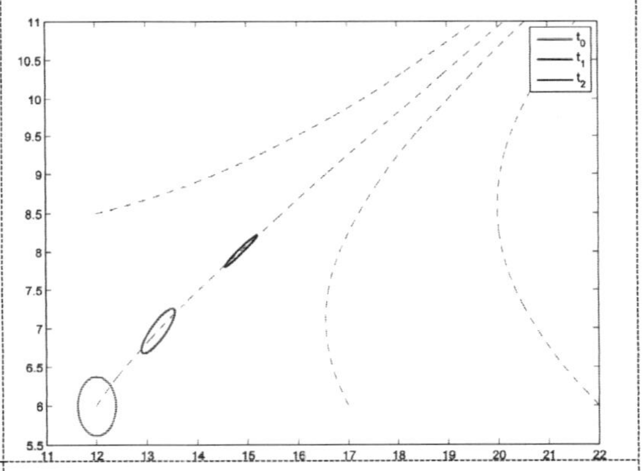

Figure 4. Hyperbolic DF $\mathbf{v}^\delta$ and iso-lines of concentration by perturbation of the NDF of figure 3 with $\mathbf{u}^\delta$, $\delta=-10^{-5}s^{-1}$.

3. Summary and concluding remarks

Standard methods to estimate the wind field from observational data yield, in general, a field $\mathbf{v}^\delta = \mathbf{v} + \mathbf{u}^\delta$ with a nonzero divergent component $\mathbf{u}^\delta$ [2,3]. The procedure suggested in the literature to make transport calculations with a DF $\mathbf{v}^\delta$ (subtracting the value of the divergence at each grid point in the transport equation [3,4]), does not consider that a perturbation of a mass-balanced flow $\mathbf{v}^0$ can yield a divergent flow $\mathbf{v}^\delta$ with a significantly different structure. Thus, the use of a DF $\mathbf{v}^\delta$ in transport calculations can yield trajectories that diverge from those given by a mass-balanced flow $\mathbf{v}^0$. According to Kitada [3], simple data assimilation methods yield a DF $\mathbf{v}^\delta$ with $\nabla \cdot \mathbf{v}^\delta$ of order 10^{-5} s^{-1}, but the examples reported in [5] show that variational methods conceived to minimize the divergence [6], can yield fields with $\nabla \cdot \mathbf{v}^\delta$ of order 10^{-3} s^{-1}. In these cases, the use of $\mathbf{v}^\delta$ in transport calculations can yield results with errors larger than those reported above. A more complete study of the effects of perturbing the mass-balance of a wind field in the flow structure and transport calculations will be given in a forthcoming work.

4. References

[1] Sthol A 1998 Computation, accuracy and applications of trajectories-A review and bibliography *Atmos. Environ.* **32** 947-966. Seaman N L 2000, Meteorological modeling for air-quality assessments *Atmos. Environ.* **34** 2231-2259.

[2] Trenberth K, Hurrel J W, Solomon A 1995, Conservation of mass in three dimensions in global analyses *J. Climate* **8** 692.

[3] Kitada T 1987Effect of non-zero divergence wind fields on atmospheric transport calculations *Atmos. Environ.* **21** pp.785-788.

[4] Carmichel G R, Peters L K , Kitada T 1986, A second generation model for regional-scale ransport/Chemistry/deposition *Atmos. Environ.* **20** 173. Kitada T, Carmichel G R, Peters L, J. Clim. Appl. Met. **23** 1153.

[5] Núñez M A 2102, Improving variational mass-consistent models of hydrodynamic flows via boundary conditions *Eur. Phys. J. Plus* **127** 40. Núñez M A, Sanchez-Sanchez J E 2102, A formulation to compute mass-consistent models of hydrodynamic flows *Eur. Phys. J. Plus* **127** 39.

[6] Ratto C F, Festa R, Romeo C, Frumento O A, Galluzzi 1994, Mass-consistent models for wind fields over complex terrain: The state of the art *Environ. Software* **9** 247-268.

5. Acknowledgments

One of us (R. M.) wants to thank to Conacyt for a scholarship 588339

Journal of Physics: Conference Series **582** (2015) 012016 doi:10.1088/1742-6596/582/1/012016

Study of new chaotic flows on a family of 3-dimensional systems with quadratic nonlinearities

S Carrillo Moreno, K Casas-García, J J Flores-Godoy, F Vázquez Valencia, G Fernández-Anaya
Department of Physics and Mathematics, Universidad Iberoamericana, Av. Prolongación Paseo de la Reforma 880, Lomas de Santa Fe, CP 01219, México DF

E-mail: guillermo.fernandez@ibero.mx

Abstract. Based on a wider systematic search on a family of 3-dimensional systems with quadratic nonlinearities, three new simple chaotic systems were found. One of them has the unusual feature of having a stable equilibrium point, and it is the simplest one of other chaotic flows with this property. The others have some interesting special properties.

1. Introduction

The search of chaotic flows began two decades ago, e.g., in [1] a systematic search of 3-dimensional autonomous systems with quadratic nonlinearities, allowed them to find 19 distinct simple chaotic flows. It is well-known that mathematically simple dynamical systems can present chaotic flows [2], [3]. Recent works have investigated simple chaotic flows with quadratic nonlinearities, and were found with either a stable equilibrium point or a line equilibrium, see [4], [5] and references there. This class of systems have hidden attractors; a hidden attractor has a basin of attraction that does not intersect with small neighbourhoods of any equilibrium points. In particular any dissipative chaotic flow with no equilibrium or with only stable equilibrium must have a hidden strange attractor [4]. Hidden attractors are important in engineering applications because they allow unexpected and potentially disastrous responses to perturbations in a structure like a bridge or an airplane wing [4]. Another class of interesting chaotic flows are those with only an equilibrium point [6]. The goal of this work is to present 3 new chaotic flows, one of them with a stable equilibrium point which is, to our best knowledge, the simplest mathematical system in the literature today with this property. The last is one of the few known chaotic systems in the literature with only one equilibrium point. These systems are found via a systematic computer search for chaos in 3-dimensional autonomous systems with quadratic nonlinearities.

2. The new chaotic flows

The 3-dimensional family of systems with quadratic nonlinearities considered in this work can be expressed as

$$\dot{X} = A_0 + M \times X + N \times R + Q \times S \quad (1)$$

where

$$X = \begin{pmatrix} x \\ y \\ z \end{pmatrix}; \; R = \begin{pmatrix} yz \\ xz \\ xy \end{pmatrix}; \; S = \begin{pmatrix} x^2 \\ y^2 \\ z^2 \end{pmatrix} \tag{2}$$

and parameters as

$$A_0 = \begin{pmatrix} a_1 \\ a_2 \\ a_3 \end{pmatrix}; \; M = \begin{pmatrix} m_{11} & m_{12} & m_{13} \\ m_{21} & m_{22} & m_{23} \\ m_{31} & m_{32} & m_{33} \end{pmatrix}; \; N = \begin{pmatrix} n_{11} & n_{12} & n_{13} \\ n_{21} & n_{22} & n_{23} \\ n_{31} & n_{32} & n_{33} \end{pmatrix}; \; Q = \begin{pmatrix} q_{11} & q_{12} & q_{13} \\ q_{21} & q_{22} & q_{23} \\ q_{31} & q_{32} & q_{33} \end{pmatrix} \tag{3}$$

The parameters are searched in a space determined as follow:

- A_0 parameters were set to zero.
- M parameters were set within the interval [–50, 50] with 2.5 steps and including {–1, 1}.
- N and Q parameters were set only to {–1, 0, 1}.

We use Monte Carlo methods to generate random numbers and assign values to the corresponding parameters. Our intention is to have few parameters in N and Q. Each generated new set of equations are solved using a Runge-Kutta non stiff fourth-order method with fixed step size, and stored if it does not diverge. We calculate Lyapunov exponents using [7] for each stored system and keep it if it's largest Lyapunov exponent is greater than 0.5.

We present 3 new systems with chaotic flows founded with this method. The criteria to determine whether the system shows a chaotic behavior or not is based on the bifurcation diagram, the Lyapunov exponents, and the state space diagram.

2.1 First chaotic flow

The first system is given by

$$\begin{aligned} dx/dt &= y(z-50) \\ dy/dt &= 15(z-y) \\ dz/dt &= x(25-y) \end{aligned} \tag{4}$$

Where the Lyapunov exponents are {2.2, 0.014, –17.2}. In consequence the system is dissipative and with the following two equilibrium points: {0, 0, 0) and {0, 50, 50}. The eigenvalues for {0, 0, 0} are {–32.6, 8.8 – 22.3i, 8.8 + 22.3i} and the eigenvalues for {0, 50, 50} are $\{25\sqrt{2}i, -25\sqrt{2}i, -15\}$. Note that the last equilibrium point is stable. This system has a hidden attractor although the other equilibrium point is not stable. In the Figure 1 the bifurcation diagram and in Figure 2 the state space diagram are shown where the attractor is presented. This system is interesting particularly because is very simple, it is more simple than the 23 systems in [4], since this system has only 6 terms and none of them are pure quadratic. Therefore this system is, to our best knowledge, the simplest mathematical system in the literature today with a stable equilibrium point.

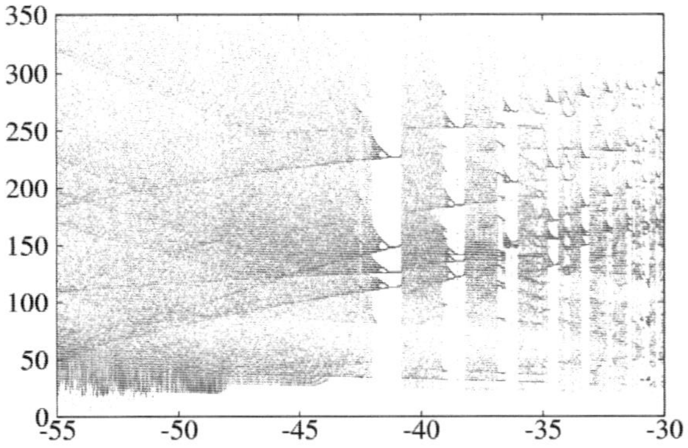

Figure 1. Bifurcation diagram for equations (4). Varying parameter –50 of first equation.

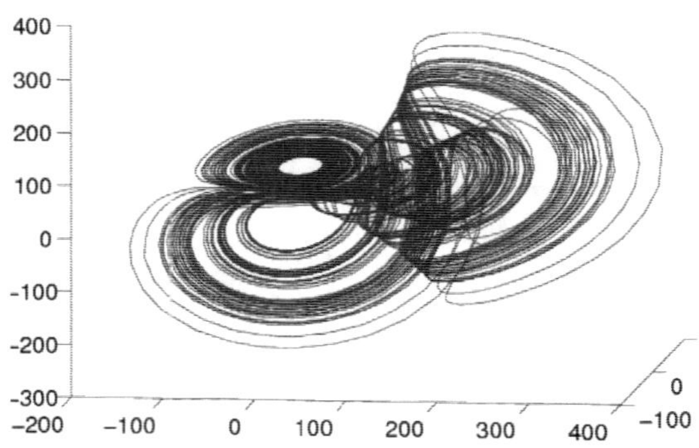

Figure 2. State space diagram for equations (4)

2.2 Second chaotic flow

The second system is given by

$$\begin{aligned} dx/dt &= -35x + yz + 32.5x \\ dy/dt &= -2.5y + 50z + z^2 \\ dz/dt &= -xy - 50y + 15z \end{aligned} \tag{5}$$

The Lyapunov exponents are {8.1, 0.009, –19.4}. Also, the system is dissipative and with the following 3 equilibrium points: {–98.14, 15.82, –50.77}, {–18.32, –23.11, –48.81} and {0, 0, 0}. The corresponding eigenvalues for {–98.14, 15.82, –50.77} are {8.93 + 57.96i, 8.93 – 57.96i, –30.36}, for {–18.32, –23.11, –48.81} the eigenvalues are {55.34, –33.92 + 22.99i, –33.92 – 22.99i} and the eigenvalues for {0, 0, 0} are {6.25 + 49.22i, 6.25 – 49.22i, –25}. In Figure 3 the bifurcation diagram and in Figure 4 the state space diagram are shown. This system shows a special mismatch over the trajectories as can be seen from the state space diagram. Also, this system has only 9 terms and three are quadratic, clearly it is a simple system with three hyperbolic equilibrium points.

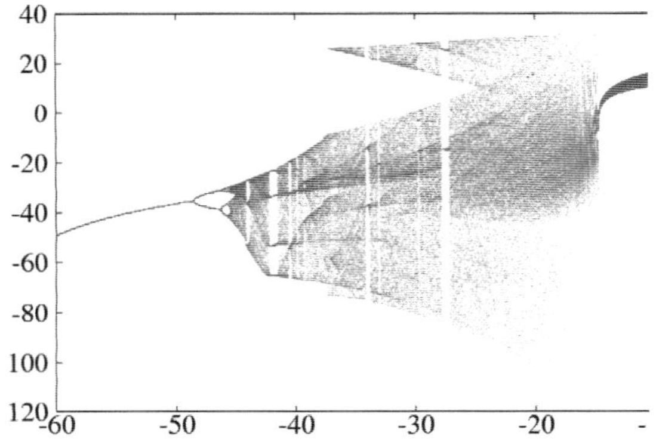

Figure 3. Bifurcation diagram for equations (5). Varying parameter –25 of first equation.

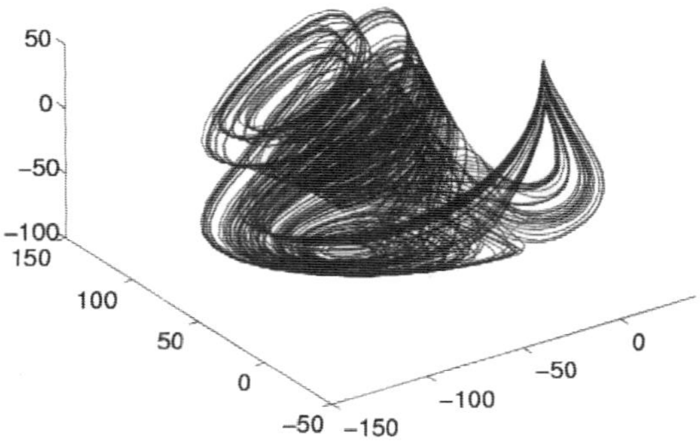

Figure 4. State space diagram for equations (5)

2.3 Third chaotic flow

The third system is given by

$$dx/dt = -30y$$
$$dy/dt = 25x - 47.5z \qquad (6)$$
$$dz/dt = -xy + 27.5y - y^2 - yz - 37.5z$$

The Lyapunov exponents are given by {1.43, 0.006, –31.0}. Again, the system is dissipative and with only one equilibrium point in {0, 0, 0}. The eigenvalues are {9.38 + 20.29i, 9.38 – 20.29i, –56.26}. In Figure 5 the bifurcation diagram and Figure 6 the state space diagram are shown. This system is linear in the first and second equations and quadratic in the third equation with 5 terms. It is not common to find chaotic flows with only one equilibrium point, for example in [7] a chaotic system with only one stable equilibrium point was recently reported. In this case we have one hyperbolic equilibrium point.

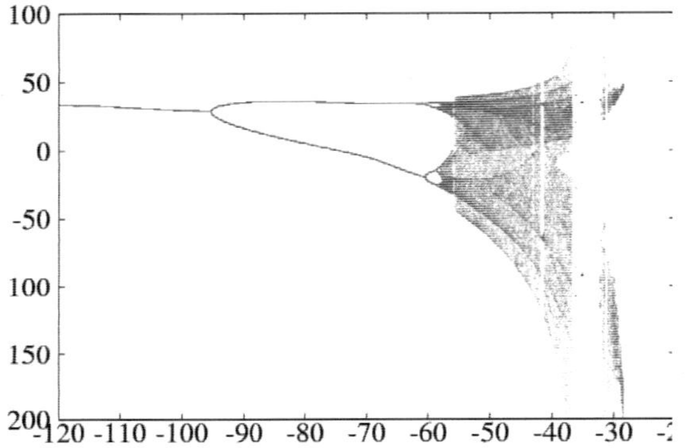

Figure 5. Bifurcation diagram for equations (6). Varying parameter –30 of first equation.

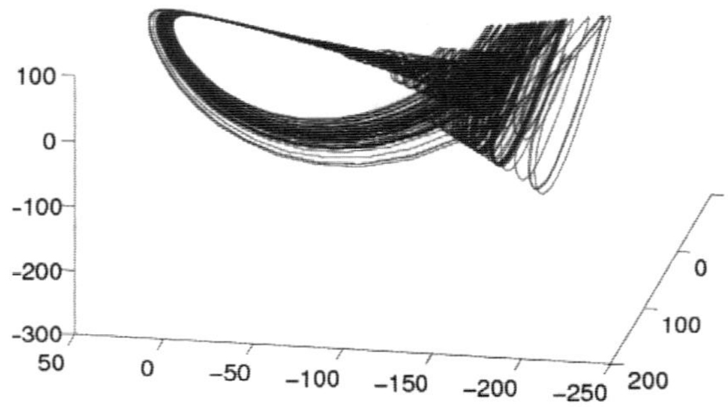

Figure 6. State space diagram for equations (6)

3. Conclusions

We present three new simple chaotic systems with the following properties: the first is a very simple system with a stable equilibrium point and a hyperbolic equilibrium point. The second shows a special mismatch over the trajectories. The third is a chaotic flow with only one hyperbolic equilibrium point. Based on the number of searches performed, we agree with [4] and [5] in the sense that these classes of systems seem to be more common than usually thought.

References

[1] Sprott J C 1994 Phys. Rev. E **50** R647(R)
[2] Lorenz E N 1993 *The Essence of Chaos* (University of Washington Press, Seattle) p 148
[3] Sprott J C 2010 *Elegant Chaos: Algebraically Simple Chaotic Flows*(World Scientific)
[4] Molaie M, Jafari S, Sprott J C and Golpayegani S M R H 2013 *Int. J. Bifur. and Chaos***23** 11 1350188
[5] Jafari S and Sprott J C 2013 *Chaos, Solitons& Fractals***57** 79-84
[6] Wang X and Chen G A 2012*Commun. Nonlinear Sci. Numer. Simul.***17** 3 1264-1272
[7] Wolf A 1986 Quantifying chaos with Lyapunov exponents In: *Chaos*, V. Holden (Edit.), Manchester University Press, Manchester, 273-290

Journal of Physics: Conference Series **582** (2015) 012017 doi:10.1088/1742-6596/582/1/012017

Bioelectric Signal Measuring System

A Guadarrama-Santana[1], L Pólo-Parada[2], A García-Valenzuela[1]

[1] Centro de Ciencias Aplicadas y Desarrollo Tecnológico
Universidad Nacional Autónoma de México
Ciudad Universitaria, Apartado Postal 70-186, Distrito Federal, 04510, México
[2] Department of Medical Pharmacology and Physiology, University of Missouri
John M. Dalton Cardiovascular Research Center
134 Research Park Drive Columbia, Missouri 65211, USA

E-mail: asur.guadarrama@ccadet.unam.mx

Abstract.
We describe a low noise measuring system based on interdigitated electrodes for sensing bioelectrical signals. The system registers differential voltage measurements in order of microvolts. The base noise during measurements was in nanovolts and thus, the sensing signals presented a very good signal to noise ratio. An excitation voltage of $1V_{rms}$ with 10KHz frequency was applied to an interdigitated capacitive sensor without a material under test and to a mirror device simultaneously. The output signals of both devices was then subtracted in order to obtain an initial reference value near cero volts and reduce parasitic capacitances due to the electronics, wiring and system hardware as well. The response of the measuring system was characterized by monitoring temporal bioelectrical signals in real time of biological materials such as embryo chicken heart cells and bovine suprarenal gland cells.

1. Introduction

Interdigitated Capacitance Sensors (IDCS) have been used in different fields of industry as well as in physical, chemical, biochemical and medical research. They offer practical advantages for materials characterization by means of measuring the electrical properties of a Material Under Test (MUT) such as impedance and capacitance due to the frequency dependent dielectric function. An important advantage of plane sensors is that only one surface of the MUT can be directly in contact with the electrodes allowing the sensing of the electric field perturbations on the MUT [1]. The dynamic range and sensitivity of the sensor can be improved depending on the number and dimensions of the electrodes that form the Interdigitated Capacitor (IDC), the electric field applied to it, the substrate and the MUT thickness principally. Finite Element Method is a very useful tool to obtain a good approximation of the physical parameters mentioned above [2]. These devices can sense temporal changes of the electric field in the MUT due to physical or chemical variables affecting biological samples. Temporal variations of the dielectric function of a MUT are manifested as perturbations to an electric field applied to it and can be sensed as temporal electric signals. This sensing principle makes it suitable for bio-sensor applications [3].

Bio-sensors are used to detect and analyze temporal cellular response to electrical or bio-chemical conditions. The resulting bio-signals can be sensed as voltage or current signals by means of impedance changes of the cell or biological tissue. The cell impedance technique is used to sense and monitor the behavior and growth of cell population when cells are in touch with conductive electrodes deposited on a plain substrate [4]. Measurements with many cells use superficial electrodes as sensors, the electric field generated by the action of many cells distributed in the neighbourhood of the electrode constitute the bioelectric signal. The electric field traverses the biological substance, so the electric potential difference can be sensed in suitable locations across the biological tissue, eliminating the need of invading the system. Bioelectrical signals require a relatively simple transducer for its registering, as the case of the electrocardiogram ECG, electroencephalogram EEG, electromyography EMG and others. With these techniques, information can be obtained only in a collective way to a given physiological condition on an electrode array pattern, usually plane or coplanar electrodes.

Patch-Clamp commonly used in Electro-Physiology is another measuring technique used to characterize a single cell response that cannot be made in a collective way [5]. Ion currents that flow through the ion channels of the cellular membrane due to physiological conditions are sensed with this technique. Patch-Clamp technique is suitable for studying a single cell response to a determined stimulus (electrical, optical, chemical or spontaneous) [6,7]. Other bioelectrical signals are generated by nerve and muscular cells. Their source is the membrane potential, which under certain conditions can be excited to generate it. Single cell measurements use specific microelectrodes as sensors and the same excited electric potential is the bioelectrical signal. With the measuring system proposed microelectrodes could be used as sensors, for electrophysiology local measuring techniques. Although, measuring work must be done in order to characterize the system response for a particular electrophysiology application.

IDCSs can detect small electric field perturbations present near its sensing surface due to small bioelectrical signals of some biological sample as in signal measuring techniques mentioned above. These electrical signals can be very small and be immersed in surrounding noise. In common situations, noise becomes a real problem to the point of impeding electric signal registering. Therefore, a low noise measuring system is required for bioelectric other sensing techniques in order to obtain a convenient SNR like those obtained with electrophysiology techniques [8].

The low noise measuring system proposed here is based on interdigitated capacitive sensors, a home- made electronic conditioning stage and a digital Lock-in amplifier. The measuring system allows to reduce noise and parasitic capacitances present in the system hardware and secondary glass substrates where the biological samples are placed. The measuring system detects electric field perturbations on the sensitive surface of the sensing device, the perturbations are registered as differential voltage measuring due to the impedance of the biological sample and its bioelectrical activity with a very suitable SNR with nanovolts base noise.

The response and sensitivity of the system were obtained characterizing several measurements of bioelectrical activity from different cell samples. The experiments proposed are related to the study of the release of neurotransmitters of adrenal medulla cells which can be determined with an independent bioelectrical measurement to capacitance changes in the cell that represent the fusion of the vesicles. Also, cardiac cells were used to study the evaluation of the possible changes in capacitance due to spontaneous contractions, which are reflected in the biosignal activity.

The results obtained indicate that it is possible to apply this measuring system in the electrophysiology field.

2. Low Noise Capacitive Measuring System

The low noise capacitive measuring system proposed is based on interdigitated capacitive sensors, an electronic conditioning stage, an analog to digital converter (ADC) and a digital Lock-in amplifier, figure 1.

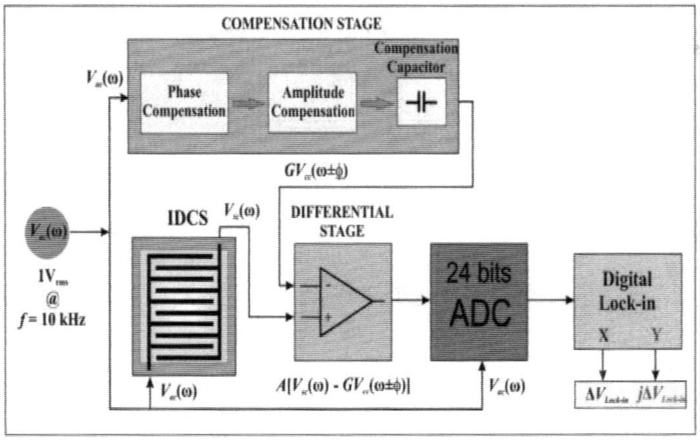

Figure 1. Block diagram of the Low Noise Capacitive Measuring System.

The sensing method proposed is based on phase changes of a sine excitation voltage $V_{ac}(\omega)$ due to electric field perturbations of biological samples placed over 1cm^2 sensitive area of the interdigitated capacitive sensor (IDCS). The excitation voltage $V_{ac}(\omega)$ with amplitude and frequency used in the sensing device were 1V$_{rms}$ and 10 kHz respectively resulting a voltage signal $V_{sc}(\omega)$. The same excitation voltage $V_{ac}(\omega)$ was applied simultaneously to a compensation capacitor (mirror device) as well.

The resulting $GV_{cc}(\omega\pm\varphi)$ signal from the mirror device was compensated in amplitude and phase in order to equalize it with $V_{sc}(\omega)$ obtaining an initial minimal reference voltage from a differential stage. The differential signal can be amplified up to a maximum gain factor of A=1x10^6 to obtain a suitable voltage for digitalizing it with the 24 bits ADC.
The sensitivity primary limit of the digital Lock-in is in its dynamic range, which in order is determined principally for the National Instruments 4431 ADC. In this case the dynamic range was 100 dB and 102.4kS/s sample rate. The differential output signal $A[GV_{cc}(\omega\pm\varphi) - V_{sc}(\omega)]$ is digitalized by the ADC with approximately 1μV minimum voltage level. This signal is processed by the digital Lock-in stage implemented with LabView to the reference frequency mentioned before.
The signal goes through phase detection and digital filter stages in order to obtain a very narrow width band around the reference central frequency to reduce surrounding noise to a minimum level.
The registered data are displayed in real time in graphs generated with the same software as a part of a virtual instrument. Besides, statistics data are obtained which are necessary to calculate the base noise of the system in the width band determined.
Data samples in study can also be stored to be analyzed in its real and complex resulting components. The reference base noise obtained with the system was in nanovolts range with ambient conditions.

3. Experimental

Experimental measurements were made with the system proposed in order to characterize its sensitivity and resolution response. When a MUT is placed on the sensitive surface of the capacitive sensor impedance and current are generated when a potential is applied across to it.
Thus, current and voltage across the MUT can be sensed. When an ac potential is applied to the MUT a dissipative current due to the resistive impedance and a displacement current due to the capacitive impedance are generated [9].
Performing an admittance analysis of the system the capacitance as a function of the displacement current across the material under test can be obtained. This would be useful for impedance characterization of biological samples, but it is out of the scope of this work.

Journal of Physics: Conference Series **582** (2015) 012017 doi:10.1088/1742-6596/582/1/012017

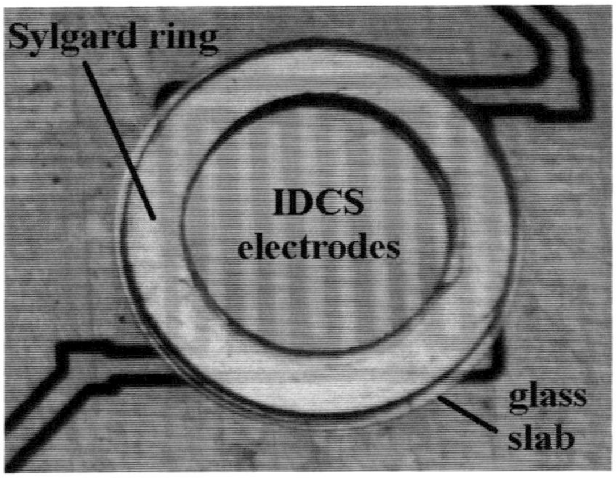

Figure 2. Small Petri dishes formed with circular glass slabs and Sylgard rings placed over $1cm^2$ sensitive surface of the Interdigitated Capacitive Sensor (IDCS).

Small Petri dishes were fabricated with 100 μm thick glass substrates and 1cm diameter placed over $1cm^2$ IDCS sensitive surface with 50μm wide and 50μm gaps between electrodes. The edges of the circular glass slabs were surrounded with Sylgard rings in which biological samples immersed in culture medium were deposited in order to sense its bioelectrical signals in diverse experiments, figure 2.

Before performing any measurements a minimal reference voltage was set in the electronic conditioning stage, a $3mV_{p\text{-}p}$ noise voltage was present at the 24 bit ADC output. After the digital Lock-in stage the noise output voltage was reduced to $3nV_{p\text{-}p}$. This value was the system base noise as shown in figure 3.

Experimentally, the real and imaginary components of the voltage signal were registered by the digital Lock-in stage.

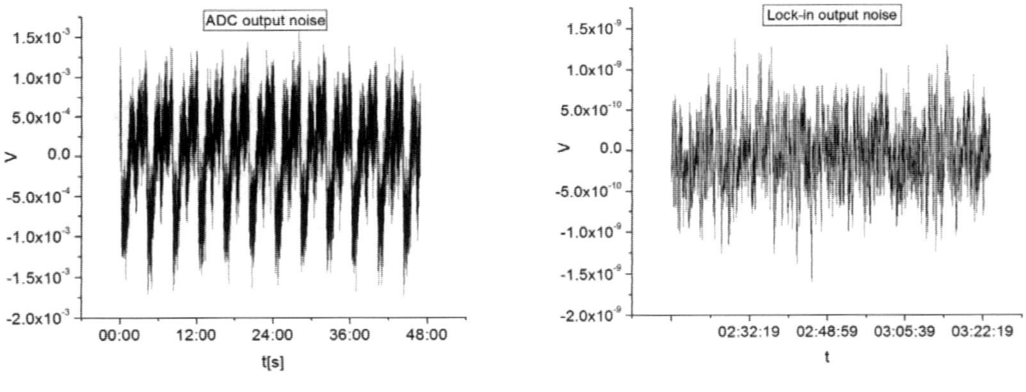

(a) (b)

Figure 3. a) $3mV_{p\text{-}p}$ ADC output base noise, b) $3nV_{p\text{-}p}$ digital Lock-in output base noise.

The differential stage is based on an instrumentation amplifier that has an offset voltage of 50μV max. The sensing signal from the sensing device and a reference signal compensated in amplitude and phase enter the differential stage in order to get a minimum offset reference signal before place any biological sample, as can be seen in figure 3b.

We will refer to the sensitivity as the smallest signal change that the measuring system can detect. In this case the sensitivity is twice the Lock-in stage output base noise.

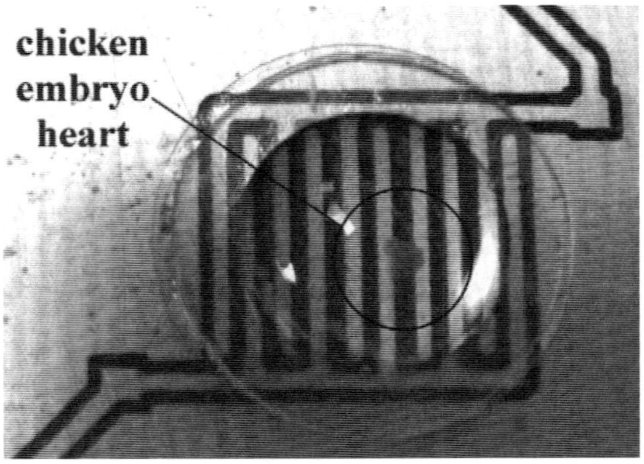

Figure 4. Chicken embryo heart immersed in 70μl culture medium over an Interdigitated Capacitive Sensor with $1 cm^2$ sensitive surface.

Different biological samples were monitored in order to register their bioelectrical signals with the measuring system proposed. First, a chicken embryo heart of around 100 microns in diameter immersed in 70 μl culture medium was placed in the Petri dish over the sensing device as it is shown in figure 4.

Chicken embryo heart cells and bovine suprarenal cells were also monitored following the same methodology.

4. Results

The measuring system proposed was able to register temporal bioelectrical signals of the chicken embryo heart, as it can be seen in figure 5. Electrical activity is represented as amplitude voltage variations over an offset around -0.305 mV for X(t) and 0.489 mV for Y(t) corresponding to the real and imaginary components of the signal respectively. The signals show certain periodicity and 6μV minimum variation showing SNR with respect $3nV_{p-p}$ system base noise despite the offset presented in the graphs.

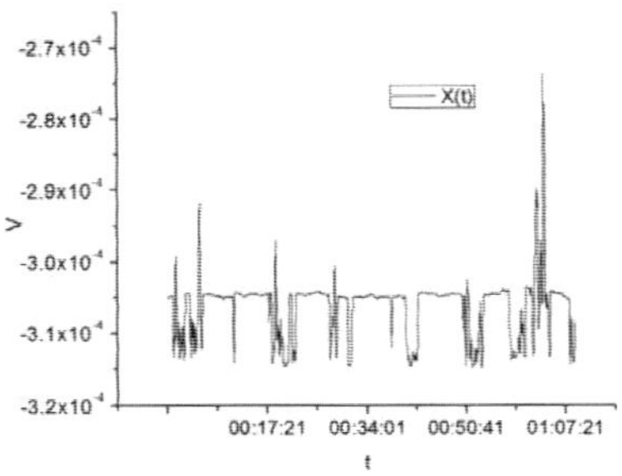

(a)

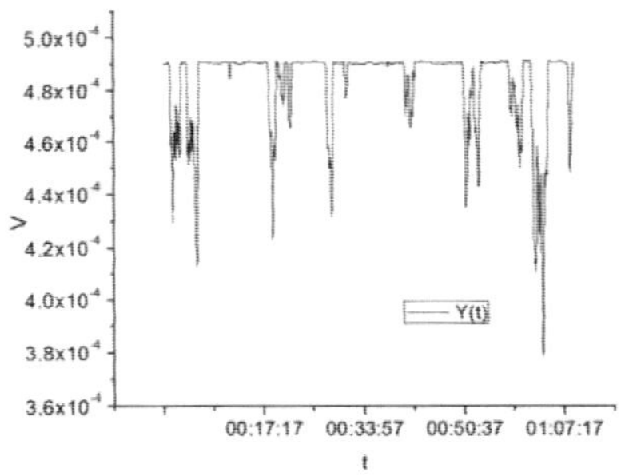

(b)

Figure 5. Signals registered from a chicken embryo heart, a) for X(t) and b) for Y(t).

The same measurement methodology was applied with chicken embryo heart cells of around 10-12 μm in diameter immersed in 50 μl culture medium, used as a second biological sample. Temporal bioelectrical signals registered are shown in figure 6. The graphs show an offset voltage around 5μV for X(t) and 3.6μV for Y(t) and maximum signal variation around 1μV showing a good signal to noise ratio with respect $3nV_{p-p}$ system base noise.

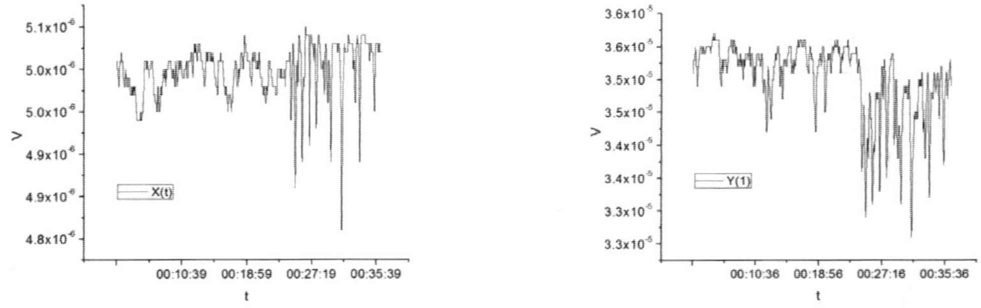

Figure. 6. Signals registered from a bunch of chicken embryo hart cells, a) for X(t) and b) for Y(t).
t is in a h:m:s format.

Finally, bovine suprarenal gland cells of about 14-16 μm in diameter immersed in 50 μl culture medium were used as a third biological sample following the same measurement methodology. The resulting bioelectrical signals registered are shown in figure 7.

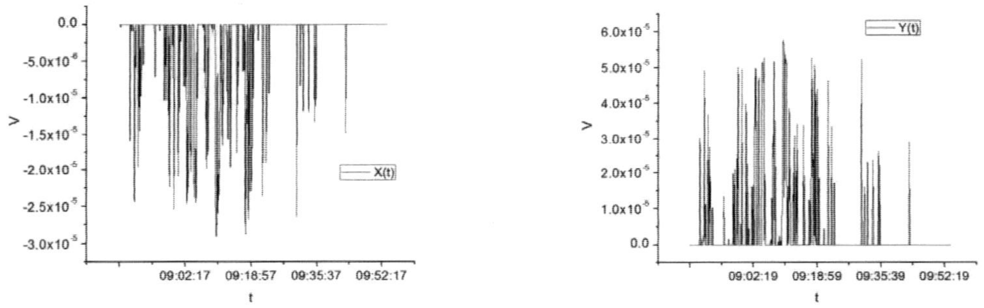

(a) (b)

Fig. 7. Bioelectrical signals registered from a bunch of suprarenal gland cells, a) for X(t) and b) for Y(t). t is in a h:m:s format.

In this case the graphs do not show a significant offset voltage as in the cases above, this is the main difference between bioelectrical signals of chicken embryo hart cells and bovine suprarenal gland cells. Clearly it can be noticed the minimal offset voltage of the signal and the maximum signal variation registered, around -3μV for X(t) and 6μV for Y(t) showing a good SNR, with respect 3n$V_{p\text{-}p}$ system base noise.

Graphs obtained showed bioelectrical signals from some nanovolts with SNR up to 100 as in the case of chicken embryo heart cells, figure 6a, and up to 10000 as in the cases of chicken embryo hart and bovine suprarenal gland cells, figures 5a and 7a respectively.

The offset shown in figures 5 and 6 for chicken heart embryo and chicken heart embryo cells could be due to the culture cell and the same biological samples as well. This is not the case with suprarenal gland cells where offset seems to be lower. In all cases bioelectrical activity is well defined with a good SNR.

5. Conclusions

The measurements obtained with the system proposed in this work are useful to estimate its sensitivity and response to different kind of bioelectric signals with no controlled ambient conditions. The measuring system presented very good stability with a base noise around 3nV peak-peak, this characteristic allowed us to register biosignals for cellular electrophysiology applications. The results obtained with the proposed measuring system show that capacitance sensors can be used for bioelectrical signal characterization from different kind of cells. The analog bioelectrical signals were electronically conditioned, digitalized and registered in real time in order to monitor biologic material for its analysis. Up to three samples can be monitored in real time with the measuring system due to the characteristics of the ADC. The measuring technique is practical because only it is necessary to change the small Petri dish with the biological sample of interest using the same capacitive sensor. Finally, this measuring system can offer additional information compared with established measuring techniques used in electrophysiology as the Patch Clamp technique [5]. One possible application for this measuring technique is to characterize the bioelectrical response of the cells to different drugs in order to determine the effectiveness of a particular treatment by comparing bioelectrical signals activity. Even though, more measurements must be done in order to characterize different bioelectric signals for different cells in determined conditions, the present work shows the potential of the measuring system proposed and IDCS for applications in electrophysiology research.

6. References

[1] Starzky F 2008 Parametrisation of interdigit comb capacitor for dielectric impedance spectroscopy, Archives of Materials Science and Engineering, Vol. **34**, Issue 1, 31-34.
[2] Hu X and Wuqiang 2010 Planar capacitive sensors – designs and applications, Sensor Review, **301** 24-39.
[3] Ibrahim M, Claudel J, Kourtiche D and Nadi M 2013 Geometric parameters optimization of planar interdigitated electrodes for bioimpedance spectroscopy, Journal of Electrical Bioimpedance, vol. **4** pp. 13-22.
[4] Berggren C, Bjarnason B and Johanson G 2001 Review Capacitive Biosensors, Electroanalysis, 13, No. **3** 173-180.
[5] Zhao Y, Inayat S, Dikin D A, Singer J H, Ruoff R S, and Troy J B 2009 Proc. IMechE Vol. **222** Part N: J Nanoengineering and Nanosystems.
[6] Asphahani P, Wang K, Thein M, Veiseh O, Young S, Xu J and Zhang M 2011 Phys Biol 015006 (11pp).
[7] Shapiro M G, Homma K, Villarreal S, Richter C P and Bezanilla F 2012 nature communications DOI:10.1038/ncomms1742.
[8] Koester P J, Mortiz Buehler S, Stubbe M, Tautorat C, Niendrof M, Baumann W and Gimsa J 2010 Modular glass chip system measuring the electric activity and adhesión of neuronal cells-application and drug testing with sodium valproic acid, Lab Chip **10**, 1579-1586.
[9] Baxter L K 1997 Capacitive Sensors Design and Applications, IEEE Series on Electronics Technology, ISBN o-7803-1130-2.

7. Acknowledgments

Authors wishing to thank the Dirección General de Asuntos del Personal Academico, Universidad Nacional Autonoma de Mexico for financial support of AGS's postdoctoral work

Journal of Physics: Conference Series **582** (2015) 012018 doi:10.1088/1742-6596/582/1/012018

Kinetics modeling of the drying of sunflower stem (Helianthus annuus L.) in a forced convection tunnel

R López, M Vaca, H Terres, A Lizardi, J Morales, J Flores, S Chávez
Universidad Autónoma Metropolitana-Azcapotzalco. Distrito Federal, México.

E-mail: rlc@correo.azc.uam.mx

Abstract. The sunflower is an annual plant native to the Americas. It possesses a large inflorescence (flowering head), and its name is derived from the flower's shape and image, which is often used to capture the sun. The plant has a rough, broad, hairy stem, coarsely toothed, with rough leaves, and circular flower heads. The sunflower seeds are appreciated for their oil , which has become a widespread cooking ingredient. Leaves of the sunflower can be used as cattle feed, while the stems contain a fiber that may be used in paper production. Recently this flower has been used in phytoremediation of soils, contaminated with heavy metals. Sunflower has been probed as an efficient phytoextractor of chromium, lead, aluminum, zinc, cadmium from soil.
In this work we present the experimental results of the drying of the sunflower stem, cut in 100 mm longitudinal sections, with diameters in the range of 11-18 mm. The aim was to obtain a dry and easy-to-handle final product, since these plants were originally cultivated in order to extract heavy metals from a polluted soil. The dried stems could then be easily confined or sent to recycle premises to concentrate the metals.
The drying process was done in forced convection within a hot air tunnel. The used temperature was 60 °C, the velocity of air was 3 m/s and the required times were 8 hours. The initial average wet mass was 28 g and the final value was 5 g, resulting in the aimed product.

1. Introduction

Traditionally, the oilseed sunflower plant is used to extract from its seeds oil and flour, the rest of the plant is used as fodder, namely animal feed. The kinetics of the flower drying has been studied. Syaref *et al.* [1], used hot air at 75 ° C and 0.3 m/s to dry 200 g of sunflower seeds with initial humidity contents of 57%, to 12% on dry basis in 40 min. The structure, the morphology and the chemical composition are factors that influence the drying kinetics due to changes caused by the different moisture levels during this process [2]. According to Rovedo *et al.* [3], the complete sunflower seed is viewed as a spherical core, and the kernel is surrounded by a single hull. The solution of the Fick's second law for moisture diffusion from a sphere in terms of the average moisture contents was used to adjust the drying curves and to evaluate the diffusion coefficient at the same temperature of 75 °C. Gely and Santalla [4], found that the equation of Page was the one that better described the kinetics of drying, when they obtained a value of $r^2 = 0.99$ and a diffusion coefficient of 7.14×10^{-8} cm^2/s. Kundu *et al.* [5], used a fluid bed dryer to dry sunflower seeds with air at 70 ° C and 1.36 m/s velocity in 20 min, with initial and final moisture contents of 22% and 8%, respectively. The sorption characteristics of sunflower seeds were measured and analyzed to provide a means for evaluating stability conditions preventing microbial spoilage and oil acidification. Water activities a_w were measured at three temperatures in a wide moisture content range and the data were interpreted with isotherm models [6]. Shanmugam *et al.* [7] used an oscillating-bed solar dryer with sunflower seeds to obtain in just 8 h an average moisture level of 7.9% which is in the acceptable level of moisture contents of less than 10%. The initial moisture contents were of 18%, the average thermal efficiency of the solar dryer was 30.1%. The possibility of using sunflower to produce biofuel has also been analyzed [8-11] and finally, also some research has been done to use the plant in phytoremediation of soils contaminated with heavy metals [9, 12-17]. In this work we present the experimental results of the drying of the sunflower stem in a forced convection tunnel. The aim was to obtain a dry and easy-to-handle final product, since these plants were originally cultivated in order to extract heavy metals from a polluted soil. The dried stems could then be easily confined or sent to recycle premises to concentrate the metals.

2. Mathematical modeling

The moisture ratio was calculated using $R = (M_t - M_e)/(M_0 - M_e)$, which has been simplified to M_t / M_0 by some researchers, including Kaymak-Ertekin [18], and Akpinar *et al.* [19], because of the continuous fluctuation of the relative humidity of the drying air during the process. Where M_t is the mass at time t, M_e is the mass at equilibrium and M_0 is the initial mass. For mathematical modeling, there are several thin-ayer drying equations; we tested them and select the best model to describe the drying curve equation for the sunflower stem using hot air.

The correlation coefficient (R^2), the reduced chi-square (χ^2), and the root mean square error (RMSE) were used as criteria for adequacy of the fit. The lower the values of the reduced χ^2, the better is the goodness of the fit. The RMSE gives the deviation between the predicted and experimental values, and it is required to approach zero. These statistical analysis values can be calculated as follow,

$$R^2 = \frac{\sum_{i=1}^{N}\left(\mathrm{MR_i} - \mathrm{MR_{pre,i}}\right)\sum_{i=1}^{N}\left(\mathrm{MR_i} - \mathrm{MR_{exp,i}}\right)}{\sqrt{\left[\sum_{i=1}^{N}\left(\mathrm{MR_i} - \mathrm{MR_{pre,i}}\right)^2 \cdot \sum_{i=1}^{N}\left(\mathrm{MR_i} - \mathrm{MR_{exp,i}}\right)^2\right]}} \qquad (1)$$

$$\chi^2 = \frac{\sum_{i=1}^{N}\left(\mathrm{MR_{exp,i}} - \mathrm{MR_{pre,i}}\right)^2}{\mathrm{N - n}} \qquad (2)$$

$$RMSE = \sqrt{\frac{\sum_{i=1}^{N}\left(\mathrm{MR_{pre,i}} - \mathrm{MR_{exp,i}}\right)^2}{N}} \qquad (3)$$

where stands $MR_{exp,i}$, for the experimental moisture ratio found in any measurement, $MR_{pre,i}$ is the predicted moisture ratio for this measurement and n are the number of observations and the number of constants, respectively [20].

3. Instrumentation and measurement.

Air and product-surface temperatures were measured using calibrated K-type thermocouples (0.5 °C exactitude); the relative humidity of the environment was determined using an EA25-model EXTECH digital hygro-thermometer, with 0.1 % resolution. A 451112-model EXTECH anemometer (0.1 m/s resolution) was used to measure the velocity of air. Mass was quantified using a BL1505-model SARTORIUS scale, with a 0.001 g span. Data acquisition was programmed using LABVIEW software.

4. Experimental procedure.

Sunflower seeds were sown on soil contaminated with heavy metals, which was previously characterized [21] . When the plants grew to a height of approximately 1.60 m, they were cut and the stem was divided into pieces of 0.10 m length; diameter was 0.018 m on average, with an approximate weight of 0.028 kg, initial moisture content was of 65%. The temperature of air was 60 °C at a velocity of 3.0 m/s. Tests were done in a continuous drying process, to guarantee that no perturbation could hinder the course of the experiment [22]. The electrical resistances were operated to raise the surface temperature of the probes up to 60 °C. Mass, air temperature, probe-surface temperature and relative humidity were automatically monitored and registered in a computer. The measurements were done every 10 minutes.

5. Experimental apparatus

The scheme of the drying tunnel used in this study is presented in figure 1. Air is conducted by means of a radial-flow fan, controlled by a variable-velocity motor (a). The velocity of air can be varied between 1.0 and 10.0 m/s; in this study the velocity applied was 3.0 m/s. There is a panel of electrical resistances (b); it is formed by two resistances of 2 kW each, two resistances of 1 kW and two more of 500 W, which together provide a maximum power of 7.0 kW. All of them are managed independently, which allows the selection of the energy to be provided. This section is cover with mineral wool as insulator (c); the length of the test chamber is 30 cm, the width,30 cm, and the height, 40 cm (d); electronic scale (e); the computer for registration all the values (f), and finally the controls system (h).

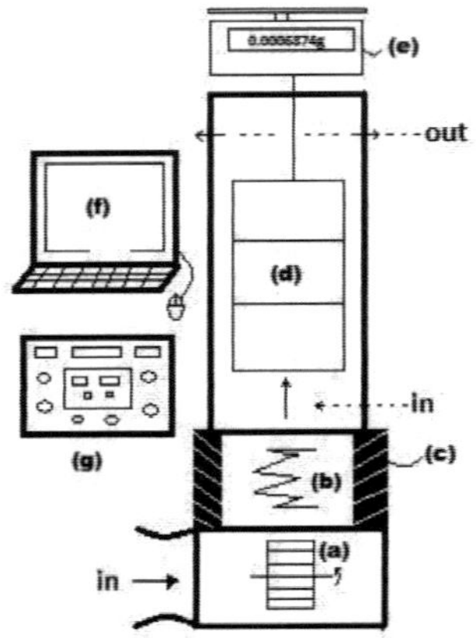

Figure 1. Vertical drying tunnel.

6. Results.

The initial average mass of the sun flower stem pieces was 28 g and the final value was 5 g, resulting in the aimed product. The temperature of the air at the entrance of the drying chamber was about 62 °C, and the temperature measured on the surface of the sample was 60 °C. At the beginning of the experiment the sample presented a moisture content of 4.82 kg water/kg dm and after 8 hours it was 0.1 kg water/kg dm. This was the minimum value that could be obtained, since the experiment continued for 30 minutes more but humidity remained unchanged. The variation of moisture against time is shown in figure 2. This experiment was carried out on a continuous basis and was repeated five times, the results were consistent.

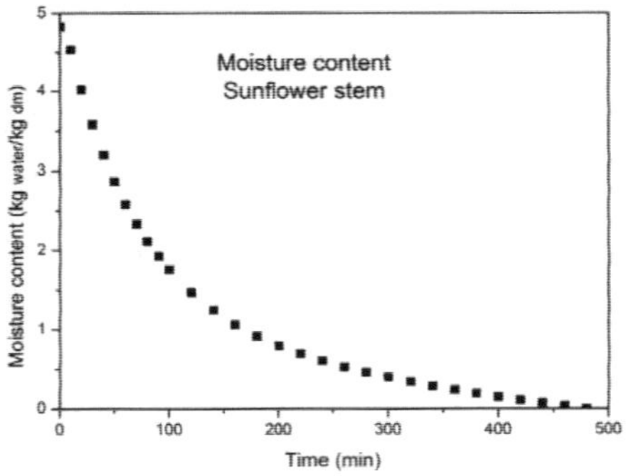

Figure 2. Moisture contents vs time for sunflower stem, the air temperature was of 60 °C on its surface, and the velocity of air was 3.0 m/s.

Modeling the drying behavior of different agricultural products often requires statistical methods of regression and correlation analysis. Linear and non-linear regression models are important tools to find the relationship between different variables, especially, for those for which no established empirical relationships exist. In this study, we determined relationships for constants of the best suitable model for the drying air temperature by multiple regression techniques, using Arrhenius, exponential and power regression models [19]. Regression analysis was performed using the Origin computer program.

The mathematical model that best described the drying process under study was the logarithmic one, with R^2 close to one, χ^2 close to zero and RMSE close to zero too (figure 3). For the sunflower stem the equation is:

$$\mathrm{MR} = 0.118 + 0.5071\exp(-0.00016\,t)$$
$$R^2 = 0.996$$
$$\chi^2 = 0.000012$$
$$\mathrm{RMSE} = 0.00053$$

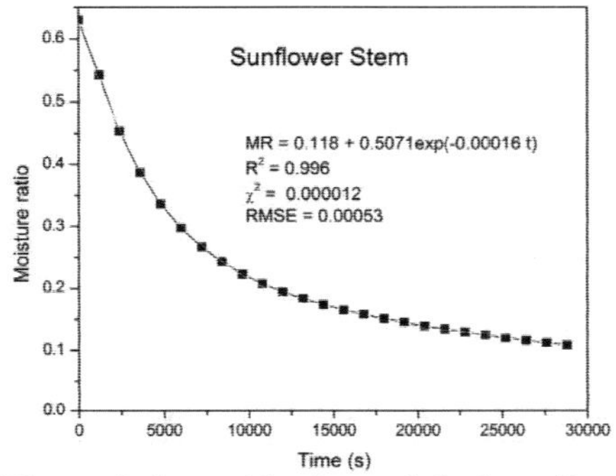

Figure 3. Logarithmic model of sunflower stem.

A fast variation in the velocity of drying of the specimens at the beginning of the process was observed, reaching values of 0.348 kg water/kg dm min on average during the first 40 minutes. Afterwards it remained practically constant at a value of 0.13 kg water/ kg dm min, until the last 20 minutes, in which it increased slightly. These results are summarized in figure 4.

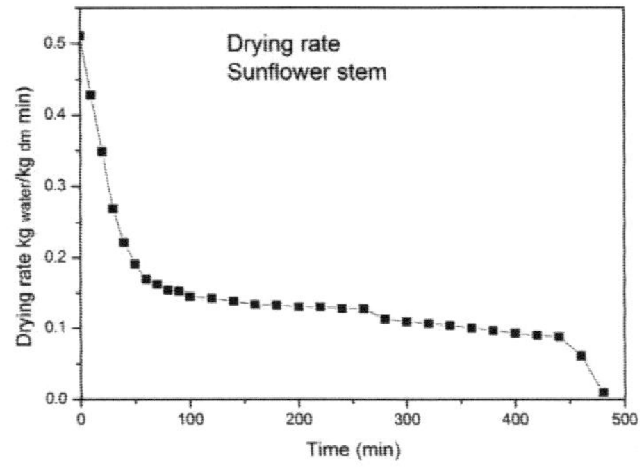

Figure 4. Drying rate of sunflower stem.

The equation of the drying of the sunflower stem was obtained by the dimensionless plot of the data in figure 5, through the Marquardt- Levenberg's non-linear optimization, and it is as follows:

$$\mathrm{Y} = 0.0211 + 1.3088 * \mathrm{X} - 1.7594 * \mathrm{X}^2 + 1.3243 * \mathrm{X}^3$$

with $R^2 = 0.979$ and SD $= 0.0127$, values that reflect good precision in the modeling of the physical phenomena.

This curve is known as *"characteristic curve of drying"* of the product [23]. Thus, the characteristic curve of drying of the sunflower stem, according to the last equation, is:

$$\mathrm{f} = 0.0211 + 1.3088\mathrm{MR} - 1.7594\mathrm{MR}^2 + 1.3243\mathrm{MR}^3$$

where f is the dimensionless drying relationship.

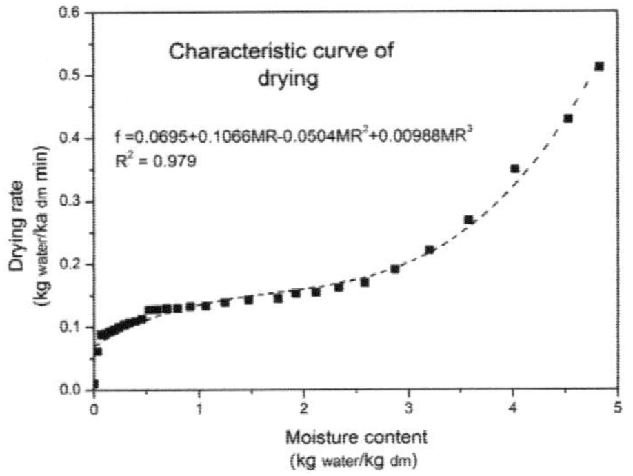

Figure 5. Characteristic curve of drying of the sunflower stem.

Finally, from the environmental point of view, since the stems contained concerning amounts of heavy metals [21], the resulting reduction of 82.12% of the total mass implied that the aimed final product, an easy- to- handle waste which could be economically sent to further treatment, confinement or metals recycling, was obtained.

7. Conclusions

In this investigation the sunflower stem could be dried within 8 hours with air at a temperature of 60 °C and a velocity of 3 m/s. The initial and final moisture contents were 4.82 kg water/kg dm and 0.1 kg water/kg dm, respectively. The numerical model that best represented the kinetics of drying of the product was the exponential one, according to the imposed conditions. The model was $MR = 0.118 + 0.5071\exp(-0.00016\,t)$. The characteristic drying curve turned out to be a third degree equation, namely $f = 0.0211 + 1.3088MR - 1.7594MR^2 + 1.3243MR^3$.

References

[1] Syaref A M, Morey R V and Gustafson R J 1988 Thin-layer drying rates of sunflower seeds, *Trans, of the ASAE.* pp. 195-200.

[2] Brooker D B, Bakker-Arkema F W and Hall C W 1992 Drying and storage of grains and oilseeds, *AVI Book Ed.*

[3] Rovedo C A, Aguerre R J and Suarez C 1993 Moisture diffusivities of sunflower seed components. *International Journal of Food Science and Technology.* **28** pp. 159- 168.

[4] Gely M C and Santalla E M 2000 Effect of Some Physical and Chemical Properties of Oilseeds on Drying Kinetics Parameters. *Drying Technology* **18** pp. 2155-2166.

[5] Kundu K M, Datta A B, Chatterjee P K 2001 Drying of oilseeds. *Drying Technology* **19** pp. 343–358.

[6] Giner S A and Gely M C 2005 Sorptional Parameters of Sunflower Seeds of Use in Drying and Storage Stability Studies. *Biosystems Engineering* **92** pp. 217–227.

[7] Shanmugam S, Kumar P and Veerappan A R 2011 Evaluation of thermal efficiency of oscillating-bed solar dryer through drying of sunflower seeds. *Journal of Renewable and Sustainable Energy.* **3** pp. 1-8.

[8] Figueiredo A K, Baümler E, Riccobene I C, Nolasco S M 2011 Moisture-Dependent Engineering Properties of Sunflower Seeds With Different Structural Characteristics. *Journal of Food Engineering.* **102** pp. 58–65.

[9] Llorent-Martínez E J, Ortega-Barrales P, Fernández-de Córdova M L, Domínguez-Vidal A, Ruiz-Medina A 2011 Investigation by ICP-MS of trace element levels in vegetable edible oils produced in Spain. *Food Chemistry.* **127** pp. 1257–1262.

[10] Ozge Cepeliog ullar and Ays E Putun 2013 Thermal and kinetic behaviors of biomass and plastic wastes in co-pyrolysis. *Energy Conversion and Management* **75** pp. 263–270.

[11] Siqueira Pantoja S, da Conceição Rafael V, da Costa Carlos E F, Zamian J R, da Rocha Filho G 2013 Oxidative stability of biodiesels produced from vegetable oils having different degrees of unsaturation. *Energy Conversion and Management.* **74** pp. 293–298.

[12] Lievens C, Yperman J, Vangronsveld J, Carleer R 2008 Study of the potential valorisation of heavy metal contaminated biomass via phytoremediation by fast pyrolysis: Part I. Influence of temperature, biomass species and solid heat carrier on the behaviour of heavy metals. *Fuel*. **87** pp. 1894–1905.

[13] Mendil D, Uluözlü O D, Tüzen M, Soylak M 2009 Investigation of the levels of some element in edible oil samples produced in Turkey by atomic absorption spectrometry. *Journal of Hazardous Materials*. **165** pp. 724–728.

[14] Ansari B R, Kazi T, Jamali M, Arain M, Wagan M, Jalbani N, Afridi H, Shah A 2009 Variation in accumulation of heavy metals in different verities of sunflower seed oil with the aid of multivariate technique. *Food Chemistry*. **115** pp. 318–323.

[15] Stals M, Thijssen E, Vangronsveld J, Carleer R, Schreurs S, Yperman J 2010 Flash pyrolysis of heavy metal contaminated biomass from phytoremediation: influence of temperature, entrained flow and wood/leaves blended pyrolysis on the behaviour of heavy metals. *J. Anal. Appl. Pyrolysis*. **87** pp. 1–7.

[16] Priya P, Shivendra V S 2011 Suppression subtractive hybridization reveals differential gene expression in sunflower grown in high P. Plant Physiology and Biochemistry **49** pp. 584-591

[17] Rojas-Tapias D, Bonilla R, Dussan J 2012 Effect of inoculation with plant growth-promoting bacteria on growth and copper uptake by sunflowers. *Water Air Soil Pollut*. **223** pp. 643–654.

[18] Kaymak-Ertekin, F2002 Drying and rehydrating kinetics of green and red peppers, *Journal of Food Science*, Vol. **67** pp. 168-75.

[19] Akpinar, E, Bicer Y and Yildiz C 2003 Thin layer drying of red pepper, *Journal of Food Engineering*, Vol. **59** pp. 99-104.

[20] Yaldiz O and Ertekin C 2001 Thin layer solar drying some different vegetables, *Drying Technology*, Vol. **19** pp. 583-596.

[21] Basabe V, Vaca M, Rojas N, López R 2013 Effectiveness of treatment phytoextracción of chromium, with sunflower (*Helianthus annuus*) and EDTA. *Proceedings of the VI Latin Environmental Sciences Network*, Costa Rica.

[22] Simal S, Deya E, Frau M and Rossello C 1997 Simple Modelling o fair drying curves of fresh and osmotically pre-dehydrated apple cubes. *Journal of Food Engineering*, Vol. **33** pp. 139-150

[23] Lahsasni S, Kouhila M, Mahrouz M, Jaouhari J T 2004 Drying kinetics of prickly pear fruit, *J Food Eng*, **61** pp. 173–179.

Journal of Physics: Conference Series **582** (2015) 012019 doi:10.1088/1742-6596/582/1/012019

Heat and mass transfer analysis of convective drying of chickpea (*Cicer arietinum*)

R López, M Vaca, H Terres, A Lizardi, J Morales, J Flores, S Chávez
Universidad Autónoma Metropolitana-Azcapotzalco. Distrito Federal, México.

E-mail: rlc@correo.azc.uam.mx

Abstract. The objective of this article is to describe the modelling and simulation of the dehydration of chickpea in a complex drying system process, using COMSOL Multiphysics Program. A model, based on mass and energy balances, was developed for the simulation of unsteady convective drying with air (3.0 m/s and 60 °C). The program predicted an 8 hours-dehydration time, with an effective moisture diffusivity of $3.1 * 10^{-10}$, which was experimentally obtained. The empirical model that best represented the process was the exponential one.

1. Introduction

Chickpeas, (*Cicer arietinum*), also known as *garbanzos* are grown in plants between 20-50 cm high that have small leaves on either side of their stems. Chickpeas are a type of pulse, with one seedpod containing two or three peas. There are two main kinds of chickpeas: *Desi*, cultivated mostly in the Indian subcontinent, Ethiopia, Iran, and Mexico; and *Kabuli*, which has lighter colored, larger seeds and a smoother coat, mainly grown in Southern Europe, Northern Africa, Pakistan, Afghanistan, India and Chile. In 2011, 132 000 tons were produced in Mexico for human consumption in approximately 90 000 cultured ha [9].

Drying has a negative environmental impact because the thermal energy that it requires is obtained from the combustion of fossil fuels. It has been estimated that from 10 to 25% of the national industrial energy consumption is applied to thermal dehydration in industrial countries [7]. As global economies succeed, the request for energy used in drying increase. Thus, there is need to know this operation and to guarantee that it is carried out as efficiently as possible within the economic limitations of the market. It is also important to keep abreast with the recent drying systems, as well as the emerging new ones.

Drying has been used from ancient times to preserve food products. Open-air drying is the most commonly used technique, especially in tropical and subtropical regions. However, several difficulties are usually confronted [7]: product destruction caused by some insects; product degradation due to direct exposure to sun light, rain, and morning dew; contamination with strange particles and smokes due to air pollution; potential cracking of the product and loss of germination capacity; decrease of quality during storage due to inefficient drying. Open air drying is also a common practice for chickpeas; in hot weather, it takes from 15 to 20 days to achieve complete drying of the grains. Thus, alternative drying methods, such as those based on solar energy may imply important savings and effective advantages, if these drying times can be cut off while the product quality is preserved.

Dehydration process involves the simultaneous transfer of mass, heat, and momentum in which the moisture is removed by evaporation into an unsaturated gas phase. Due to the complexity of the drying process, no generalized model exists to explain the mechanism of internal moisture movement. Although it is now recognized that in most practical conditions of air drying of products the principal rate determining step is internal mass transfer, there is no agreement on the method of internal moisture movement [1]. In the case of porous materials such as fruits and vegetables, interstitial spaces, capillaries and gas filled cavities exist within the food medium and water transport takes place via several possible mechanisms acting in various arrangements. Most of fruits and vegetables are categorized as capillary porous media or capillary porous colloids, therefore, it is often suggested that a combination of capillary flow and vapor diffusion mechanisms should be used to describe internal moisture transfer [3].

The heat and mass transfers in the bodies are determined by the structure of the substance. The agricultural products are anisotropic, inhomogeneous and complex. In natural materials structural and dimension fluctuations happen during environmental changes or artificial external conditions. These variations result hysteretic type functions (e.g. sorption isotherms) [6]. These facts can also be basically true for inorganic materials, however in biological products phenomena appear together [3]. In order to get better acquainted with these methods we have to develop the mathematical simulations describing the heat and mass transfer processes from all physical correctly points of view. In the past authors basically dealt with the modeling of heat and mass transfer in cereal grains, and consequently have literature experiences first of all in this area [2, 4]. The simplifications were: (a) the grains were considered to be isotropic and homogenous, therefore moisture concentration gradient was used as the main driving force of the mass transfer; (b) all the transfer factors were considered to be constant; (c) complex models were primarily simplified to squared ones. As the result of these conditions the models could follow only in certain accuracy the integral moisture content changes or the average temperature changes. Experimental measurements are necessary as input factors for the simulations. For a better accurate model the knowledge of the forces potentials during heat and mass transfers is necessary. Water potential gradients as the driving force are used in contrast with the old practice (where moisture gradients were used) for modeling of mass transfer in chickpea.

2. Mathematical modeling of drying processes.

The basic governing differential equation system for the drying processes is given in [6], and simplified by Husain et al. [3] in the following form:

$$\frac{\partial X}{\partial \tau} = \nabla(D\nabla X) \qquad (1)$$

$$\rho\frac{\partial T}{\partial \tau} = \nabla \mathrm{k}\nabla \mathrm{T} + \mathrm{L}\rho\frac{\partial \mathrm{X}}{\partial \tau} \qquad (2)$$

where, X is the moisture content dry base; D is the diffusion coefficient; ρ is the density; T is the temperature; τ is the time; k is the thermal conductivity; L is the latent heat of vaporization of water. The previously mentioned simplifications were taken into account by other authors, who, for instance, analyzed the problem using different physical properties of the different constituents [2, 10]. Other way was to consider the coefficients as functions instead of constants [8]. The above equation systems proved that inner moisture distribution in e.g. a single chickpea, does not agree with the calculated distribution [4]. This equations system was solved with the COMSOL Multiphysics software.

The best numerical methods that describe the drying process result from the theories of irreversible thermodynamics in which the fluxes are taken to be directly proportional to the appropriate water potential gradients [1]. These considerations make possible to obtain the moisture contents and the temperature fields inside the grain solving the partial differential equations. These equations include transport coefficients which must be determined experimentally, and are functions of moisture content.

The physical composition characteristics are different for all food products; they depend on variety, location grown, climatic conditions, etc. The best sources of thermal property models are empirical rather than theoretical, they are prediction equations based on chemical composition, temperature and physical structure (density, porosity, size and configuration of void spaces), they are based on statistical curve fitting rather than theoretical derivations involving heat transfer analysis. The values of the physical variables used in the software such as the properties of the chickpea [7] are summarized in Table 1. Some values were taken from the database that the program contains, and the diffusion coefficient value was obtained by experimentation

Table 1: Parameters used for the chickpea in the COMSOL Multiphysics software.

▾ Parámetros

Nombre	Expresión	Valor	Descripción
T_air	60[degC]	333.15 K	Oven air temperature
T0	18[degC]	291.15 K	Initial patty temperature
rho_p	1383[kg/m^3]	1383.0 kg/m³	Density of patty
h_T	30[W/(m^2*K)]	30.000 W/(m²·K)	Heat transfer coefficient
c0	0.93*rho_p/M_H2O	71455 mol/m³	Initial moisture concentration
c_b	0.02*rho_p/M_H2O	1536.7 mol/m³	Air moisture concentration
C_m	0.003	0.0030000	Specific moisture capacity
k_m	1.29e-9[kg/(m*s)]	1.2900E-9 kg/(m·s)	Moisture conductivity
h_m	1.67e-6[kg/(m^2*s)]	1.6700E-6 kg/(m²...	Mass transfer coefficient in mass units
D	k_m/(rho_p*C_m)	3.1092E-10 m²/s	Diffusion coefficient
k_c	h_m/(rho_p*C_m)	4.0251E-7 m/s	Mass transfer coefficient
D_m	9.7e-9[m^2/s]	9.7000E-9 m²/s	Surface moisture diffusivity
lda	2.3e6[J/kg]*M_H2O	41400 J/mol	Molar latent heat of vaporization
M_H2O	18[g/mol]	0.018000 kg/mol	Water molecular weight

3. **Results and Discussion**.

COMSOL Mutiphysics program was used to simulate the dehydration of chickpea in a complex drying system processes, obtaining the numerical solution of the model equations. The above system of non-linear partial differential equations had been solved by the finite element method implemented by COMSOL Multiphysics 4.4, together with the already described set of initial and boundary conditions. We built the geometry of the model, and then were fixed the boundary settings, the mesh parameters and compute the final solution (Figure 1).
The chickpea has a height of 0.016 m and 0.019 m in diameter. The selected mesh was the triangular type and 6586 nodes or degrees of freedom were formed, for its fourth part, since it was considered symmetric, the air flow incided horizontally at a speed of 3.0 m/s and a temperature of 60 °C, as shown in Figure 2.

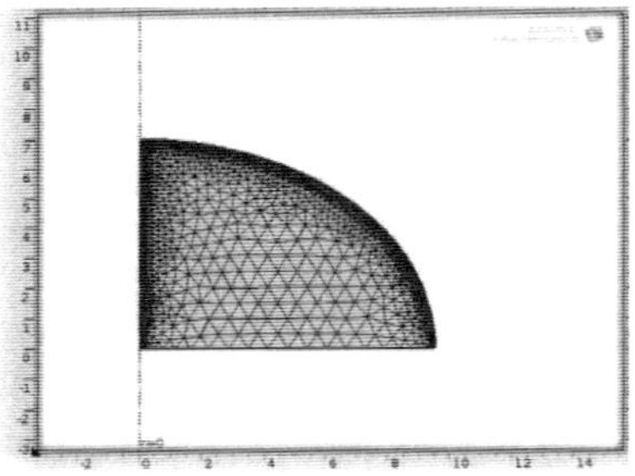

Fig. 1. Mesh selected on the grain of chickpea.

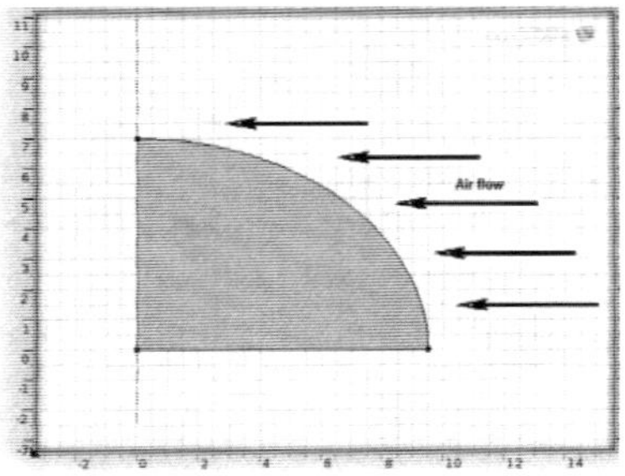

Fig. 2. Incidence of air on the grain, the air speed and temperature were 3.0 m/s and 60 ° C, respectively.

The simulation with COMSOL program of hot-air drying began with an initial concentration in the sample of 67 000 mol/cm^3 and the final was 5 000 mol/cm^3, the used time was 8 hours, the obtained graph is displayed in Figure 3. This time is almost equal to that obtained experimentally and reported in [5], which was 8. 5 hours. At the beginning the moisture lost was fast, in the first 30 minutes the mass decreased to 45% of its initial value, then the loss became slower. In the following 3.5 hours an additional 30% was lost and in the final stage, it took 4 hours to lose 7.5% of moisture content.

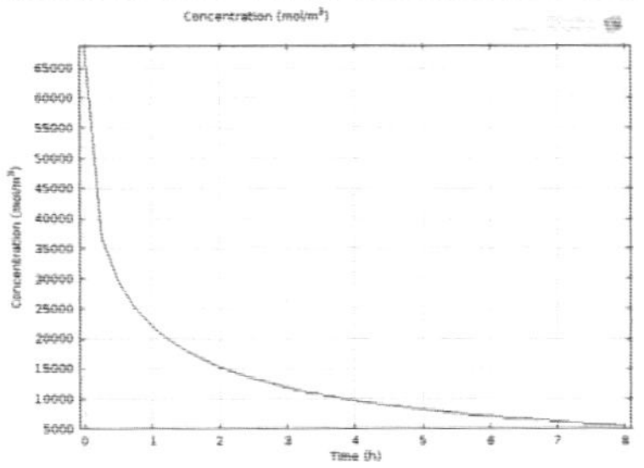

Fig. 3. Graph of the loss of concentration of mass against time.

These values were used to produce the corresponding moisture content vs time graph (Figure 4), to get the model of drying of the chickpea, which was logarithmic, the equation is:
$MR = 0.0397 + 0.5381exp(-1.4214\,t)$, with $R^2 = 0.934$, $\chi^2 = 0.00013$ and $RMSE = 0.00092$.
In the graph presented in Figure 5 the variation of the temperature against time in the center of the chickpea can be observed. It starts to gradually increase from 18 °C until it reaches the temperature of the drying air, 60 °C. The drying time was approximately 1 hour, and afterwards the temperature remained constant until the end of the simulation. This temperature was not experimentally obtained due to the technical difficulties involving the corresponding measurement.

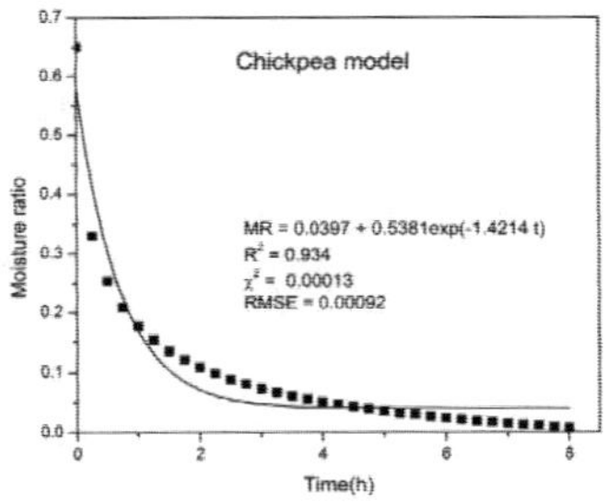

Fig. 4. Logarithm model of chickpea drying.

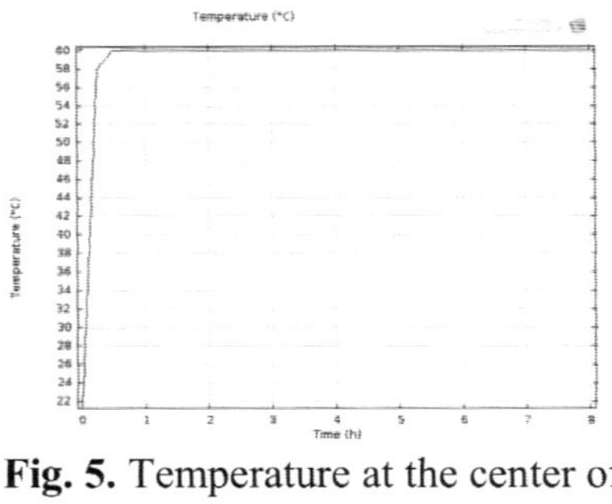

Fig. 5. Temperature at the center of the sample of the chickpea.

In the sequence shown in Figure 6, drying process of the sample is displayed. At the start the whole grain has the same concentration of mass, as the simulation moves on the blue-color strip, which represents the lowest concentration, becomes wider, which indicates that the product is losing moisture. At the end of the drying time, it is observed that the concentration of mass in the center is quite small, 5 000 mol/cm^3. The color bar that appears on the right hand side corresponds to the values of concentration.

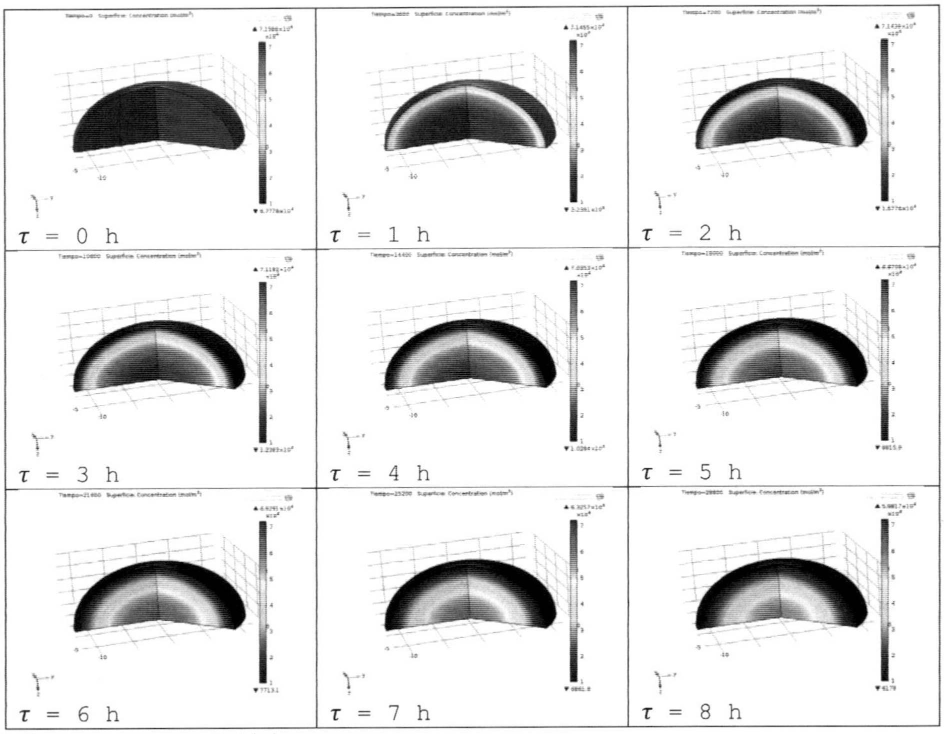

Fig. 6. Drying process of chickpea.

4. Conclusions.

The finite element method is applied to solve the heat and mass transfer equations of the drying of chickpea, regardig it as a composite biological material. The COMSOL Multiphysics software was used for the calculation of the model.

The process of drying a grain of chickpea has been simulated using air at a temperature of 60 ° C and 3.0 m/s speed. The initial temperature of the air was considered equal to the environmental one of 18 ° C. The initial concentration of mass in the sample was 67 000 mol/cm^3 and the final was 5 000 mol/cm^3, the time used in the simulation was 8 hours. This type of simulation allows the step-by-step observation of the phenomenon of drying. The drying temperature, 60 ° C, was reached in the center of the sample after an hour of the process start.

References.

[1] Ghiaus A G, Margaris D P and Papanikas D G 1997 Mathematical modelling of the convective drying of fruits and vegetables, Journal of Food Science, *An international journal of the Institute of Food Technologists*, **62** 1154-1157

[2] Gustafson R J, Thompson D R, Sokhansanj S 1979 Temperature and stress analysis of corn kernel finite element analysis. Trans. ASAE, **22** 955-960

[3] Husain A, Chen C S, Clayton J T 1973 Simultaneous heat and mass diffusion in biological materials. *J. Agric. Eng. Res*. **18** 343- 354

[4] Kovács A J, Neményi M 1999 Moisture gradient vector calculation as a new method for evaluating NMR images of maize (Zea mays, L.) kernels. *Magn. Reson. Imaging,* **17**(**7**) 1077-1082

[5] López C R 2014 Cinética de secado del garbanzo con aire caliente. Informe interno UAM-A. México

[6] Luikov A V 1966 Heat and Mass Transfer in Capillary - porous Bodies, Pergamon Press.

[7] Mujumdar A 2007 Handbook of Industrial Drying. 2nd ed. Marcel Dekker, New York

[8] Neményi M, Czaba I, Kovács A, Jáni T 2000 Investigation of simultaneous heat and mass transfer within the maize kernels during drying, Computers and Electronics in Agriculture. **26** 123-135

[9] Servicio de información y estadística agroalimentaria y pesquera con información de las delegaciones de la SAGARPA en los estados, <http://www.sagarpa.gob.mx>.
[10] Sokhansanj S, Gustafson R J 1980 In: Mujumdar, A. S. (Ed.): Prediction of Heat and Mass Transfer within a Grain Kernel – A Finite Element Application, Drying '80 Proceedings of the Second International Symposium, 2. McGraw-Hill, New York, 229-232

First-order irreversible thermodynamic approach to a non-steady RLC circuit as an energy converter

G Valencia and L A Arias
Departamento de Física, Escuela Superior de Física y Matemáticas IPN, Ciudad de México, 07738, México.

E-mail: gvalencia@esfm.ipn.mx

Abstract. In this work we show a RLC-circuit as energy converter within the context of first-order irreversible thermodynamics (FOIT). For our analysis, we propose an isothermic model with transient elements and passive elements. With the help of the dynamic equations, the Kirchhoff equations, we found the generalized fluxes and forces of the circuit, the equation system shows symmetry of the cross terms, this property is characteristic of the steady state linear systems, but in this case phenomenological coefficients are function of time. Then, we can use these relations, similar to the linear Onsager relations, to construct the characteristic functions of the RLC energy converter: the power output, efficiency, dissipation and ecological function, and study its energetic performance. The study of performance of the converter is based on two parameters, the coupling parameter and the "forces ratio" parameter, in this case as functions of time. We find that the behavior of the non-steady state converter is similar to the behavior of steady state energy converter. We will explain the linear and symmetric behavior of the converter in the frequencies space rather than in the time space. Finally, we establish optimal operation regimes of economic degree of coupling for this energy converter.

1. Introduction

It is well-known that in actual dissipative heat engines, the maximum power output and maximum efficiency points do not coincide and their separation can be managed by some phenomenological parameter which depend on the engine's design and the material employed in its construction [1,2].

In 1931 [3], Lars Onsager published his linear theory of irreversible processes, also known as first-order irreversible thermodynamics, this theory is able to explain various phenomena that occur out of equilibrium within a framework of flows, forces and generalized coefficients. Onsager relations are symmetric with regard to the cross coefficients corresponding to the contributions of generalized forces to generalized flows. The most basic version of this converter posses two generalized fluxes and two forces in steady state, besides it is in contact with an ideal bath at fixed temperature. The converter can operate in several working regimes, so we can study the energetic performance of the converter.

This paper is organized as follows: in section 2 we obtain the relation between fluxes and forces, as in the Onsager theory, by using the Kirchhoff laws [4]. In section 3, we will construct the characteristic functions and will make the energetic performance analysis of the electromagnetic converter. Finally, we write some conclusions about of our results.

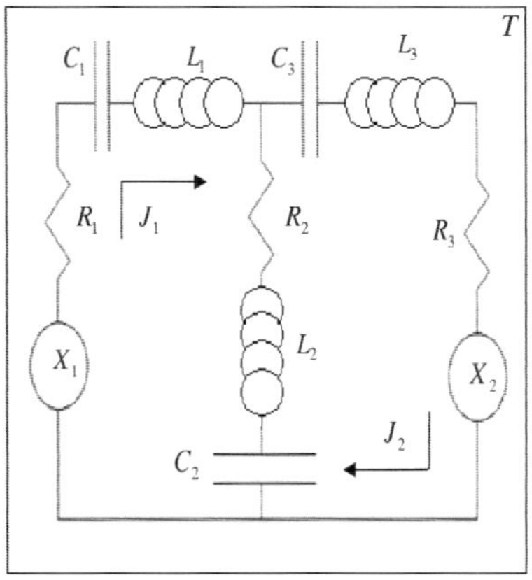

Figure 1. RLC-circuit of two meshes as non-steady state electromagnetic energy converter.

2. FOIT fundamentals

In this section, we first review the assumptions from which the linear Onsager relations for a power converter are built. Next, we will apply the Kirchhoff law of meshes to the RLC-circuit, in order to find the mesh currents depending on the resistances, capacitances. inductances and voltages of direct current sources of the circuit. With these equations we will try to write a kind of Onsager relations, identifying the generalized forces and the phenomenological coefficients of the system.

2.1. Constitutive equations

If we consider the simplest version of an energy converter, let J_1 and J_2 be two coupled generalized fluxes (J_1 being the driven flux and J_2 the driver flux); X_1 and X_2 are the conjugate generalized potentials associated with the fluxes. For the linear case, fluxes and potentials are given by the equation (1):

$$J_1 = \sqrt{L_{11}}\left(\sqrt{L_{11}}X_1 + q\sqrt{L_{22}}X_2\right) \tag{1}$$

$$J_2 = \sqrt{L_{22}}\left(q\sqrt{L_{11}}X_1 + \sqrt{L_{22}}X_2\right) \tag{2}$$

with $L_{ij} = L_{ji}$, the Onsager symmetry relation between crossed coefficients. In these equations we use the coupling coefficient defined by

$$q(t)^2 = \frac{L_{12}^2(t)}{L_{11}(t)L_{22}(t)} \geq 0, \tag{3}$$

This parameter gives us a measure of the coupling of spontaneous and nonspontaneous fluxes [5]. On the other hand, we can define a parameter of operation in terms of the forces, as follows

$$x(t) = \sqrt{\frac{L_{11}(t)}{L_{22}(t)}\frac{X_1}{X_2}} = \sqrt{\frac{L_{11}(t)}{L_{22}(t)}}x', \tag{4}$$

where x' is the ratio between the forces (where L_{ij} are the FOIT-phenomenological coefficients).

2.2. RLC circuit as energy converter

In the RLC-circuit (see figure. 1) there are two fluxes and two forces, we denote by J_1 and J_2 the coupled fluxes (currents), as well as X_1 and X_2 the forces (voltages). To resolve this electric circuit we use the Kirchhoff laws [4]; if we apply the corresponding Kirchhoff equation to the mesh 1 (figure.1), we obtain the following

$$X_1 = \left(\frac{1}{C_1} + \frac{1}{C_2}\right)\int J_1 dt + (L_1 + L_2)\frac{dJ_1}{dt} + (R_1 + R_2)J_1 - \frac{1}{C_2}\int J_2 dt - L_2\frac{dJ_2}{dt} - R_2 J_2, \quad (5)$$

while for the mesh 2 the equation is

$$X_2 = \left(\frac{1}{C_2} + \frac{1}{C_3}\right)\int J_2 dt + (L_2 + L_3)\frac{dJ_2}{dt} + (R_2 + R_3)J_1 - \frac{1}{C_2}\int J_1 dt - L_2\frac{dJ_1}{dt} - R_2 J_1. \quad (6)$$

It should be noted that (5) and (6) are integral equations, then if we apply a linear transformation such that these equations are sent to the frequency space (Laplace transformation). The new equations system can be resolved easily, if we consider constant in time voltage sources in time, the new equations have the following algebraic form:

$$\frac{X_1}{s} = \left(\frac{1}{C_1} + \frac{1}{C_2}\right)\frac{I_1(s)}{s} + s(L_1 + L_2)I_1(s) + \frac{q_o}{sC_1} + (R_1 + R_2)I_1(s) - \frac{I_2(s)}{sC_2} - sL_2 I_2(s) - R_2 I_2(s), \quad (7)$$

and

$$\frac{X_2}{s} = \left(\frac{1}{C_2} + \frac{1}{C_3}\right)\frac{I_2(s)}{s} + s(L_2 + L_3)I_2(s) + \frac{q_o}{sC_3} + (R_2 + R_3)I_2(s) - \frac{I_1(s)}{sC_2} - sL_2 I_1(s) - R_2 I_1(s), (8)$$

where R_1, R_2 and R_3 are the resistive elements, C_1, C_2 and C_3 are the capacitive elements, L_1, L_2 and L_3 are the inductive elements and q_0 is the initial charge of the capacitors. We note that (7) and (8) can be inverted, such that the fluxes $I_1(s)$ and $I_2(s)$ are expressed in terms of the forces X_1 and X_2. In order to obtain the fluxes J_1 and J_2 in the real solutions space, to invert the Laplace transformations shown in (7) y (8). In (9) the parameters $a_k = a_k(R_1, R_2, R_3, L_1, L_2, L_3, C_1, C_2, C_3)$ are the roots of Q(s). In this case the fluxes can be expressed by using a matrix as follows:

$$\begin{pmatrix} J_1 \\ J_2 \end{pmatrix} = \begin{pmatrix} (L_2+L_3)\sum_{k=1}^{2}\frac{\left[a_k^2 + \frac{R_2+R_3}{L_2+L_3}a_k + \frac{1}{L_2+L_3}\left(\frac{1}{C_2}+\frac{1}{C_3}\right)\right]e^{a_k t}}{Q'(a_k)} & L_2\sum_{k=1}^{2}\frac{\left(a_k^2 + \frac{R_2}{L_2}a_k + \frac{1}{L_2 C_2}\right)e^{a_k t}}{Q'(a_k)} \\ L_2\sum_{k=1}^{2}\frac{\left(a_k^2 + \frac{R_2}{L_2}a_k + \frac{1}{L_2 C_2}\right)e^{a_k t}}{Q'(a_k)} & (L_1+L_2)\sum_{k=1}^{2}\frac{\left[a_k^2 + \frac{R_1+R_2}{L_1+L_2}a_k + \frac{1}{L_1+L_2}\left(\frac{1}{C_1}+\frac{1}{C_2}\right)\right]e^{a_k t}}{Q'(a_k)} \end{pmatrix}\begin{pmatrix} X_1 - \frac{q_o}{C_1} \\ X_2 - \frac{q_o}{C_3} \end{pmatrix}, \quad (9)$$

with

$$Q'(a_k) = 4[L_1(L_2 + L_3) + L_2L_3]a_k^3 + 3[R_1(L_2 + L_3) + L_1(R_2 + R_3) + L_3R_2 + L_2R_3]a_k^2$$
$$+ 2\left[\frac{(L_2 + L_3)}{C_1} + L_1\left(\frac{1}{C_1} + \frac{1}{C_2}\right) + R_1(R_2 + R_3) + R_2R_3 + \frac{L_3}{C_2} + \frac{L_2}{C_3}\right]a_k$$
$$+ \left[R_1\left(\frac{1}{C_1} + \frac{1}{C_2}\right) + \frac{R_2 + R_3}{C_1} + \frac{R_2}{C_3} + \frac{R_3}{C_2}\right]. \tag{10}$$

From the matrix (9) we can see that the Onsager relations for the RLC energy converter also exhibit symmetry of the coefficients associated with cross-flow contributions, if the generalized forces are taken as

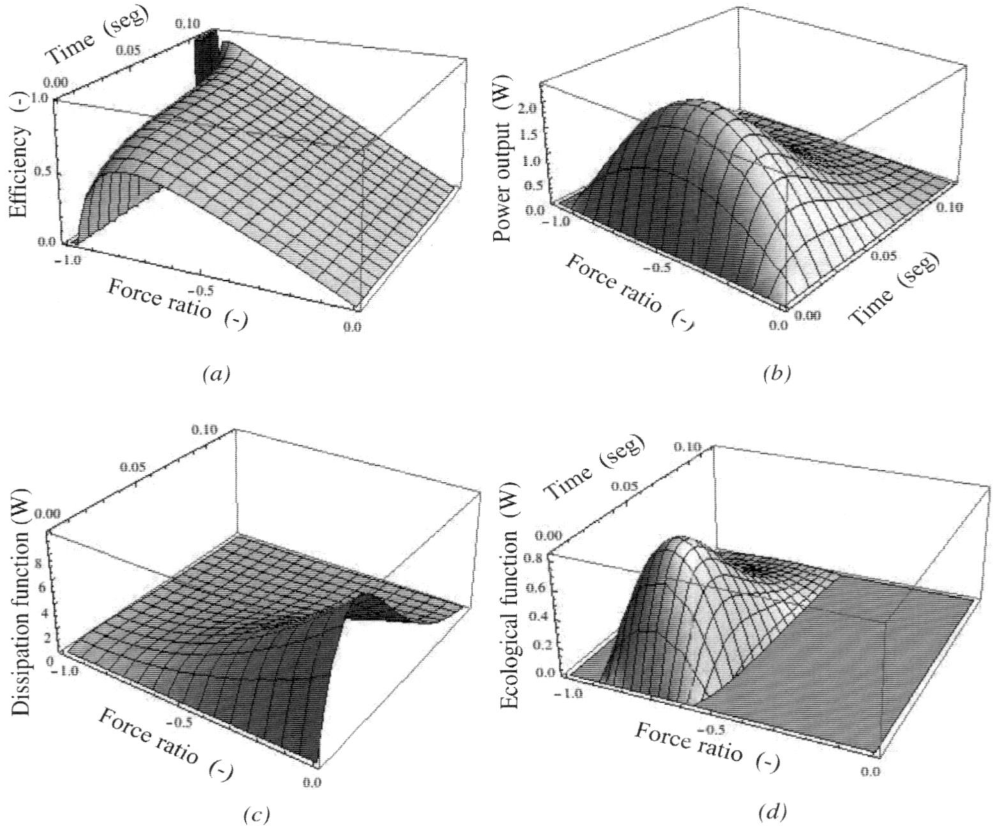

Figure 2. Characteristic functions in terms of the forces ratio and the time, for a given coupling parameter, a) efficiency and b) power output. Also c) dissipation function in terms of the forces ratio and the time, for a fixed coupling parameter and d) ecological function.

3. Performance of the RLC-converter

The energy analysis of the converter can start associating the summands of the entropy production with input power and output power. First, we construct the dissipation function multiplying the entropy production by the temperature T of the converter, which is constant. The dissipation function for an isothermal system can be expressed as [6]

$$\Phi(t) = T\sigma(t) = TJ_1(t)X_1 + TJ_2(t)X_2 = \left[(x')^2 L_{11}(t) + 2x'L_{12}(t) + L_{22}(t)\right]X_2. \tag{11}$$

From above equation the power output of the RLC-converter is given by

$$P(t) = -TJ_1(t)X_1 = -T\left[(x')^2 L_{11}(t) + x'L_{12}(t)\right]X_2^2, \tag{12}$$

this function measures the energy interchange between the system and the surroundings.

We can define also the efficiency following Caplan and Essig, the converter using the coupling parameter and the forces ratio we have

$$\eta(t) = \frac{power\ output}{power\ input} = -\frac{TJ_1(t)X_1}{TJ_2(t)X_2} = \frac{-x(t)[q(t)+x(t)]}{x(t)q(t)+1} = -\frac{(x')^2 L_{11}(t) + x'L_{12}(t)}{x'L_{12}(t) + L_{22}(t)}. \tag{13}$$

This function shows the influence of parameter design and operating parameter in the energy behavior of the converter.

Some authors [7] have found that the graphic output power versus efficiency of real machines draw loops, these loops have two conspicuous points, a maximum power output and a maximum efficiency, between these two operating points can be found a region. The most commonly used objective function to find the operating points that are in the region of compromise is called ecological function, whose definition we will write down, [8]

$$Ec(t) = P(t) - \Phi(t) = -T\left[2(x')^2 L_{11}(t) + 3x'L_{12}(t) + L_{22}(t)\right]X_2^2. \tag{14}$$

In Fig. 2 we show the time evolution of these two parameters, considering the conventional resistances and capacitances values reported in the literature about electric circuits; in this case we take the followings values: R_1=10Ω, R_2=680Ω, R_3=12Ω, L_1=0.2H, L_2=4H, L_3=0.1H C_1=3μF, C_2=500μF, C_3=5μF. From these values, we can see that the graphs in Fig. 2 show the temporal behavior well known, and the same behavior as for steady states.

In 1980 Stucki [9] proposed a thermodynamic model for the oxidative production of ATP in the mitochondria. He found that this process could be carried out under different operating modes, among them those that correspond to maximum efficiency so as to some other condition of optimization, he called this modes economic degrees of coupling. We can find the operation modes that correspond to economic degree of coupling for the RLC-converter.

4. Conclusions

This RLC circuit was taken as an example, because it is one of the simplest energy converters in its construction and exhibits dependence over time. We can see the same symmetry of the cross coefficients as in the steady state, perhaps the Curie theorem could be valid in the space of frequencies.

On the other hand, the converter time evolution is dominated for the coupling parameter, and its behavior depends of the transient elements. With these considerations, we can minimize the entropy production in electric power generating plants or reduce power dissipation in transmission networks.

References

[1] Ferguson Colin R 1986 Internal Combustion Engines: Applied Thermosciences, Wiley, New York.

[2] Gordon J M, Huleihil M 1992 General performance characteristics of real heat engines, *J. Appl. Phys.* **72**, 829.
[3] García-Colín Scherer L, Goldstein-Menache P 2003 La fisica de los procesos irreversibles (The physics of irreversible processes), El Colegio Nacional (in spanish).
[4] Edminister J A 1965 Electric Circuits. Mc. Graw Hill
[5] Caplan S R, Essig A 1983 Bioenergetic and Linear Nonequilibrium Thermodynamics, Harvard University Press, Cambridge
[6] De Vos A 1995 Endoreversible thermoeconomics, Energy Convers. Mgmt. **36** (**1**) 1-5.
[7] Arias-Hernandez L A, Angulo-Brown F, Paez-Hernández R T 2008 First-order irreversible thermodynamic approach to a simple energy converter. Phys. Rev. E **77** (**1**): 011123 1-8.
[8] Arias-Hernández L A 2000 Ph.D. thesis, Instituto Politécnico Nacional, México (in spanish).
[9] Stucki J W 1980 The optimal efficiency and the economic degrees of coupling of oxidative phosphorylation. Eur. J. Biochem. **109(3)** 269-83.

Journal of Physics: Conference Series **582** (2015) 012021 doi:10.1088/1742-6596/582/1/012021

Optical evaluation on the setting of cement paste

H A De León Martínez, J J Soto Bernal, R González Mota and I Rosales-Candelas.
Instituto Tecnológico de Aguascalientes. Av. Adolfo López Mateos # 1801 Ote. Fracc. Bona Gens, Aguascalientes, México

E-mail: asa_15@hotmail.com

Abstract. In the construction area, one of the most widely used cement is the CPC 30R, it is a hydraulic binder consisting of CaO, SiO_2, Al_2O_3 and Fe_2O_3, when mixed with water forms cement pastes and its four crystallographic phases start to hydrate. The diffuse reflection on cement paste can give an indication of the behaviour on optical properties on the hydration of the cement and early formation products. In this study, Portland cement (CPC) pastes were prepared with 0.45 a water to cement ratio (w/c). This work is aimed to evaluate the optical properties of cement pastes on the hydration reaction during the first 24 hours by measuring the intensity of diffuse reflection changes.

1. Introduction

Portland cement is a hydraulic binder [1, 2], composed mainly of CaO, SiO_2, Al_2O_3 and Fe_2O_3, which is called clinker, after mixing with water form cement pastes and its four crystallographic phases start to hydrate: alite (C_3S), belite (C_2S), aluminate (C_3A) and the ferrite (C_4AF) [3, 4, 5, 6]. The reaction of C_3A and C_4AF phases predominate in the stages 1 and 2 of the hydration [6, 7, 8, 9] and the reaction of tricalcium silicate (C_3S) and dicalcium silicate (C_2S) phases predominate in the third and fourth cement hydration stages, forming hydrates of calcium (CH) and calcium hydrosilicate (CSH) [6, 10].

To predict the behavior of concrete, it is important to model the hydration of cement pastes incorporating their properties. Powers-Brunauer model describes cement paste as a low-crystalline gel layer [11]; in this model, the behavior during hydration is explained in four stages. In the first stage, immediately after the cement has contact with water, there is a rapid evolution of heat, the space occupied by water between the particles of cement begins to fill whit hydration products, forming a C-S-H film around cement particles; this phase ceases in about 15 minutes [7, 12, 6]. In the second stage (15 min-4 hrs), the reaction rate slows [18], temperature decreases and there is less formation of hydrates but CH crystals grow and make pressure on the C-S-H film until they break it, leaving newly exposed to water the unhydrated cement particles. In the third stage (4-8 hrs) the reaction rate increases and the cement temperature rises [13, 11, 14]; the initial setting occurs in the early stage, when the reaction rate increase and accelerates over time; final set occurs at the end of it. In the fourth stage (8-24 hrs) there is a slow but continuous formation of hydration products [11, 12] and the temperature of the cement begins to stabilize.

Currently, the techniques that exists to monitor setting of cement are conventional, discrete and their outcomes are influenced due to the human factor [15, 16]; so, it is necessary to develop more reliable techniques; in this regard, laser technology has reached an important role due to the ability to make measurements and accurate image reproduction, besides a surface analysis of the cement [17, 18].

The Kubelka-Munk model is used to monitor the optical properties of the materials, as it establishes that the color of any turbid medium results in the absorption and scattering of light [19] and is expressed by the absorption (K) and dispersion (S) coefficient of the light reflected from an inhomogeneous medium according to Eq. 1 [20, 21].These coefficients describe the amount of radiation that a material absorbs or reflects according to wavelength of light emitted by the source, which is reflected by the material to be measured; these coefficients vary for each material and specifically for cement pastes vary with the changes in its microstructure.

$$f(R_\infty) = \frac{(1-R_\infty)^2}{2R_\infty} = \frac{K}{S}, \quad (1)$$

where R_∞, is the diffuse reflection

The diffuse reflection of cement pastes can give us information about the behavior of the microstructure; so, the changes in the cement paste are monitored to obtain the initial an final setting time. In this paper, the diffuse reflection was used to continuously monitor the micro-structural changes in a cement paste sample during hydration; initial and final setting time were obtained with this technique. The temperature of the cement paste was measured in order to compare the behavior of it with diffuse reflection.

2. Materials and Methods

2.1 Materials

A cylinder of 20 mm diameter by 40 mm in height were prepared with cement CPC 30R, with w/c = 0.45 according to ASTM-C187-04 [23] standard, at an ambient temperature of 25 ° C and a relative humidity of 70%; in Table 1 the composition of the cement is shown.

Table 1. Phase composition of portland cement clinker [6].

Phase Composition	Mass (%)
C_3S	55
C_2S	20
C_3A	6
C_4AF	9

2.2 Temperature measurements

Measurement of temperature during the hydration process of Portland cement samples was performed using an infrared thermometer FLUKE model 63.

2.3 Optical arrangement

In the optical arrangement, a He-Ne laser (633nm) was used as a radiation source with an output of 5mW, it was used a beam splitter in order to obtain two beams and we use the 30% as a reference one, using a 22.2% filter to attenuate the beam and prevent the OPT sensor is saturated; The 70% beam was impinge on the sample for diffuse reflection.

3. Results and Discussion

3.1 Temperature measurement

At the beginning of the first stage there is a fast generation of hydration heat as shown in Figures 1a and 1b, according to Carmel Jolicoeur [12]; this hydration heat is produced due to the initial hydrolysis reaction with the aluminate and ferrite phases [12, 24]; hydration process generates heat and increases the temperature inside and outside of the cement paste, so that the temperature rises in the first stage to 28.3 ° C [25]. Hydration products formation occurs, creating a protective layer around the cement particles in the final part of the stage.

After two hours the temperature drops, recording the minimum of the curve due to a slow hydration reaction due to cement particles are coated by a layer of hydration products, this layer hinders diffusion of water with the cement, reducing the rate of reaction of the components [12];

so that, less heat is released; furthermore, the formation of calcium hydrosilicate (C-S-H) in the second stage is lower than in the first stage and the temperature depends on the formation of the C-S-H [11]. At the end of the second stage CH crystals grow causing a breakdown of the protective film, exposing the unhydrating cement particles to water.

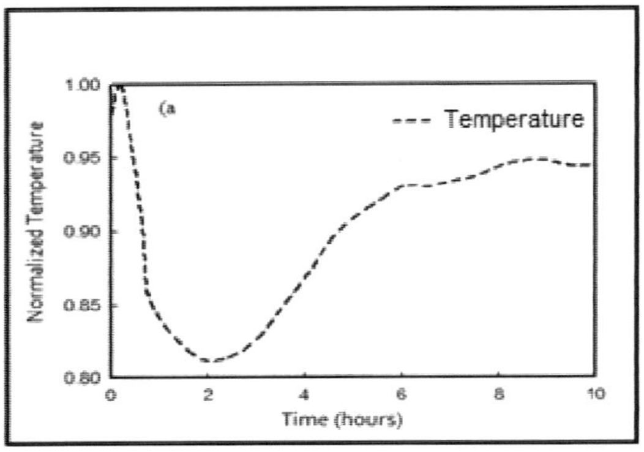

Figure 1a. Temperature behavior during hydration of cement paste.

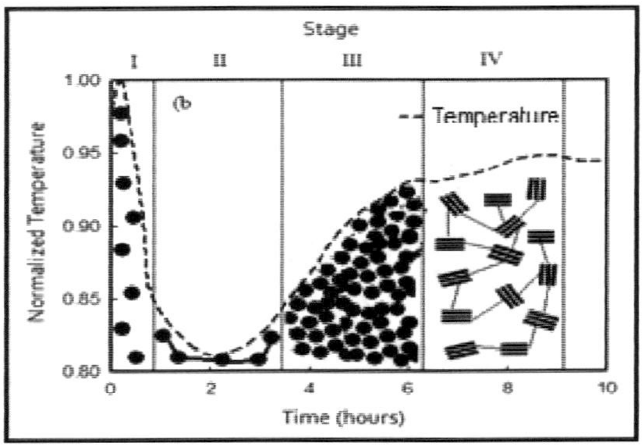

Figure 1b. Temperature behavior during hydration correlated with the stages of Powers-Brunauer model.

At the start of the third stage the initial setting is given. After the breakdown of the protective layer, the unhydrated cement is exposed to water, so that, the hydration reactions generate greater heat release; the silicate phases predominates from this stage forward and hydration products continue to form until the system has resistance to deformation, giving place to the final set [12]. In the fourth stage, there is a lower reaction rate and a slow but continuous formation of hydration products [8, 13]; besides, concrete temperature starts to stabilize, so changes in the temperature of the cement paste are small [14].

3.2 Diffuse reflection

At the first stage, the cement particles are more exposed to the incident light since the cement paste is an agglomerate of small particles, interlaced and intertwined; so, it forms a porous mass [26] in which the incident light may penetrate due to the separation of cement particles; monomeric silicates are unified reaching an amorphous C-S-H gel form and the film formed around the cement particles is small at the beginning and grow with time. The diffuse reflection has a maximum at this stage, before the film grows.

At the second stage, hydrates layer becomes larger, coating the cement particles, in addition, the cement paste is agglomerated and the light cannot penetrate the cement grains; thus, there is a greater absorption of incident light, causing decreasing of the diffuse reflection; the hydration rate is very slow and the loss of water determines the agglomeration of C-S-H in sheets [11] and, this effect is manifested in the contraction of the material; This stage can last several hours [11], and finally crystallization processes of hydration products will be accelerated.

At the third stage, the cement has a more homogeneous but rough arrangement, breakage of the film leaves the particles exposed to the incident beam, increasing the diffuse reflection. At the end of this stage the cement begins to lose plasticity and hydration products begin to crystallize [27, 28]. At fourth stage, the formation of hydration products continue to grow and the cement paste is crystallized [11], causing diffuse reflection decreases slightly. After this stage, the diffuse reflection decreases. During hardening of the cement paste, the diffuse reflection begins to decrease; this may be due to crystallization of the elements, so that, the reflections are more specular. The diffuse reflection changes when the micro-structure changes; besides, it is possible to infer the time of initial and final setting from it.

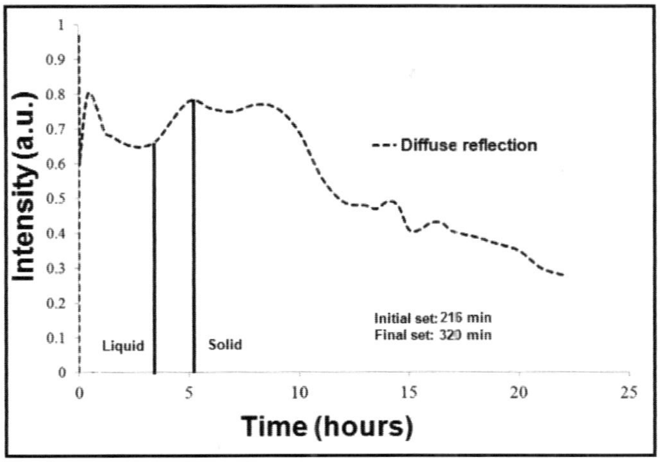

Figure 2. Diffuse reflection of cement paste for 24 hours.

During the cement setting process, the behavior of the diffuse reflection and the temperature is similar, see Figure 3; therefore, from the diffuse reflection it is possible to infer the initial and final setting time; 216 minutes and 320 minutes, respectively, the results obtained by the diffuse reflection are within the range marked in norms C 150-02 and ASTM C 191-01 and our results can be compared with those obtained by D. Mikulić *et al.*,2005 and Yimen *et al.*, 2009. With these results, it is possible to distinguish the stages of hydration and define the start and end of each stage. Therefore it is possible to eliminate human error in the measurement for obtaining the setting times. These results are only from one cylinder and in future work we contemplate more cylinders.

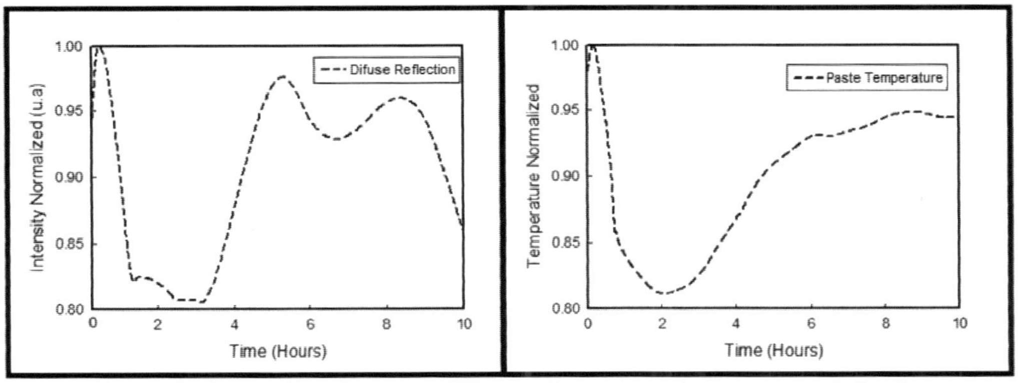

Figure 3. Comparison between diffuse reflectance and temperature behavior

4. Conclusions

During the setting process of the cement paste, the behavior of the diffuse reflection changes as the microstructure changes; as a result, it can be inferred the initial and final setting time of the cement paste by means of diffuse reflection. The setting times obtained by the diffuse reflection are: 216 minutes for initial and 320 minutes for the final setting time, respectively.

By optical monitoring with laser technology, the dynamic behavior of cement pastes was measured, and therefore, it can be seen in real time the changes in the behavior of cement pastes at different ages and at different stages. The behavior of the diffuse reflection curve is similar to the behavior of the temperature curve, so one can infer the setting times from diffuse reflection, due to one of the techniques to measure the properties of cement, is the temperature.

References

[1] Alexandre G S 2002 *XV Brazilian Symposium* **138**.

[2] Rodríguez J E, Ahumada L M, Bustamante J M and Ruiz de Murgueito R 2007 *Boletín de la sociedad española de cerámica y vidrio* **44** 421.

[3] Emanuelson A, Hansen S and Viggh A 2003 *Cement and Construction Research* **33** 1613.

[4] Emanuelson A, A R, Cánovas L and Hansen S 2003 *Cement and Concrete Research* **33** 1623.

[5] Pang X, Bentz D P, Meyer C, Funkhouser G P and Darbe R.

[6] Heikal M, Morsy M and Aiad I 2005 *Cement and Concrete Research* **35** 680.

[7] D and Damidot N A 1991 *Proceedings of the International RILEM Workshop on hydration and setting* 23.

[8] Rahhal V, Donza H, Delgado A, Gutiérrez J and Talero R

[9] Young J F 1972 *Cement and Concrete Research* **2** 415.

[10] Marín López C, Reyes Araiza J L, Manzano Ramirez A, Rubio Avalos J, Perez Bueno J J, Muñiz Villareal M S, Ventura Ramos E and Vorobiev Y 2009 *Inorganic Materials* **45** 1429.

[11] Ramachandran V S 1995 *Concrete Admixtures Handbook* Noyes Publications Canada.

[12] Jolicoeur C and Simard M A 1998 *Cement and Concrete Composites* **20** 87.

[13] H P C Lea's Chemistry of Cement and Concrete, Oxford, England Butterworth-Heinemann, 2001.

[14] C. Camp, "Department of Civil Engineering," 11 03 2013. [Online]. Available: http://www.ce.memphis.edu/1101/notes/concrete/everything_about_concrete/04_hydration.html. [Accessed 22 05 2014].

[15] R J, P J S. and S K V 2009 *ACI Mat. J* **97** 675.

[16] Christian U G. and Hans Wolf R 2003 *International Symposium*.

[17] Lawrence J and Li L 2000 *Journal of Material Science and Engineering* **287** 25.

[18] Frías M, de Luxan M P and Sánchez de Rojas M I 1988 *Materiales de Construcción* **38** 37.

[19] Barron V and Monte Alegre L 1986 *Amerincan Journal of Science* **286** 792.

[20] Kubelka P and Munk F 1931 *Z. Tech. Physik* **12** 593.

[21] Saunderson J 1942 *Optical Soc. Am. Journ* **32** 727.

[22] A C 0 A Standars 2004 *American Society for Testing and Materials* PA USA.

[23] Kjellsen K and Lagerblad B 2007 *Cement and Concrete Research* **37** 13.

[24] Christine L A, Atherton S, Roy D M and Pugh St S, "High Temperature Cement". United States 23 June 1981.

[25] ASTM-C39/C39M-01 2001 *American Society for Testing and Materials* **04.02**.

[26] Donald H P. Campbell 1999 Microscopical Examination and Interpretation, DodgeVille, WI: PCA, 1999.

[27] Torréns Martín D, Fernández Carrasco L and Martínez Ramirez S 2013 *Cement and Concrete Research* **47** 43.

[28] Pierre Henri J V C 2007 *Cement and Concrete Research* **37** 1321.

[29] Khalid M, Karl J B and Reza Z 2001 *Ieee Transactions on Instrumentation and Measurement* **50** 1225.

[30] A C 9 A Standars 2001 *American Society for Testing and Materials* PA USA.

[31] Pincigallo A, Lura P, Van Breugel K and Levita G 2003 *Cement and Concrete Research 33* 1013.

[32] S I 1979 *Portland Cement Paste and Concrete*.

Journal of Physics: Conference Series **582** (2015) 012022 doi:10.1088/1742-6596/582/1/012022

Alternate model of Chladni figures for the circular homogenous thin plate case with open boundaries

H A Trejo-Mandujano[1], G Mijares-Bernal[1], E G Ordoñez-Casanova[2]
[1] Departamento de Física y Matemáticas, Universidad Autónoma de Ciudad Juárez, Chihuahua, México
[2] Departamento de Industrial y Manufactura, Universidad Autónoma de Ciudad Juárez, Chihuahua, México

E-mail: htrejo@uacj.mx

Abstract. The wave equation is a direct but a complex approach to solve analytically for the Chladni figures, mainly because of the complications that non-smooth and open boundary conditions impose. In this paper, we present an alternate solution model based on the principle of Huygens-Fresnel and on the ideas of Bohr for the hydrogen atom. The proposed model has been implemented numerically and compared, with good agreement, to our own experimental results for the case of a thin homogenous circular plate with open boundaries.

1. Introduction

The Chladni experiment is commonly used as an undergraduate visualization of acoustical stationary waves in a plate [1], but the models that describes this figures also serves in other wave-like phenomena like quantum billiards [2]. The classical demonstration consists of harmonically exciting an elastic plate that has been sprinkled with sand or other granular media. The excitation will propagate in the plate and under certain conditions; it will form a stationary wave which is commonly represented mathematically as a separation of the wave equation solution into a purely temporal, and a purely spatial product. The effect on the plate is that there will be constant regions of zero amplitude and others zones vibrating at some intensity. The vibrant regions will propel the granular media in all directions until it falls in a region of zero amplitude maintaining that position. Because we are dealing with a stationary wave, the media will maintain a fixed pattern: the Chladni figure.

The standard way to analyze this type of figures is to solve the inhomogeneous Helmholtz equation using the appropriate boundary conditions. However, this approach is hard to implement, especially in the case of vibrating plates with non-smooth (irregular) open boundaries. The difficulty of this problem is reflected in the literature where most methods only solve for the plate irregularity. For instance, Amore presents a method for solving the Helmholtz equation using what he call “little sinc functions”; his method works for irregular and/or inhomogeneous membrane with fix boundary conditions only [3]. Muller [4] and Gander [5] utilizes finite element methods to find the eigenvalues of the homogenous Helmholtz equation; this approach excellent reproduce the Chladni figures for the rectangular plate case but the final product requires the use of a computer to make it useful.

Journal of Physics: Conference Series **582** (2015) 012022 doi:10.1088/1742-6596/582/1/012022

The above examples are to show the difficulty of finding an analytical expression that describes Chladni figures. Therefore in this paper we present an alternate method based on the Huygens principle of wave propagation and on ideas from the Bohr postulate of quantized angular momentum for the hydrogen atom. The method was applied to the well-known case of the circular plate with excitation at the origin; then compared to familiar experimental results obtaining an acceptable level of agreement. We expect that this method can be extended to other geometries, irregularities and excitation point location.

The distribution of this paper is the following: In section 2, we give a brief view of the Huygens-Fresnel principle and the Bohr atom postulates. In section 3, we describe the postulates of the proposed model and apply them to describe the Chladni figures of the circular plate with excitation at the center. In section 4, we show the numerical implementation of the model and compare it to our own experimental results. Finally, in section 5, both conclusions and future work are presented.

2. Huygens-Fresnel Principle and Bohr postulate

In order to better expose the alternate model; in this section we concisely describe the principles and postulate that serve as a basis for our prototptype.

2.1. Huygens-Fresnel Principle[6]

In an effort to explain reflection, refraction, interference and diffraction of light, Huygens proposed in 1678 that every point in a wave-front can be considered as a source of secondary waves that propagate forward at speed of light; the following wave-front is constructed from the envelop of all these secondary waves.

Afterwards in 1816, Fresnel introduced an obliquity factor to the Fresnel principle that gives preponderant amplitude to the secondary waves in the direction of propagation. In order to achieve better experimental verification, Fresnel also made some assumptions about the amplitude and phase of the secondary waves. Summing all this up, the mathematical expression of the Hugens-Fresnel model can be written:

$$\psi(\vec{r}) = \frac{ik}{2\pi}\int_S \frac{e^{ik|\vec{r}-\vec{r}'|}}{|\vec{r}-\vec{r}'|} F(\theta)\, dr', \tag{1}$$

where ψ represents the amplitude of the wave at $\vec{r}$, k is the wave number, $F(\theta)$ is the Fresnel factor and A is the amplitude at the origin $\vec{r}_0$. It is interesting to note that the now called Huygens-Fresnel model predicted a bright spot of light in front of an opaque circular plate when illuminated from the back, this prediction caused the rejection of the model by the scientific community until was proved experimentally.

2.2. The Bohr atom[7]

At the beginning of the past century, it was known that matter is composed of negative and positive charged particles. Rutherford found experimentally that the positive charges where gather in dense clusters named nuclei. What followed was several models to explain how the negative charge is distributed respect the nuclei in order to conceive a minimum unit of matter called atom.

Bohr proposed a model where the electrons spin around the nuclei analogous to the planets around the sun. Even that this was a simple and logic model constructed from the Coulomb and Newton laws; it had an error based on Maxwell equations: an accelerated electron should emit radiation, making the electron lose energy and fall in to the nucleus.

In order to skip this problem, Bohr postulated that the electron could only spin around the nuclei in certain orbits in which it does not irradiate energy. These orbits are such that the angular momentum of the electron is an integer number of the Planck constant h. If we express this postulate in terms of the linear moment p and substitute the Debroglie relation $\lambda = 2\pi h/p$, then the permitted orbit of radius r follows

$$2\pi r = n\lambda, \tag{2}$$

where λ is the electron wavelength and n an integer number. From the expression above, we can see that the Bohr postulate implies that the permitted orbits are the ones that have a circumference equal to an integer of the electron wavelength. This is the idea that we are going to use in the proposed model.

3. The alternate model

As described before, Chlandi figures are stationary solutions to the wave equation, which is linear. This implies that the sum (superposition) of independent solutions can be used to construct a general solution. Therefore, we can propose a model as a linear combination of solutions that obey the following postulates:

1. The solution is composed in part of a divergent (principal) and convergent circular waves. The first wave travels from the excitation point up to the outer edge of the plate and the second wave, travels back with a phase shift proportional to the acoustical length path.
2. The principal wave excites the plate border creating secondary waves at every point around its perimeter. Every secondary wave reflects back in a direction perpendicular to the plate border following the Huygens-Fresnel principle.
3. The only frequencies that produce Chladni figures are those that form stationary waves around the perimeter of the plate in similar way to the stationary waves in the Bohr atom. This stationary wave also modulates the amplitude of the secondary waves of postulate 2. The sum of all these waves are what forms the Chladni figures.

The first postulate comes from the phenomenon of wave propagation and reflection by change of media. This is the case for Chladni plates, where a portion of the wave is transmitted to the air and the rest reflected back to the source.

The second postulate is a direct application of the Huygens-Fresnell principle with the consideration that the secondary waves are traveling back, perpendicular to the border perimeter and phase shifted in concordance to the distance from the excitation point.

The first two postulates are not sufficient because they would allow the formation of Chlandi patterns for any excitation frequency. Consequently, the third postulate limits the formation of Chladni patterns to excitation frequencies that yield stationary waves around the plate. In addition, the stationary wave modulates the amplitude of the secondary waves of postulate 2 at the border.

3.1 Circular plate

In this paper, we are going to apply the above postulates to a thin circular metal plate of radius R and excitation at the center. Let ψ represents the amplitude at any point inside the plate, then the mathematical expression in polar coordinates (r, θ) of the proposed model for this case reads

$$\psi(r,\theta) = \frac{A}{r}e^{ikr} + \frac{A}{r}e^{-ik(r+\varphi)} + A\int_0^{2\pi} \frac{e^{ik|\vec{r}-R\hat{\theta}|}}{|\vec{r}-R\hat{\theta}|} cos(kR\theta)d\theta, \tag{3}$$

where A is the amplitude of the wave excitation point, r_0, k is the wave number permitted by postulate 3, φ is the phase change mentioned in postulate 2. The first and second term represents the outgoing and reflected wave respectively. Observe that the reflected wave have a

Journal of Physics: Conference Series **582** (2015) 012022 doi:10.1088/1742-6596/582/1/012022

phase shift φ, and the third term represents the infinite sum of waves around the plate modulated by an stationary wave represented by the cosine function.

4. Numerical implementation and experimental validation

The model presented in equation (3) was numerically implemented in MATLAB software and the code is available at the minimal request. The algorithm starts by first constructing a generalized circular wave function that represents a complex circular wave having the wave number, phase and origin as inputs. Second, we deploy and add two of these functions into a matrix, both with the same wave number and origin at the center, but with a phase difference φ between them.

Third, we use a cycle to sum multiple circular wave functions each having as origin a point in a circle subscribed on to the matrix. The amplitude of these outer waves is modulated by a cosine function with an argument that complies with postulate 3 with an angle measured respect the center of the matrix. The Fresnel factor was not implemented at this moment.

4.1. Experimental setup

The experimental setup consisted of a 15 inches radius (R) 20ga laminated plate placed over a mechanical vibrator (PASCO SF-2394), which was driven by a function generator (PASCO PI-9587). The procedure consisted in sprinkling common salt in the plate and sweeping the frequency of the function generator until a constant pattern was obtained.

4.2. Results

The literature describes that the Chladni figures for the circular plate consist on a combination of radial and concentric nodes. The following figure shows the case of one of these possible combinations; other cases are similar. The simulation are for $\lambda = 0.45R$, $n = 2$ and the correspondent experiment is shown to the right.

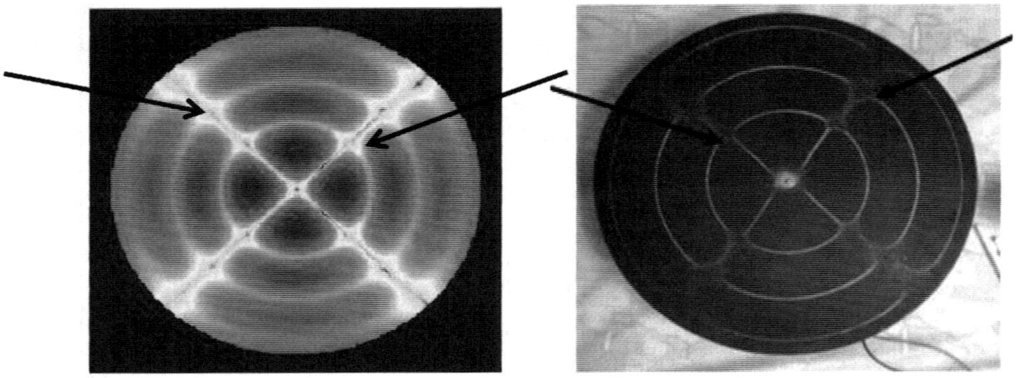

Figure 1. Simulation (left) and experiment (right) comparison, observe the similarity of the features where the radial and concentric nodes meet.

We can observer that the model reproduces with good level of accuracy the radial and concentric nodes. Furthermore, notice how the model also replicates the small features (arrows) that appear on the intersection of the nodes.

On figure 2, we observe a ripple effect on the experiment that we can reproduce on the simulation, in a comparable way, if we consider a non-integer value for n. If this is the case, we can consider this anomaly as a quasi-stationary case.

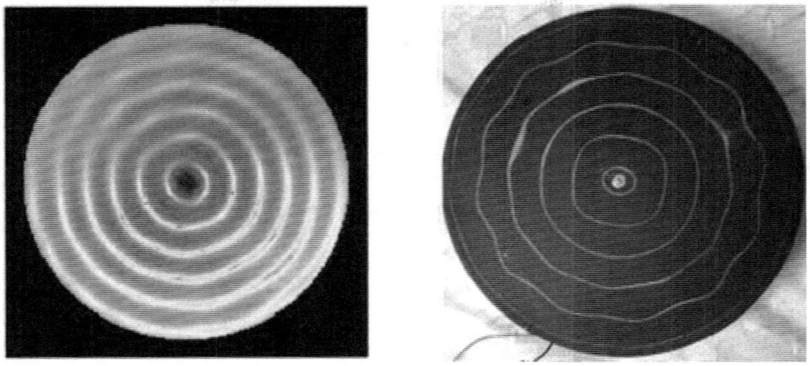

Figure 2. A possible explanation of the experimental ripple pattern (right) is a quasi-stationary state for $n = 1/5$ and $\lambda = 0.3R$, as shown in the simulation (left).

5. Conclusion and future work

We present a new model that reproduces the Chladni figures for the thin homogenous circular plate. The model uses a superposition of Huygens principles and ideas of the Bohr atom to obtain a mathematical expression. This expression was numerical implemented and confirmed experimentally with an adequate level of accuracy as it exhibited the radial and concentric nodes characteristic of the Chladni circular plate, and even reproduced the features at the intersection of the two types of nodes. The model also seemed to replicate a ripple patter observed experimentally.

In a future work, we are going to test the model for other geometrical figures such as conic sections and also for non-homogenous plates. In addition, the implementation of the Fresnel factor on the numerical simulation will be addressed.

References

[1] Rossing T 1981 *Am. J. Phys.* **50** 271
[2] Libisch, Rotter and Burgdörfer *Eur. Phys. J. Special Topics* **145** 245
[3] Amore P 2008 *J. Phys. A. Math. Theor.* **41** 265206
[4] Müller T 2013 *Eur. J. Phys.* **34** 1067
[5] Gander and Kwok, *SIAM Rev.* **54** 573
[6] Born M and Wolf E 1989 *Principle of Optics* (Pergamon Press) p.370
[7] Beiser A 1995 *Concepts of Modern Physics* (McGraw-Hill, Inc) p.129
[8] Ventsel and Krauthammer *Thin plates and shells Theory Analysis and Application* (Marcel Deker, Inc.) p.106

Journal of Physics: Conference Series **582** (2015) 012023 doi:10.1088/1742-6596/582/1/012023

Heat Transfer Convection in The Cooking of Apple Using a Solar Cooker Box-Type

H Terres, S Chávez, A Lizardi, R López, M Vaca, J Flores, A Salazar
Universidad Autónoma Metropolitana, Unidad Azcapotzalco, Departamento de Energía, Area de Termofluidos, San Pablo 180, Col. Reynosa Tamaulipas, Azcapotzalco, 02200, México, D. F.,México, D.F.

E-mail: tph@correo.azc.uam.mx

Abstract. In this work, experimental results to determine the convection heat transfer coefficient in the cooking process of apple using a solar cooker box-type are presented. Experimental data of temperatures for water, surface and central point of the apple were used. To determine the convection coefficient, the apple was modelled as a sphere. The temperatures evolution was defined using thermocouples located at water, surface and central point in the vegetables. Using heat transfer convection equations in transitory state and the temperatures measured, the Biot number and the convection coefficient were determined.

1. Introduction

For many applications in different areas the convection heat transfer coefficient must be determined for different proposes. Due to this, the variety of cases is big, such as next is exposing.

The external heat transfer coefficient in steam processing was determined experimentally in a pilot scale. Using the finite element method, heat transfer equations were established and solved. The actual retort temperature profile were used as boundary conditions. To analyse the time-variability along a retort cycle, the heat transfer coefficient values were determined for various instants [1].

In foods and liquids like beverages, the cooling and freezing times are required. For this purpose, the design of food refrigeration equipment are necessary. Also, refrigeration loads are required. Using the surface heat transfer coefficient for the freezing or cooling operation the refrigeration loads can be estimated. For this purpose, heat transfer coefficients from cooling curves were obtained by means of a iterative algorithm. For various food items heat transfer coefficients were determined [2].

In an experimental study for potato, the heat transfer properties in terms of its convective heat transfer coefficient were calculated. For potato cylinders drying experiments were performed. The convective heat transfer coefficient was evaluated for combined effects of heat capacities in food products. According results the convective heat transfer coefficient for potato cylinders was varying from 11.73 to 16.23 W/m^2 °C [3].

A rectangular metal tank filled with an alcoholic solution was used to determine the spatial distribution of h values inside tunnels. As a model system, a temperature sensor inserted in its interior was implemented. The h values for different positions inside the tunnels were calculated. The experimental method itself is the main result, due to it is reproducible to determining heat transfer coefficients inside industrial tunnels [4].

Journal of Physics: Conference Series **582** (2015) 012023 doi:10.1088/1742-6596/582/1/012023

In the food preservation, the thermal processing is very usefull due to its easy methodology. In this case, steam, air or water are used and the convective heat transfer coefficient must be determined. By this reason, a method to determine the convective heat transfer coefficient was implemented for thermal process of foods [5].
In an article, the behaviour of heat and mass transfer relation during khoa making was investigated. Considering a thermal model to estimate the maximum evaporation during heating of milk, indoor experiments were done. It was observed that with the varying voltage, the convective and evaporative heat transfer coefficients decrease when the rate of heating is increased.
It was also observed that when the operating temperature is increased, the convective and evaporative heat transfer coefficients also are increased. The rate of increment of evaporative heat transfer coefficient was higher than the convective heat transfer coefficient [6].
In this work experimental results for the convection heat transfer coefficient for apple are presented. Agree to results, the highest value to the convection coefficient is 90 W/m^2 °C for a specific solar radiation of 1150 W/m^2. The results obtained can be useful in numerical simulation when an estimation for heat convection coefficient is required.

2. Experimental procedure

In the experimental procedure, a solar cooker box-type was used. The solar cooker is integrated by the following elements: 1. a cover with two flat glasses with a clearance between them. 2. Reflectors placed on the cover outer of the cooker 3. Internal reflectors made in commercial aluminium paper placed to different tilt angles, 4. Thermal insulator placed in the lateral part of the same one, and 5. Recipient contains the product to cook. The solar cooker is locked tightly; this allows reaching considerable temperatures in the water. In figure 1 the solar cooker is shown.

Figure 1 Solar cooker box-type: with exterior reflectors and interior reflectors

The solar radiation (total) was measured using an Eppley piranometer model 8-48. Multimeters were used to measure the temperature. Thermocouples k-type at surface, central point of the apple and water were placed. For the tests the temperature was measured each 5 minutes. The interval for the time cooking process was 2.5 hours. The average value of the mass for the eggplant was 250 g and the average value for the water was 250 ml. In figure 2, temperature data measured for apple after to the experimental procedure are shown. This data were selected among ten experimental tests. In all cases, the behavior was similar.

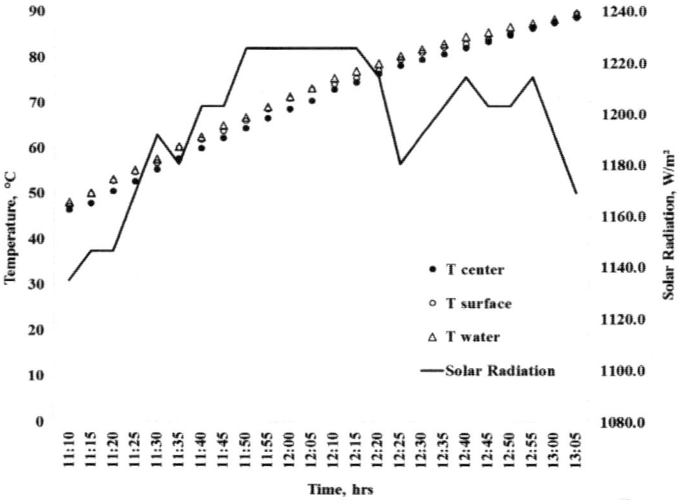

Figure 2 Experimental temperatures for the apple

3. Convection coefficients results

When the temperature data is obtained, heat transfer equations are used to determine the convection heat transfer coefficient. Some considerations for this purpose were done: 1. Thermal conductivity constant. 2. One-dimensional flow. 3. Without heat generation. 4. Spherical coordinates.
Agree to data measured, is possible to determine the temperatures difference ratio in function of water, surface and center by using [7]

$$\boldsymbol{\theta} = \frac{T_0 - T_\infty}{T_i - T_\infty} \quad (1)$$

This equation is related with correlations. The reference to determine the convection heat transfer coefficient is [8]

$$\boldsymbol{\theta} = A_1 e^{-\lambda_1^2 F_0} \quad (2)$$

Where A_1 and 1_1 must be determined by using tables. Biot number (Bi) and the Fourier number (Fo) are defined as

$$F_0 = \frac{\alpha t}{D^2} \quad (3)$$

$$Bi = \frac{hD}{k} \quad (4)$$

The Biot number is unknown, for this reason an iterative method is used to determine A_1 and 1_1.
Finally, the convection heat transfer coefficient can be determined solving for *h*. So, when equation 1 is determined, equations 2 and 4 can be solved for Bi. After this, h can be calculated by using equation 4. This procedure is applied for all the time during the cooking process.
In figure 3, results for the Biot number corresponding to apple are shown and the convection heat transfer coefficient values for apple are shown in figure 4

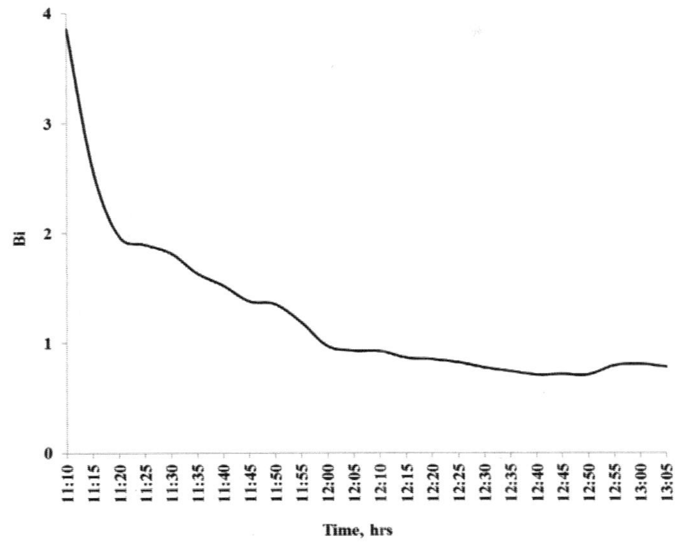

Figure 3 Experimental results for Biot number

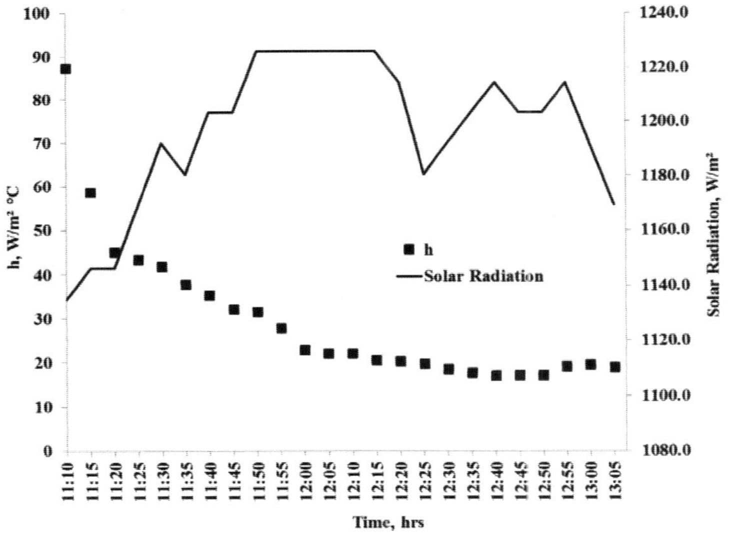

Figure 4 Experimental results for h, convection heat transfer coefficient

4. Discussion

Agree to results, temperatures follow next behavior Tf > Ts > Tc (fluid, surface and center respectively) for all cases. This is consistent for the heating process. The difference temperatures between surface and center inside the apple were approximate 0.06.

The temperature inside of the apple was 88.9 °C which is enough to reach a complete cooking in the same ones.

The maximum Biot value was 3.86. This value indicates that the conduction resistance is greater than the convection resistance for the apple. The convection heat transfer coefficient value was 87.47 W/m² °C.

5. Conclusion

An experimental procedure to determine the convection heat transfer coefficient was exposed. The results show how the convection happens in the heating process for apple.

The resistance values are useful, because it allows identify what is the most important thermal resistance in the heating process for apple in their heating process using a solar cooker. The convection coefficient can be used in numerical simulation, particularly, in solar energy applications due to this information is unknown and there is not works about it.

References

[1] Varga S, Oliveira C 2000 Determination of the heat transfer coefficient between bulk medium and packed containers in a batch retort *Journal of Food Engineering*, **44** 191-198

[2] Fricke B, Becker B 2002 Calculation of heat transfer coefficients for foods *Int. Comm. Heat Mass Transfer* **29** 731-740

[3] Tripathy P, Abhishek S, Bhadoria P 2013 Determination of convective heat transfer coefficient and specific energy consumption of potato using an ingenious self tracking solar dryer *Food Measure* **30** 1-10

[4] Santos A, Carciofi M, Dannenhauer, Hense H, Laurindo J 2007 Determination of heat transfer coefficient in cooling-freezing tunnels using experimental time–temperature data *Journal of Food Process Engineering* **30** 717–728

[5] Esteves P, Augusto D, Cristianini M 2011 Determining the Convective Heat Transfer Coefficient (h) in Thermal Process of Foods *International Journal of Food Engineering* **7** 1-22

[6] Kumar M, Prakash O, Kasana S 2012 Experimental investigation on natural convective heating of milk *Journal of Food Process Engineering* **35** 715–726

[7] Cengel Y. A. and Ghajar A. J. *Heat and mass transfer 4th edition McGraw-Hill Co.* 2011.

Journal of Physics: Conference Series **582** (2015) 012024 doi:10.1088/1742-6596/582/1/012024

First and Second Law Efficiencies in the Cooking Process of Eggplant using a Solar Cooker Box-Type

H Terres, S Chávez, A Lizardi, R López, M Vaca, J Flores, A Salazar

Universidad Autónoma Metropolitana, Unidad Azcapotzalco, Departamento de Energía, Area de Termofluidos, San Pablo 180, Col. Reynosa Tamaulipas, Azcapotzalco, 02200, México, D. F.,México, D.F.

E-mail: tph@correo.azc.uam.mx

Abstract. In this work an experimental procedure and the determination of first and second law efficiencies for the cooking process of eggplant using a solar cooker box-type are shown. The eggplant was modelled as cylinder. In the experimental process a NI Compact Field Point was used as acquisition data system which allows measure temperatures in simultaneous form. The temperatures evolution was defined using thermocouples located at water, surface and central point of the eggplant. After to measure the evolution temperatures in a solar cooker thermodynamics principles were applied to determine the first and second laws. The results obtained indicates what is the numerical difference between the first and second laws in the cooking process of eggplant. The results allow to understand how the inlet energy that impacts on solar cooker is converted in energy useful in the cooking process of eggplant. This work be used in future designs of solar cookers.

1. Introduction

In the solar energy field is important determine which is the real value or magnitude that is used for the several solar devices. For this reason, useful parameters are the first and second laws, which are obtained in the thermodynamics analysis and they are aimed to the different processes where heat or work have an important contribution. Some references are cited to illustrate applications of the exergy and efficiency in solar energy.

Petela [1], works in equations for heat transfer between three surfaces: cooking pot, reflector and imagined surface making up the system, were derived. He obtained a model which allowed for theoretical estimation of the energy and exergy losses: unabsorbed insolation, convective and radiative heat transfer to the ambient, and additionally, for the exergy losses: the radiative irreversibilities on the surfaces, and the irreversibility of the useful heat transferred to the water. The exergy efficiency of a solar parabolic cooker, was found to be relatively very low (1% aprox.), and to be about 10 times smaller than the respective energy efficiency which is in agreement with experimental data from the literature.

Kamthania et al. [2] In their paper an attempt to analyze the performance of semi-transparent hybrid PVT double pass air collector is shown. Considering a quasi-steady state condition, analytical

expression for air temperature at the outlet in a hybrid PVT air collector was obtained. Energy balance equations applying the thermodynamical first law were formulated.
Their paper shows a detailed analysis of energy and exergy for a semi-transparent hybrid with double pass air collector. This device is compared with a single pass air collector.

Different conditions of weather for five different cities (New Delhi, Bangalore, Mumbai, Srinagar, and Jodhpur) of India were condidered. When the compare is done, is possible to see that hybrid double pass air collector have better performance than the single pass air collector.

Candau [3] proposes a derivation of the exergy of radiation, based solely on classical thermodynamics notions, making thus possible a ready check of the validity of the results. Results are given first for the blackbody case, then extended to a radiation with arbitrary spectrum. Finally, application of the notion of radiation entropy to a few simple examples shows the consistency of this notion with some well-known physical laws, and can even give some insight into the real signification of these laws.

Fudholi et al. [4] designed, constructed and tested a solar drying system for drying of seaweed. Seaweed is a potential source of renewable energy, and it can be converted into energy such as biofuel oil, biodiesel and gas. Drying kinetics of red seaweed were investigated and obtained. The nonlinear regression procedure was used to fit three different drying models. The Page's model clearly showed a better fit to the experimental data between Newton's model and Henderson and Pabis model. The Page's model was resulted in the highest value of R^2 and lowest values of MBE and RMSE. At average solar radiation of about 500 W/m^2 and air flow rate 0.05 kg/s, the collector, drying system and pick-up efficiencies were found about 35, 27 and 95%, respectively. This study was performed with energy analysis and exergy analyses of the solar drying process of red seaweed. The specific energy consumption (SEC) of 2.62 kWh/kg was obtained. Moreover, the exergy efficiency of solar drying ranged from 1% to 93%, with an average of 30%.

Gunerhan and Arif Hepbasli [5] worked on a solar water heating. In Izmir, Turkey, the device was evaluated. Exergy destructions as well as exergy efficiency relations are determined for the Solar heating system. Exergy efficiency values on a product/fuel basis are found to range between from 2.0 to 3.3 %, and 3.2 to 4.3 % at a dead state temperature of 32.7 C.

In this work, a study for first and second law applied to the cooking process of eggplant using a solar cooker box-type is shown. The results obtained can be used to evaluate the process and they indicate what is the real energy quantity used in the cooking process of eggplant.

2. Experimentation

In the experiment a solar cooker box-type as energy source is used. The solar cooker is integrated by the following elements: 1. a cover with two flat glasses with a clearance between them. 2. Reflectors placed on the cover outer of the cooker 3. Internal reflectors made in commercial aluminium paper placed to different tilt angles, 4. Thermal insulator placed in the lateral part of the same one, and 5. Recipient contains the product to cook. The solar cooker is locked tightly; this allows reaching considerable temperatures in the water. In figure 1 the solar cooker is shown.

The solar radiation (total) was measured using an Eppley piranometer model 8-48. Multimeters were used to measure the temperature. Thermocouples k-type at surface, central point of the eggplant and water were placed. The temperature measuring was realized each 5 minutes. The time cooking process considered was 2.5 hours. The average value of the mass in the eggplant was 250 g. The average value for the water was 250 ml. The geometrical characteristics and the thermocouples colocation at eggplant, are shown in figures 2.

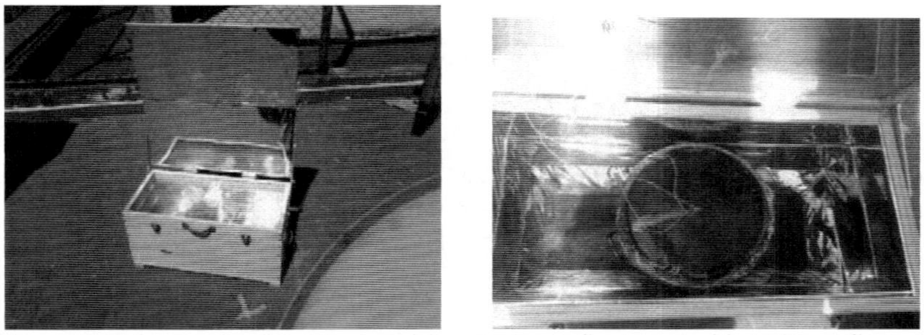

Figure 1 Solar cooker box-type: with exterior reflectors and interior reflectors

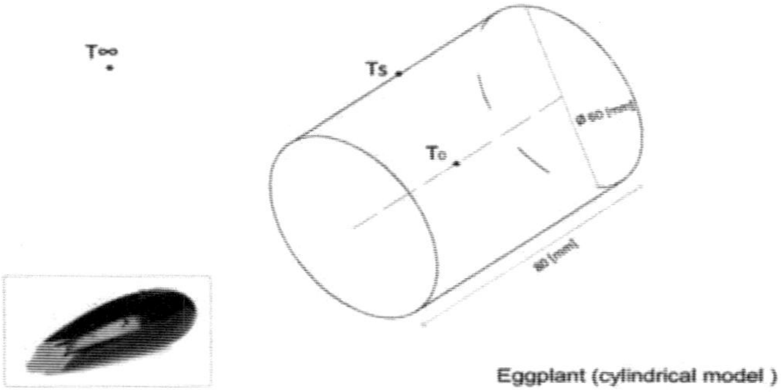

Figure 2 Thermocouples position and geometry for eggplant

Once temperature data is obtained, first and second law are defined to calculate the values for the process. For this purpose, some considerations were done: 1. Thermal conductivity constant. 2. One-dimensional flow. 3. Without heat generation. 4. Cylindrical coordinates.
According to thermodynamics theory states [6]

First law Efficiency

$$\eta_I = \frac{Q_{out}}{Q_{in}} = \frac{m(u_f - u_i)}{\frac{A_v G(t_f - t_i)}{1000}} \quad (1)$$

Second law Efficiency

$$\eta_{II} = \frac{\psi_{out}}{Q_{in}} = \frac{mcT_0\left[\frac{T_f}{T_0} - 1 - ln\left(\frac{T_f}{T_0}\right)\right]}{\frac{A_v G(t_f - t_i)}{1000}} \quad (2)$$

The heat specific was determined using correlations in function of temperature [7].

$$C = C_{H_2O}x_i + C_{protein}x_i + C_{fat}x_i + C_{carbohidrates}x_i + C_{fiber}x_i \quad (3)$$

Where

$$x_i = Mass\ component\ of\ the\ product\ (eggplant) \quad (4)$$
$$C_{H_2O} = 4176.2 - 0.090864T - 0.0054731T^2 \quad (5)$$

$$C_{protein} = 2008.2 + 1.2089T - 0.0013129T^2 \quad (6)$$
$$C_{fat} = 1984.2 - 1.4733T - 0.0048008T^2 \quad (7)$$
$$C_{carbohidrates} = 1548.8 + 1.9625T - 0.0059399T^2 \quad (8)$$
$$C_{fiber} = 1845.9 + 1.8306T - 0.0046509T^2 \quad (9)$$

Q is the heat in kJ, m is the mass of the eggplant in kg, u is the internal energy kJ/kgK, T is the absolute temperature, in K, A_v is area of the cover glass in m^2, G is the solar radiation in W/m^2, t is the time in seconds. In subscripts, *out* represents the out energy, *in* is the out energy and *f*, *i* and *0* correspond to final, initial and dead state respectively.

3. Results and discusion

After the experimental procedure, results for the thermal process are obtained. In figure 3, temperatures for water, surface and centre of the eggplant are shown.

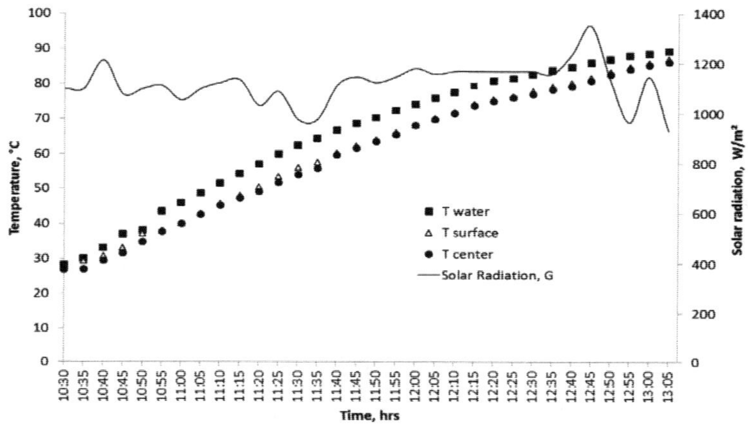

Figure 3 Thermocouples position and geometry for eggplant

With this data and using equations (1) and (2) first and second law efficiencies are determined, in figure 4 are shown.

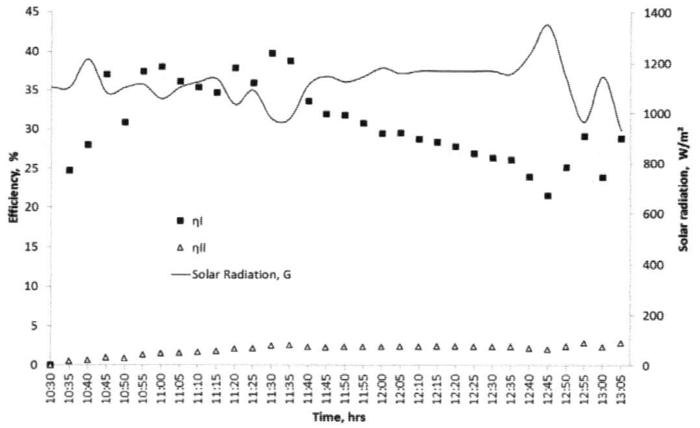

Figure 4 Thermocouples position and geometry for eggplant

The experimental data allow to appreciate the temperature behavior in the cooking process which is consistent for the heating process of eggplant, this is $T_{water} > T_{surface} > T_{center}$.
According to the efficiencies data, the second law values are not bigger than 5 % while the first law values reaches 40 %. Those results allow to evaluate the quantity of energy not used in the cooking of eggplant.
The first law efficiency indicates the cooking process in an ideal interpretation.
By other hand, the second law efficiency shows how the inlet energy is really used during the cooking process. This efficiency is called real efficiency because it allows to view the irreversibility in a process like this case.
It is identified in other works [8] the loss of energy occurs in the cover of the solar cooker and it is associated to the conduction and convection heat transfer.
This work can be useful to identify what is thermal behavior in the cooking process of eggplant. The results also can be useful to establish how and where the solar cooker could be improved.

4. Conclusions

Experimental procedure to determine the evolution of temperatures was shown. The implement of instrumentation in solar energy was illustrated. Thermodynamics principles were applied to determine the first and second laws. The results obtained indicates what is the numerical difference between the first and second laws in the cooking process of eggplant. The results obtained can be used in future designs of solar cookers

References

[1] Petela R 2005 Exergy analysis of the solar cylindrical-parabolic cooker *Solar Energ.* **79** 221-233
[2] Kamthania D, Nayak S, Tiwari G. N 2011 Energy and Exergy Analysis of a Hybrid Photovoltaic Thermal Double Pass Air Collector *Applied Solar Energy* **47** 199–206
[3] Candau Y 2003 On the exergy of radiation *Solar Energy* **75** 241–247
[4] Fudholi A, Sopian K, Yusof O, Hafidz M 2014 Energy and exergy analyses of solar drying system of red seaweed *Energy and Buildings* **68** 121–129
[5] Gunerhan H, Hepbasli A 2007 Exergetic modeling and performance evaluation of solar water heating systems for building applications *Energy and Buildings* **39** 509–516
[6] Russell L, Adebiyi G 1993 Classical Thermodynamics *Saunders College Publishing* USA
[7] Choi Y, Okos M 1986 Effect of temperature and composition on the thermal properties of foods, Food Engineering and Process Applications *Elsevier Applied Science Publisher*
[8] Terres H, Morales J, Lizardi A, López R, Portillo R 2013 Estudio numérico de eficiencias en estufas solares tipo caja con reflectores internos *Revista Internacional de Métodos Numéricos para Cálculo y Diseño en Ingeniería* **29(2)** 122–128

Journal of Physics: Conference Series **582** (2015) 012025 doi:10.1088/1742-6596/582/1/012025

System of acquisition and processing of images of dynamic speckle

F Vega[1], C Torres[1]
[1]Laboratorio de Óptica e Informática, Universidad Popular del Cesar, km 4 Via patillal,Valledupar-Colombia

E-mail: fabiovega@unicesar.edu.co

Abstract. In this paper we show the design and implementation of a system to capture and analysis of dynamic speckle. The device consists of a USB camera, an isolated system lights for imaging, a laser pointer 633nm 10mw as coherent light source, a diffuser and a laptop for processing video. The equipment enables the acquisition and storage of video, also calculated of different descriptors of statistical analysis (vector global accumulation of activity, activity matrix accumulation, cross-correlation vector, autocorrelation coefficient, matrix Fujji etc.). The equipment is designed so that it can be taken directly to the site where the sample for biological study and is currently being used in research projects within the group.

1. Introduction

At present, the acquisition and processing video in real time is a field of research open. The biological tissue is one of the most complex of those present in nature. This situation is aggravated by the inherent variability present in one sample to another, which makes it more difficult the comparison of results between different samples even for the same stimulus.

In biological materials, dynamic speckle phenomenon is also known as biospeckle and allows analysis of biological activity or image rendering in various applications such as seeds, Mushrooms, fruits, blood flow, parasites, bacteria, organic films, ice cream, roots, semen, etc. Speckle dynamic applications in non-biological materials include various, including the analysis of the dried paint, control of gels, foams, corrosion, efflorescence, etc.

The dynamic speckle has applications in many fields of current research [1], as referents have Research where they developed the first descriptors such as [2], which was obtained image of local activity in fruits, fruit bruising study [3] in 2003, evaluating viability in seeds [5], characterization paint drying processes [6]. Also been used as statistical descriptors co-occurrence matrix [4] typical of this type of study. Work is currently being developed as the measurement of blood flow velocity [9], reconstruction of audio starting biospeckle image [8] etc.

2. Software acquisition and processing video

The acquisition system and video processing, was designed with the following features: must allow the capture of video from different sources such as analog video inputs (CCD camera), digital cameras with USB output to the computer; because Nowadays high resolution USB cameras are the order of the day. Then the system should allow the processing of images, video and apply different filters to images. Various algorithms were developed as: Filters images, color histograms, histogram equalization, line profile, conversion of pixels to micrometers according objective microscope, saved images, streaming video, histograms and export files for your particular MATLAB processing time. The graphical interface of user (GUI) of the software can observe the Fig. 1 (b).

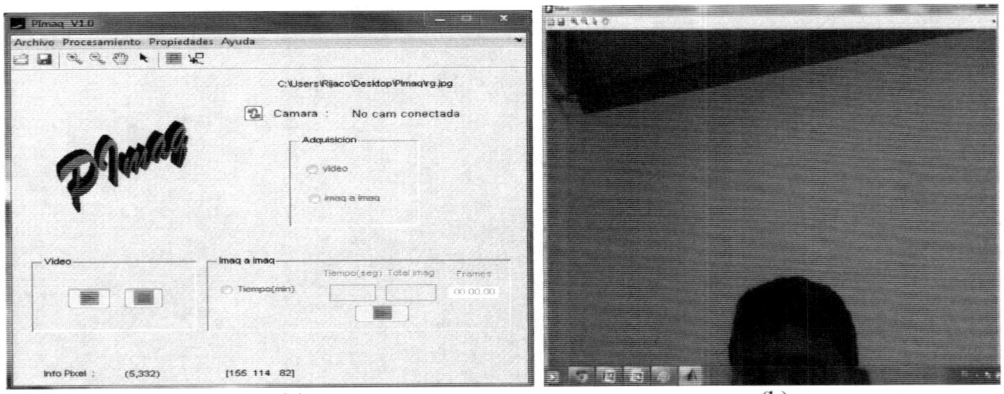

(a) (b)

Figure 1. Graphical interface software acquisition and processing of video.

The developed software can be run on notebook computers with USB port for video capture. The software has two windows one for control and one for video view.

2.1. Algorithms developed

Were implemented algorithms for filtering the image as seen in Fig. 2. Specialized algorithms are also developed in the microparticle diameter measurement view Fig. 3 where you select the source calculation, cores optical fiber which are observed by microscopy with different objectives of different lateral raises shown in Fig. 4, statistical processing for dynamic speckle; Among these are the autocorrelation coefficient matrices see equation (1).

Figure. 2. Example filters, In order a) Original image b) Edges c) grayscale d) Gaussian e) unsharp

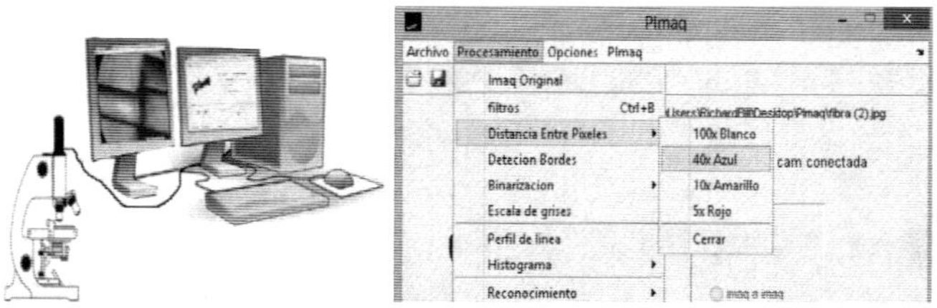

Figure 3. Selection of microscope objective for diameter calculation

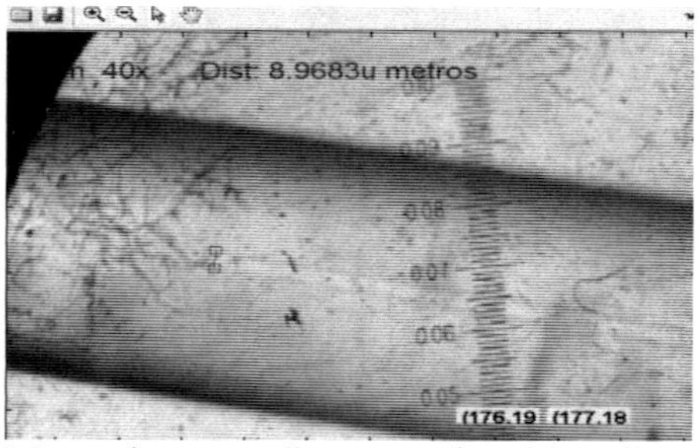

Figure 4. Measuring core diameter of multimode optical fiber

Now for compute the correlation coefficient between an image and the same image size is:

$$r = \frac{\sum_m \sum_n (A_{mn} - \bar{A})(B_{mn} - \bar{B})}{\sqrt{\left(\sum_m \sum_n (A_{mn} - \bar{A})^2\right)\left(\sum_m \sum_n (B_{mn} - \bar{B})^2\right)}} \tag{1}$$

Where $\bar{A}$ and $\bar{B}$ is Average or mean of matrix elements of (A) and (B). And (m,n) are pixel position. In this work, we used correlation coefficients for study in feeds.

3. Hardware for capturing dynamic speckle

The hardware required for applications biospeckle is specially designed to work with small objects of study, such as fruits, seeds and any biological object that can contain hardware designed.

The specific characteristics to this project were summarized as follows:

- ✓ The equipment is portable
- ✓ It has USB interface for PC and Laptops capture
- ✓ You can adapt different coherent laser sources
- ✓ The closed system prevents interference from other sources of light in the visible range
- ✓ Possesses graphical interface to capture and save
- ✓ The system can record video or save image sequence

Taking into account the required characteristics was developed a closed box, of black inner insulation for optical wavelengths in the visible range, in which to place the object of study. Also has a system movable sheets for location of the object of study, the prototype has holes to usb camera for location capture images or video, the implemented prototype can be seen in Fig. 5.

Journal of Physics: Conference Series **582** (2015) 012025 doi:10.1088/1742-6596/582/1/012025

Figure 5. Prototype for capturing images of objects of study

3.1 Preliminary Catches

This work is part of a larger project wingspan which is a study of seed germination in grasses, work which is in development; In Fig. 6.a is observed the capturing a tender corn seed, in good condition, messed and their capture speckle. Where you can clearly shown the quality of images obtained in the equipment and their dynamic speckle images of maize seeds.

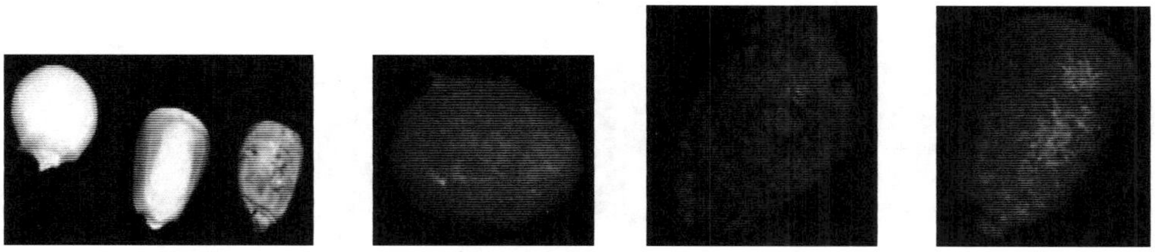

a) (b) (c) (d)

Figure. 6. Capturing Images with the equipment: a) Original Seeds, b) Tender Speckle seed, c) speckle seed disrepair. d) Seed Perfect condition

4. Prototype application

The equipment was initially applied in a research paper to estimate the percentage of germination of rice in Colombia, where it counted with the collaboration of Colombia Fedearroz Valledupar headquarters, which supply the seeds under study.

This work is important for the Colombian region because, in Colombia only can sell rice paddy it to germinate above 90% (Article 12 of Resolution 3034 of 1999, the Colombian agricultural institute ICA) [10]. The certification process consists initially on tests the feasibility with tetrazolium, where work has been done. Which indicates that the seed can germinate, but the final proof is the germination test samples of seed process that takes about 15 days, from sowing of the seed until the counting seeds germinated successfully. In this work is to develop a system for determining of percentage of seed germination using techniques biospeckle where will decrease the time for obtaining results from days to just minutes.

The experimental setup used for cross-correlation methods is shown in Fig. 7. and Fig. 8. It used an expanded low power He–Ne LASER (10mW, 633nm) as coherent light source, a diffuser, attenuator for decrease to avoid saturating the camera sensor; images were registered by a USB camera with focus len. Then, the images were digitized and stored as files in a host computer with data acquisition software developed in MATLAB IDE. Care was taken so that the speckles were well resolved by the camera sensor. We used very low illuminating intensity so that the effect of the irradiation on the sample could be negligible. Laser illumination was adjusted to illuminate a broader region of the fruit during all measurements. The focus of camera was adjusted with perfect visualization of a new paper for focused capture of frame image.

The object of study was used seeds of rice variety Fedearroz 2000, the rice seed Fedearroz 2000 is a new variety that present similar to Fedearroz 50 characteristics in terms of their resistance to lodging and certain types of diseases, but their productivity levels are much higher. Fedearroz 50 yields 7 to 8 tons on average, although in certain regions of the country as the terrace of Ibague Colombia, Fedearroz 2000 has reached 13 tonnes per hectare.

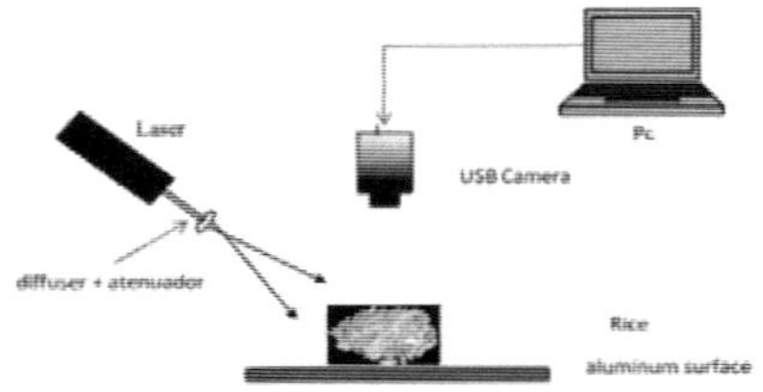

Figure. 7. Experimental setup for the Biospeckle activity analysis.

Figure. 8. Final prototype implemented for study

It is also developing a portable prototype for capturing and processed dynamic speckle images which can be observed in Fig. 2; equipment allows the capture of image on equal terms for all samples, not require a special location for its functioning; another of its main functions is to isolate samples of wavelengths in the visible range to not interfere the Laser Source thus improving the quality of the procedure; There was also the need to place seeds on an aluminium surface for better statistical processing.

For the experiment, the recording time was equal to 60 seconds with frame rate equals to 2 seconds for a total 30 images. The objective speckle patter was obtaining by free propagation between the sample and the camera sensitive area. Data were registered and recorded on files rice seeds at different stages of latency, were obtained sample by sectional Fedearroz Valledupar of seed freshly harvested in the countryside (green paddy), Freshly processed for storage (paddy) and low latency stage (maximum germination). The samples selected by groups of 100 seeds in the same latency and grouped as shown in Fig. 9. After measurement of dynamic speckle proceeded to make them germinate and manually count how many seeds were born and how many not. These three groups of data samples, the conclusions of this work are obtained.

Journal of Physics: Conference Series **582** (2015) 012025 doi:10.1088/1742-6596/582/1/012025

Figure. 9. Rice seed sample

Every point of image was essentially representative of the phenomenon and at the same instants. So, each column of these images could be used to generate THSPs that were all instantiations of the same experimental conditions. The images captured are of 640x480 pixels. Then, we proceed to convert image to grey scale 256 level and to apply a segmentation algorithm to obtain matrix NxN for analysis of interest area. The matrix of NxN is required for computing algorithm cross-correlation. The autocorrelation coefficients were obtained.

5. Segmentation and image processing

The segmentation process consist of four step, the first step is original image capture, the second step is convert the image to grayscale, the third step is to use the MATLAB function "bwlabel", this function returns a binary matrix, of the same size as image, containing labels for the connected objects in same image, then the object are removed from a binary image all connected components (objects) that have fewer than 1000 pixels, producing another binary image, BW2 using the function "bwareaopen"; next properties.

('Area','Centroid','FilledArea','MajorAxisLength','MinorAxisLength','ConvexHull') are obtained using the "regionprops" function. Now the mayor area is obtained in the properties object with your central position pixel (x, y) and its maximum row and column in the image. The final step is the segmented image obtained the original grayscale image; this image is used for biospeckle image processing. The image segmentation process is shown Fig.10.

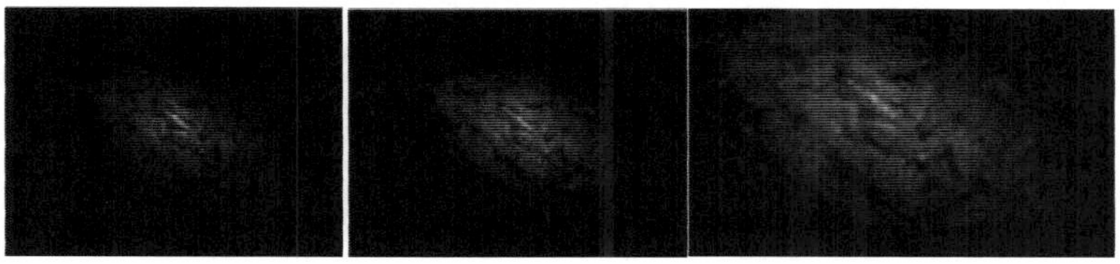

(a) (b) (c)

Figure. 10. Image segmentation. a) Original image, b) Grayscale image c) Segmented image

The segmented image is the correlation coefficient calculation with the Equation (1) and was obtained the graphs of autocorrelation coefficients of the different groups of seeds, which can be seen in Fig. 11.

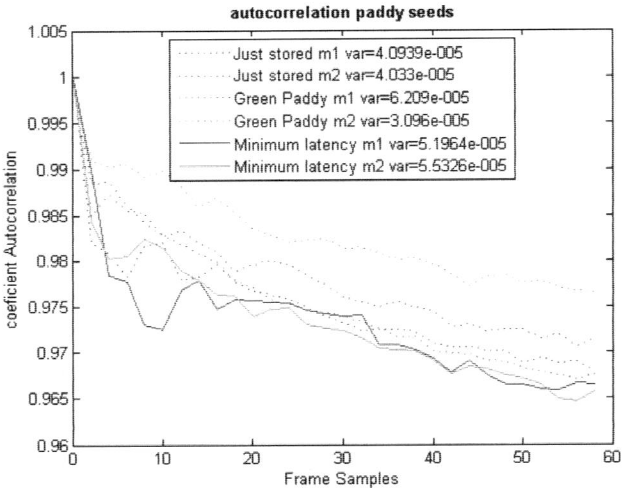

Figure. 11. Autocorrelation paddy rice seeds

In the graph we can see that the seeds have minimum latency greater slope, and having a higher cross correlation with respect to the group of seeds (green and freshly stored paddy). Table of physical evidence germination of the seeds used in the experiment can be seen in Table I.

Table 1. Percentage of seed germination

Group of sedes	% Germination	Average size cm.
Minimum latency Sample 1	98	2,7
Minimum latency Sample 2	95	2,8
Just stored Sample 1	70	1,5
Just stored Sample 2	65	1,07
Green Paddy Sample 1	80	1,4
Green Paddy Sample 2	75	1,6

Now as we have no equation of graphic for calculating the tangent line that is the speed with which the sample is decorrelates. For this resorted to use of equation (2), in which we calculated the slope between two points; for our case y_n values are the autocorrelation coefficients and the values of x_n are the time lag between samples, which are spaced every two seconds in the design of the experiment.

$$\mathbf{m} = \frac{y_i - y_f}{x_i - x_f} \tag{2}$$

Then these data proceeds to average them to obtain the average velocity decorrelation. This is summarized in equation (3)

$$\mathbf{m}_{\mathbf{prom}} = \frac{A}{N-2}\sum_{n=3}^{N}\left|\frac{y_n - y_{n-1}}{x_n - x_{n-1}}\right| \tag{3}$$

Where N is the total number of samples (30), n is the umpteenth sample this starts at position n=3, because the first point the coefficient of autocorrelation is 1 and could affect the average overall slope and the value A is a gain value of the slope. The linearized coefficient cross correlation is shown in fig. 12.

Table 2. Slope values, with A = 1000, N = 30

Group of sedes	Average slope
Just stored Sample 1	0.4747
Just stored Sample 1	0.4738
Green Paddy Sample 1	0.3948
Green Paddy Sample 2	0.3814
Minimum latency Sample 1	0.7250
Minimum latency Sample 2	0.5076

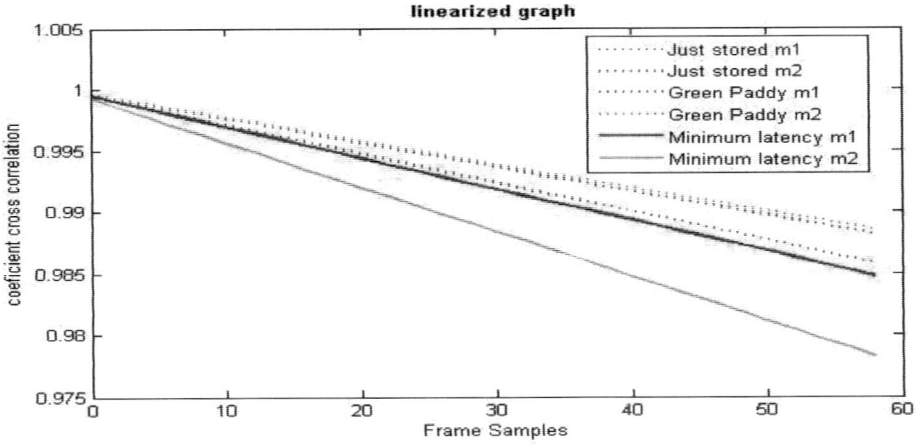

Figure. 12. Linearized coefficient cross correlation graph

As we can see in Table II and relating it to the germination percentage of the real test shown in table I, it can be seen when the cross correlation slope was above 0.5, with amplification A = 1000 and N = 30 factor, there was a 90% higher germination of seeds in physical proof of birth of the same seeds.

6. Conclusions

It was possible to develop portable equipment for acquisition and processing video in real time, the program allows capturing a number of images desired with a programmable time (tenths of a second, seconds and minutes) and then carrying out the relevant statistical processing. It was used, developed to a prototype system for estimating percentage of germination in seeds of rice, earning decorrelation slope as an indicator of germination percentage. As we can see in Table II and relating it to the germination percentage of the real test shown in table I, it can be seen when the cross correlation slope was above 0.5, with amplification A = 1000 and N = 30 factor, there was a 90% higher germination of seeds in physical proof of birth of the same seeds.

Acknowledgment

The authors give acknowledgment to the Popular University of Cesar by funding of the research project, Dr. Marcelo Trivi in CIOP in Argentina for kindly advise, to Fedearroz in valledupar colombia by providing seeds to develop the study.

References

[1] Rabal, HJ; Braga, RA 2008 Dynamic laser speckle and applications Taylor and Francis NY 304p

[2] R. A. Braga, C. M. B. Nobre, A. G. Costa, T. Sáfadi and F. M. Costa 2011 Evaluation of Activity through Dynamic Laser Speckle Using the Absolute Value of the Differences Optics Communications **284** 646- 650

[3] M. Pajuelo, G. Baldwin, H. Rabal, N. Cap, R. Arizaga, M. Trivi 2003 Bio-speckle assessment of bruising in fruits *Optics and Lasers in Engineering* **40** 13-24

[4] Ricardo Arizaga, Marcelo Trivi, Héctor Rabal 1999 Speckle time evolution characterization by the co-occurrence matrix analysis *Optics & Laser Technology* **31** 163-169

[5] Roberto A. Braga, Inácio M. Dal Fabbro, Flávio M. Borem, Giovanni Rabelo, Ricardo Arizaga, Héctor J. Rabal, Marcelo Trivi 2003 Assessment of Seed Viability by Laser Speckle Techniques *Biosystems Engineering* **86** 287-294

[6] Mavilio, Margarita Fernández, Marcelo Trivi, Héctor Rabal, Ricardo Arizaga 2010 Characterization of a paint drying process through granulometric analysis of speckle dynamic patterns *Signal Processing* **90** 1623-1630

[7] Kausik Basak, Goutam Dey, Manjunatha Mahadevappa, Mahitosh Mandal, Pranab Kumar Dutta 2014 in vivo laser speckle imaging by adaptive contrast computation for microvasculature assessment *Optics and Lasers in Engineerin* **62** 87-94

[8] Ziyi Chen, Cheng Wang, Chaohong Huang, Hongyan Fu, Huan Luo, Hanyun Wang 2014 Audio signal reconstruction based on adaptively selected seed points from laser speckle images *Optics Communications* **331** 6-13

[9] Chong Hui Li, Hong Dong Wang, Jian Jun Hu, Xin Lan Ge, Ke Pan, Ai Qun Zhang, Jia Hong Dong 2014 The monitoring of microvascular liver blood flow changes during ischemia and reperfusion using laser speckle contrast imaging *Microvascular Research* **94** 28-35

[10] Resolution 3034 of 1999 the Colombian agricultural institute ICA Online http://www.ica.gov.co/

Journal of Physics: Conference Series **582** (2015) 012026 doi:10.1088/1742-6596/582/1/012026

Effect of ultraviolet radiation in the photo-oxidation of High Density Polyethylene and Biodegradable Polyethylene films

A Martínez-Romo[1], R González Mota[1], J J Soto Bernal[1], C Frausto Reyes[2] and I Rosales Candelas[1].

[1]Instituto Tecnológico de Aguascalientes. Av. Adolfo López Mateos 1801 Ote. Fracc. Bona Gens, Aguascalientes, México.

[2]Centro de Investigaciones en Óptica A.C. Av. Prol. Constitución 607 Fracc. Reserva Loma Bonita, Aguascalientes, México.

E-mail: ahiza_2323@hotmail.com

Abstract. One of the most widely used plastics in the world is the High density polyethylene (HDPE), it is a stable material due to its carbon-carbon bonds, causing their slow degradation; which is why we are looking for alternative ways to accelerate the degradation process of this polymer. An alternative is the addition of oxidized groups in its molecular structure, which results in the development of polymers susceptible to biodegradation (PE-BIO). In this paper, HDPE and PE-BIO films were exposed to UV-B radiation (320-280 nm) at different exposure times, 0-60 days. The effects of UV radiation in samples of HDPE and PE-BIO were characterized using infrared spectroscopy with attenuated total reflectance (ATR). The results show that the exposed materials undergo changes in their molecular structure, due to the infrared bands formed which corresponds to the photo-oxidation of HDPE and PE films when submitted to UV-B radiation.

1. Introduction

Nowadays, plastics are products widely used and manufactured in large quantities, their difficult process of degradation causes these materials are the most frequent contaminants in soils and oceans [1, 2]. The high density polyethylene (HDPE) is one of the most commonly used plastics in everyday life [3], its widespread use is due to its low manufacturing cost, in addition to its excellent mechanical and chemical properties; its low rate of degradation is one of the reasons that new alternatives for disposal reduction are studied. One of the proposed alternatives is the manufacture of self-biodegradable polyethylene (PE-BIO), which is an oxidized material, containing carbonyl groups (C=O) in its molecular structure; whose important feature is that they are susceptible to microbial attack. Another alternative is the use of UV-B radiation, in order to degrade and to incorporate plastics as harmless material to the environment [3].

2. Theory

Plastics are synthetic materials derived from petroleum, so the problems caused by their indiscriminate use and their persistence in the environment, has resulted in a worldwide established system of coding to identify them. Plastic products have a distinctive symbol to identify each type of polymer. The HDPE is identified with the number two.

The high density polyethylene (HDPE), is a semicrystalline thermoplastic, which is obtained by the addition polymerization mechanism of the monomer ethylene; its structure is a linear chain of CH_2-CH_2 with few branching, its density is 96 g/cm^3 and its melting point is 135 °C. The HDPE is not toxic, has a lower stiffness compared with the polypropylene, its chemical resistance is acceptable, also, has good mechanical strength; these properties have made from HDPE an attractively material for the production of plastic bags, toys, cases (elbow, knee), household items; also is widely used in the automotive and food packaging industry. The widespread use and high persistence in the environment of HDPE, have led to seek solutions to reduce the presence of this polymer in the environment. The biodegradable polyethylene (PE-BIO) has been proposed as a solution to the problem of environmental pollution by HDPE. The PE-BIO is a synthetic polymer including carbonyl groups in its molecule, so that, microorganisms are able to metabolize the chains of this materials; however for microorganisms perform their function is necessary for the PE-BIO exposed to weather with temperature and humidity controlled [4].

The degradation of the polymers is mainly due to photo-oxidation r [5, 6, 7] and thermo-oxidation reactions [8, 9]; causing breakage and cross-linking of the polymer chains, the formation of carbonyl (C=O) and vinyl ($CH=CH_2$) groups, as well as changes in the conformation of the polymer [10, 11, 12]. These effects have been characterized by other authors with techniques such as differential scanning calorimetry (DSC), electron microscopy (SEM), X-ray diffraction (XRD) and Infrared Spectroscopy (FTIR) [13, 14, 15, 16, 17]. In the present work, the effect of UV-B radiation on the photo-oxidation of HDPE and PE-BIO films were characterized using FTIR spectroscopy.

3. Materials and methods

3.1 Material
High density polyethylene (HDPE) and biodegradable polyethylene (PE-BIO) films of dimensions 40x50 mm with a thickness of 0.5 mm.

3.2 Exposure to UV-B radiation
The samples of HDPE and PE-BIO, were exposed to UV-B radiation for a period of time from 0 to 60 days, at a distance of 5 cm between the sample and the radiation source, using a UV-B BYTEK lamp model Ultraviolet Multiprom Eraser in a 320-280 nm range emission.

3.3 FTIR spectroscopy
The effects of UV-B radiation in the films of HDPE and PE-BIO, were characterized by infrared spectroscopy (FTIR), using a spectrophotometer Thermo Nicolet iS10, equipped with ATR, Germanium mirror and a Potassium bromide beam splitter.

4. Results

In figure 1, the infrared spectra of HDPE and PE-BIO unirradiated are shown, which have the same set of IR absorption bands; stretching vibration of the carbon-hydrogen bonds (CH) of the main chain occurs at 2772-3038 cm^{-1}; the regions of 1440-1490 cm^{-1} and 700-750 cm^{-1} correspond to the vibration of wagging and rocking methylene (CH_2), respectively [18]. The IR spectrum of PE-BIO, figure 1, has an absorption band at 1640 cm^{-1} which is due to stretching vibration of the carbonyl group (C=O), indicating that this material is oxidized from manufacturing.

Photo-oxidation of PE films exposed to UV-B radiation is developed according to the reaction mechanism *Norrish Type II* [19, 20], which is considered a homolytic process consisting of four stages: initiation, propagation, chain branching and termination [21]. The stage of initiation is mainly caused by hydroperoxides formed during the polymerization reactions of PE.

Hydroperoxides when decomposed by the effect of UV-B radiation, allow the formation of functional groups such as hydroxyl, anhydrides and carbonyl. As the dose of UV-B radiation increases in the films of HDPE and PE-BIO, the breaking bond between tertiary carbons and branch leading to the formation of vinyl groups [22, 23]; when increasing dose of UV-B radiation, the carbonyl and vinyl groups releasing from main chains causing crosslinking of the same [24].

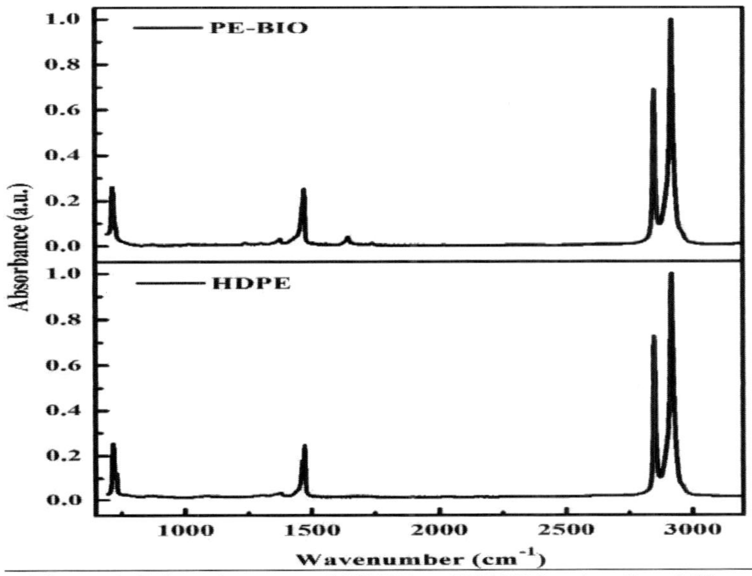

Figure 1. Infrared spectra of HDPE and PE-BIO unirradiated films

In figures 2a and 2b, the IR spectra of HDPE and PE-BIO films are shown, respectively, from 0 to 60 days of exposure to UV-B radiation. The infrared absorption band located in the range of 1600-1800 cm^{-1} corresponding to stretching vibration of the carbonyl functional group (C=O). During the first 30 days of exposure of the films of HDPE to UV-B radiation, it is observed the increase in band intensity of IR absorption of carbonyl, due to the photo-oxidation of plasticizers and stabilizers that HDPE contains [16]. From 30 to 60 days of exposure to UV-B radiation, the photo-oxidation of the chains of HDPE occurs and IR absorption band tends to wide, due to the simultaneous formation of groups such as esters, ketones, anhydrides and vinyl groups (see table 1).

Table 1. Infrared vibration of photo-oxidation products of HDPE and PE-BIO films exposed to UV-B radiation.

Photo-oxidation products	Wavenumber (cm^{-1})
Vinylidene	888
Terminal vinyl	909
Trans-vinylene	965
Ketones	1715
Esters	1730
Anhydrides	1770

PE-BIO films are in oxidized state from its manufacture and, during the first 5 days of exposure to UV-B radiation, the presence of oxidized groups decreases, since the carbonyl group is more susceptible to experiment cleavage by action of UV-B radiation [23]. However, after 5 days of exposure to UV-B radiation, the intensity of the IR absorption band is increased due to photo-oxidation of PE-BIO stabilizers and plasticizers.

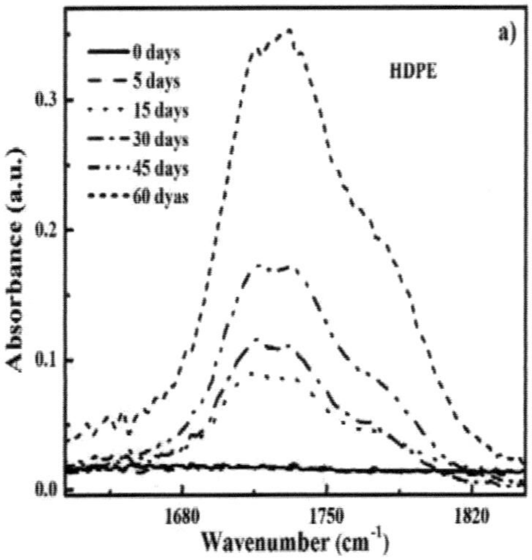

Figure 2a. IR absorption region of the carbonyl group of HDPE films exposed to UV-B radiation from 0 to 60 days.

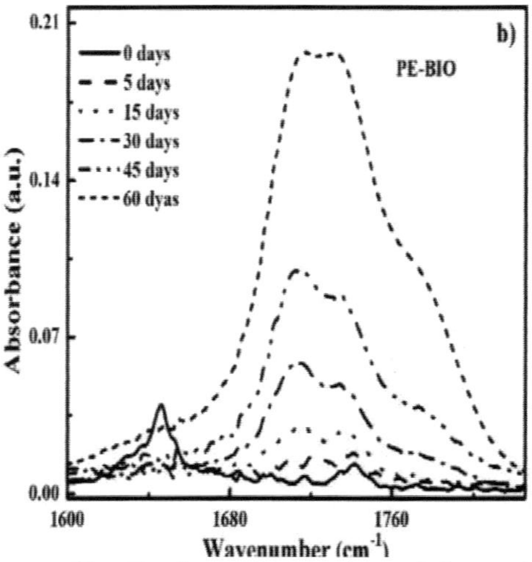

Figure 2b. IR absorption region of the carbonyl group of PE-BIO films exposed to UV-B radiation from 0 to 60 days.

Figures 3a and 3b shows the IR region from 850-1300 cm^{-1} which corresponds to different types of vinyl groups of HDPE and PE-BIO films exposed to UV-B radiation from 0 to 60 days. For both types of polymeric materials, the intensity of IR absorbance is increased as the exposure time of the material to UV-B radiation progresses, due to the formation of vinyl groups from the reaction of tertiary carbons by effects of UV-B radiation [21].

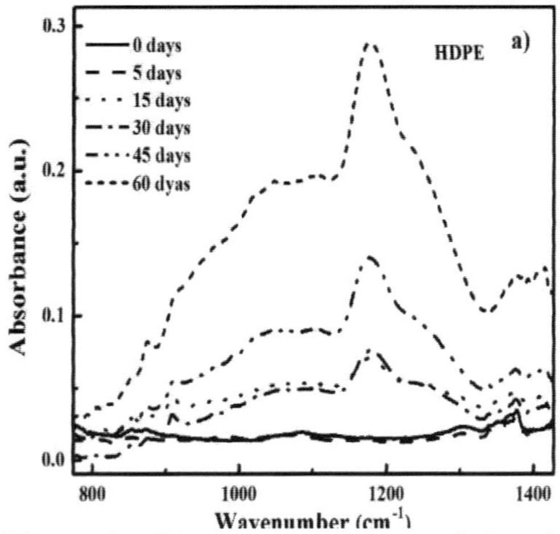

Figure 3a. IR absorption region of the vinyl group of HDPE films exposed to UV-B radiation from 0 to 60 days.

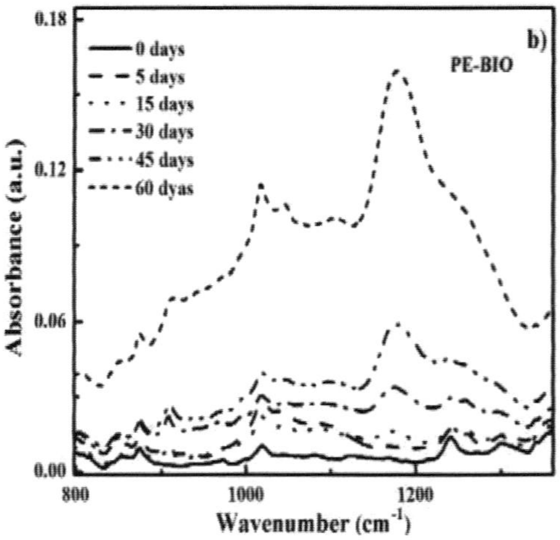

Figure 3b. IR absorption region of the vinyl group of PE-BIO films exposed to UV-B radiation from 0 to 60 days.

Journal of Physics: Conference Series **582** (2015) 012026 doi:10.1088/1742-6596/582/1/012026

5. Conclusions

In the presence of UV-B radiation and oxygen, chains of HDPE and PE-BIO experiment structural changes at molecular level, forming vinyl and carbonyl groups, which are considered the main products of photo-oxidation.

The broadening of IR absorption band of carbonyl and vinyl groups is due to the simultaneous formation of different products of photo-oxidation such as ketones, esters, anhydrides, vinylidenes, terminal vinyl and *trans*-vinylenes groups during exposure of the HDPE and PE-BIO films to UV-B radiation.

The formation of oxidation products occurs mainly in HDPE films, because the presence of the carbonyl group in the PE-BIO from manufacturing slows down the photo-oxidation reaction of this polymer; so that, it is found that the final disposition of PE-BIO must be satisfied with the conditions of disposal of solid waste, temperature and humidity, stipulated for this type of polymeric material.

References

[1] Satoto R, Wiwik S Subowo, Retno Yusiasih, Yoshimitsu Takane, Yasushi Watanabe and Tatsuko Hatakeyama 1997 *Polymer Degradation and Stability* **56** 275
[2] Lofgren, Jabarin Saleh A, and Elizabeth A 1994 *Journal Applied Polymer Science* **53**
[3] Hokestra H D 1997 *Stabilization and Degradation of Polymer* **247** 91
[4] Prasun K. Roy, Minna Hakkarainen and Indra K. Varma 2011 *Environmental Science and Technology* **45** 4217
[5] Tidjani Adams 2000 *Polymer Degradation and Stability* **68** 465.
[6] Ayako Torikai, Hirose Hasegawa 1998 *Polymer Degradation and Stability* ***61*** 361.
[7] Mingguang Liu, Richard Horrocks A and Michel E. Hall 1995 *Polymer Degradation and Stability* **49** 1
[8] Jeffery D Peterson, Sergey Vyazovkin, Charles A Wight 2001 *Macromolecular Chemistry and Physics* **202** 775
[9] Farideh Khabbaz, Ann Christine Albertsson, Sigbritt Karlsson 1999 *Polymer Degradation and Stability* **63** 127
[10] Hoekstra 1997 *Die Angewandte Makromolekulare Chemie* **247** 91
[11] Khoylou Farah 2005 *Iranian Polymer Journal* **14** 875
[12] Scott Gerald 2000 *Polymer Degradation and Stability* **68** 1
[13] Gallo R, Bramilla L, Castiglioni Chiara and Severini F 2006 *Journal of Macromolecular Science* **43** 535
[14] Gulmine J V 2003 *Polymer Degradation and Stability* **79** 385
[15] Klaeboe P. 1995 *Vibrational spectroscopy* **9** 3
[16] Gugumus F 1990 *Polymer Degrdation and Stability* **27** 19
[17] Liberata Guadagno 2001 *Polymer Degradation and Stability* **72** 175
[18] Bonilla José. 2006, Handbook of Plastic Analaysis Marcel Dekker cap 7
[19] Arnau R, Tidjani A 1995 *Polymer* 36 2841
[20] Klein, Jalma María 2013 *Progress in Ruber, Plastic and Recycling Technology* **29** 39.
[21] Gugumus F 1990 *Die Angewandte Makromolekulare Chemie* ***3051*** 27
[22] Telmo Ojeda, Ana Freitas, Kátia Birck, Emilene Dalmolin and Rodrigo Jacques 2011 *Polymer degradation and stability* **96** 703
[23] Wypych George *Handbook of Material Weathering* 2003 ChemTec Publishing
[24] Amin S H, Hamid and M.B 1995 *Journal of Applied Polymer Science* 55 1385
[25] Xingzhou Hu 1997 *Polymer Degradation and Stability* 55 13

Journal of Physics: Conference Series **582** (2015) 012027 doi:10.1088/1742-6596/582/1/012027

Photopyroelectric Techniques for thermo-optical characterization of gold nano-particles

B E Chávez-Sandoval,[1-2] J A Balderas-López,[1] G Padilla-Bernal,[1] M A Moreno-Rivera,[1] M O Franco-Hernández,[1] A Martínez-Jiménez[2] and F García-Franco[2]

[1]Unidad Profesional Interdisciplinaria de Biotecnología IPN,Av. Acueducto S/N, Barrio la Laguna Ticomán, México D.F. CP 07340.
[2]Universidad Autónoma Metropolitana-Azcapotzalco, Departamento de Ciencias Básicas.
Avenida San Pablo No. 180. Colonia Reynosa Tamaulipas, México D.F. CP 02200.

E-mail: abrahambalderas@hotmail.com

Abstract. Since the first methodology, proposed by Turkevich, to produce gold nanoparticles (AuNPs), improvements have been made as to allow better controllability in their size and shape. These two parameters play important role for application of gold nanoparticles since they determine their optical and thermal properties. Two photopyroelectric techniques for the measurement of the thermal diffusivity and the optical absorption coefficient for nano-particles are introduced. These thermo-physical properties were measured for the colloidal systems at different nano-particle´s sizes and, for optical properties, at three different wavelengths (405 nm, 488 nm and 532 nm). No significant difference, on thermal properties, was found in the range of nano-particles` sizes studied in this work; in opposition optical properties shown more sensitive to this parameter.

1. Introduction

Metallic nano-particles have shown potential applications especially in biomedicine (photothermal therapy and radiofrequency hyperthermia) due mainly to their optical properties, which can be conveniently changed by changing its size and morphology. At this respect gold nano-particles are very important since gold is a noble metal, not toxic for human been [1]. Simple and precise experimental techniques for optical and thermal characterization of nanofluids are then relevant since these two physical properties are important factors for such applications. In this work two photothermal techniques, for thermal diffusivity and optical absorption coefficient measurements for liquids, are used for thermal and optical characterization of gold nano-particles. Gold-nanofluids at different sizes, ranging between 5 to 60 nm in diameter, were used for this goal. The optical absorption coefficient at three different wavelengths (405 nm, 488 nm and 532 nm) was obtained and it was shown the strong dependence of this optical parameter on the nanoparticle´s size.

Journal of Physics: Conference Series **582** (2015) 012027 doi:10.1088/1742-6596/582/1/012027

2. Theoretical bases

The mathematical model of the photopyroelectric methodologies for thermal and optical characterization of liquids involves the heat diffusion problem through a three layer system [2,3]. The heat source, for each case, consists of modulated radiation at fix modulation frequency, f, which, for the case of thermal diffusivity measurements [2], is absorbed on the surface of a strong absorbing material (Fig. 1) and, for the case of optical absorption coefficient measurements [3], is absorbed through the liquid sample under study (Fig. 2) by following the Beer-Lambert´s model for light absorption, with optical absorption coefficient β. The heat generated on each case travels through the liquid sample under study to reach the pyroelectric sensor which generates, on response, a voltage which is measured by means of a lockin amplifier.

The voltage difference on the pyroelectric sensor can be written as a complex expression that depends on the thermal and optical properties of the involved media and on the modulation frequency and thickness of the liquid sample as well.

However, if the modulation frequency is keep constant and the sample´s thickness, L, is taken as the only variable it can be shown that, under the appropriate experimental conditions, this voltage difference, $V(L)$, for the case of thermal diffusivity measurements, can be written in the very simple form $V(L)=C\exp(-\sigma L)$, where $\sigma=(1+i)(\pi f/\alpha)^{1/2}$; meanwhile that for the case of optical absorption coefficient measurements it can be written as $V(L)=D\exp(-\beta L)$. C and D are complex expressions which are independent on the sample´s thickness. The photopyroelectric phase in this last case is constant and this fact can be used as an experimental criterion as to validate the optical parameter to be measured.

The experimental procedure to obtain thermal diffusivity involves plotting the pyroelectric´s signal phase, as a function of the sample´s thickness, and carried out a linear fitting procedure. This thermal property can be obtained from the fitting parameter $B=(\pi f/\alpha)^{1/2}$.

The experimental procedure to obtain the optical absorption coefficient, on the other hand, involves plotting the pyroelectric´s signal amplitude (in semi-log scale), as a function of the sample´s thickness, and carry out a fitting procedure. The optical absorption coefficient, in this case, is the slope of the best linear fit.

3. Experimental

3.1 Samples´ preparation

Colloidal gold nano-fluids, at different nano-particle´s sizes (Table I, first column), were obtained starting from a Stock solution of $HAuCl_4$ $3H_2O$, 25 mM in deionized water, at different pH´s. 500 µl of this Stock solution were poured in a 250 ml volumetric flask, with 50 ml of deionized water inside it.

The flask, covered with a watch glass, was put on an electric stove and heated with agitation to bring the mixture to boil, then 1.25 ml of sodium citrate solution ($Na_3C_6H_5O7{\cdot}2H_2O$), at 1%, was added very fast, and the mixture was keep heating with agitation up to the nano-particle´s synthesis was complete (about 15 minutes). Sodium citrate worked in this case as a reduction agent and stabilizer at the same time. The pH for each solution was controlled with HCl and NaOH. Nanoparticles´ sizes were characterized by means of atomic force microscopy.

3.2 Photopyroelectric technique for Thermal Diffusivity measurements

The experimental photopyroelectric (PPE) set-up for thermal diffusivity measurements, shown schematically in Fig. 1, consisted of a diode laser with collimation optics (Thorlabs, L785PO90, λ=785 nm), whose light-intensity was modulated by means of a diode current driver (Thorlabs, model LDC202C) controlled by the TTL (Transistor-Transistor-Logic) lockin's output (Stanford Research, model SR 830).

The diode laser with optics was put inside an aluminum cylindrical body and sealed, with epoxy resin, by a thin silicon slab on its bottom (0.02 cm thickness). This set was attached on a micrometer stage with 5 microns of minimum step. The pyroelectric sensor consisted of a PVDF film, 52 microns thickness, with metal electrodes (Ni–Al) on both sides, and 1.5 cm diameter.
The pyroelectric sensor was attached, with a thin layer of conductive epoxy, with the metalized surface of a thin silicon slab (about 200 microns thickness) as a protective layer. The metalized surface of the silicon slab connects the PVDF´s top surface to the metallic body which was the pyroelectric´s electrical ground. The PPE voltage signal generated in the sensor was pre-amplified (Stanford Research systems, model SR-550) and sent to the lock-in amplifier for further amplification and demodulation. The PPE signal was monitored as a function of the sample's thickness *L* (a sample's thickness scan) in steps of 50 microns and a modulation frequency of 1 Hz, setting the lockin's time constant at 3 s, 15 data points were taken for the analysis.

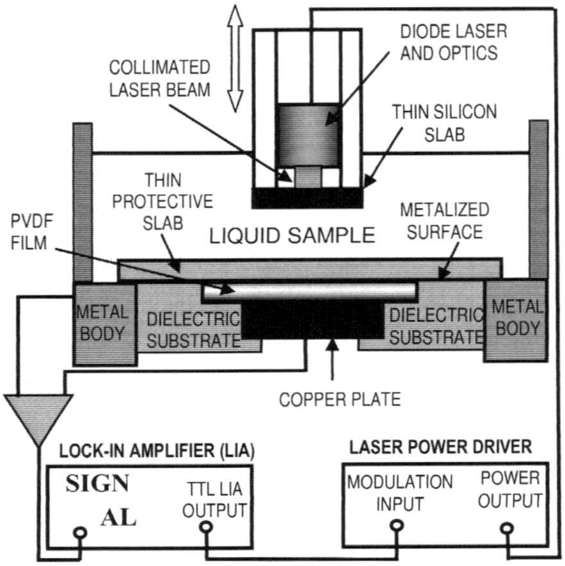

Fig. 1. Cross section of the experimental photopyroelectric set up for thermal diffusivity measurements.

Journal of Physics: Conference Series **582** (2015) 012027 doi:10.1088/1742-6596/582/1/012027

3.3 Photopyroelectric technique for Optical Absorption Coefficient measurements

The experimental photopyroelectric (PPE) set-up for Optical absorption coefficient measurements, shown schematically in Fig. 2, was essentially the one described in section 3.1.

The difference was on the thin silicon slab which was replaced by an optical window, to allow light goes through the liquid sample to be absorbed on its volume.

Three different monochromatic light sources, from lasers, were used this time (commercial laser pointers at 405 nm and 532 nm and an argon ion laser (Spectra Physics), at 488 nm). 30 experimental data points, in steps of 50 microns, were taken for the analysis in this case.

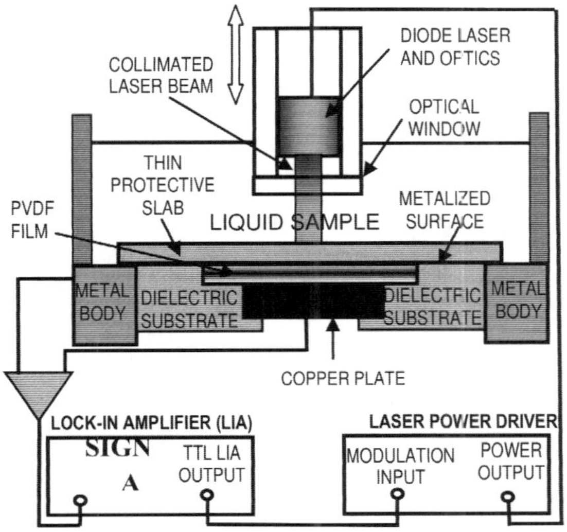

Fig. 2. Cross section of the experimental photopyroelectric set up for optical absorption coefficient measurements.

4. Results and Conclusions

Figure 3 shows typical photopyroelectric signal phases, as function of the sample´s thickness, for two of the gold nanofluids studied in this work (5 nm (Fig. 3a) and 7 nm (Fig. 3b)). The continuous line on each figure is the best linear fit as to obtain the sample´s thermal diffusivity, as it was described before. Table I (column 2) summarize the thermal diffusivity values obtained for all samples.

Figure 4 shows the corresponding photopyroelectric (PPE) signal (a) PPE phase and b) PPE amplitude), as function of the sample´s thickness, for gold nano-particles at 20 nm of dimeter (pH 7). It is evident in Fig. 4a) the constant phase behavior, which validates the corresponding linear fit in Fig. 4b) for the PPE amplitude (in semi-log scale) for obtaining the sample´s optical absorption coefficient at 405 nm, in this case, as it was described in the theoretical section. Similar results are shown on Fig. 5a) for the PPE phase and Fig. 5b) for the PPE amplitude, for the same sample but at a wave-length of 488 nm. Table I, columns 3 to 5, summarizes the corresponding optical absorption coefficients at the three wave-lengths used in this work (405 nm, 488 nm and 532 nm, respectively).

Table 1. Thermo-optical properties for the gold nano-particles studied in this work. α stands for the thermal diffusivity and β for the optical absorption coefficient.

Gold nanoparticle´s diameter (nm)	$\alpha(10^{-2})$ (cm^2/s)	β (405 nm)	β (488 nm)	β (532 nm)
5 (pH 3)	0.1470±0.0008	0.79±0.03	1.56±0.02	1.33±0.03
10 (pH 5)	0.1460±0.0004	1.55±0.04	1.46±0.03	1.13±0.02
20 (pH 7)	0.1466±0.0006	1.75±0.03	1.96±0.02	1.74±0.03
30 (pH 9)	0.1464±0.0007	1.47±0.02	1.55±0.02	1.16±0.04
60 (pH 11)	0.1458±0.0009	1.42±0.05	2.64±0.03	2.07±0.02

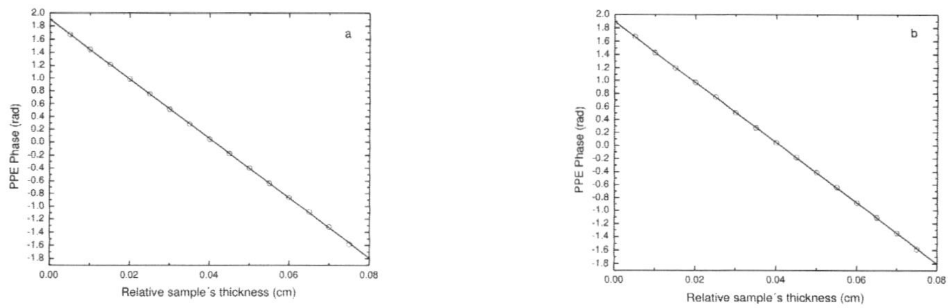

Fig. 3. Photopyroelectric (PPE) phase, as a function of the sample´s thickness, for colloidal solution of gold nano-particles. a. 5 nm (pH 3) and b. 20 nm (pH 7). The continuous lines on each figure is the best linear fit as to obtain the sample´s thermal diffusivity.

By looking at the thermal diffusivity values obtained for all samples (Table I, column 2) it is evident that there is no significant difference between these samples in terms of this physical parameter, at least in the range of sizes used in this work. Different conclusions can be reached by looking at the optical absorption coefficients (Table I, columns 3 to 5) since the optical absorption coefficient are clearly different for all samples. It is interesting to note that the greater optical absorption coefficient was measured for the gold nano-fluid at 60 nm and for 405 nm wave-length.

Journal of Physics: Conference Series **582** (2015) 012027 doi:10.1088/1742-6596/582/1/012027

This fact is interesting since the wave-length for strongest absorption for these systems is usually reported around 530 nm, using conventional spectroscopy [4].

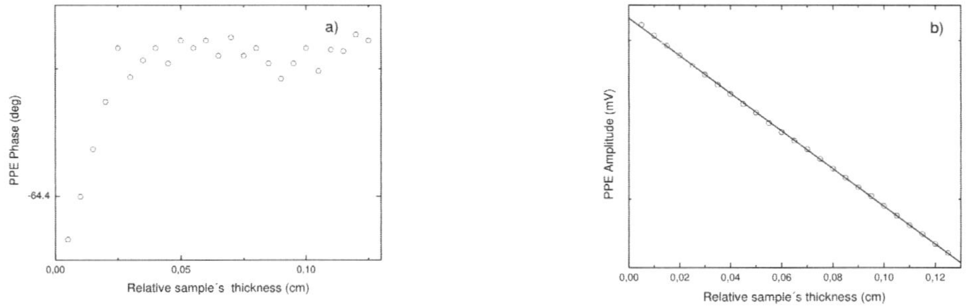

Fig. 4. Photopyroelectric (PPE) signal (a) phase and b) amplitude), as a function of the sample´s thickness, for colloidal solution of gold nano-particles at pH 7. The continuous line on b) is the best linear fit as to obtain the optical absorption coefficient at 405 nm.

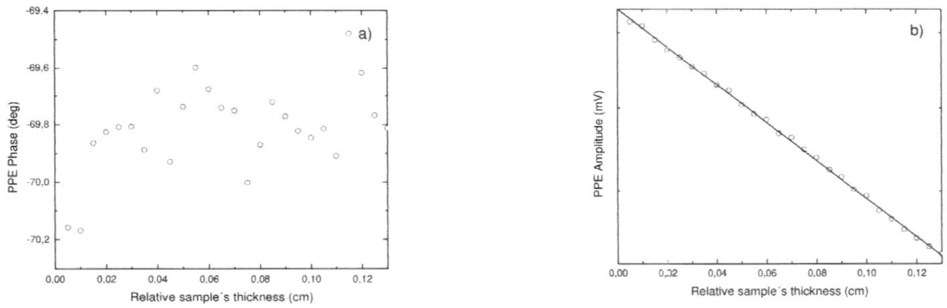

Fig. 5. Photopyroelectric (PPE) signal (a) phase and b) amplitude), as a function of the sample´s thickness, for colloidal solution of gold nano-particles at pH 7. The continuous line on b) is the best linear fit as to obtain the optical absorption coefficient at 488 nm.

References

[1] Turkevich J., Stevenson P.C., Hillier J. 1951 A study of the nucleation and growth processes in the synthesis of colloidal gold *Discuss Faraday Soc* **11** 55–75

[2] Balderas-López J. A, Mandelis A. and García J. 2000 Thermal-wave resonator cavity design and measurements of thermal diffusivity measurements of liquids *Rev of Sci Instruments* **71** 2933-2937

[3] Balderas-López J. A. 2012 Generalized 1-D Photopyroeledctric technique for optical and thermal characterization of liquids *Meas. Sci Technology* **23** 10

[4] Haiss W., Thanh N. T. K., Aveyard J.,Ferning D. G. 2007 Determination of size and control of gold nanoparticles fron UV-Vis spectra *Analytical Chemistry* **79** 4215-4221

VII International Congress of Engineering Physics IOP Publishing
Journal of Physics: Conference Series **582** (2015) 012028 doi:10.1088/1742-6596/582/1/012028

Assessment of high performance concrete containing fly ash and calcium nitrite based corrosion inhibitor as a mean to prevent the corrosion of reinforcing steel

P Montes-García[1], V Jiménez-Quero[1], H López-Calvo[2]

[1]Instituto Politécnico Nacional, Grupo de Materiales y Construcción del CIIDIR Oaxaca. Hornos # 1003 Sta. Cruz Xoxoxcotlán Oaxaca, Mexico.
[2]Universidad Autónoma Benito Juárez de Oaxaca. Av. Universidad S/N, ExHacienda Cinco Señores Oaxaca, México

vg.jimenezquero@gmail.com

Abstract. This research analyses the effectiveness of the water-to-cement ratio (w/c), fly ash and a calcium nitrite based corrosion inhibitor to prevent the corrosion of reinforcing steel embedded in high performance concrete. The interactive effect between the inhibitor and fly ash was evaluated because the occurrence of a negative effect when both ingredients are added together in a concrete mixture has been reported. All the concrete mixtures studied in this investigation had 8.2% of silica fume. Twenty seven prismatic concrete specimens were fabricated with dimensions of 55 x 230 x 300 mm each containing two steel rods embedded for the purpose of corrosion monitoring. The specimens were exposed to a simulated marine environment with two daily cycles of wetting and drying for one year. To evaluate the deterioration of the specimens corrosion potentials and linear polarization resistance tests were carried out. The results indicate that the use of a low w/c, the addition of fly ash and the addition of the corrosion inhibitor contributed to the reduction of the corrosion of steel in the concrete specimens. The results further suggest that the combination of fly ash and corrosion inhibitor does not promote the deterioration of the concrete matrix.

1. Introduction

High performance concrete (HPC) can be made with cement alone or any combination of cement and mineral components, such as, ground granulated glass blast-furnace slag (GGBS), fly ash (FA), silica fume (SF), metakaolin, and fillers, such as limestone powder. Ternary systems are increasingly used to take advantage of the synergy of some mineral components to improve concrete properties in the fresh and hardened states, and to make high performance concrete more economical and ecological [1-2]. Corrosion of reinforcing steel in ordinary and High performance concrete has become a major problem world-wide, especially in buildings, bridges, parking decks, tunnels and other buildings exposed to a seawater or de-icing salts. As a result of this deterioration process, the repair costs nowadays constitute a major part of the current spending on infrastructure [3].

The results reported in the literature on the use of the combination of Calcium Nitrite based corrosion inhibitor (CNI) and fly ash (FA) in high performance concrete, when used to prevent corrosion, differ considerably. According to Li et al., [4] the addition of CNI into the mix resulted in the deterioration of the resistance to chloride diffusion which had been previously enhanced by the incorporation of fly ash. Through the mercury intrusion porosimetry test, the influence of CNI on chloride diffusion was studied further. The results showed that the addition

Journal of Physics: Conference Series **582** (2015) 012028 doi:10.1088/1742-6596/582/1/012028

of CNI leads to the increase of micropore diameter. In another research, the test results demonstrate that the incorporation of CNI accelerates the formation of calcium hydroxide and ettringite crystals, and weakens the pore refinement effect caused by the secondary hydration reaction of fly ash and microsilica [5]. Montes [6] mentioned that the inhibitor and fly ash alone do not provide sufficient protection against corrosion, but apparently the combination of the two provides a significant benefit in high performance concrete, increasing the resistance to chloride diffusion.

Because of the contrasting results more research is needed, especially on the interactive effects of fly ash and calcium nitrite based corrosion inhibitor when they are used as corrosion prevention methods.

In view of this, this paper presents the results of research aimed to investigate the effects of fly ash and CNI on the corrosion potentials and corrosion densities of reinforcing bars embedded in three different water-to-cement ratio concretes.

1. Experimental program

1.1. Materials and Methods

Maximum size of the coarse aggregate was 12.5 mm and its relative density was 2.69. Natural river sand was used as fine aggregate, which has a fineness modulus of 2.65 and a relative density of 2.62. . CSA Type 10 Portland cement, which already has 8.2% silica fume as cement replacement, was used for all the concrete mixtures. The fly ash used belongs to Type F, according to the ASTM classification. The commercial Calcium Nitrite based corrosion inhibitor used in this study contains a minimum of 30% of Nitrite ions. The solution (simulated salt water) was prepared as recommended by the ASTM D 1141-80 Standards.

For the development of this project, a 33 factorial design was performed. Twenty seven concretes considering three water-to-cement ratios (0.29, 0.37 and 0.45), three fly ash replacement levels (0%, 20% and 40%) and three dosages of Calcium Nitrite based corrosion inhibitor (0, 12.5 and 25 l/m3) were prepared. To evaluate the effect of artificial seawater on the corrosion process of the reinforcing bars, twenty seven small-scale slabs were cast. Each slab had embedded two black steel bars.. A marine environment simulation setup was utilized to test the specimens made from each mixture. The setup for accelerated corrosion testing consists of two chambers where it is possible to simulate wetting and drying cycles in concrete specimens in marine exposure conditions. Two complete cycles are performed within a 24-h period as compared with two tidal cycle in a natural marine environment. Also, the effect of temperature ranging from 26 ± 2°C in the wet cycle to 55 ± 2°C in the dry cycle is considered.

When there is active corrosion, current flow through the concrete between anodic and cathodic sites, the current is accompanied by an electric potential field surrounding the corroding bar. The equipotential lines intersect the surface of the concrete and the potential at any point can be measured using the half-cell potential method [7]. The polarization resistance can be measured by direct current measurements, impedance spectra, and transient techniques [8].

In this research, the corrosion current density (Icorr) was estimated from results of Linear Polarization Resistance (LPR) using a CMS 105 equipment (Gamry instruments Inc.) was used. The parameters to carry out this test were: scan rate of -20 to 20 mV. For the corrosion potentials test (Ecorr) [9]a calomel(mercuric chloride) reference electrode and a high impedance voltmeter were used. Measurements for both corrosion potentials and corrosion current density were taken every other month during one year.

Figure 1 shows results of the corrosion potentials. For their analysis and interpretation the threshold values proposed by the ASTM 876 standard [10] can be used. The standard divides the corrosion condition in low risk, intermediate risk, high risk and severe corrosion.

In this research, the measurements indicate that the water-to-cement ratio has a significant effect on the probability of corrosion. Results also show that the lower the water-to-cement ratio the lower the probability of corrosion. However, the 0.45 water-to-cement ratio concrete was not able to protect the steel reinforcement because at the end of the experiment had severe corrosion. Results are consistent with findings related to the use of a w/c lower than 0.4 when durable reinforced concrete is required [11]. The addition of 20% fly ash decreased the probability of corrosion and this effect was more evident for 0.29 w/c concrete. For higher w/c ratios seems that 20% fly ash is not good enough to prevent severe corrosion. Nevertheless, 40%fly ash was able to prevent severe corrosion, especially for 0.29 and 0.37 w/c concretes. Similar effect than that observed for fly ash can be observed when 12.5 L/m^3 of CNI was added to the concrete, this is, it was more effective in lower w/c concretes. In this case, even the addition of 25 L/m^3 of CNI was not enough to prevent the corrosion of 0.45 w/c concrete.The combination 1, consisting of 20%fly ash and 12.5 L/m^3 of CNI, was effective to decrease the corrosion potential of 0.29 and 0.37 w/c concretes to a low risk; however, the 0.45 w/c concrete showed severe corrosion at the end of the experiment.

The combination 2, consisting of 40%fly ash and 12.5 L/m^3 of CNI, did show only a slight difference reducing the corrosion risk of the steel reinforcement when compared to the combination of 20%fly ash and 12.5 L/m^3 of CNI. Nevertheless, the corrosion risk of the 0.45 w/c concrete containing this combination was less than for combination 1. The combination 3, consisting of 20%fly ash and 25 L/m^3 of CNI, had similar effect than combination 1 showing a reduction of the corrosion risk of the steel bars for 0.29 and 0.37 w/c concretes and not significant reduction for the 0.45 w/c concrete. Finally, the combination 4, consisting of 40%fly ash and 25 L/m^3 of CNI, was the most effective of all maintaining a intermediate corrosion risk for all the w/c ratio concretes studied.

Figure 2 shows the corrosion current densities recorded for all the specimens tested. In this case, the criterion proposed by Broomfield [12] was used for the classification of the corrosion condition. In such a work, similarly to the corrosion potentials technique, Broomfield proposes that the corrosion condition can be passive, low, intermediate and severe. The tendencies of the corrosion current densities agree in a great deal with those from the potentials. This corroborate the positive observed effects of the different combinations of fly ash and Calcium Nitrite Inhibitor on the reduction of corrosion of the steel reinforcement embedded in high performance concrete. of inhibitor and the addition of 40% fly ash values of 50% probability of corrosion were presented. To the water/cement ratio of 0.45 is observed that, at higher amounts of fly ash and corrosion inhibitor, values lower risk probability corrosion are obtained. Severe corrosion values (greater than 90% risk) was reduced to values of 10% risk of corrosion using a combination of 40% ash and 25 L/m^3. Is also observed in the graph h) the water/cement ratio of 0.29 and 0.37, in combination with 20% fly ash and 25 L/m^3 inhibitor, values lower risk of corrosion were obtained. Based on the criteria established by Broomfield [10] and the analysis of the graphs of Figure 2 in which it is observed that the combination of fly ash and corrosion inhibitor decreases the corrosion rate for the water/cement ratios studied, and that higher amounts of fly ash and corrosion inhibitor generate lower rates corrosion of the reinforcing steel, it is concluded that the combination of fly ash and corrosion inhibitor decreases the reinforced concrete corrosion when subjected to a chloride contaminated environment.

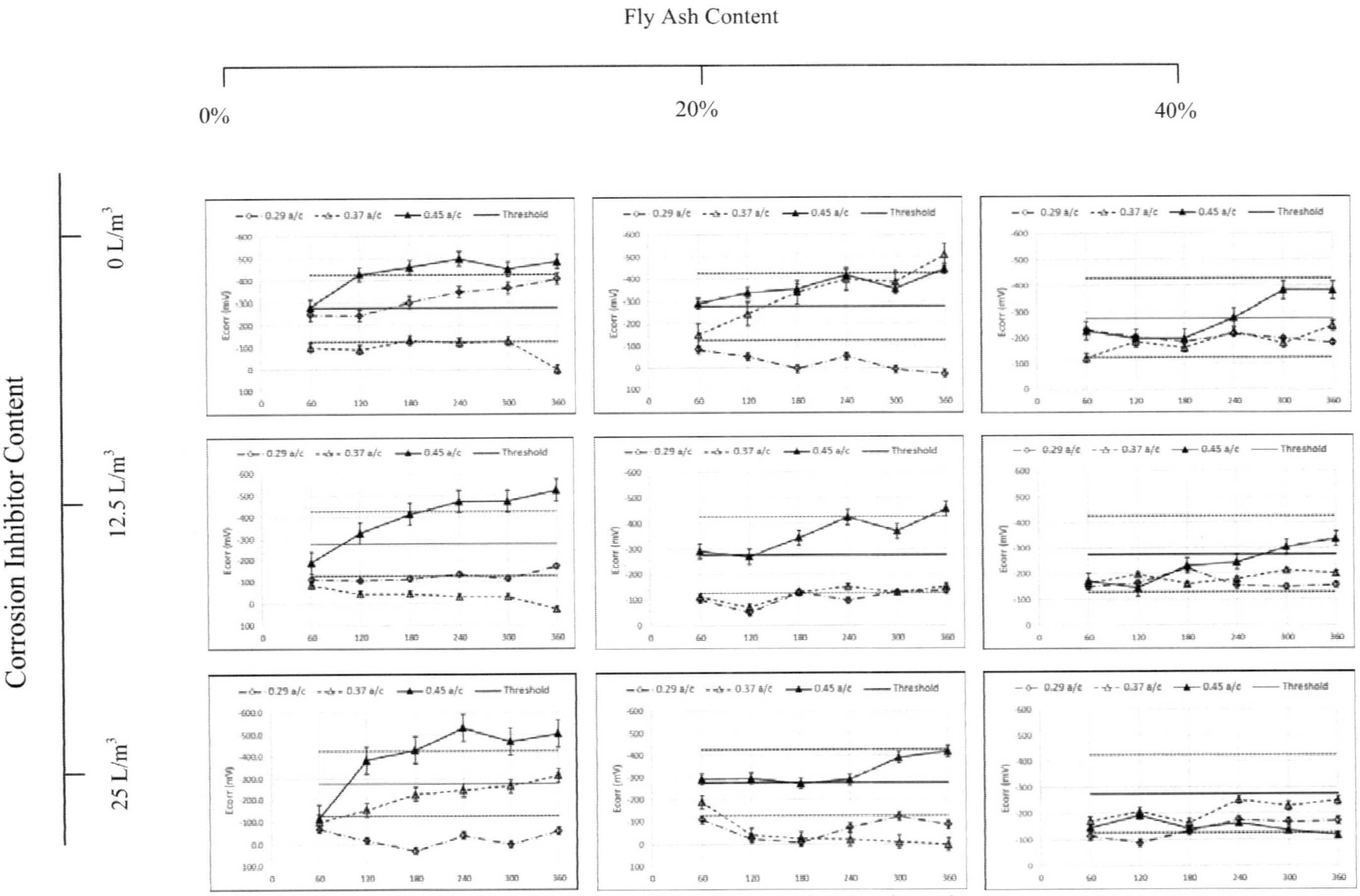

Figure 1 -. Corrosion potentials vs Reference electrode

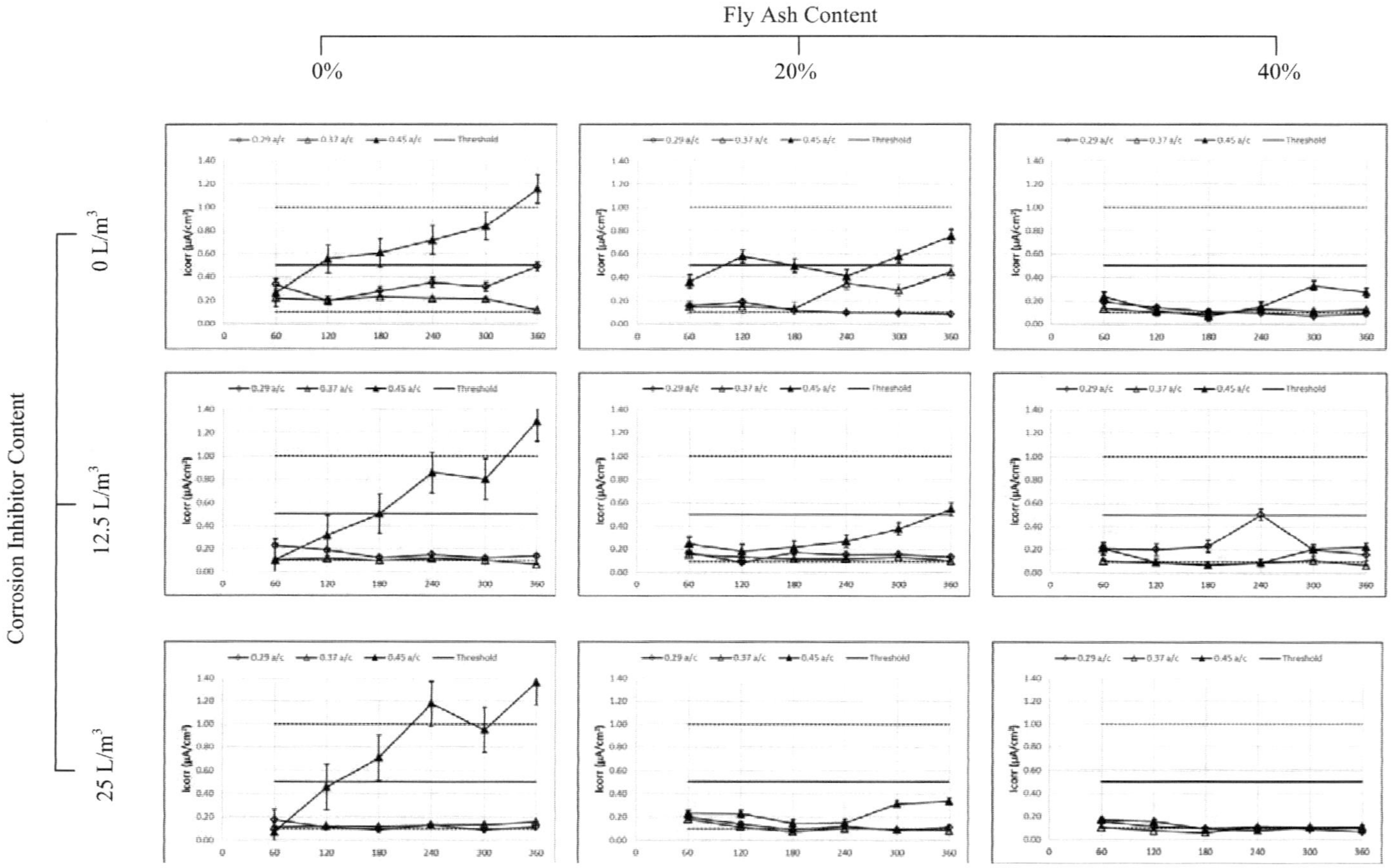

Figure 2 -. Corrosion density

3. Conclusions

Based on the analysis of the results the following conclusions can be drawn:

1.-It was demonstrated that the incorporation of fly ash or CNI into concrete alone is useful to decrease the level of corrosion; however, this is a temporary effect as they do not provide full protection to the steel reinforcement because only a delay the corrosion process is experienced. Also, it was corroborated that reinforced concrete exposed to a chloride contaminated environment should be prepared with a w/c lower than 0.4.

2.-The addition of 20%fly ash and 12.5 L/m^3 of CNI is effective decreasing the corrosion of steel reinforcing embedded in low w/c concretes.

3.-The addition of 40%fly ash and 12.5 L/m^3 of CNI is also effective decreasing the corrosion of steel reinforcing embedded in low w/c concretes; however, the increase from 20 to 40%fly ash shows only a marginal effect decreasing corrosion.

4.-The addition of 20%fly ash and 25 L/m^3 of CNI is effective decreasing the corrosion of steel reinforcing embedded in low w/c concretes.

5.-The addition of 40%fly ash and 25 L/m^3 of CNI is effective decreasing corrosion of the concretes evaluated in this research maintaining an intermediate level of corrosion during the testing period.

Acknowledgements

The authors acknowledge the financial support of the National Council for Science and Technology (CONACYT), the Instituto Politécnico Nacional (IPN), the COFAA-IPN, and the CIIDIR Oaxaca.

References

[1] Aïtcin, P. C 1998 Ed E & FN SPON *High–Performance Concrete*

[2] Aïtcin P.C. 2003 The durability characteristics of high performance concrete: a review. Cement & Concrete Composites **25** 409–420

[3] Raupach M, Schießl P 2001 Macrocell sensor systems for monitoring of the corrosion risk of the reinforcement in concrete structures *NDT&E International* **34** 435–42

[4] Li Z., Ma B., Peng J. and Qi M. 2000 The Microstructure and Sulfate Resistance Mechanism of High-Performance Concrete Containing CNI *Cement and Concrete Composites* **22** 369-377

[5] Li, Z., Peng J. and Ma, B. 1999 Investigation of Chloride Diffusion for High-Performance Concrete Containing Fly Ash, Microsilica and Chemical Admixtures *ACI Materials Journal* **96** 391-396

[6] Montes P., Bremner T.W. and Lister D. 2004 Influence of calcium nitrite inhibitor and crack width on corrosion of steel in high performance concrete subjected to a simulated marine environment *Cement and Concrete Composites* **26** 243-253

[7] Wojtas H. 2004 Determination of corrosion rate of reinforcement with a modulated guard ring electrode; analysis of errors due to lateral current distribution *Corrosion Science* **46** 1621–32

[8] Millard S.G., Law D., Bungey J. H. and Cairns J. 2001 Environmental influences on linear polarization corrosion rate measurement in reinforced concrete *NDT & E International* **34** 409-417

[10] ASTM C876 – 09 .Standard Test Method for Corrosion Potentials of Uncoated Reinforcing Steel in Concrete

[11] ACI. Corrosion Committee

[12] Broomfield, J. P. 1997 Corrosion of Steel in Concrete, Understanding, Investigation and Repair *E & FN SPON*

Journal of Physics: Conference Series **582** (2015) 012029 doi:10.1088/1742-6596/582/1/012029

Thermocouples calibration and analysis of the influence of the length of the sensor coating

M Noriega, R Ramírez, R López, M Vaca, J Morales, H Terres, A Lizardi, S Chávez
Universidad Autónoma Metropolitana-Azcapotzalco. Distrito Federal, México.

E-mail: rlc@correo.azc.uam.mx

Abstract. This paper presents the design and construction of a lab prototype, with a much lower cost compared to the ones commercially sold, enabling the manufacture of thermocouples which are then calibrated to verify their functionality and acceptance. We also analyze the influence of the external insulation over the wires, to determine whether it influences temperature measurement. The tested lengths ranged from 0.00 m up to 0.030 m. The thermocouple was compared against the behavior of a thermocouple of the same type that was purchased with a commercial supplier. The obtained measurement showed less than 1 °C difference in some points. This makes the built thermocouple reliable, since the standard allows a difference of up to 2.2 °C.

1. Introduction

Temperature is an important factor that should be taken into account in the majority of energy exchange processes, thermometers are instruments specifically designed to measure temperature, there is a wide variety of thermometers today and their operating principles are equally extensive.

The thermometers are the instruments that are used for temperature measurement, a type of them, contact thermometers, should be placed in physical contact in the middle of the medium from which you want to know the temperature, for example: in production processes or exchange of energy, the medium can consist of an electric oven, in exhaust gases of a combustion engine, water at the entrance of a cooling tower, bioreactors, etc. These contact thermometers base its operation in thermal equilibrium that is established by direct contact of sensor and the medium of interest.

Thermocouples are a type of contact temperature gauges; they are widely used in the industrial branch due to their capacity of measuring a wide range of temperatures, their reliability, easy operation and the possibility of registration of the measurement automatically. A thermocouple is composed of two conducting wires made of different material; these cables have a point in common. This union works as measuring point (sensing joint), which comes in contact with the medium in which you want to determine the temperature, at the other end, wires are connected to an electronic circuit that delivers a reference voltage (the voltage is usually associated with a temperature of 0 °C). When thermocouple wires are subjected to a difference in temperature between the measuring joint and the end connected to the reference circuit, there is a potential difference based on the change of temperatures (Seebeck effect). One of the most widely used thermocouples is the K-type, because of its low cost compared to other types of thermocouples.

The commercial cost of these devices is high, although the material to build them is usually inexpensive; the cost of the commercial product rises considerably, up to eight times. The punctual

measurement of temperature has different key point applications in laboratory work, especially when it comes to evaluating the thermodynamic behavior of some systems as for example in solar energy, heat transfer, thermal machines, etc. There are some prototypes which require up to 50 measurements in different sites at the same time, which makes the cost of this kind of experience rather high, considering the use of one or more thermocouples for each point.
In order to reduce costs, to the extent possible, in the area of Thermo-fluids, at UAM-A, it was decided to design and build an apparatus that would manufacture the thermocouples used in the instrumentation of the prototypes. We subsequently performed the calibration of the produced thermocouples. A question that arises when these instruments are evaluated is if the distance of coating that is removed to the thermocouple measuring wires influence on the precision of the measurement or not. The objectives of this work were to present the design and construction of the prototype for the thermocouples manufacturing and to investigate whether the distance that is removed from the wire coating had a significant impact on the temperature measurements.

2. Description of the manufacturing prototype and calibration of thermocouples

For the manufacture of thermocouples, type K (Chromel-Alumel) cable for thermocouple was used, the joint between the two metals that is required for the measurement of temperature was made producing a short circuit; this merged both metals forming an amalgam, and therefore the principle of operation of the prototype is to produce this phenomenon. The components of the device are: (A) a mechanical arm, (H) a ceramic rod, (D) a tip to join the thermocouple, (B, C) thermocouple attachment lugs, and (E, F) the power switch box.
One end of the lever arm is firmly fixed to a base (D) at the other end is placed a piece of charcoal which helps to close the electrical circuit with the tip of the thermocouple (C). This material was selected because it was the only one that would not leave residues and also it would not stick to the fused union of the temperature sensor; the ceramic bar (H) is fixed and its function is to serve as support and insulation when the short circuit between the carbon and the tip of the thermocouple occurs; the tip of the thermocouple to be joined is the type K-thermocouple cable. The power switch (F) is located in the switch box; it is a 10 A thermomagnetic switch, which must be activated before making any temperature measurement (F). Finally, there is a plug for connection to the Electric power supply. The diagram of the device is presented in Figure 1 and the built prototype, in Figure 2.

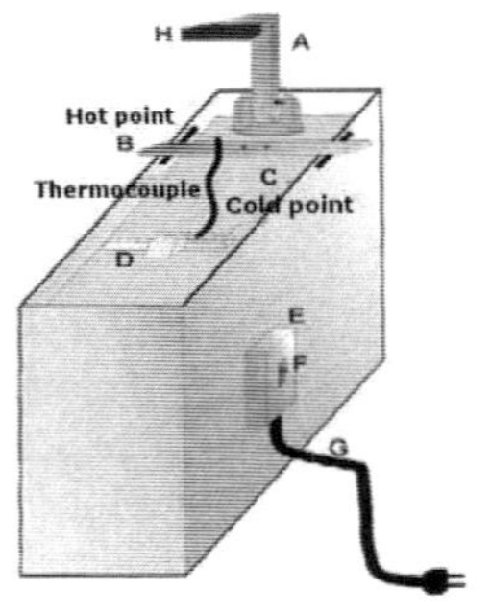

Fig.1. Prototype diagram.

Fig. 2. Image of the finished prototype.

LabView software in which the instrumentation (ADAM modules) is integrated, was used in the registration of measurements. The proper programming for reading in real time and data storage for their subsequent management was performed.

A virtual instrument (program) which allows the measurement, registration and presenting the graphs for four temperatures on screen at the desired time, was created. The main window of this LabView program is shown in figure 3.

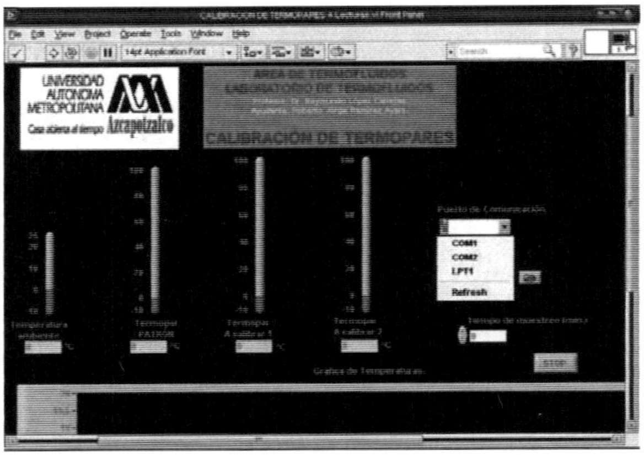

Fig. 3. Main window of the program.

3. Manufacture of thermocouple

The thermocouple manufacturing was carried out with a section of 0.40 m type K-cable, a length of 0.02 m of the insulating material at one end was withdrawn, this side should be placed in the hold-down shoe of the prototype for the manufacture of thermocouples; a length of 0.005 m of the insulation material is removed from the end of wires, then both wires are united with a rat-tail type lashing; this would be the temperature sensor. This end was placed in the ceramics holding of the positive pole of the apparatus and then the soldering thermocouple wires by the action of short-circuit proceeded. The thermocouple was withdrawn from the device making sure that the molten tip was perfect, i.e. that a small ball with both wires, free of any kind of fracture, had been formed. For details on the design and use of the available equipment, [2] can be consulted. For the first test the length of the insulation removed at the thermocouple was considered null, i.e., the thermocouple had two wires covered completely, then lengths of 0.003 m, 0.005 m, 0. 010 m, 0.015 m, 0.020 m 0.025 m and 0.030 m were withdrawn, and each test was repeated four times to verify that there was no significant variation in the measurements, or in its case, to obtain an average of them. A commercial thermocouple of the same type was employed as reference, and served as a pattern for measurements.

4. Experimental procedure.

The experiment was conducted by placing 750 mL of water at 0 °C, in a beaker of 1 liter of volume, where two thermocouples were inserted into the center of the bowl and exactly at the same depth, to ensure that both were in the same isothermal region and to avoid, to the extent possible, variations in temperature due to the flow currents. The vessel was warmed on an electrical thermal plaque until the water reached boiling point. At the same time both ends of thermocouples were connected to the ADAM module and the LabView program was started on the computer and the registration of measurements, performed every second, begins. The employed array is shown in Figure 3.

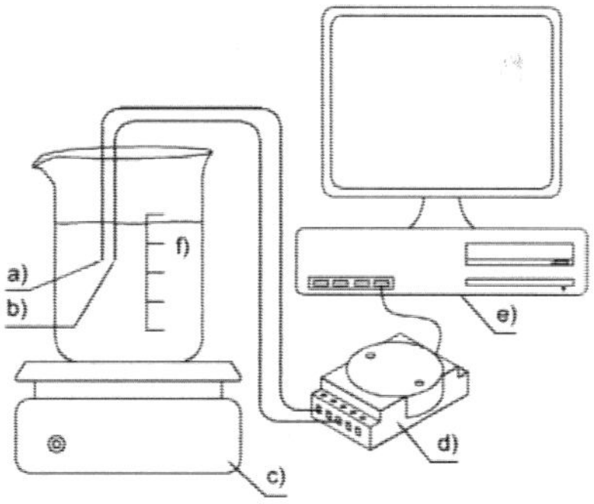

Fig. 3. Measurement arrangement: a) pattern thermocouple, b) thermocouple, c) electric heater, d) ADAM unit, e) computer, and f) water.

5. Results

All the tests that were conducted began at 0 °C and ended at approximately at 92 °C, in each experiment the average of the four tests was obtained, most of the results were almost equal, and the duration of each time was approximately 26 minutes. The graph of temperature against time obtained when the two wires of the thermocouple were completely covered with plastic is presented in Figure 4. There are some differences of maximum 1 °C in some spots. In all tests no greater value difference was observed.

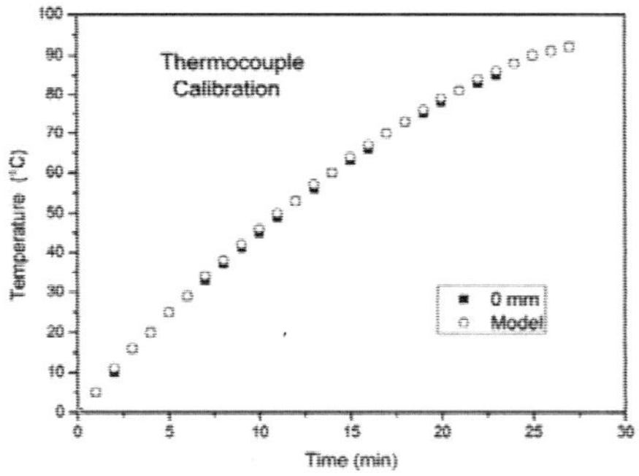

Fig. 4. Graph of temperature against time, with both wires of the thermocouple fully covered with insulating material.

In the case of the thermocouple for which 0.030 m of the insulating cover was withdrawn, the variation of temperature for some points did not exceed 1 °C, as shown in the graph of Figure 5. According to the ASTM E230/E230M-12 standard [1], in the temperature range of 0 °C to 760 °C a difference of up to 2.2 °C is allowed, therefore the built thermocouple was considered acceptable for measurements in the specified temperature range. Subsequently, the described procedure was repeated several times to this same thermocouple, i.e., a length of 0.003 m of plastic was removed and four tests were performed, until it a final length of 0.030 m of cover was withdrawn.

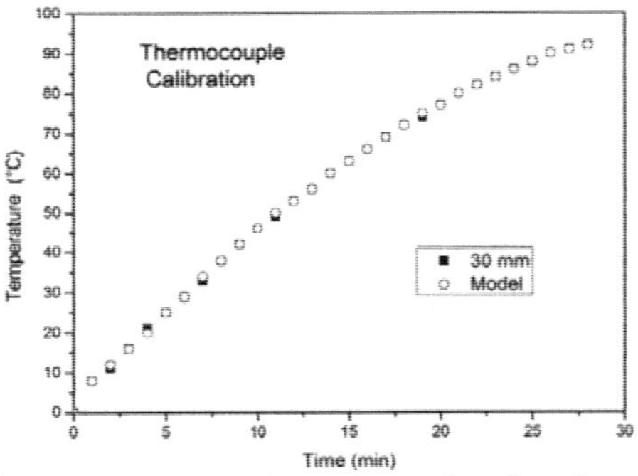

Fig. 5. Graph of temperature against time, for the thermocouple without 0.030 m insulating material.

Figure 6, shows the graph of the temperature of the thermocouple to be calibrated against the temperature of the reference thermocouple, in order to find the percentage of error between two measurements and the value of R^2 of each of the performed experiences. In the case of both wires completely covered the R^2 = 0.9996, the percentage of average error found was 1.2%, which makes it acceptable.

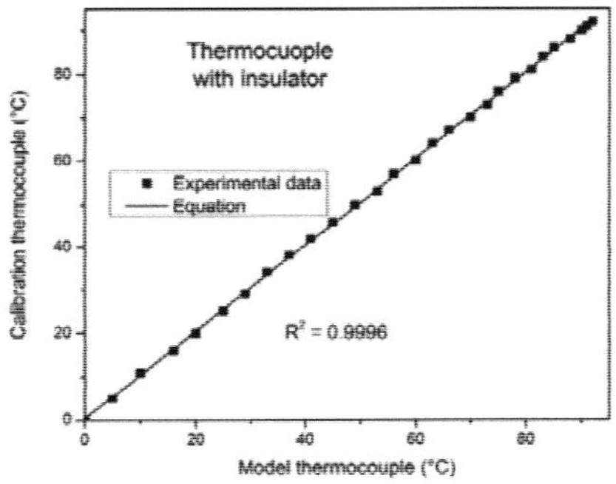

Fig 6. Fit graph for the insulated thermocouple, the value of R^2 is very close to the unit.

The values of the percentage of error and the value of R^2 for every experienced thermocouple are presented in Table 1. The maximum value of the percentage of retrieved error was observed when we withdrew a length of 0.030 m of the insulating material and the value of R^2 in all cases was very close to the unit.
The conclusion from these results is that the insulating material of a thermocouple can be removed up to 0.025 m length and readings obtained will be within the recommendation mentioned in the corresponding ASTM standard.

Journal of Physics: Conference Series **582** (2015) 012029 doi:10.1088/1742-6596/582/1/012029

6. Conclusions.

An economical lab prototype that enables the production of thermocouples which are subsequently calibrated for its functionality and acceptance has been built. The influence of the insulation covering of the wires in temperature measurement was analyzed. The experienced lengths of the uncovered portion of the wire ranged from 0.00 m up to 0.030 m, and were compared against the behavior of a thermocouple of the same type that was purchased with a commercial distributor. The obtained measurements presented a variation on some points of less than 1 °C, which confirms the built thermocouple as reliable, since the standard allows a difference of up to 2.2 °C.

Table 1. Error percentage and value of $\boldsymbol{R^2}$ p for each evaluated condition of the thermocouple.

Non-insulated length [m]	% Error	R^2
0.000	1.10	0.99967
0.003	0.71	0.99972
0.005	1.27	0.99967
0.010	0.43	0.99969
0.015	0.65	0.99977
0.020	1.90	0.99976
0.025	0.60	0.9997
0.030	2.90	0.99944

References

[1] ASTM E230/E230M-12 2012

[2] López C.R., Morales G. J., Vaca M. M., Lizardi R. A., Terres P. H, Lara V. A. 2013. Equipo para Laboratorio: Fabricación y Calibración de Termopares *UAM-Azcapotzalco*

Development of low weight self-levelling mortars

A Padilla, I Panama, A Toledo and A Flores
Material Department, Universidad Autónoma Metropolitana
Av. San Pablo No. 180 Col. Reynosa Tamaulipas, Azcapotzalco, México D.F

E-mail: ajpr@correo.azc.uam.mx

Abstract. This work shows the development of self levelling mortars, using micro bubbles based on aluminium silicate with a density of 0.25 g/cm^3. Mortars formulations are composed by 8 different components in order to achieve properties balance between fresh and solid state. The mean objective is development light weight mortars with high fluidity and compression strength using micro bubbles and some additives. Formulations were designed employing Taguchi DOE of 8 variables and 3 states. Result analysis according to Taguchi method lets indentify the preponderant effect of each variable on the cited properties. Several formulations reached fluidity higher than 250%, with compression strength around 100 kg/cm^2 and a low volumetric weigh. Obtained volumetric weights are 20% less than commercial self levelling mortars weight. Finally some relations are presented such: as relation water/cement with fluidity, and micro bubble content versus mortars volumetric weight, and finally compression strength versus the volumetric weight of mortars.

1. Introduction

Self-levelling or self compacting mortars are used due to their ability to flow easily under their own weight and levelling without any mechanical helping. This property lets mortar to full all spaces between bars even in the presence of congest reinforcement and achieving full compaction. Also this kind of mortars reduces the process time due to mechanical vibration is omitted. When vibration is omitted in casting process, there are other advantages such as: segregation phases and mortar damage is reduced and vibration and noise are eliminated. By other hand there is the tendency to reduce de specific weight of building materials in order to decrease building death weight. Particularity density of commercial self compacting mortars is between 2,000 to 2,200 kg/m^3. Therefore, this work was focussed to development light weight self compacting mortars using micro bubbles based on aluminium silicate, which have density around 0.25 g/cm^3. Previous experience [1] on addition of low density glass micro bubbles specifically in normal hydraulic mortar has shown that with addition of glass micro bubble in only 6% wt, it is possible to reduce 25 % the weight of the final piece, without secondary effects on paste viscosity and mechanical compression properties.

2. Experimental section

Formulation of this type of mortars is more complex than normal mortar formulation [2,3,4], due to it is necessary to get equilibrium between properties on the fresh state and solid state. In the fresh state is requested high flow without arid segregation. Rapid and high strength is requested in the solid state. Properties depend of composition variables and the interaction between them. Employed materials in this work were: cement Portland 40R ; calcium oxide; silica sand mesh 30 (0.584 mm) as arid compound; micro bubbles, which are based on aluminum silicate, with a true density of 0.25 g/cm^3, and nominal particle size corresponding to mesh 200 (0.076 mm); Wannin®, as flow additive based on lignocelluloses; Joincel®, which chemical structure, based on cellulose ethers, is composed of linear chains of etherified cellulose and it is usually employed as dispersant and plastizer; and calcium formate used as cure accelerant additive.

Due the high number of variables, formulations were done based on Taguchi design for 8 variables and 3 states, which are shown in Table 1. In this way, it allows identification of key parameters that have the most effect on the performance characteristic value so that further experimentation on these parameters can be performed and the parameters that have little effect can be ignored [5,6].

The evaluated properties of mortars were: fluidity which was tested for all formulation in the fresh state. Fluidity was determinate using fluidity table equipment. In the solid state, volumetric weight and compression resistance were evaluated for all formulations after 7 days of cure.

Table1. Self compacting mortars formulations. Percentages are in weight

Formulation No	Cement	Silica sand	Micro bubble	Water	Flow additive	Dispersant	Calcium oxide	Accelerant additive
1	30%	41%	4%	22%	0.1%	0.1%	2.2%	0.8%
2	29%	39%	5%	23%	0.2%	0.2%	2.6%	1.0%
3	28%	38%	6%	23%	0.3%	0.4%	2.9%	1.2%
4	29%	41%	3%	22%	0.2%	0.2%	3.0%	1.2%
5	29%	40%	5%	23%	0.3%	0.4%	2.1%	0.8%
6	28%	39%	6%	23%	0.1%	0.1%	2.5%	1.0%
7	32%	39%	3%	21%	0.1%	0.3%	2.4%	1.1%
8	32%	38%	4%	22%	0.2%	0.1%	2.8%	0.7%
9	31%	37%	5%	24%	0.3%	0.2%	1.9%	0.9%
10	33%	37%	3%	23%	0.3%	0.2%	2.4%	0.7%
11	32%	36%	4%	24%	0.1%	0.3%	2.7%	0.9%
12	33%	37%	6%	21%	0.2%	0.1%	2.0%	1.2%
13	36%	36%	3%	21%	0.3%	0.1%	2.7%	0.9%
14	36%	36%	4%	21%	0.1%	0.2%	1.9%	1.1%
15	33%	33%	5%	27%	0.2%	0.3%	2.1%	0.6%
16	36%	37%	3%	21%	0.2%	0.3%	1.9%	0.9%
17	36%	37%	4%	19%	0.3%	0.1%	2.3%	1.1%
18	33%	35%	5%	24%	0.1%	0.2%	2.5%	0.6%

2. Results and discussion

Table 2 shows the mean properties of the different formulations. As one can see it is difficult to establish a relationship between properties and composition of self leveling mortars. Making analysis according to Taguchi method, it is possible to identify the preponderant effect of each variable in the properties of the mortars. In this way, Table 3, shows the effect of each formulation variable on the mortar properties.

Table2. Properties of each formulation

Formulation No.	Fluidity	Volumetric weight kg/m^3	Compression strength kg/cm^2	Formulation No.	Fluidity	Volumetric weight kg/m^3	Compression strength kg/cm^2
1	310%	1.664	86	**10**	215%	1.536	62
2	141%	1.547	56	**11**	113%	1.525	64
3	63%	1.389	29	**12**	86%	1.581	78
4	169%	1.587	71	**13**	350%	1.715	91
5	98%	1.432	22	**14**	70%	1.624	110
6	350%	1.544	66	**15**	268%	1.469	103
7	104%	1.597	62	**16**	55%	1.597	59
8	293%	1.611	107	**17**	55%	1.672	83
9	221%	1.488	62	**18**	95%	1.507	110

So one can see fluidity property for example, is reduced by accelerant and Joincel®; but is increased by calcium oxide and silica sand. The others components have no o little effect on fluidity. Volumetric weight is reduced by Wannin®, Joincel®, silica sand and micro bubbles but is increased by cement and silica sand. Finally compression strength is increased by calcium oxide, accelerant and cement, but is reduced by Wannin®, Joincel® and silica sand.

Table3. Effect of each formulation variable on the mortar properties.
Variable increases its value

Variable	Fluidity	Volumetric weight	Compression strength
Wannin®	No change	Reduce	Reduce
Joincel®	Reduce	Reduce	Reduce
Calcium oxide	Increase	No change	Increase
Accelerant	Reduce	Increase	Increase
Cement	No change	Increase	Increase
Silica sand	Increase	Reduce	Reduce
Micro bubble	No change	Reduce	No change
Water	No change	No change	No change

By other hand, for mortars is very usual to define the effect of water/cement relation on fluidity. According to experimental design, relation water/cement (w/c) change from 0.54 to 0.84. As one expect fluidity increases with w/c. Data are reported on Figure 1, and there is large data dispersion, as the points inside circles shown. One of them has a high fluidity with a very low w/c relation, and other has low fluidity with high w/c relation. This behavior could be due to the presence and interaction whit other components. However, there are mortar formulations that achieve fluidity values upper to 250%. So we can consider these mortar formulations are self compacting mortars.

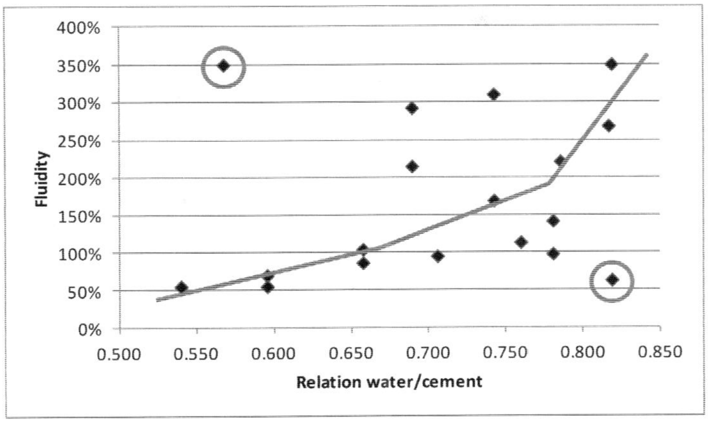

Figure 1. Effect of relation w/c on mortar fluidity

As micro bubbles are the mean compound, results important evaluate the effect of micro bubbles on volumetric weight. As one expect, content increase of micro bubbles reduces volumetric weight of the mortar. Obtained data are reported in Figure 2. Volumetric weights change from 1,400 to 1,700 kg/m^3 when micro bubbles content goes from 3% to 6%wt. This represents a interesting weight reduction if we consider that weight of normal self mortars is around 2,100 kg/m^3. Unfortunately, there is not possible to establish a relation between these parameters.

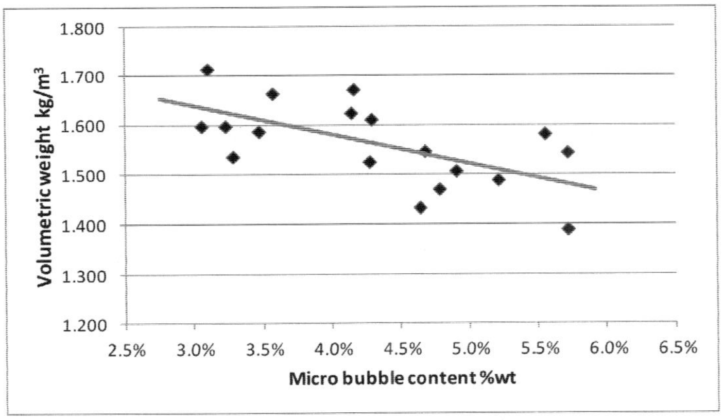

Figure 2. Effect of micro bubble content on mortar volumetric weight

Also, there is a good correlation between volumetric weight and compression resistance of the mortars, as it shown in Figure 3. Compression strength higher than 100 kg/cm^2 can achieve with volumetric weight between 1,600 to 1,750 kg/m^3.

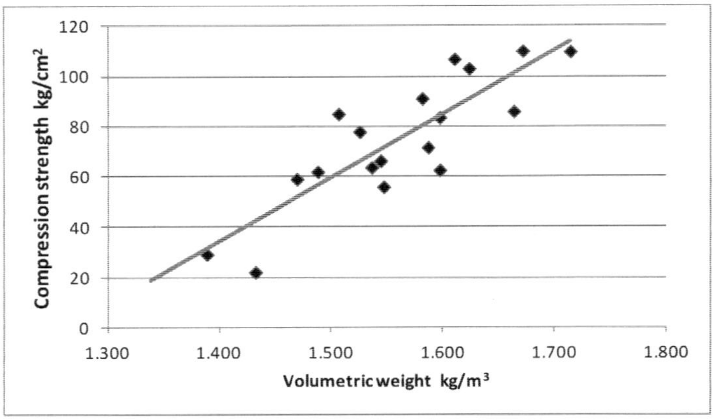

Figure 3. Correlation between volumetric weight and compression strength

3. Conclusions

Development formulations reduce volumetric weight to value around 1,600 kg/m^3, with fluidity higher than 200% and with compression strength upper 100 kg/cm^2. These values together with a low weight of 1,600 kg/m^3 open the possibility to use this material for casting and elaboration of light weight structural elements.

References

[1] Padilla A, Flores A., Rocha L., Panama I. 2010 Efecto de microesferas en morteros hidráulicos, Proceedings *5° Congreso Internacional de Ingenería Fisica,* MP-CIIF-060

[2] Okamura H., Ouchi M. 2003 Self Compacting Concrete *J of Adv. Technol*

[3] Ouchi-Kochi M., Hibino-Nagaoka M. 2000 Development, Applications and Investigations of Self-Compacting Concrete

[4] Sedran, T., de Larrard F., Horst F.1996 Contamines Mix design of self compacting concrete Proceedings of RILEM International Conference on Production Methods and Workability of Fresh Concrete *E&FN Spon* 439-450

[5] Cesatrone J. 2001 The Power of Taguchi *IIE Solutions* 2001 36-40

[6] Sullivan Laurenz P 1987 The Power of Taguchi Methods *Quality Progress* 77-83

Journal of Physics: Conference Series **582** (2015) 012031 doi:10.1088/1742-6596/582/1/012031

Photoluminescence Emission and Structure Diversity in ZnO:Ag Nanorods

E Velázquez Lozada[1], G M Camacho González[2], T Torchynska[3]

[1]SEPI – ESIME – Instituto Politécnico Nacional, México D. F. 07738, México.
[2]ESIME – Instituto Politécnico Nacional, México D. F. 07738, México.
[3]ESFM – Instituto Politécnico Nacional, México D. F. 07738, México.

E-mail: evlozada5@yahoo.com.mx

Abstract. The photoluminescence, its temperature dependences, as well as structural characteristics by methods of Scanning electronic microscopy (SEM) have been studied in ZnO:Ag nanorods prepared by the ultrasonic spray pyrolysis (USP). The PL spectra of the ZnO:Ag NRs over the temperature range from 10 K to 300 K are investigated. Three types of PL bands have been revealed: i) the near-band-edge (NBE) emissions, ii) defect related emission and iii) IR emissions. It is shown the IR emission corresponds to the second-order diffraction of the near-band-edge (NBE) emission bands. The study of NBE PL temperature dependences reveals that the acceptor bound exciton (ABE) and its second-order diffraction peak disappeared at the temperature higher than 200 K. The attenuation of the ABE peak intensity is ascribed to the thermal dissociation of ABE with appearing of a free exciton (FE). The PL bands, related to the LO phonon replica of FE and its second-order diffraction, dominate in the PL spectra at room temperature that testify on the high quality of the ZnO:Ag films prepared by the USP technology.

1. Introduction

Zinc oxide (ZnO) nanocrystals (NCs) with wide band gap energy (3.37 eV) have attracted great attention due to exceptional exciton properties (high exciton binding energy equal to 60 meV at 300K) and a number of deep levels that emit in the whole visible range and, hence, can provide intrinsic "white" light emission. ZnO NCs are promising candidates for the different optoelectronic applications such as light emitting diodes [2-7]. The control of the ZnO defect structure in these nanostructures is a necessary step in order to improve the device quality. Since the structural imperfection and defects generally deteriorate the exciton related recombination process, it is necessary to grow the high quality films for efficient light-emitting applications. The ultrasonic spray pyrolysis (USP) method is a simple, inexpensive, non-vacuum and a low temperature technique for the film synthesis [8]. It will be interesting to study the optical emission of the USP produced ZnO nanostructures doped with Ag versus temperature in order to identify the best regimes for obtaining bright emitting NCs and the nature of optical transitions.

2. Experimental details

ZnO:Ag thin solid films were prepared by the USP technique on the surface of soda-lime glass substrate for the substrate temperatures 400^0 C and different deposition times of 3, 5 and 10 min.

Journal of Physics: Conference Series **582** (2015) 012031 doi:10.1088/1742-6596/582/1/012031

Using this technique the nanoparticle's size can be easily controlled by changing a concentration of starting solution and the atomization parameters. The deposition system includes a piezoelectric transducer operating at variable frequencies up to 1.2 MHz and the ultrasonic power of 120 W. ZnO:Ag thin solid films were deposited from a 0.4 M solution of zinc (II) acetate [Zn(O2CCH3)2] (Alfa), dissolved in a mix of deionized water, acetic acid [CH3CO2H] (Baker), and methanol [CH3OH] (Baker) (100:100:800 volume proportion). Separately, a 0.2 M solution of silver nitrate [Ag(NO3)] (Baker) dissolved in a mix of deionized water and acetic acid [CH3CO2H] (Baker) (1:1 volume proportion) was prepared, in order to be used as doping source. A constant [Ag]/[Zn] ratio of 2 at. % was applied at the ZnO Ag film preparation. The morphology of ZnO:Ag films has been studied using the scanning electron microscopy (SEM) Dual Beam, FEI brand, model Quanta 3D FEG with field emission gun. PL spectra were measured in the temperature range 10-300K at the excitation by a He-Cd laser with a wavelength of 325 nm and a beam power of 20 mW at 300K using a PL setup on a base of spectrometer SPEX500 described in [9-11].

3. Experimental results and discussion

SEM images of the typical ZnO:Ag nanorods (NRs) obtained at the deposition times of 3 and 10 min and the substrate temperatures 400^0 C are presented in figure 1. It is clear that the ZnO nanoroads have the hexagonal cross section and the road orientation along the c axis. The cross section size of ZnO nanoroads increases with the duration of UPS process from 50-70 nm (for the duration of 3min), 100-150 nm (at 5min) and 150-200 nm (at 10 min). PL spectra of ZnO:Ag NRs are shown in figure 2 for all studied samples.

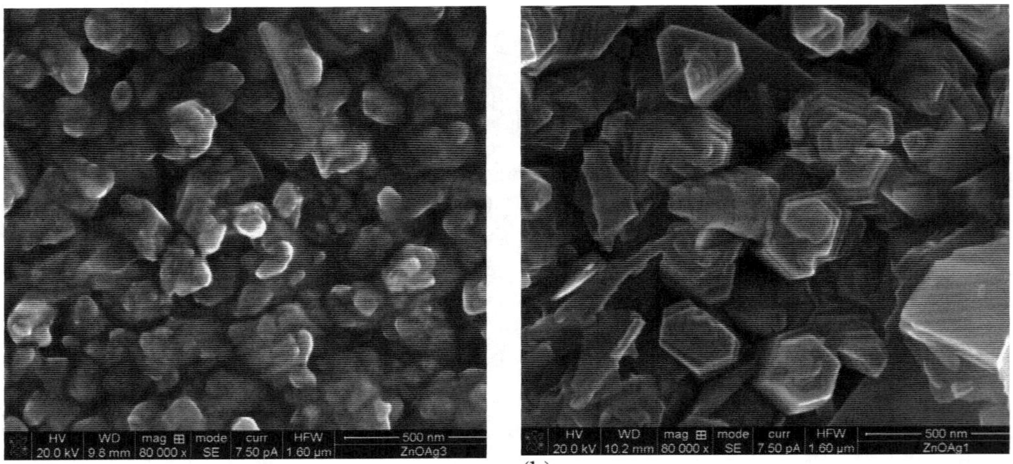

(a) (b)

Figure 1. SEM images of the samples prepared at the substrate temperatures 400^0 C and the durations of 3 (a) and 10 (b) min.

It is clear that the PL spectra are complex and can be represented as a superposition of elementary PL bands with the peaks in the spectral ranges: 2.90-3.25 eV (I, II), 2.00-2.50 eV (III, IV) and 1.45-1.61 eV (V,VI). The deconvolution procedure has been applied to the PL spectra with the aim to separate the elementary PL bands (Fig.3). The analysis of figures 3 permits to distinguish six elementary PL bands with the peaks at: 3.25 (1), 2.92 (2), 2.75 (3), 2.10 (4), 1.62 (5) and 1.46 (6) eV at 10K (Fig.3).

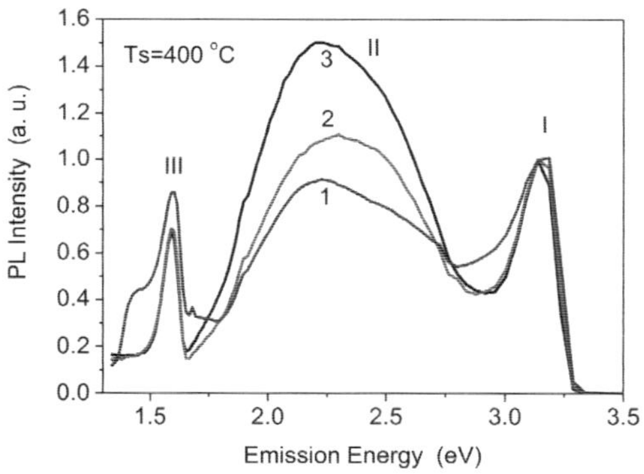

Figure 2. PL spectra of samples prepared on the substrate with T = 400° C at the deposition times: 1-3 min, 2-5 min, 3-10 min.

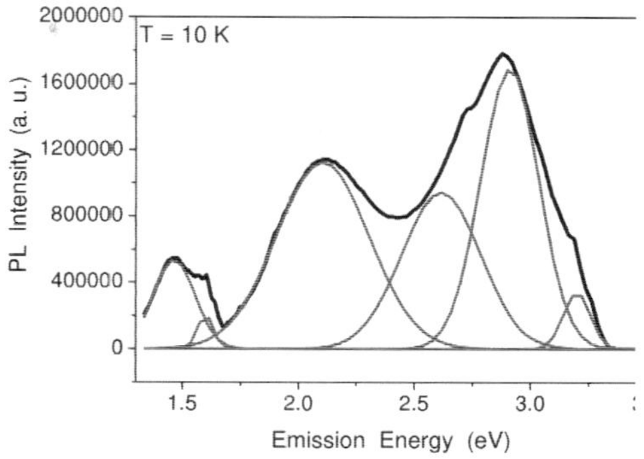

Figure 3. PL spectrum deconvolution on six elementary PL bands for the film obtained at 400^0 C at the duration 3 min.

It well known that the UV-visible PL bands (I) in ZnO are near-band-edge (NBE) or exciton emissions [12]. The PL band in the spectral range 2.00-2.50 eV related to the defect emission [12] and the nature of IR PL bands has to be clarified. With increasing the USP duration the intensity of defect (II) related PL bands rose mainly in comparison with the intensity of NBE PL bands (Fig.2).

PL spectra of ZnO:Ag NRs measured at different temperature in the range of 10-300K are shown in Fig.4. The same rates of the variation of PL peak intensities versus temperature have been revealed for the PL bands 1 and 7, as well as for the emission bands 2 and 8 (Fig.4). This fact, as well as the PL peak positions, permits to assign the IR PL bands 7 and 8 to the second order diffraction of the NBE PL bands 1 and 2, respectively [13].

A great variety of luminescence bands in the UV and visible spectral ranges have been detected in ZnO crystals [12]. The near-band-edge (NBE) emission at 3.0-3.37 eV is attributed to the free (FE) or bound (BE) excitons, their LO phonon replicas, such as FE-1LO or FE-2LO, to optical transition between the free to bound states, such as the shallow donor and valence band, or to donor-acceptor pairs [14]. However, the position of the near-band-edge emission at room temperature can vary significantly due to the variation of relative contributions of free exciton emission and phonon replicas [15].

The defect related green PL band in the spectral range 2.40-2.70 eV in ZnO is assigned ordinary to oxygen vacancies [16], Cu impurities [17] or surface defects [18]. The orange PL band with the peak at 2.02-2.10 eV was attributed earlier to oxygen interstitial atoms (2.02 eV) [19] or to the hydroxyl group (2.10eV) [20, 21]. Taking into account that the PL intensity of this PL band increased with raising the USP duration (Fig.2) the assumption that the corresponding defects are related to oxygen interstitial atoms (or to the hydroxyl group) looks very reliable.

Finally, we need to discuss the nature of the near-band-edge PL bands at 2.92 and 3.25 eV. Studied ZnO films were doped by Ag and, therefore, have the acceptor type defects, Ag_{Zn}, which were formed when the Ag atoms substitute Zn atoms in the ZnO crystal lattice. The intensity of 2.92 eV PL band decreases essentially with temperature and this PL band and its IR second-order diffraction peak (1.46eV) disappeared completely at 200K (Fig.4).

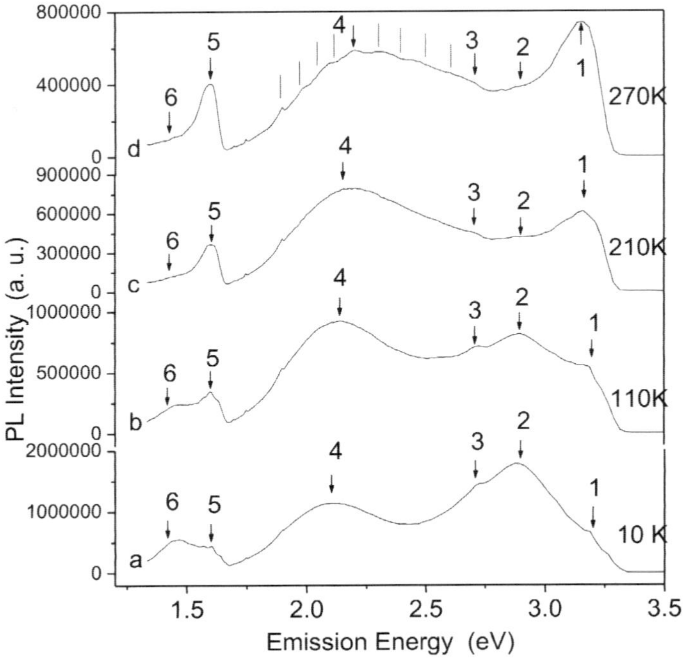

Figure 4. PL spectrum variation with temperature for the simple prepared at the duration 5 min.

Thus it is possible to assume that this PL band (2.92 eV) owes to emission of the acceptor bound excitons (ABE), involving the acceptor Ag_{Zn}, or its complexes such as donor-acceptor pairs, in the ZnO:Ag nanorods. The thermal decay of the 2.92 eV PL intensity is assigned to the thermal dissociation of ABE to free exciton. In this case the 3.25 eV PL band, which dominates in the room temperature PL spectrum (Fig.4) can be attributed to the LO phonon replica of FE emission.

4. Conclusions

ZnO:Ag nanorods with hexagonal structures have been successfully synthesized by the USP method. The PL spectra of the ZnO:Ag NRs over the temperature range from 10 K to 300 K have been investigated. Temperature-dependent PL spectra show that the IR emissions correspond to the second-order diffraction of the NBE emissions. The study of NBE PL temperature dependences reveals that ABE and its second-order diffraction peak disappeared when the temperature higher than 200 K. The attenuation of the ABE peak intensity is ascribed to the thermal dissociation of ABE to free exciton. The PL bands, related to the LO phonon replica of free exciton and its second-order diffraction, dominate in the PL spectra at room temperature that testifies on the high quality of the ZnO:Ag films prepared by USP.

5. Acknowledgement

The authors would like to thank the CONACYT (project 130387) and SIP-IPN, Mexico, for the financial support, as well as the CNMN-IPN for the SEM measurement.

References

[1] S.J. Pearton, D.P. Norton, K. Ip, Y.W. Heo, T. Steiner 2005 Prog. Mater. Sci. **50**, 293
[2] M.H. Koch, P.Y. Timbrell, R.N. Lamb 1995 Semicond. Sci. Technol. **10**.
[3] K. Vanheusden, C.H. Seager, W.L. Wareen, D.R. Tallant, J. Caruso, M.J. Hampden-Smith, T.T.

Kodas, J. Lumin. 1997 **75** 11

[4] R. Scheer, T. Walter, H.W. Schock, M.L. Fearheiley, H.J. Lewerenz 1997 Appl. Phys **63** 3294

[5] Y. Chen, D.M. Baghall, H. Koh, K. Park, K. Hiraga, Z. Zhu, T. Yao, J.1998 Appl. Phys. **84** 3912

[6] Y.B. Li, Y. Bando, D. Golberg, 2004 Appl. Phys. **84** 3603

[7] J. Ding, T.J. McAvoy, R.E. Cavicchi, S. Semancik 2001 Sens. Actuat. **77** 597

[8] W. Tang, D.C. Cameron 1994 Thin Solid Films **238** 83

[9] M. Dybic, S. Ostapenko, T.V. Torchynska, E. Velazquez Lozada 2004 Appl. Phys. **84** 5165

[10] T. V. Torchynska, A.I. Diaz Cano, M. Dybic, S. Ostapenko, M. Mynbaeva, Physica B, 2006 Condensed Matter 376-377

[11] T.V. Torchynska, J. Palacios Gomez, G.P. Polupan, F.G. Becerril Espinoza, A. Garcia Borquez, N.E. Korsunskaya, L.Yu. Khomenkova 2000 Appl. Surf. Science **167** 197-204

[12] A. B. Djuris, A.M.C. Ng, X.Y. Chen. 2010 Progress in Quantum Electronics **34** 191-259

[13] J. Lv, Ch. Liu, W. Gong, Zh. Zi, X. Chen, K. Huang,T. Wang, G. He, Sh. Shi, X. Song, Zh. Sun, 2012 Optical Materials **34**

[14] T. Mahalingam, K.M. Lee, K.H. Park, S. Lee, Y. Ahn, J.Y. Park, K.H. Koh, 2007 Nanotechnology **18**

[15] T. Voss, C. Bekeny, L. Wischmeier, H. Gafsi, S. Borner, W. Schade, A.C. Mofor, A. Bakin and A. Waag 2006 Appl. Phys. **89**

[16] M.K. Patra, K. Manzoor, M. Manoth, S.P. Vadera, N. Kumar, J. Lumin. 2008

[17] N.Y. Garces, L. Wang, L. Bai, N.C. Giles, L.E. Halliburton, G. Cantwell, Appl. Phys. 2002 **81**

[18] A.B. Djurišic, W.C.H. Choy, V.A.L. Roy, Y.H. Leung, C.Y. Kwong. K.W. Cheah, T.K. Gundu Rao, W.K. Chan, H.F. Lui, C. Surya, 2004 Adv. Funct. Mater

[19] X. Liu, X. Wu, H. Cao, R.P.H. Chang, J. 2004 Appl. Phys. **95**

[20] J. Qiu, X. Li, W. He, S.-J. Park, H.-K. Kim, Y.-H. Hwang, J.-H. Lee, Y.-D. Kim, 2009 Nanotechnology **20**

[21] T.V. Torchinskaya, N.E. Korsunskaya, B. Dzumaev, B.M. Bulakh, O.D. Smiyan, A.L. Kapitanchuk, S.O. Antonov, 1996 Semiconductors **30**

Journal of Physics: Conference Series **582** (2015) 012032 doi:10.1088/1742-6596/582/1/012032

Graphene-like carbon synthesized from popcorn flakes

D Mendoza[1], C B Flores[1], R Y Sato Berrú[2]
[1]Instituto de Investigaciones en Materiales
[2]Centro de Ciencias Aplicadas y Desarrollo Tecnológico, Universidad Nacional Autónoma de México, 04510, Coyoacán D. F., México.

E-mail: doroteo@unam.mx

Abstract. The synthesis of graphene-like carbon using popcorn kernels as a renewable resource is presented. In a first step popcorn kernels were heated to produce popcorn flakes with a spongy appearance consisting of a polygonal cellular structure. In a second step, the flakes were treated at high temperature in an inert atmosphere to produce carbonization. Raman spectroscopy shows graphene-like structure with a high degree of disorder.

1. Introduction

Since the first report on graphene, a one layer of carbon atoms arranged in a hexagonal structure, there is plenty of reported work on this system and related materials. One of the main characteristics is its high electrical mobility, which makes graphene a good candidate for high frequency electronics [1]. Other related materials, such as graphene oxide or porous carbons have great interest due to their possible applications in the purification of water [2], or for electrodes in supercapacitors and batteries [3].

One of the main concerns is the use of renewable resources as the precursor of carbonaceous materials, such as porous carbon [4] and graphene [5]. In the present work we explore the possibility of the use of corn as a renewable precursor to synthesize graphene-like carbon. Specifically we use popcorn kernels to make popped flakes which have a foamed appearance with a volume roughly 30 times the volume of the original kernel. The visualized physical process is as follows. When the popcorn kernel is heated, the temperature and the internal pressure increase due to the evaporation of water moisture; at a temperature around 180°C the pressure is around 9 atmospheres and the hull of the kernel breaks producing the explosion ("popping") of the endosperm. The final flake is a foam with polygonal cellular structure [6]. The majority of the flake is expanded starch and a careful examination of the foam reveals that the cells are separated by thin planar walls. The main idea of the present work is that the planar structure of the cell walls may transform to graphitic structures under thermal treatments.

2. Experimental Section

Popcorn kernels were heated in a home microwave oven (650 W) during several minutes until the popping process is concluded. The white foam is separated from the residual hull and loaded into a fused quartz tube in the interior of a horizontal furnace. In an atmosphere of nitrogen, the temperature was raised to 1000°C or 800°C and maintained during a period of one hour, the resultant material has a black color and is named CORNBON1000 and CORNBON800, respectively. Originally we selected 1000°C for the heat treatment because this is the temperature for the synthesis of graphene using the Chemical Vapor Deposition (CVD) technique [7], but a lower temperature was also used for comparison. Samples were characterized by scanning electron microscopy (SEM), X-ray diffraction and Raman spectroscopy.

3. Results

In figure 1 SEM images of the worked material are presented. The pristine popcorn flake shows an overall porous morphology (figure 1a) and in a close up a polygonal cellular structure is clearly revealed with an average diameter of 20 microns (figure 1b). After the heat treatment the morphology changes, such as is observed in figure 1c and figure 1d for temperatures of 800°C and 1000°C, respectively.

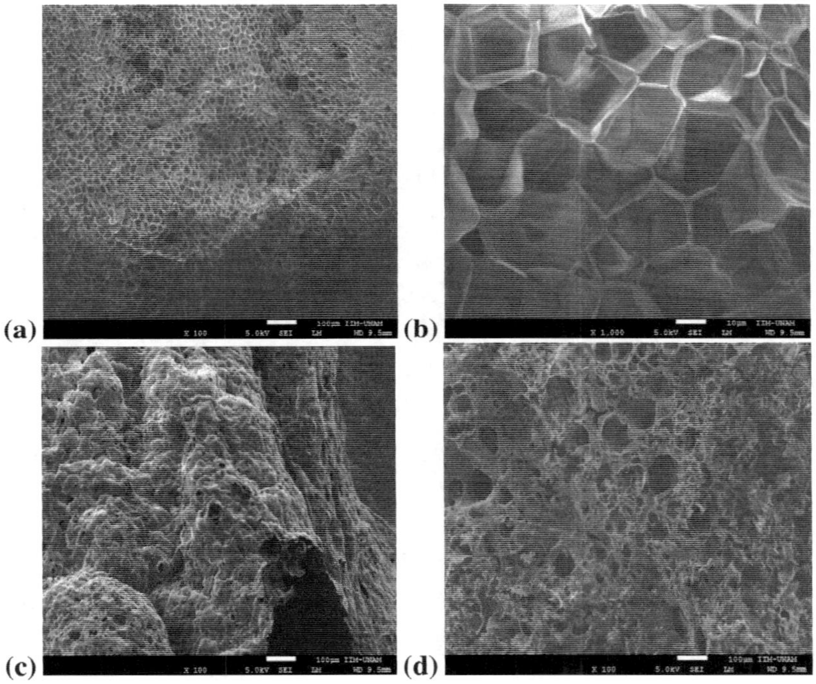

Figure 1. (a), (b) SEM images of the popcorn flakes, in the close up in (b) the cellular polygonal structure is observed, note the lamellar form of the intercell walls. Carbonaceous material after the heat treatment at 800°C (c) and 1000°C (d).

X-ray diffraction indicates that the carbonaceous samples are highly disordered (figure 2), showing a broad peak around 27° which is related to (002) diffraction peak of graphite and other broad peak around 43° that is associated to turbostratic graphite, in which the planes stack in a very disordered manner [8].

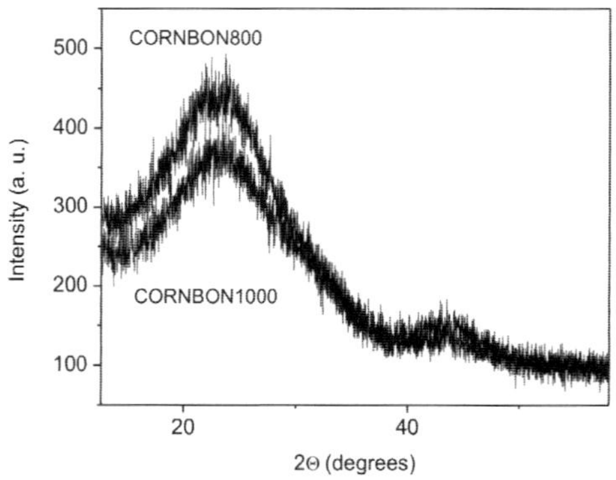

Figure 2. X-ray diffractograms of samples with heat treatment at 800°C (CORNBON800) and at 1000°C (CORNBON1000). The carbonaceous material shows two broad bands centered around 27° and 43°. The first band is more noticeable in sample treated at 1000°C as a shoulder, the highest broad band centered around 23° is due to the glass container of the carbonaceous powder.

Raman spectroscopy of the samples shows two intense bands centered around 1330cm^{-1} and 1583cm^{-1} for the sample treated at 800°C, and 1320cm^{-1} and 1580cm^{-1} for sample treated at 1000°C; both samples show a very broad band between (2400-3200)cm^{-1} (see inset in figure 3a). The main difference between these two kinds of samples is that a higher stage of graphitization at the higher temperature is observed.

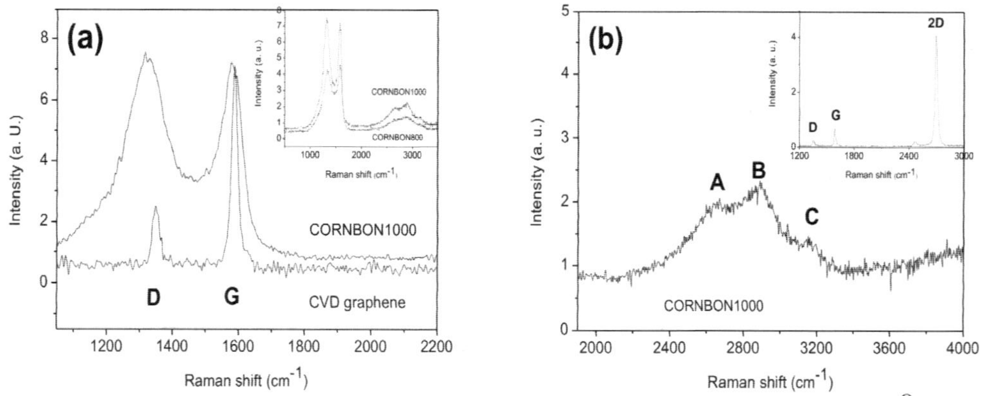

Figure 3. Raman spectra of the graphene-like material with heat treatment at 800°C (CORNBON800) and at 1000°C (CORNBON1000). (a) Comparison with graphene obtained by CVD technique showing the D and G bands, in the inset the overall spectra of the synthesized samples are shown. (b) Close up of the zone around the 2D band showing three bands labeled A, B and C discussed in the main text, the inset show the total spectrum of graphene obtained by CVD.

Journal of Physics: Conference Series **582** (2015) 012032 doi:10.1088/1742-6596/582/1/012032

4. Discussion

Raman spectroscopy is a very useful technique to study the great variety of carbon structures, in particular, the graphene-based materials [9, 10]. Graphene is characterized by three bands: G band around 1580cm^{-1} related with in plane stretching carbon-carbon bonds, D band ~1350cm^{-1} associated with structural disorder, and the 2D band located ~2700cm^{-1} and considered as an overtone of the D band. In the process to excite the 2D band, two phonons with opposite wave-vectors are involved, hence it is not necessary the existence of disorder or defects and always is present in the graphene spectrum [10]. It is also possible to quantify the number of graphene layers using the intensity ratio I(2D)/I(G) and the local structure of the 2D band which can be modeled with different sub-bands depending on the number of graphene layers. Respect to the peaks intensity, the general rule is that for graphene (monolayer), I(2D)/I(G)>3 to 4, and for many graphene layers the peaks intensity ratio is less than one [9].

For comparison purposes, in the inset of figure 3b the overall spectrum of graphene synthesized by CVD technique [11] is presented. The characteristic of this kind of material is the existence of the D band compared to graphene obtained by exfoliation of graphite crystals where D band is almost absent [12]. In figure 3a the comparison of CVD graphene and the material synthesized at 1000°C is presented, clearly we can associate the prominent bands located at 1320cm^{-1} and 1580cm^{-1} with D and G bands, respectively. Although the two bands are well defined, they are very broad and we propose it is due to structural disorder (recall: X rays diffraction shows high degree of disorder).
In a systematic study on ion bombardment of graphene and few layer graphene, Martins, et al. [13] show the evolution of the Raman spectrum. In the case of monolayer graphene, G and 2D are the original bands, as the dose of ions increases D band appears, and finally the three bands deform. At the final stage, Raman spectra of graphene and few layers graphene are very similar to that presented in the inset of figure 3a of the present work. This process is explained in terms of disorder induced by ion bombardment [13]. It should be noted that at the higher ions dose (high disorder), the 2D band changes to a very broad and complex band, very similar to that shown in figure 3b for our material. The broad bands labeled A, B and C are identified as overtones: 2D, D+G and D'+G, respectively; where D' band (around 1620cm^{-1}) is also related to disorder but involucres intravalley scattering processes.

The carbonaceous material obtained by our method has a high degree of disorder but it also resemble the structure of graphene and few layers of graphene, this is the reason for which we denominated this material as graphene-like carbon. We suppose there are small zones or crystallites that are similar to graphene with defects. We can estimate the size L of these crystallites using the following relation [10]: I(D)/I(G) ~5nm/L, valid for wave length excitation of 532 nm, which is our case. From figure 3a we can see that D and G bands have similar intensities, hence the average size of the graphene nanocrystals is around 5 nm, which explains the absence of crystalline peaks in the X-ray pattern and also the overall structure of the Raman spectrum.

Finally, taking advantage of the foam structure of the popcorn flake, this material can be used as a template to generate other materials. As a preliminary result, in figure 4 the formation of silver particles is shown. In this case, the white flakes were immersed into a silver nitrate-water solution which is readily absorbed by the popcorn flake. After drying, the material is heated to 500°C during one hour in a nitrogen atmosphere. We believe that choosing the appropriate conditions, it would be possible to control the size of the silver particles.

Figure 4. Silver particles grown on popcorn flake after a heat treatment at 500°C using silver nitrate as the precursor.

5. Conclusion

Using corn kernels as a renewable resource we synthesized carbon with a graphene-like structure with a high degree of disorder. We envisage that the obtained material may be useful as electrodes for supercapacitors and batteries due to its porous and spongy structure.

Acknowledgements. We thank A. Tejeda (X-rays) and O. Novelo (SEM) of IIM-UNAM for their valuable assistance. We also thank Laboratorio Universitario de Caracterización Espectroscopica (LUCE), CCADET-UNAM for the use of Raman equipment.

References

[1] Geim A K and Novoselov K S 2007 *Nature Mater.* **6** 183
[2] Joshi R K, Carbone P, Wang F C, Kravets V G, Su Y, Grigorieva I V, Wu H A, Geim A K and Nair R R 2014 *Science* **343** 752
[3] Kim T Y, Jung G, Yoo S, Suh K S and Ruoff R S 2013 *ACS Nano* **7** 6899
[4] White R J, Budarin V, Luque R, Clark J H and Macquarri D J 2009 *Chem. Soc. Rev.* **38** 3401
[5] Ruan G, Sun Z, Peng Z and Tour J M 2011 *ACS nano* **5** 7601
[6] Sweley J C, Rose D J and Jackson D S 2013 *Food Reviews International* **29** 157
[7] Li X, Cai W, An J, Kim S, Nah J, Yang D, Piner R, Velamakanni A, Jung I, Tutuc E, Banerjee S K, Colombo L and Ruoff R S 2009 *Science* **324** 1312
[8] French B L, Wang J J, Zhu M Y and Holloway B C 2006 *Thin Solid Films* **494** 105
[9] Ferrari A C, Meyer J C, Scardaci V, Casiraghi C, Lazzeri M, Mauri F, Piscanec S, Jiang D, Novoselov K S, Roth S and Geim A K 2006 *Phys. Rev. Lett.* **97** 187401
[10] Ferrari A C and Basko D M 2013 *Nature Nanotechnology* **8** 235
[11] Bautista C, Mendoza D 2011 *World J. Conden. Matter Phys.* **1** 157
[12] Bautista C, Sato R, Mendoza D 2013 *Mundo Nano* **6** 29
[13] Martins E H, Moutinho M V, Stavale F, Lucchese M M, Capaz R B, Achete C A and Jorio A 2010 *Phys. Rev.* B **82** 125429

Journal of Physics: Conference Series **582** (2015) 012033 doi:10.1088/1742-6596/582/1/012033

Obtaining the wavefront phase maps of free form surfaces: using the least squares algorithm

B Villalobos-Mendoza, D Aguirre-Aguirre, F Granados-Agustín and A Cornejo-Rodríguez
Instituto Nacional de Astrofísica, Óptica y Electrónica, Luis Enrique Erro No.1, Sta. María Tonantzintla, C.P. 72840 Puebla, México.

E-mail: bvillalobosmendoza@gmail.com

Abstract. In this work is presented the validation of the least squares algorithm proposed by Morgan (1982) and Greivenkamp (1984), to obtain the wavefront phase maps of a free form surface. The validation was made by simulating the synthetic interferograms of a free form surface using a Bessel function, each interferogram was simulated with a phase-shifting of $\pi / 20$. This algorithm is applied to the experimental interferograms that are obtained in a Twyman-Green interferometer where the phase shifting is achieved by using a SLM (Spatial Light Modulator) that is placed in one of its arms; the phase shifts are achieved by displaying all the gray levels from 0 to 255 in the SLM. The phase shifts that are performed in this experimental setup are lower than $\pi / 4$, therefore the conventional algorithms cannot be applied.

1. Introduction

Many different algorithms have been published for the determination of the wavefront phase, Carré (1966), Bruning et all. (1974), Sommargren (1975), Wyant (1975), Morgan (1982), Bhushan and George (1984), Greivenkamp (1984), Wyant and Creath (1985), Reid (1986), [1-8]. Some techniques shift the phase a known amount between intensity measurements. A minimum of three measurements is necessary to determine the phase [1, 9], since there are three unknowns in the interference equation

$$I_0 = I_0\left[1+\gamma_0 \cos(\phi)\right]. \tag{1}$$

These unknowns are the intensity I_0, the modulation of the interference fringes γ_0, and the wavefront phase ϕ [1, 9]. The phase shift between intensity measurements can be any value between 0 and π as long as it is constant and lineal. The recorded intensity is

$$I_i(x, y) = a_0(x, y) + a_1(x, y)\cos(\alpha_i) + a_2(x, y)\sin\alpha_i, \tag{2}$$

where

$$a_0(x,y) = I_0(x,y),$$
$$a_1(x,y) = I_0(x,y)\gamma_0(\text{sinc}\ \tfrac{1}{2}\Delta)\cos\phi(x,y),\tag{3}$$
$$a_2(x,y) = I_0(x,y)\gamma_0(\text{sinc}\ \tfrac{1}{2}\Delta)\sin\phi(x,y).$$

The unknowns of these set of equations are $I_0(x,y)$, $\gamma_0(x,y)$ and $\varphi(x,y)$, which are represented by $a_0(x,y)$, $a_1(x,y)$ and $a_2(x,y)$ [1, 8, 9]. The least squares solution to these equations is

$$\begin{pmatrix} a_0(x,y) \\ a_1(x,y) \\ a_2(x,y) \end{pmatrix} = A^{-1}(\alpha_i)B(\text{x},\text{y},\alpha_i),\tag{4}$$

where

$$A(\alpha_i) = \begin{pmatrix} N & \sum\cos\alpha_i & \sum\sin\alpha_i \\ \sum\cos\alpha_i & \sum\cos^2\alpha_i & \sum(\cos\alpha_i)\sin\alpha_i \\ \sum\sin\alpha_i & \sum\cos\alpha_i\sin\alpha_i & \sum\sin\alpha_i \end{pmatrix},\tag{5}$$

and

$$B(\alpha_i) = \begin{pmatrix} \sum I_i(x,y) \\ \sum I_i(x,y)\cos\alpha_i \\ \sum I_i(x,y)\sin\alpha_i \end{pmatrix}.\tag{6}$$

The phase at each point in the interferogram is determined by evaluating the value of B at each point and then solving for the coefficients a_0 and a_2 :

$$\tan\varphi(x,y) = \frac{a_2(x,y)}{a_1(x,y)} = \frac{I_0\gamma_0(\text{sinc}\frac{1}{2}\Delta)\sin[\phi(x,y)]}{I_0\gamma_0(\text{sinc}\frac{1}{2}\Delta)\cos[\phi(x,y)]}.\tag{7}$$

The phase calculation assumes that the phase-shifts between measurements are known and Δ is constant for every measurement [1, 5, 8, 9].

Analysis of the phase-shifted interferograms

To make this analysis, it were simulated a set of phase-shifted interferograms with a phase shift of $\pi/20$. This phase-shifts where chosen with this value given that the algorithm is going to be applied to the experimental interferograms, which are obtained in a Twyman-Green interferometer. The experimental interferograms have phase-shifts lower than $\pi/4$ that is the minimum value used in conventional algorithms, is for this reason that the least squares algorithm was used.

In figure 1a is shown the synthetic interferogram of a spherical surface with local deformations, and figure 1b shows the synthetic interferogram of a free form surface.

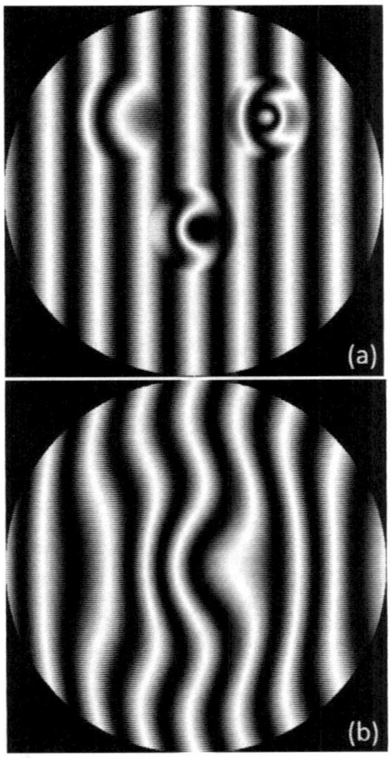

Figure 1. Synthetic interferograms. (a) Spherical surface with local deformations and (b) Free form surface

A total of ten interferograms were simulated with a phase shift of $\pi/20$ for every synthetic surface under test. In figure 2 are shown the graphics of the simulated surfaces that were simulated in Matlab®. Figure 2a shows the spherical surface with local deformations with a peak-valley of 1.7λ, and figure 2b shows the free form surface with a peak-valley of 0.7λ. The deformations introduced in the synthetic interferograms produce phase shifts of $\pi/2, \pi$ and $3\pi/4$. These phase-shifts values of the deformations can be compensated using the SLM in the experimental setup, due to the total phase shift that are achieved with this SLM is nearly to 2π [10].

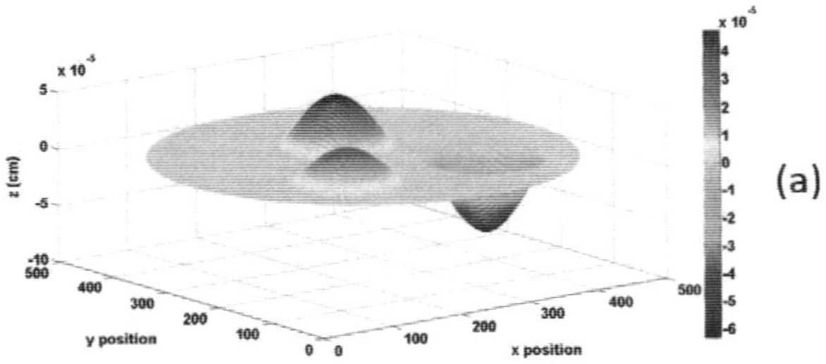

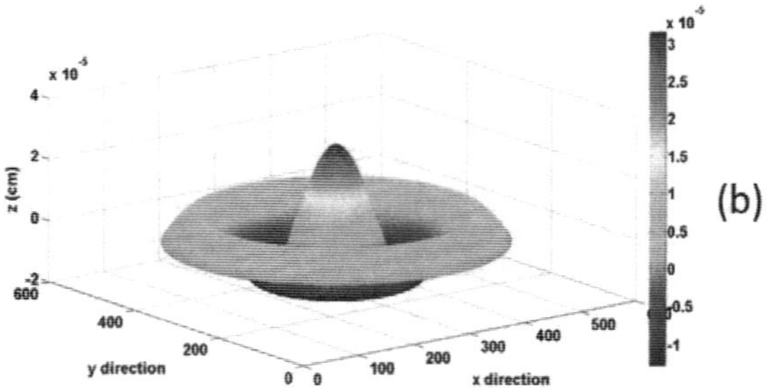

Figure 2. Surface graphics. (a) Spherical surface with local deformations and (b) Free form surface.

With this information the next step was to apply the least squares algorithm, Eq. 4, to the synthetic interferograms simulated. Next section shows the obtained results.

3. Results and Discussion

In figure 3 are shown the wavefront phase maps of both, the spherical surface with local deformations and the free form surface that were obtained applying the least squares algorithm, Eq. 4.

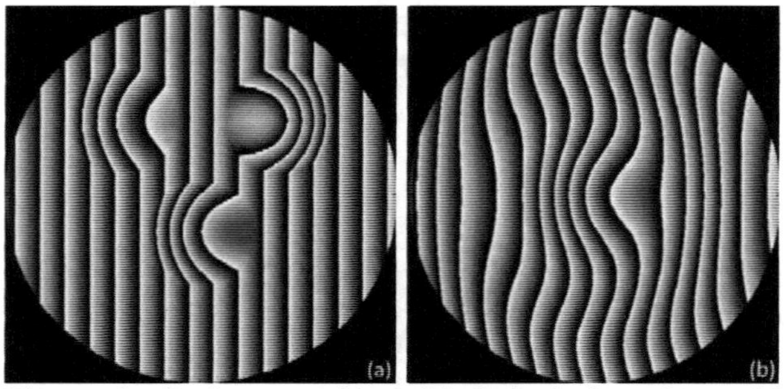

Figure 3. Surface graphics. (a) Spherical surface with local deformations and (b) Free form surface.

As can be seen, the phase values recovered by using the least squares algorithm, Eq. 7, are those that were expected (values that are well behaved), regardless the values of phase-shift introduced, ($\leq \pi / 20$). In contrast with the conventional algorithms that need that the phase-shifts have values greater than $\pi / 4$ [1, 8, 9].

4. Conclusions

In this work was described the process carried out to obtain the wavefront phase maps of an spherical surface with local deformations and a free form surface, by using the least squares algorithm. This work will be applied to the experimental interferograms that are obtained in a Twyman-Green interferometer, where the phase-shifts are produced by a SLM that is placed in one of its arms and displaying the gray levels from 0 to 255. By using the least squares algorithm allow us to analyze a phase shifts lower than $\pi / 4$ which is the minimum value that is required in almost of the conventional algorithms. This analysis enable us to apply the least squares algorithm to our experimental interferograms that have phase-shifts lower than $\pi / 4$ and this is the work we are currently working.

References

[1] Wolf E (1988) *Progress in Optics XXVI* (New York: Elsevier Science Publishers)
[2] Carré P 1966 *Metrologia* 2 13
[3] Bruning J H, Herriott D R, Gallagher J E, Rosenfeld D P, White A D and Brangaccio D J 1974 *Appl. Opt.* 13 2693
[4] Sommargren G E 1975 *J. Opt. Soc. Am.* 65 960
[5] Wyant J C 1975 *App. Op.* 14 11
[6] Morgan C J 1982 *Opt. Lett.* 7 368
[7] Greivenkamp J E 1984 *Op. Eng.* 23 4
[8] Creath K 1985 *Appl. Opt.* 24 18
[9] Malacara D (2007) *Optical Shop Testing* (New York: Wiley-Interscience John Wiley & Sons Inc. Publication)
[10] Villalobos-Mendoza B, Granados-Agustín F, Aguirre-Aguirre D and Cornejo-Rodríguez A 2013 *Proc. SPIE* 8785

Acknowledgments

B Villalobos-Mendoza and D Aguirre-Aguirre thank the CONACyT (México) the graduate studies scholarships granted, CVU: 271299, 272864, respectively.

Solar light collection from buildings façades

C García Cadena, C Sánchez-Pérez, M Tovar Rosas.
Centro de Ciencias Aplicadas y Desarrollo Tecnológico, Universidad Nacional Autónoma de México, Ciudad Universitaria, Apartado Postal 70-186, Distrito Federal 04510, México.

E-mail: garciacadenac@gmail.com

Abstract. We present an analysis of the light collection from the building's façade for interior lighting using diffuse reflective surfaces. The use of diffuse surfaces may increase the acceptance angle of such systems, in order to collect more sunlight coming from different directions according to the objects surrounding the building façade for longer periods of the day without the use of sun tracking systems. We discuss a numerical analysis and experimental results comparing light collection with the proposed system with respect to an equivalent system using specular reflecting surfaces. Results show that the collection is around 50 lx with a system of dimensions 10 cm x 15 cm and is of about of 55% higher than the collection achieved by a system of specular surfaces.

1. Introduction

The contribution of natural light available on the façades of buildings (in densely populated areas), comes mostly from reflections from the ground or nearby objects, which is non-directional light rather than light coming from the sun [1]. Light pipe lighting systems in different versions are most commonly used but they generally require modifications to the building structure [2] or to be planned previous a building's design [3]. Light pipes work more efficiently in the case of sunny days but its performance decreases considerably on cloudy days [4]. Another approach to capturing sunlight are the anidolic systems that are proposed for capturing light with good efficiency through specular reflectors installed on the façade of the building principally in sunny days [5].

Since most of the available sun light in the building's façade is not directional or diffuse, in the case of cloudy days, in this work we analyze the use of collection systems with diffuse reflective surfaces to improve the light collection that can be extended to more hours during the day in different type of weather conditions.

1. Proposed system

Most sunlight collection systems for interior lighting found in literature make use of surfaces that reflect incident light in a specular manner, as shown in Figure 1A, limiting light collection to only some specific angles of the incident light.

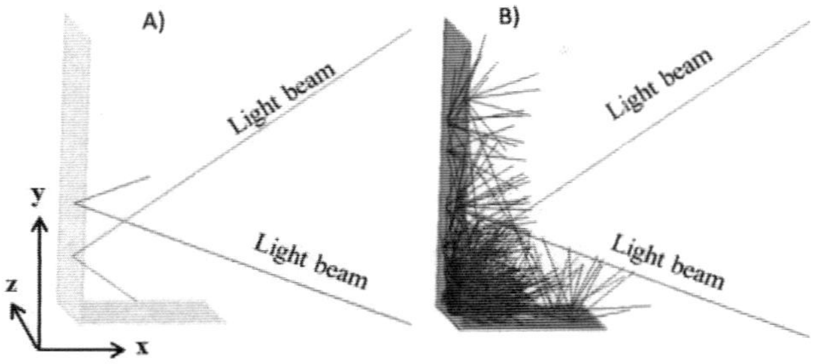

Figure 1. A) Reflection of light beams in specular surfaces. B) Reflection of light beams in diffusing surfaces.

Moreover, a uniformly diffusing surface reflects the total incident light in multiple directions with the same intensity [6] as shown in Figure 1B, it is for this reason that is of our interest to analyze the use of a system formed with diffuse reflectors. We propose a system using two diffuse reflectors joined by one of its edges as shown in Figure 2.

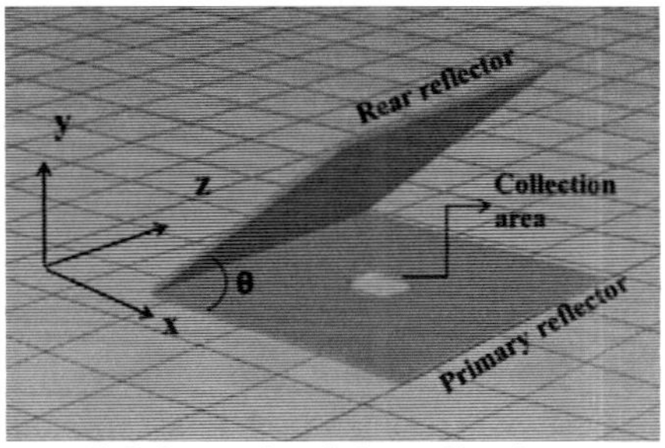

Figure 2. Schematic of the collecting system.

This system consists by a primary reflector (PR) where the day light from the sun directly enters and then is reflected by a rear reflector (RR) in which also reflects the light from the ground and surroundings objects. The reflected light will be collected in the collection area. We consider as the collected light all the radiant flux measured trough the collection area. The luminous flux emitted by the diffuse reflecting surface in the direction of the observer depends on the solid angle in the x-y-z plane as shown in Figure 3 therefore;$\mathbf{d\omega = Sen\phi d\psi d\phi}$ and depends on the light ray direction with respect to the $\boldsymbol{\phi}$ angleand the $\boldsymbol{\psi}$ angle formed between the normal to the surface and the direction of the observer with respect to the x-z plane. The proposed system surfaces reflect light according to the cosine law given by Lambert [7]. In this type of surfaces the reflected light flux (irradiance L) is given by:

$$L = \frac{dp}{dA} = L_d d\omega Cos\phi, \quad (1)$$

where $\mathbf{L_d}$ is the radiant flux per unit area and solid angle.

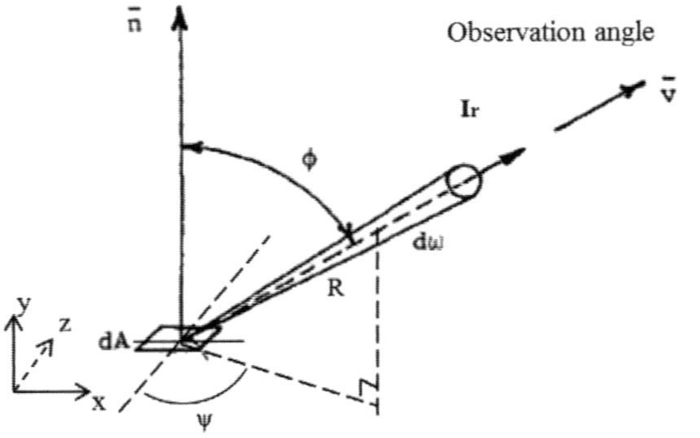

Figure 3.Projecting light reflected from a diffuse surface in the direction of the observer

It is possible to calculate the total optical power emitted by an ideal diffusing surface (Lambertian surface) when it is illuminated by an external source as follows:

$$\mathbf{M = \frac{dP}{dA} = \int L_d cos\phi d\omega = \int_0^{2\pi} \int_0^{\frac{\pi}{2}} L_d cos\,\phi sen\phi d\phi d\psi} \quad [\mathbf{Wm^{-2}}] \qquad (2)$$

Since the irradiance is not a function of φ or ψ, the solution of this double integral leads to the total power radiated by an illuminated Lambertian surface is constant as:

$$\mathbf{M = L_d \int_0^{2\pi} \int_0^{\frac{\pi}{2}} cos\,\phi sen\phi d\phi d\psi = L_d} \qquad (3)$$

The equation (3) is essential to calculate the interaction of radiant flux between diffuse surfaces, this term represents the total radiant flux reflected by a diffuse surface. The total radiant flux interaction between diffuse surfaces was estimated numerically using a ray tracing software as shown in next section.

2. Numeric simulations

To estimate the response of the proposed system we performed a series of simulations using a computer ray tracing program Zemax. For these simulations a homogeneous light source with a non-directional light distribution is used as the case of a partially cloudy day [8]. This source was placed at a distance of 1m from the collection system at a position where the emission of light flux completely covers the system.

In these simulations we tested different possible configurations to determine which should better collect light. To estimate the amount of luminous flux collected by the system a sensor was placed at the center of the PR in the collector area and we varied the angle θ formed by tilting the RR with respect to PR. We present in Table 1 results of these simulations.

Table 1 Luminous flux collected for different values of θ.

θ [°]	Luminous flux [lx]
10	3.58
20	6.41
30	9.84
40	10.4
50	8.96
60	8.42
70	6.25
80	4.73
90	4.05

From Table 1, it can be observed that the maximum light collection is found at 40° with 10.4 lx collection, below and over this angle there is a flux decrease by more than 40% of luminous flux collected by the system.

Regarding to the number of incident rays at the reflector surfaces, in Fig 4 we plotted the incidences of light rays (number of hits) at the RR and PR reflectors considering an angle θ=30°. For the RR and PR surfaces the number of incident rays is higher near the apex formed by the two reflectors where they are closer.

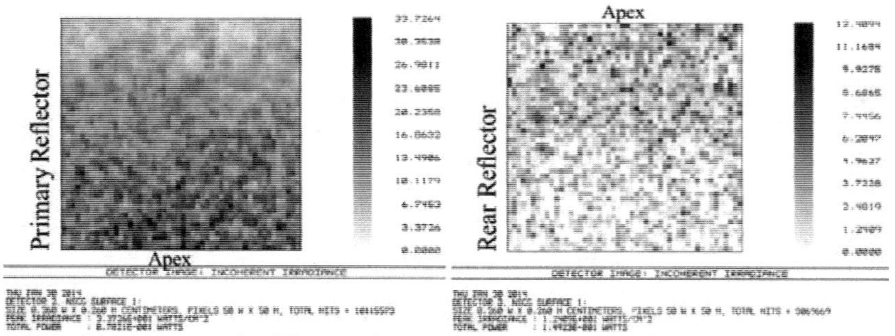

Figure 4. Incident rays at the PR (graph on the left) and RR reflector (graph on the right).

According to this we placed a sensor as the collecting element within the apex zone and calculated the luminous flux at the sensor for two systems of the same size but one formed by diffusing reflectors and the other formed by specular surfaces. The collector sensor was placed at the same distance from the light source for both of the systems. To emulate light coming from different angles at different hours of the day, we used 5 light sources configurations by placing at 5 different positions. For the five configurations we considered the same distance of 1m to the RR reflector as shown in Figure 5A. Results are shown in Figure 5B where we also plotted the flux detected by the sensor for the light source coming directly to the sensor without any reflector.

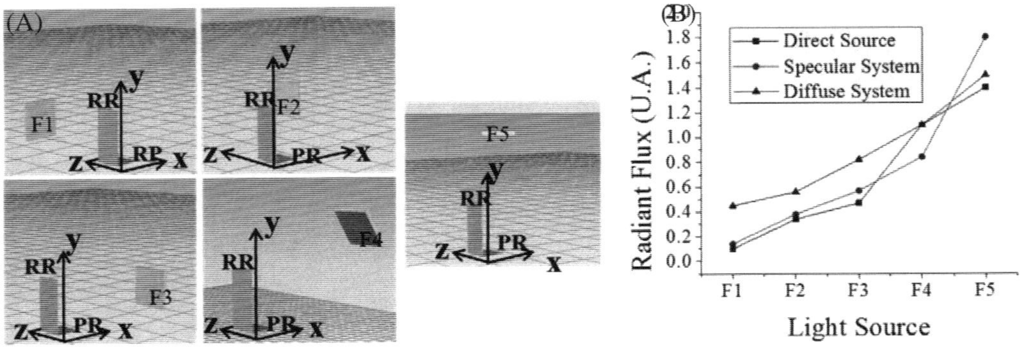

Figure 5. (A) Position of the light source with respect to PR. (B) Simulation results comparing the luminous flux collected using a diffuse system, a specular system and the light source without reflectors.

From results in Figure 5B one can see that the proposed system with diffusing reflectors collects nearly to 20% more radiant flux than the system with specular surfaces and over 25% of radiant flux coming directly from the light source. It is noteworthy that the specular system collects a larger amount of radiant flux that the diffuse system when the light source is located at 90 ° of inclination with respect to the PR, this is because at this position light reaches the sensor directly without undergoing many reflections. Results of these simulations were used to construct a collector that was tested under natural light illumination.

3. Experimental results

The diffuse system was constructed using carton reflectors of 15*cm* x 20*cm* coated with white paint, while the comparison system was made of common mirrors of the same dimensions as the diffuse system.

As shown in the numerical simulations most of the light was collected along the apex of the RR and the PR, for this reason we placed a plastic Fresnel lens covering the front of the system so it could increase light collection at the collection area. Light collected was measured by an optical fiber bundle, of 6 plastic fibers of 1mm core diameter, placed at one end collection area and the bundle output was connected to a light meter, as shown in Figure 6.

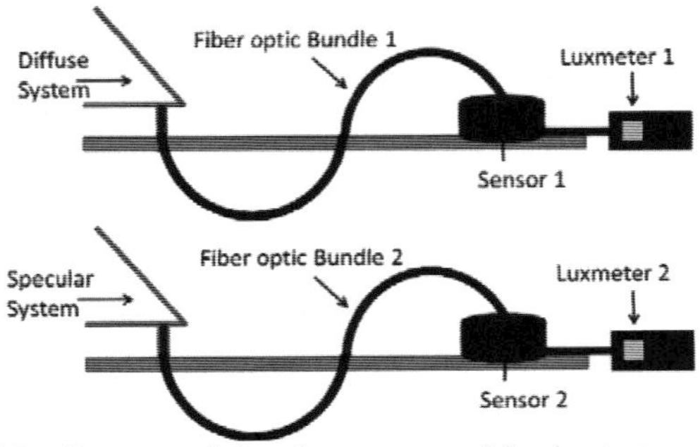

Figure 6. Assembly diagram of the elements used in the tests under natural light.

The measurements were carried out in periods of 8 hours on the north side façade of our laboratory. In figure 7 is presented the total luminous flux collected by the diffuse system compared with the collection of the specular system in a cloudy day.

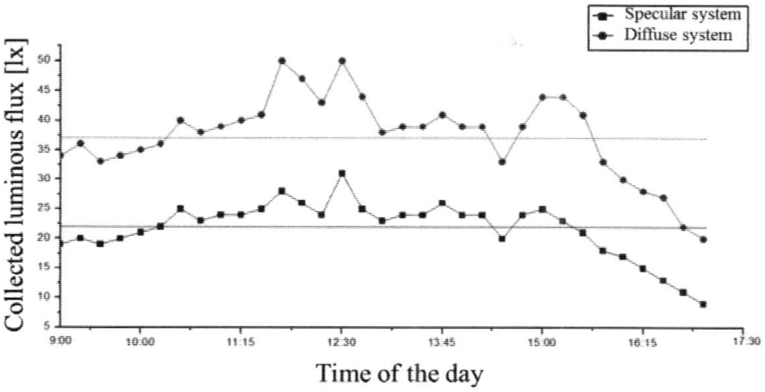

Figure 7. Luminous flux collected on a cloudy day for a system using diffuse reflectors (points) and the collected flow for a system using specular reflectors (squares).

It can be noted in Figure 7 that the proposed system collect about a 35% to 55% more light collected compared to the system formed by specular surfaces. The fluctuations in the measurements are due to changes of cloudiness along the day. It is also shown that the collection average for the diffuse system (around 40 lx) is higher than the specular system (around 25 lx) in an 8hrs period without the use of a sun-tracking system.

4. Conclusions

We propose a passive light collection system using diffuse reflectors, which can be installed in existing buildings façade. This system prove to be reliable in overcast day conditions where the diffuse system collected in average about 50% more light than the specular system in an 8 hours measurement.

It was shown in the numerical simulations which was the systems configuration that could collect a larger amount of radiant flux. This configuration was used to test the response of the proposed system experimentally. It was observed in the numerical simulations that the system formed by diffuse surfaces are able to capture about 20% more radiant flux compared to a system consisting of specular surfaces. It is also shown that the most collected radiant flux is concentrated in the apex of the PR and the RR, this is why we propose the use of a Fresnel lens to concentrate the light in the collection area. It is observed that the system, despite the small size of both the collecting system as the means of transport of light, managed to collect a significant amount of radiant flux.

This is why we can say that by adding more systems like this, we could meet the CIE interior lighting standards [9]. The possibility of a passive low-cost system is shown, this type of systems could be used to take advantage of the light available in the façades of the buildings without the need to modify the structure of them and thus have a significant reduction in the consumption of electricity [10] by increasing the actual hours of operation of this type of systems response over a day.

References

[1]Cheung H, Chung T. 2005 Calculation of the Vertical Daylight Factor on Window Façades in a Dense Urban Environment. Architectural Science Review.Vol.48 Iss 1.

[2] SolaTube international Inc. Daylighting systems Internet: **http://www.solatube.com/. Enero 2013.**

[3] Rosemann A, Kaase H. 2005. Lightpipe applications for daylighting systems.Solar Energy.Vol. 78 Iss 6.

[4] Beu D, Carter D.Lighting Design Methods for Obstructed Interiors. CIE 161:2004. CommissionInternationale de I'Eclairage, Vienna.

Journal of Physics: Conference Series **582** (2015) 012034 doi:10.1088/1742-6596/582/1/012034

[5] Ruck N, AschehougØ,Aydinli S. 2009*Daylight in Buildings*. International Energy Agency. Washington, D.C.
[6] S. K. Wittkopf, E. Yuniarti, L. Soon. Prediction of energy savings with anidolic integrated ceiling across different daylight climates. Energy and Buildings. Vol. 38, Issue 9. 2006. P.P. 1120-1129.
[7] J. Walsh. Photometry. 1958. New York. Dover Publications. pp. 136-52.
[8] Mills A. *Heat Transfer* 1999. New Jersey. Prentice Hall. pp. 538-591.
[9]Uetani Y, Aydinli S. Spatial Distribution of Daylight - CIE Standard General Sky.CIE S 011/E:2003. Commission Internationale de I'Eclairage, Vienna.
[10]CIE S 008/E: 2001/ISO 8995-1:2002(E). Lighting of Work Places - Part 1: Indoor. CommissionInternationale de I'Eclairage, Vienna.

Acknowledgments
We acknowledge financial support from DGAPA UNAM through grant PAPIIT IT100612.

Journal of Physics: Conference Series **582** (2015) 012035 doi:10.1088/1742-6596/582/1/012035

Manipulator Controlled since a Smartphone by Bluetooth

F Sánchez-Niño, G Rodríguez Pedroza, E G Castillo Baldivia
IICO-UASLP, Av. Karakorum 1470 Lomas 4a sección
San Luis Potosí, S.L.P, 78210, México

E-mail: franciscosanchezn@gmail.com

Abstract. We present the design of transmitter interface of data between a microcontroller and Smartphone to control a robot. We used a Bluetooth module to send the commands to control the moving of the manipulator. The system is formed by three parts: the first part, the program made in Android. It is run on the Smartphone that controls the position of the robot. The second part is the reception board based in the PIC18F4550 that energizes the joints of the manipulator. The last part is the driver. It use an integrated circuit L293D that is configured in two full bridges H. This system is a good tool for learning of programing, sensors, actuators, robotic, electronic and design electronic.

1. Introduction
Bluetooth is a radiofrequency technology that permits the wireless connection between devices in remote form giving an excellent security in the data transmission. The main characteristics are: sturdiness, low consumption of energy, low cost, transmission of voice and data, also can be used in fixed and mobile devices. Bluetooth is a standard IEEE 802.15.1 that works in the band since 2.4 to 2.485 GHz using the full duplex transmitter data in a rated speed of 9600 baud. Nowadays the Bluetooth technology is used in the following areas: informatics, telecommunications, automation, entertainment, laptops, system electronics and mobile technology [1, 2, 3].

2. Implemented system
In the figure 1 we show the block diagram of the implemented system It is formed basically by the Android program, the reception board based in microcontroller PIC18F4550 [4, 5, 6], the driver board and manipulation. In the next paragraphs are described each element.

Journal of Physics: Conference Series **582** (2015) 012035 doi:10.1088/1742-6596/582/1/012035

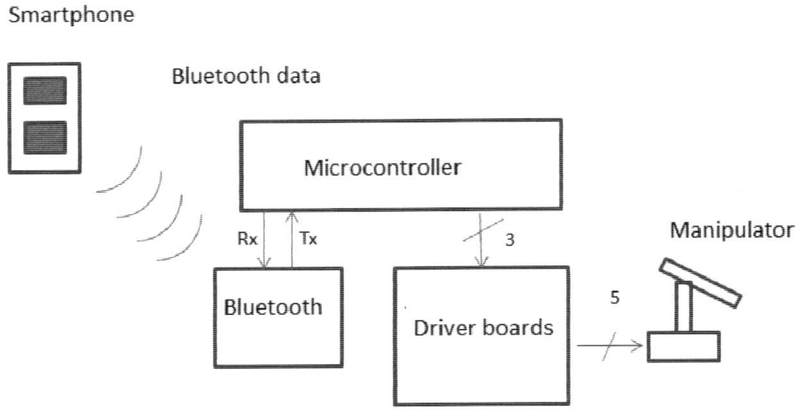

Figure 1. Block diagram of the system.

2.1 Android program

For this application, we use the libraries and tools provided by App Inventor Classic [8]. With this IDE was developed the program that runs on the Smartphone with Android system 2.3.5. In the figure 2 we show the principal screen of the application.

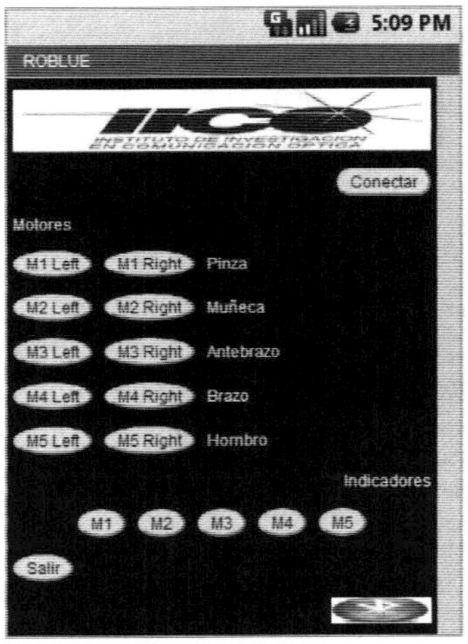

Figure 2. Control panel of the Bluetooth application.

The interface consists in a serie of buttons used to send commands to microcontroller PIC18F4550. In this form the microcontroller sends a signal to manipulator, also the application show the state or action made by the manipulator through visual indicators.

2.2 Receptor system by Bluetooth based in microcontroller PIC18F4550

The receptor system is based on the microcontroller PIC18F4550 and the main characteristics are: 5 Ports I/O, 4 timers, 20 interrupt sources, UART, USB, 13 A/D, 2 PWM and 2 EPWM [4, 5, 6]. The microcontroller is communicated through UART with Bluetooth module HC-06. It receives the commands to run by manipulator, this command is sent to PIC18F4550 by means of RX and TX Pins. The main features of the Bluetooth module are the following: voltage source since 3.6 V to 5.5 V, transmission rate since 1200 to 115200 baud, it works only like slave, current of 20 mA to 30 mA without connection and 8 mA with connection. The microcontroller receives the command and it uses a structure switch decides the action to run. The table 1 shows the commands.

Command	Function
0A	Open clamp
0B	Close clamp
0C	Up link 1
0D	Down link 1
0E	Up link 2
0F	Down link 2
0G	Up link 3
0H	Down link 3
0I	Turn left
0J	Turn right

Table1. Commands used by the microcontroller to define the action to be performed by the manipulator.

2.3 Driver board

Upon receipt of the data, the motors of the robot are energized through the driver board. In the figure 3 we show the block diagram of the driver board used to energize at the motors of the robot. This board is based in the integrated circuit L293D, it contains internally 2 full bridges H or 4 half bridges H. Three drivers boards were manufactured for apply current to the five motors of robot [7].

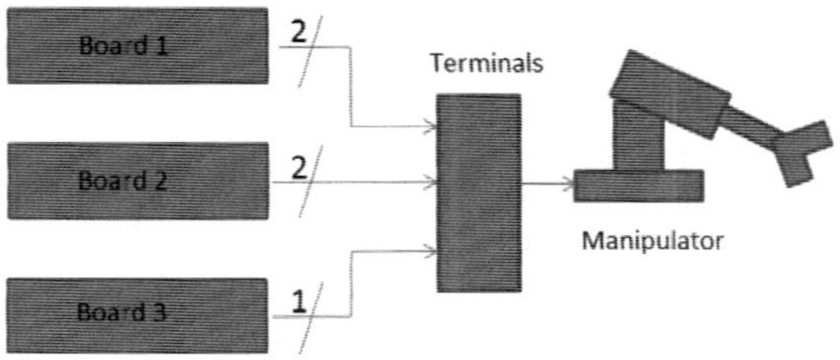

Figure 3. Block diagram of the driver board.

2.4 Manipulator

The manipulator has five degrees of freedom, three degrees for locate the object and two degrees to clamp the element. The table 2 shows the joints and their main characteristics.

Journal of Physics: Conference Series **582** (2015) 012035 doi:10.1088/1742-6596/582/1/012035

Joint	Feature	Joint	Feature
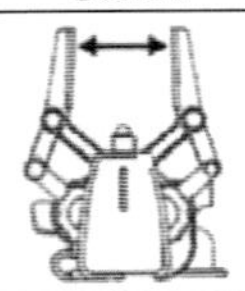	Maximum aperture of 4.5 cm.		Vertical movement of the clamp 120 °.
Joint	**Feature**	**Joint**	**Feature**
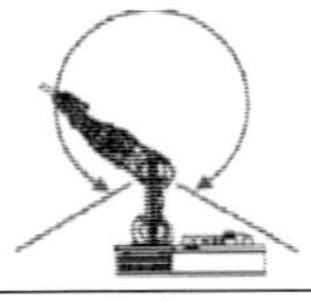	Vertical movement of the upper arm of 300 °.	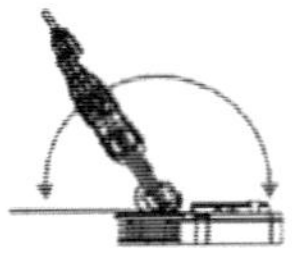	Vertical movement of the lower arm 180°.
Joint	**Feature**	The clamp has a capacity of 0.1 Kg. This arm has 5 DC motors and an LED to illuminate the work area of the clamp.	
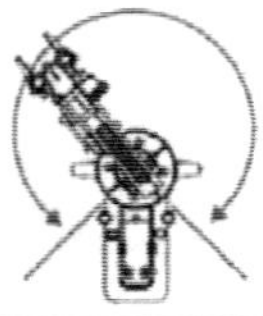	Horizontal movements of the arm mount 270°.		

Table 2. The joint features of the robot.

3. Results

We carried out several performance tests to the system developed to check reach and response. The right ranges of data are about 10 m. To lengths above 10 m the data are not reliable. Also we carried out test with obstacle and walls that interfered with the signal; we observed that the distance and the intensity signal did not change. The system response is optimal due to that the response of the motors is immediately and the movements of the manipulator are almost instantaneous.

4. Conclusions

We designed a robotics system based in microcontrollers PICs and frontal panel developed in Android to control the manipulator. This system is a good tool for learning of programing, sensors, actuators, robotic, electronic and design electronic. This system will use in courses of college and postgraduate to reinforce the knowledge of the students.

References

[1] McComb G. and Predko M 2006 *Robot Builder's Bonanza Third edition* (New York, Mc Graw Hill)

[2] Angulo J M and Romero S 2005 *Introducción a la robótica* (Thomson)

[3] Groover M P 2007 *Fundamentos de manufactura moderna: materiales, procesos y sistemas* (Mc Graw Hill)

[4] Lázaro A M 2001 *LabVIEW 6i Programación Gráfica para el Control de Instrumentación* (Paraninfo)

[5] Brey B B 2007 *Applying PIC18 Microcontrollers Architecture, Programming, and Interfacing Using C and Assembly* (Prentice Hall)

[6] Deitel P J and Deitel H M 2004 *C/C++ y Java Cómo Programar* (México, Prentice Hall)

[7] Horowitz P and Hill W 1989 *The Art of Electronics* (Cambridge University Press)

[8] http://appinventor.mit.edu/explore/ai2/tutorials.html

Journal of Physics: Conference Series **582** (2015) 012036 doi:10.1088/1742-6596/582/1/012036

Performance evaluation of a digital mammography unit using a contrast-detail phantom

J Elizalde-Cabrera and M-E Brandan

Instituto de Física, Universidad Nacional Autónoma de México, Ciudad Universitaria, México DF 04510, Mexico

E-mail: brandan@fisica.unam.mx

Abstract. The relation between image quality and mean glandular dose (MGD) has been studied for a Senographe 2000D mammographic unit used for research in our laboratory. The magnitudes were evaluated for a clinically relevant range of acrylic thicknesses and radiological techniques. The CDMAM phantom was used to determine the contrast-detail curve. Also, an alternative method based on the analysis of signal-to-noise (SNR) and contrast-to-noise (CNR) ratios from the CDMAM image was proposed and applied. A simple numerical model was utilized to successfully interpret the results. Optimum radiological techniques were determined using the figures-of-merit $FOM_{SNR}=SNR^2/MGD$ and $FOM_{CNR}=CNR^2/MGD$. Main results were: the evaluation of the detector response flattening process (it reduces by about one half the spatial non-homogeneities due to the X-ray field), MGD measurements (the values comply with standards), and verification of the automatic exposure control performance (it is sensitive to fluence attenuation, not to contrast). For 4-5 cm phantom thicknesses, the optimum radiological techniques were Rh/Rh 34 kV to optimize SNR, and Rh/Rh 28 kV to optimize CNR.

1. Introduction

In order to optimize the benefits of a mammographic study, mammography systems must operate under highly controlled technical conditions, providing the necessary image quality and the minimum dose consistent with the clinical goal. Figures of merit have been defined in order to quantify the ratio between benefit and risk [1-4]. For mammographic studies, the benefit is expressed by image quality parameters, while risk is associated with mean glandular dose.

In this work we have studied image quality parameters (determined using the CDMAM phantom) and mean glandular dose delivered by a flat panel digital mammography system for different polymethyl methacrylate (PMMA) thicknesses within a range of clinically relevant radiological techniques. We´ve proposed a method to quantify image parameters which extends the contrast-detail relation offered by the phantom design. The aim has been to evaluate the system performance, find the optimum operational conditions, and compare with those selected by the system. This evaluation is needed as a benchmark for the reliable use of the unit in the current laboratory research projects.

Journal of Physics: Conference Series **582** (2015) 012036 doi:10.1088/1742-6596/582/1/012036

1. Materials and methods

The analyzed system was a GE Senographe 2000D, with a dual-track (Mo-Rh) anode, and Mo (0.030 mm) and Rh (0.025 mm) filters. The digital detector consists of a 19x23 cm^2 columnar CsI(Tl) scintillator coupled to an a-silicon plate, with nominal pixel size equal to 100 μm. Raw ("for processing") images were used, unless specified. Two types of phantoms were used: homogeneous PMMA plates of various thicknesses and sufficiently large to completely cover the detector, and the CDMAM (version 3.4) [6]. The CDMAM phantom, shown in Fig 1, consists of a 0.5 mm thick aluminum base (99.5% pure Al), with 410 gold disks (99.999% pure Au) having diameters between 0.06 and 2.00 mm and thicknesses between 0.03 and 2.0 μm. A 5 mm thick, 16 x 24 cm^2 PMMA plate covers the base, and the CDMAM structure is equivalent to 1.0 cm of PMMA for a standard Mo/Mo/28 kV mammography X-ray spectrum. The phantom includes same area additional PMMA plates to simulate various breast thicknesses.

Figure 1. CDMAM phantom [5]. The 205 cells contain two gold disks each, one at the center and one placed randomly in a corner.

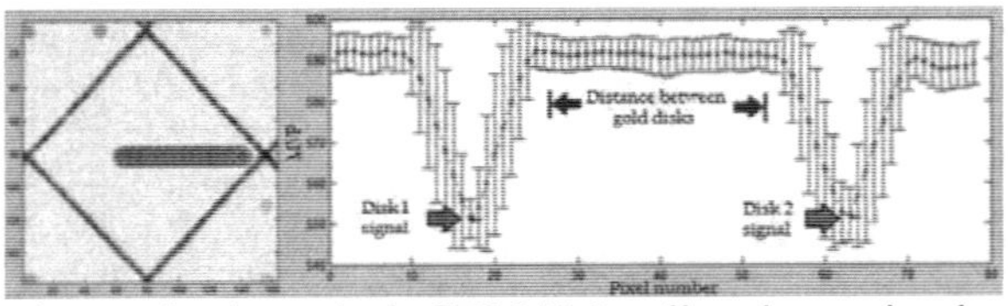

Figure 2. A typical CDMAM cell, where circular ROIs are created along the line that connects the disks. MPV and standard deviation are measured in the ROIs.

2.1 Detector response, and uniformities of the detector and X-ray field

The detector response was evaluated as the relation between mean pixel value (MPV) and X-ray tube mAs in images of a 4.5 mm thick PMMA phantom for Mo/Mo 28 kV X-rays at manually selected mAs values. MPV was obtained in squared 4 cm^2 regions-of-interest (ROIs) centered 6 cm from the torax edge of the image.The relation should be linear (squared correlation coefficient R^2> 0.99).

The detector uniformity was evaluated [5], acquiring raw images of a 4.5 cm thick PMMA phantom. The MPV was measured in squared ROIs at the center and corners of the phantom image. The relative MVP deviation of any ROI from the average should be < ±15%.

The uniformity of the X-ray field was determined exposing radiochromic film (EBT3 GrafChromic) to the appropriate air-kerma to induce a response (about 0.5 Gy) under the same X-ray conditions as the previous test. The film was located 15 cm above the breast support base and normal to the beam direction. The film was digitized before and after the exposure in order to extract the net response by subtraction [7-8]. A flatbed HP Scanjet 7650 with 300 ppm resolution was used.

2.2 Measurement of image quality and mean glandular dose

CDMAM images were acquired with the phantom placed on top of 2 cm thick PMMA in order to keep a constant distance between phantom and detector. Total phantom thicknesses were 4.0, 4.5 and 5.0 cm. All anode/filter combinations, voltage between 25 and 34 kV, and three mAs values for each combination were analyzed; 8 images were acquired for each setting. Also, images for each CDMAM thickness were obtained under automatic exposure control (AEC).

Journal of Physics: Conference Series **582** (2015) 012036 doi:10.1088/1742-6596/582/1/012036

The evaluation of image quality was performed in the traditional way, using the *CDMAM Analyser v1.5.3* [9] code, and a custom-made Matlab-based software. The first provided contrast-detail curves where the threshold disk thickness visible in the phantom image was evaluated for each disk diameter. The second code evaluated standard image quality parameters from the CDMAM images. These indicators were signal, noise, signal-to-noise ratio (SNR), contrast, and contrast-to-noise ratio (CNR), measured in individual phantom cells chosen by the user. The code locates the center of each disk in the cell, draws a straight line between them, and generates circular ROIs of the same diameter as the analyzed disks. Inside the ROIs, MPV and standard deviation are calculated. A MPV profile is created, as shown in Fig. 2, and the average MPV in both disks and the cell noise are obtained from the profile. From these, SNR and CNR are also calculated.

Mean glandular dose MGD was calculated for each PMMA thickness following Dance [10, 11],

$$MGD = K_{ES} g c s \,, \tag{1}$$

where K_{ES} is air-kerma at the phantom entrance surface, g is a factor that converts air-kerma into dose for 50% glandularity, c is a factor that corrects for differences in glandularity, and s is a factor that corrects for anode/filter combinations other than Mo/Mo. The radiological technique was the one chosen by the unit AEC.

2.3 Figures-of-merit (FOM)

Two figures-of-merit, FOM_{SNR} for SNR and FOM_{CNR} for CNR, were calculated as follows.

$$FOM_{SNR} = \frac{SNR^2}{MGD}\,,\; FOM_{CNR} = \frac{CNR^2}{MGD}\,. \tag{2}$$

In principle, these are independent of the photon fluence and depend on beam quality only.

2.4 Calculation of signal and noise

Detector signal and noise were numerically calculated following Lemacks' formalism [12] which includes knowledge of the spectrum, attenuation, detector efficiency and transformation of X-rays into light. From signal and noise, SNR and CNR were obtained. Fig 3 shows the calculation geometry.

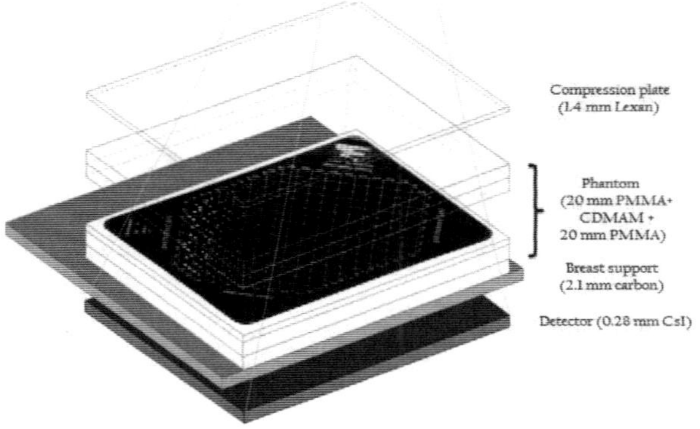

Figure 3. Geometry for numerical calculations. The focal spot is 66 cm from the detector, towards the top.

3. Results

3.1 Detector response, and uniformities of the detector and X-ray field

The detector response was lineal, R^2= 0.99991. Figure 4(A) shows the X-ray intensity distribution for Mo/Mo at 28 kV, and Fig 4 (B), the pixel value distribution for the same technique. The radiation field showed a maximum next to the thorax edge, slightly to the right, and a minimum at the superior left corner. The difference between the extreme values was ≈ 18%.The flat field pixel image showed a maximum at the center, slightly to the left and a minimum at the top right corner. The difference between extrema was ≈ 8%. Similar X-ray and detector distributions were measured for Rh/Rh.

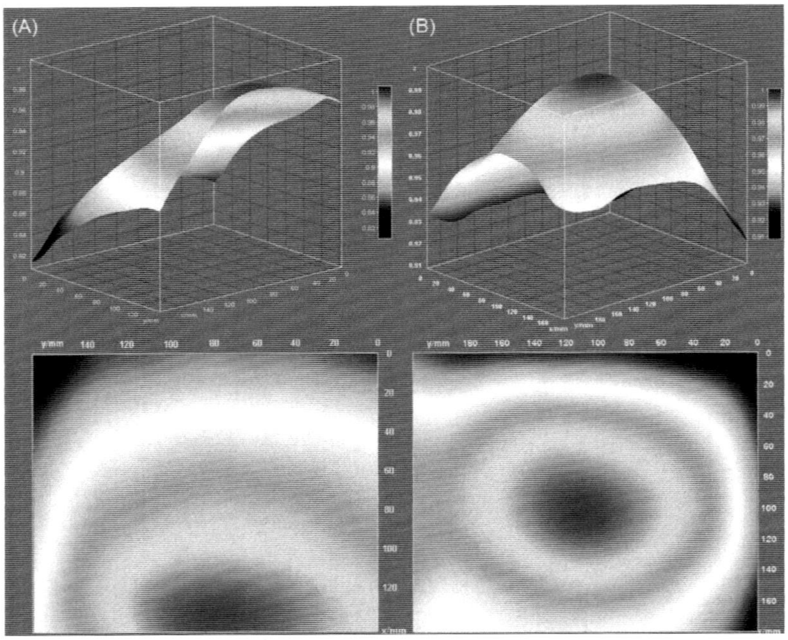

Figure 4. (A) X-ray intensity distribution measured with radiochromic film, normalized to its maximum for Mo/Mo, 28 kV. (B) Raw image pixel value normalized distributions for the same technique.

3.2 Measurement of image quality and mean glandular dose

Measured mean glandular doses were 1.2, 1.1 and 1.1 mGy for PMMA thicknesses of 4.0, 4.5 and 5.0 cm, equivalent to 4.5, 5.3 and 6.0 cm thick compressed breasts. These values comply with international recommendations [6]. The radiological techniques chosen by the AEC for 4.0, 4.5 and 5.0 cm PMMA were Mo/Mo, Mo/Rh, and Rh/Rh, all at 28 kV, respectively.

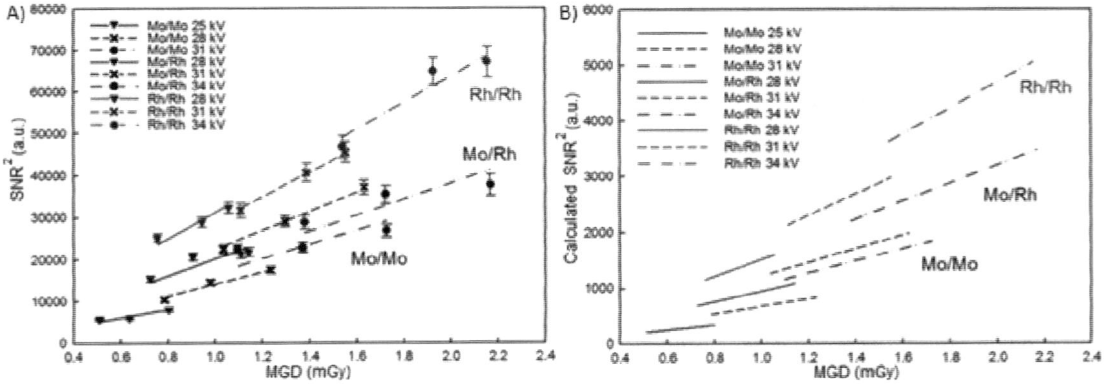

Figure 5. A) Measured SNR^2 vs MGD for a 4.5 cm thick PMMA phantom. All the analyzed anode/filter combinations and voltages are shown. B) Calculations for the same magnitudes.

Figure 5, left, shows representative values of SNR^2 and MGD measured for a 4.5 cm thick PMMA phantom. At the right, the corresponding calculations, which describe well the measurements.

3.3 Figures-of-merit

For 4.0 and 5.0 cm PMMA thicknesses, optimum FOM_{SNR} were obtained for the Rh/Rh beams and for 4.0 and 4.5 cm, changes in kV between 31 and 34 changed FOM_{SNR} little (only 1 to 4 %). For 5.0 cm, the optimum FOM_{SNR} was obtained at 34 kV. On the other hand, FOM_{CNR} optimization at each thickness required Rh/Rh beams at 28 kV. The unit AEC selected the optimum parameters (in terms of FOM_{CNR}) only for 5 cm PMMA.

4. Discussion and conclusions

Figures-of-merit have predicted optimum operation parameters. In general for the analyzed thicknesses, the optimum values were Rh/Rh at 34 kV for SNR and Rh/Rh at 28 kV for CNR. This result is consistent with independent conclusions such as those in [1-4]. These beams were harder (more energetic) than those automatically chosen by the unit; the AEC selected Rh/Rh only for 5 cm PMMA. The flat-fielding process preformed as a pre-process by the Senographe 2000D system reduced by about one half the non-homogeneities created by the heel effect, but the image of the homogeneous phantom still showed differences up to 6% between central and corner pixels. The MGD values for all PMMA thicknesses, which were equivalent to 4.5, 5.3 and 6.0 cm compressed breasts, respectively, were within international recommendations [5]. Analysis of both figures-of-merit indicated that the automatic control has been set to keep SNR constant, without consideration of contrast.

This overall evaluation has characterized the beam properties, delivered dose, and detector performance of this mammography unit which is currently used in research projects at our laboratory.

Acknowledgements

Authors thank C. Ruiz-Trejo and E. López-Pineda for continuous technical support during the project. We acknowledge partial financial support from DGAPA-UNAM Grant PAPIIT IN105813. JEC acknowledges a CONACyT scholarship for M.Sc.(Medical Physics) studies. We thank Miguel Rodríguez-Ponce, INCan, for facilitating the realization of this work.

References

[1] Baldelli P, Phelan N and Egan G 2008 Effect of Anode/Filter Combination on the Dose and Image Quality of a Digital Mammography System Based on an Amorphous Selenium Detector, *IWDM LNCS* **5116** 716-23

[2] Williams M B, Raghunathan P, More M J, Seibert J A, Kwan A, Lo J Y, Samei E, Ranger N T, Fajardo L L, McGruder A, McGruder S M, Maidment A D, Yaffe M J, Bloomquist A and Mawdsley G E 2008 Optimization of exposure parameters in full field digital mammography *Med. Phys.* **35** 2414-23

[3] Young K C, Cook J J H and Oduko J M 2006 Use of the European Protocol to Optimise a Digital Mammography System *IWDM LNCS* **4046** 362-9

[4] Young K C, Oduko J M, Bosmans H, Nijs K and Martinez L 2006 Optimal beam quality selection in digital mammography *The British Journal of Radiology* **79** 981-90.

[5] SEFM, SEPR and SERAM 2011 (in Spanish) Protocolo Español de Control de Calidad en Radiodiagnóstico

[6] Karssemeijer N and Thijssen M A O 1996 Determination of contrast-detail curves of mammography systems by automated image analysis *Digital Mammography* **96** 155-60

[7] Fuss M, Sturtewagen E, De Wagter C and Georg D 2007 Dosimetric characterization of GafChromic EBT film and its implication on film dosimetry quality assurance *Phys. Med. Biol.* **52** 4211-25

[8] Massillon-JL G, Chiu-Tsao S T, Domingo-Muñoz I and Chan M F 2012 Energy Dependence of the New Gafchromic EBT3 Film: Dose Response Curves for 50 KV, 6 and 15 MV X-Ray Beams *International Journal of Medical Physics, Clinical Engineering and Radiation Oncology* **1** 60-5

[9] European Reference Organisation for Quality Assured Breast Screening and Diagnostic Services, *http://www.euref.org*

[10] Dance D R 1990 Monte Carlo calculation of conversion factors for the estimation of mean glandular breast dose *Phys. Med. Biol.* **35** 1211-9

[11] Dance D R, Skinner C L, Young K C, Beckett J R and Kotre C J 2000 Additional factors for the estimation of mean glandular breast dose using UK mammography dissymmetry protocol *Phys. Med. Biol.* **45** 3225-40.

[12] Lemacks M R, Kappadath S C, Shaw C C, Liu X and Whitman G J 2002 A dual-energy subtraction technique for microcalcification imaging in digital mammography- A signal-to-noise analysis *Med. Phys.* **29** 1739-51

Journal of Physics: Conference Series **582** (2015) 012037 doi:10.1088/1742-6596/582/1/012037

Numerical study of remote detection outside the magnet with travelling wave Magnetic Resonance Imaging at 3T.

M López[1], F Vázquez[2], S Solís-Nájera[2], A O Rodriguez[3]

1. Departamento de Ciencias Basicas, Universidad Autónoma Metropolitana Azcapotzalco, México DF, México.
m.albert.ph@gmail.com
2. Facultad de Ciencias, Universidad Nacional Autonoma de Mexico, México DF, 04510, México.
jfva@ciencias.unam.mx, solisnajera@ciencias.unam.mx
3. Departamento de Ingeniería Eléctrica, Universidad Autonoma Metropolitana Iztapalapa, México DF, 09340, México.

E-mail: arog@xanum.uam.mx

Abstract. The use of the travelling wave approach for high magnetic field magnetic resonance imaging has been used recently with very promising results. This approach offer images one with greater field-of-view and a reasonable signal-to-noise ratio using a circular waveguide. This scheme has been proved to be successful at 7 T and 9.4 T with whole-body imager. Images have also been acquired with clinical magnetic resonance imaging systems whose resonant frequencies were 64 MHz and 128 MHz. These results motivated the use of remote detection of the magnetic resonance signal using a parallel-plate waveguide together with 3 T clinical scanners, to acquired human leg images. The cut-off frequency of this waveguide is zero for the principal mode, allowing us to overcome the barrier of transmitting waves at lower frequency than 300 MHz or 7 T for protons. These motivated the study of remote detection outside the actual magnet. We performed electromagnetic field simulations of a parallel-plate waveguide and a phantom. The signal transmission was done at 128 MHz and using a circular surface coil located almost 200 cm away for the magnet isocentre. Numerical simulations demonstrated that the magnetic field of the principal mode propagate inside a waveguide outside the magnet. Numerical results were compared with previous experimental-acquired image data under similar conditions.

1. Introduction

In the last three decades, the acquisition of magnetic resonance imaging (MRI) has been widely used for the diagnosis of diseases and various pathologies. The magnetic resonance systems commonly used in the clinic environment have intensities ranging from 1.5 to 3 T. However, MRI research systems employ magnetic field intensities greater than 3 T and up to 22 T. The used of waveguides systems at high-magnetic fields is still a very promising area of research for both clinical and preclinical applications. This is mainly due to the possibility of acquiring images with a larger field-of-views with improved signal-to-noise ratio (SNR) and good B1 homogeneity. An important feature is that the waveguides were initially used in high-field systems (7 T and greater). Brunner et.al. used a cylindrical waveguide to acquire the very first images with this novel approach

[1]. An important feature is the use of a patch antenna like excitation and reception [2]. The radiofrequency (RF) signal travels a long the waveguide until it reaches the phantom. The parallel-plate waveguide compromises two parallel conductive plates and is the simplest structure for the transmission and propagation of electromagnetic waves. This waveguide compromises two parallel conductive plates and it has been used to acquire images at 3 T with a whole-body clinical imager [3, 4]. Waveguides propagate frequencies as function of their cut-off frequency for the different M-modes which is given by the equation:

$$f_{cut-off} = m\left(2a\sqrt{\mu\varepsilon}\right)^{-1} \qquad (1)$$

Where m = 0, 1, 2, 3 ... and the values of the permeability, μ=4πx10^{-7} H/m, and permittivity $\varepsilon=\varepsilon_0\varepsilon_r$, with ε_0 = 8.85410x10^{-12} F/m, and ε_r = 65 for muscle [2]. The cut-off frequency for the principal mode (TM_0) of a parallel-plate waveguide is $f_{cut-off}$=o since m = 0. The TM_0 mode can propagate at any frequency. This particular characteristic of the parallel-plate waveguide makes it ideal for acquiring MR images with no restriction of the resonant frequency at remote points away from the bore isocenter.

2. Method

To investigate the magnetic field generated by the principal mode for this particular waveguide, numerical simulations were done with a magnetic field strength of 3T whose wavelength is 2.34 m. The waveguide consisted of two parallel copper plate 200 cm long and 12 cm wide and a 16 cm separation, plate thickness was 0.005 cm. All numerical simulations of the electromagnetic fields were conducted with the finite integral method implemented in the commercial tool, CST (CST Microwave Studio, Darmstadt, Germany). This tool solves electromagnetic systems using the integral evolution of Maxwell's equations instead of the differential form [5]. The cylindrical phantom was assumed to fill with a saline-solution and ad had 5 cm radio and was 15 cm long. The centre of the phantom was placed 10 cm inward of the waveguide: this point was considered as the origin for the numerical simulations. The distance between the phantom and the RF coil was varied with fixed step. Figure 1 (a) shows the setup used to numerically measure the magnetic field of the principal mode. A circular coil design was developed using the graphical interface of CST and shown in Figure 1 (b). To tune our coil at 128 MHz, fourteen capacitors of 100 pF were evenly placed around, and two parallel 0.7 pF capacitors, C_m were added for 50 Ohm-matching, and for fine tuning, an extra capacitor, C_t was inserted whose capacitance value varied from in 9.5 pF to 10.5 pF. The design of the parallel-plate waveguide is shown in Figure 1 (c). One of the ends of the waveguide is within in a hollow cylinder with external diameter of 60 cm and 40 cm long to mimic the magnet bore. Phantom was placed at the isocenter of the simulated magnet.

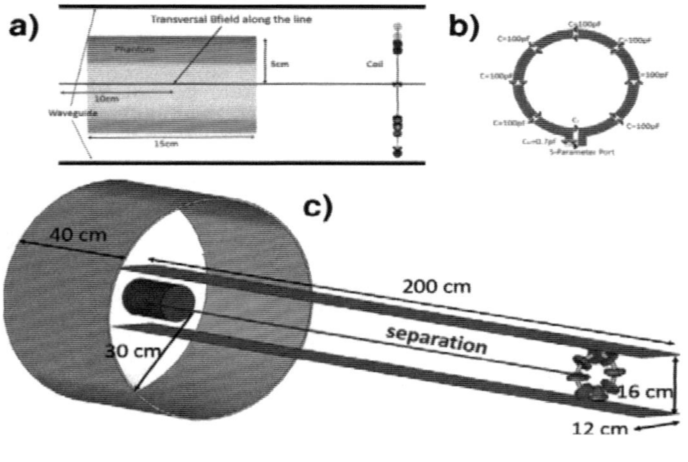

Figure 1. (a) Schematic diagram of the cylindrical phantom showing dimensions and location inside the waveguide. (b) Design of the circular antenna showing passive components for tuning and matching. (c) Setup for numerical simulations of the magnetic field of TM_0.

The simulated return loss of the parameter S11shows a good coil performance since values of around
- 35 dB can be reached. Figure 2 shows the simulated-acquired S11 for the circular coil of Figure 1.b).

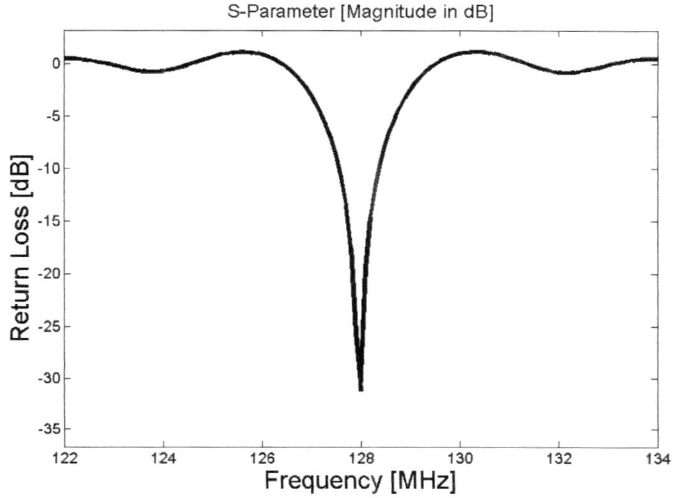

Figure 2. Simulated return loss S_{11} of the circular coil with no load in Fig. 1.

RF waves emitted by the antenna interact reach the phantom due energy conduction of the waveguide, representing the magnetic field of the principal mode, TM_0. Figure 3 shows sagittal (a) and coronal (b) views of the magnetic field inside the waveguide. An important decrease in the intensity of magnetic field can be appreciated at the centre of the waveguide.

Figure 3. Simulated magnetic field of the principal mode, TM_0 in sectional on the phantom in the parallel-plate waveguide for a specific distance between the phantom and the RF coil.

The numerical evaluation of the magnetic field at the isocenter was carried out using a longitudinal line, positioned along the waveguide, which due to the fact that the isocenter of the phantom is a separation of 10 cm measured from the edge of the guide, there is the magnetic field strength at this distance on the line, as shown in Figure 4.

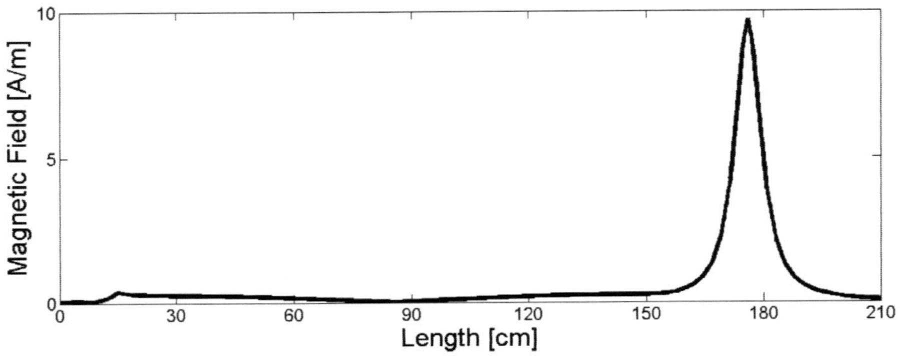

Figure 4. Variation of the magnetic field along the waveguide z-direction.

The data obtained are shown graphically in Figure 5, where we graph the separation distance against the magnetic field strength at the isocenter of the phantom vertical and horizontal

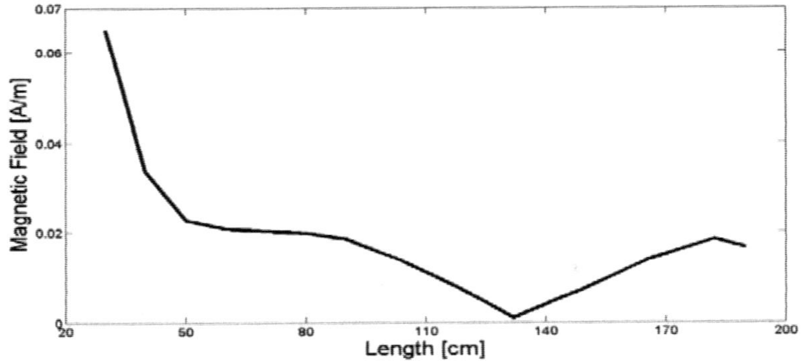

Figure 5. Behaviour of the magnetic field along the waveguide as described in Fig. 4.

The plot in Figure 6 shows experimentally obtained data of phantom images acquired at 3 T with a clinical system [3].

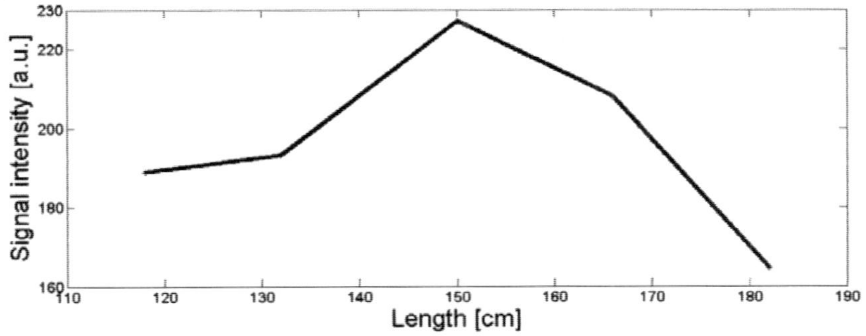

Figure 6. Separation Distance vs. Experimental signal intensity [3].

3. Discussion

An RF circular coil was successfully tune to 128 MHz and matched to 50 Ohm using the finite integral method as demonstrated by the simulated S11 plot of Figure 3. Numerical RF penetration also shows a very good coil performance to assure maximum energy transfer into the waveguide. Numerical results of the magnetic field can propagate along the parallel-plate waveguide for a RF coil positioned far away from simulated saline-solution phantom. Figure 4 shows that a higher magnetic field can be generated in the vicinity of the RF coil, while a lower magnetic field is also produced around the cylindrical phantom. However, an important decrease can be appreciated at the centre of the waveguide as depicted in Figure 4, too. At this point, a simulation using one single separation was done. A more complete study demands to numerically compute the magnetic field of TM0 for a number of separations between the RF coil and the phantom. There is a discrepancy between the experimental and numerical results that remain to be fully explained. At this moment, our guess suggests that our simulations may not be performed at greater values of magnetic field, resulting in an important decrease of the TM0 magnetic field. The electromagnetic results showed the feasibility to numerically simulate the magnetic field of the principal mode of parallel-plate waveguide.

References

[1] D. O. Brunner, et al., 2009, *Nature*, **457**, 994–998.
[2] A. Ibrahim, et al., 2010, *ISABEL*, 1-5.
[3] F. Vazquez, et. al., 2011 *ISMRM.* 19.
[4] F. Vazquez, et. al., 2013 *J. Appl. Phys.* **114** 064906.
[5] RF&Microwave Simulators -From Component to System Design; Thomas Weiland; CST GmbH, Bad Nauheimerstr. 19, 64289 Darmstadt, Germany.

Journal of Physics: Conference Series **582** (2015) 012038 doi:10.1088/1742-6596/582/1/012038

A Hydrogen detectors Wireless network for monitoring the gas ambient in a laboratory.

F Sánchez-Niño[1], F J. De Anda Salazar[1]
[1]IICO-UASLP, Av. Karakorum 1470 Lomas 4a sección
San Luis Potosí, S.L.P, 78210, México

E-mail: franciscosanchezn@gmail.com

Abstract. A system to monitor the hydrogen contents in the atmosphere present in the various sections of a crystal growth laboratory has been built. When hydrogen is used in the processes carried out in a laboratory it is important to monitor its content because the oxygen-hydrogen mixture is highly flammable and/or explosive. The system emits a warning when the Hydrogen concentration increases and if the concentration rises above 1 % in vol. it shuts down automatically any process that is using hydrogen. The design and performance of the system is reported here.

1. Introduction.

Hydrogen is necessary in many industrial and laboratory processes but there is the problem that when its concentration in the atmosphere is higher than normal the mixture is highly flammable and if its concentration reaches 4 % in vol. the mixture explodes if its temperature reaches 80 °C., the explosion can be triggered by a spark produced by a short circuit or the opening of a switch in an electric circuit [1, 2, 3, 4, 5].

It is then highly desirable to have a system to permanently monitor the hydrogen concentration in the ambient atmosphere if a small rise in the concentration is detected it is a sign that a Hydrogen is leaking somewhere in the process and if it is detected at an early stage the leakage can be found and stopped before the Hydrogen concentrations reaches dangerous values. Also, if somehow the Hydrogen concentration rises above a certain threshold an automatic shut-off of any hydrogen flow or electric circuit could prevent an explosion [1, 2, 3, 4, 5].

2. Design of the detectors wireless network.

A block diagram of the network is shown in Fig. 1 it consists of three boards containing a Hydrogen sensor, the signal of the sensor is connected directly to one of the internal ADC converters of a PIC16F877A microcontroller [6]. Depending on the value of the signal the microcontroller sends one of three different commands and an identification signal trough a radiofrequency transmitter. There are two reception boards one located outside the laboratory and the other, inside the laboratory, is connected to a PC that controls the experiment where hydrogen is being used.

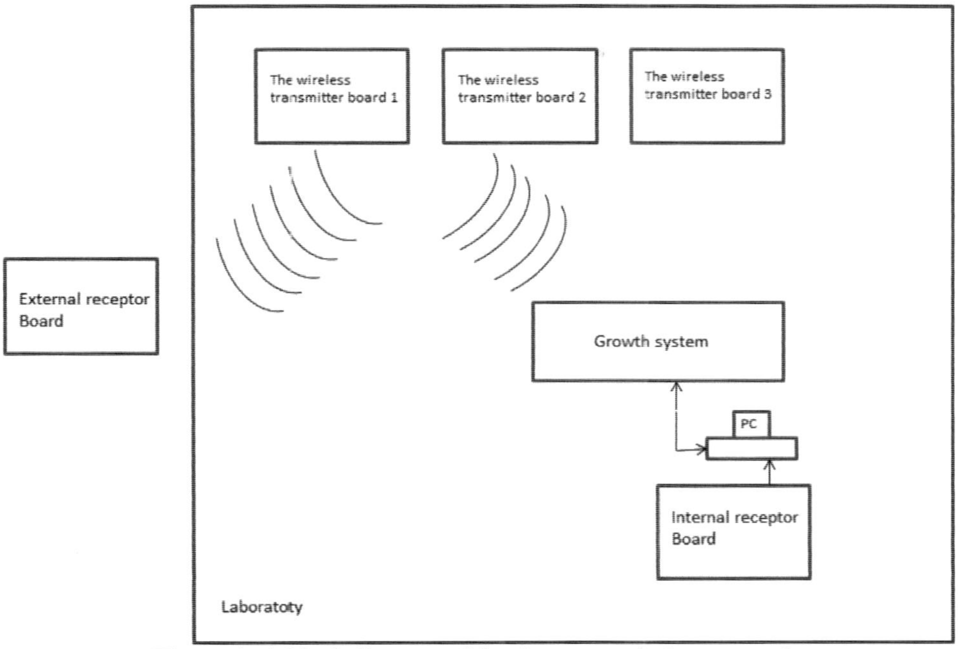

Figure 1. A block diagram of the detectors wireless network.

The external reception board uses the transmitted signals to identify the sender and to perform three different tasks depending on the command received. One command indicates that the hydrogen concentration is normal and constant, in this case there is a green LED lighted on in the reception board, other command indicates that the hydrogen concentration is above normal but below a certain limit the receptor module then emits a continuous audible and a low frequency intermittent visual yellow warning and on the third command, issued when the Hydrogen level reaches a dangerous limit, the receptor board emits an audible alarm and a high frequency intermittent red LED light signal.

The internal receptor board communicates with a PC and is used only to shut off the process when the hydrogen concentration exceeds certain limit corresponding to the third command as above.

Each detector board emits audible and visible signals according to the command it is transmitting.

Fig. 2 shows a block diagram of each Hydrogen detector board. It contains the hydrogen sensor a signal conditioning module, the microcontroller, the RF transmitter and audible and visual Indicators, each part will be described in detail below.

Journal of Physics: Conference Series **582** (2015) 012038 doi:10.1088/1742-6596/582/1/012038

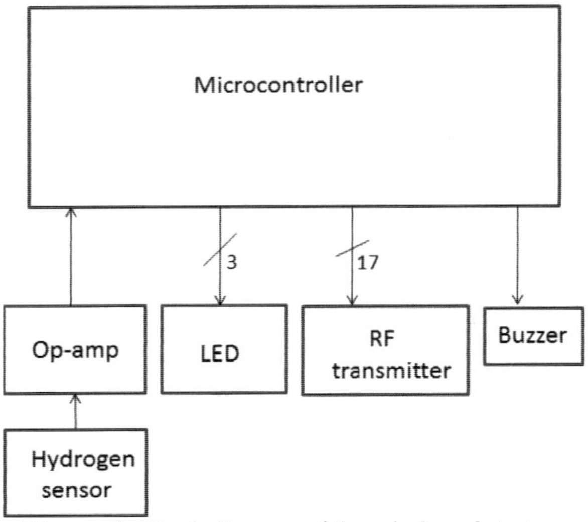

Figure 2. Block diagram of the wireless detector.

The external reception board is shown in the block diagram of Fig. 3. The internal reception board is not shown it consists simply of a RF receiver and a microcontroller that send a signal to a DAQ board in the PC, when this signal is detected the PC shuts off the process.

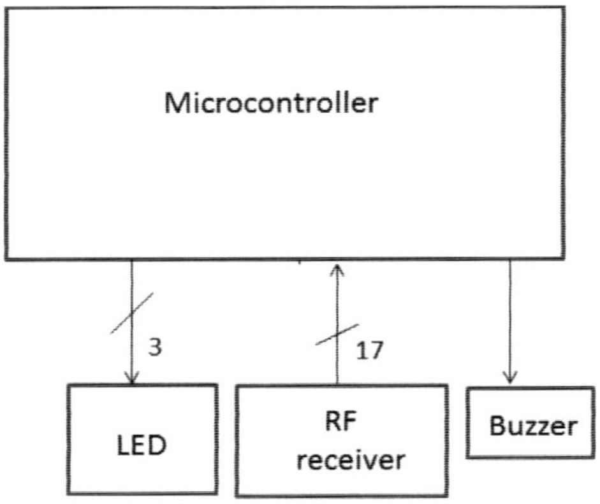

Figure 3. Block diagram of the wireless receiver.

2.1. The Hydrogen sensor

The hydrogen concentration is measured with a sensor of the company Synkera, model H2-PN-701 it can measure Hydrogen concentrations in the range 10-1000 ppm. The internal resistance of the sensor changes according to the hydrogen concentration. The characteristics of the sensor [3] are shown in Fig. 4.

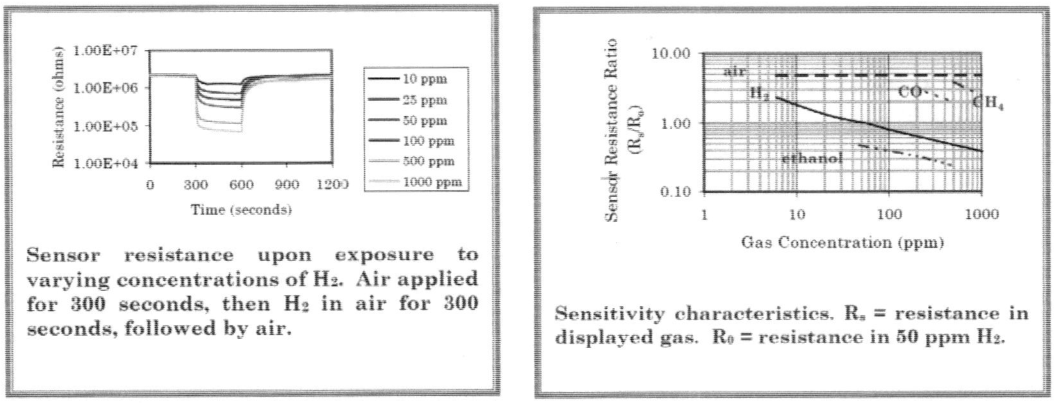

Figure 4. Sensor response in the presence of hydrogen.

The bias circuit of the sensor is shown in Fig. 5 (8). It utilizes a heating voltage (VH) of 5.4 V. The voltage supply to the divider (VC) is 5.0 V. The hydrogen concentration in contact with the sensor can be estimated from the internal resistance of the sensor (RS) and Fig. 4b above. The value of RS can be deduced from V_{Out} [3].

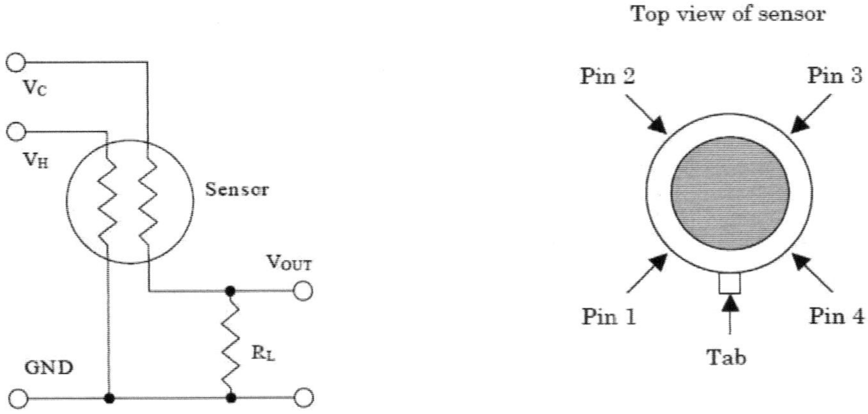

Figure 5. Basic measurement circuit is connected to pin 1 to 3 and heating voltage to pins 2-4 to the supply voltage.

2.2. The wireless transmitter board

In our case there are three transmitter boards. The block diagram of the wireless transmitter board is shown in Fig. 2. The board is designed around a PIC16F877A microcontroller [6]. It measures the output of the sensor (VOut) using one of its ADC converters, depending on its value it codifies the data that will be send by the RF transmitter and sets the corresponding state of the LEDs and buzzer. The signal was transmitted using a TWS-434 amplitude modulated (AM) radiofrequency transmitter working at 433.92 MHz with a maximum transmission rate of 8 Kbps [8, 9]. The corresponding circuit is shown in Fig. 6.

The data transmitted consist of a sequence of 17 bits with the following structure:

- A starting bit

Journal of Physics: Conference Series **582** (2015) 012038 doi:10.1088/1742-6596/582/1/012038

- 8 address bit used to verify that the signal comes from a transmitter that that belongs to system.
- 8 data bits, the first 4 bits are the data bits and the other bits identify the transmitter board sending the signal.

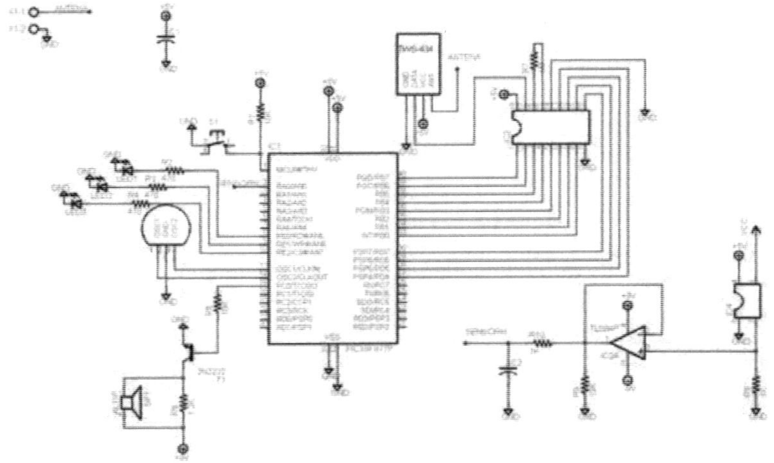

Figure 6. Transmitter circuit board.

Between each 17 bits transmission there is a time period of the same length as the 17 bits where no data is transmitted.

2.3. The wireless receptor Board

This board was designed using a PIC16F877A microcontroller [3] and a RWS-434 RF receptor module its block diagram is shown in Fig. 3. The data received is fed to the microcontroller which decodes the information to identify the transmitter board and to indicate the hydrogen level in the area where that transmitter board is located [10, 11]. The corresponding circuit is shown in Fig. 7.

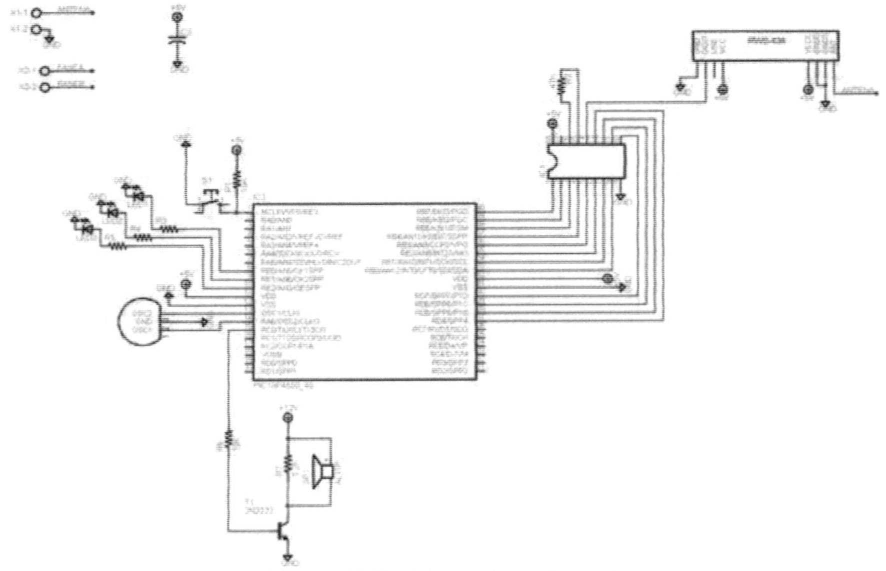

Figure 7. Receiver circuit board.

2.3.4 Network operation.

Since there are three transmitter boards and only one reception board it is necessary to synchronize the transmitters because if any two of them were transmitting at the same time they would interfere and the receptor could not decode the signal transmitted. So there is a transmission sequence as shown in Fig. 8. The transmission of each board lasts 5 s and is followed by a 2 s dead period.

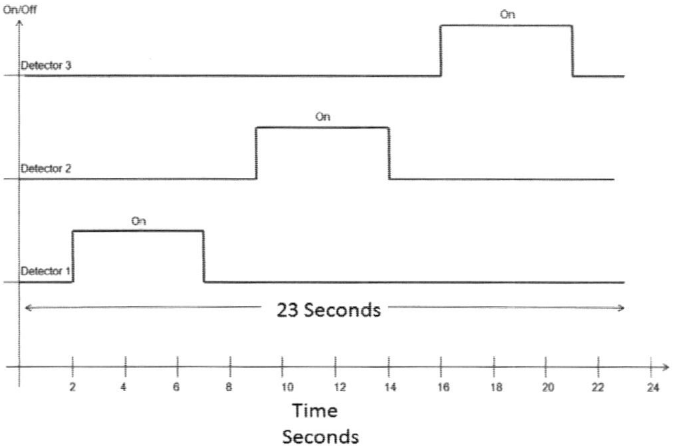

Figure 8. Diagram of wireless network operation.

3. Results

The system has been tested to check for the signal variation when there is no extra Hydrogen in the atmosphere it was found that with a Load Resistance (RL) of 1.00 K□ the signal remains between 50 and 70 mV. Leakages of Hydrogen were simulated by blowing the detector board from below at several distances with a 0.7 l/min flow of a gas mixture containing Hydrogen (5 % H_2 -95 % Ar). The time response depends mainly on the arrival time of the Hydrogen to the sensor. And the system gives either an alarm or warning signals depending on the distance of the blowing nozzle and on how long the flow lasts.

4. Conclusions

A wireless system for Hydrogen detection has been developed. It can detect dangerous concentrations of Hydrogen in different locations and send a warning signal to a receptor unit that can be located up to 50 m apart from the transmitter boards. Other receptor boards can communicate with a PC to shut down the hydrogen flow if necessary.

Acknowledgments

We appreciate the skillful help of the technical staff at IICO-UASLP. This work was partially supported by CONACYT, PROMEP and FAI at UASLP.

References

[1] Mishournyi V A, Hernández I C, Gorbatchev A Y and Lastras A 2002 Tecnologías epitaxiales de crecimiento de cristales semiconductores *Avance y Perspectiva* **21** 21
[2] Manipulación de Hidrógeno, Abelló Linde S.A, www.abellolinde.es
[3] Trace Hydrogen Sensor (P/N 701), Datasheet, Synkera Technologies Inc.
[4] Somov A, Baranov A and Spirjakin D 2014 A wireless sensor-actuator system for hazardous gases detection and control *Sensors and Actuators A: Physical* **210** 157

Journal of Physics: Conference Series **582** (2015) 012038 doi:10.1088/1742-6596/582/1/012038

[5] Automatización del Sistema de Crecimiento de Cristales por Epitaxia en Fase Líquida, IICO-UASLP, Agosto 2009.
[6] Microchip, Datasheet PIC16F877A/74A/76A/773A, 2001, www.microchip.com
[7] Microchip, Datasheet PIC18F2455/2550/4455/4550, 2004, www.microchip.com
[8] Transmitter Module: TWS-BS-3(433.92 MHz), Datasheet, www.wenshing.com
[9] HT12A/HT12E 212 series of encoders, Datasheet, April 11, 2000, www.wenshing.com
[10] Receiver Module: RWS-374-6(433.92 MHz), Datasheet, www.wenshing.com
[11] HT12D/HT12F 212 series of decoders, Datasheet, November 18, 2002, www.wenshing.com

Study of absorption, emission and EDS properties of Pr^{3+} incorporated in a SiO_2 matrix by sol-gel method

M Gómez-Miranda[1], C Velásquez Ordoñez[2] and R Sosa Fonseca[1]
[1]Physics Departement Universidad Autónoma Metropolitana- Iztapalapa, Distrito Federal, México PO BOX 55-532.
[2]Universidad de Guadalajara Centro Universitario de los Valles Centro de Investigación en Nanociencia y Nanotecnología de la Depto. de Ciencias Naturales y Exactas Ameca Jal. México

E-mail: marisol.gomez.miranda@gmail.com

Abstract
The optical response of the Pr^{3+} ions doped monolith of SiO_2 prepared by sol-gel method was investigated by absorption and emission spectra. The absorption spectrum shows some typical lines of Pr^{3+}, with these data was made the diagram levels corresponding to SiO_2: Pr^{3+}. The emission spectra were taken at 225 nm and 240 nm of wavelength of excitation was measured. In that there are reabsorption line at 423 of the Pr^{3+} ions in the emission host, that means that some kind of host-ion interactions exist. Red and NIR emission at 612 nm, 652 nm, 711 and 728 nm, respectively, from Pr^{3+} ions in SiO_2 was observed. The transition $^1S_0 \rightarrow {}^3P_2$ at 425 nm is observed because an effective radiative transfer between the silicate glasses host and the praseodymium ions is observed and discussed.

1. Introduction

Luminescent SiO_2 activated with trivalent rare earth are attractive due to their potential in technological applications. Narrow lines characterize the intra-4f emission spectra of rare earth ions due to the 4f electrons of rare earth ions are shielded by the outer 5s and 5p electrons from external forces, [1]. The positions of the 4f configuration energy levels are only slightly dependent on the host matrix, and are roughly the same as the free-ion levels. These unique properties are of interest in development of new materials with the capability to produce visible light with narrow lines for red, green, and blue phosphors. Trivalent praseodymium (Pr^{3+}) has been used as an activator in various host matrices to prepare phosphors that can be used in different types of light emitting devices. Pr^{3+} ion has unique features, one of which is the ability to emit efficiently in the visible (VIS) to the infrared (IR) spectral regions depending on the host matrix and the ion concentration, [2,3]. Photoluminescence resulting from intra-4f-configurational transitions of rare earths ions embedded in inorganic glasses has been essential of a practical fibre amplifier, though some non-crystalline materials (including glass–ceramics sometimes) doped with transition metal ions or heavy metal ions are known to exhibit fluorescent emissions in the near-infrared wavelengths, [4]. On the other hand the sol-gel process had to be a good method to synthetize metallic oxides. In this were synthetized SiO_2 monoliths with praseodymium ions by sol gel technique. This material was optically characterized by absorption, emission and EDS measurements, which are presented and discussed here.

2. Experimental

All metal alkoxides, as well as, all solid reagents employed in this work, were acquired from Aldrich, while all solvents were purchased from Fluka. Sol-gel methods were used to produce monolithic silica xerogels doped with trivalent praseodymium ions. Continuous fluorescent spectra were obtained with a Perkin-Elmer model 650-10S spectrofluorometer equipped with a 150 W Xenon lamp, and the absorption was obtained from Cary 5 equip. The photoluminescence spectra emission and absorption of SiO_2: Pr^{3+} monoliths were analyzed in the range of 230 – 800 nm. And the Electron dispersion Scattering of X-Ray (EDS) measurements of the composition of the SiO_2: Pr^{3+} sample, was taken with a Scan Electron Microscope of high resolution HRSEM Jeol 7600F.

3. Results and Discussion

In figure 1(a), it is observed the absorption spectrum of SiO_2: Pr^{3+} from 400 nm to 800 nm, to RT. The inset shows details from 700 nm to 800 nm ranges. Absorption lines in 444nm, 449.5 nm, 468 nm, 481.5 nm and 589.5 nm are observed and they corresponding to $^3P_2$, $^3P_1$, $^3P_0$, $^1D_2$ transitions, from $^3H_4$ level; and the $^3F_4 \rightarrow {}^3P_0$ (at 732 nm) transition, respectively. With these data we make the diagram level and it appear in Figure 1(b). Radiative transitions (solid line) and no radiative transitions (dash line) observed in emission spectra are marked in figure 1b, too.

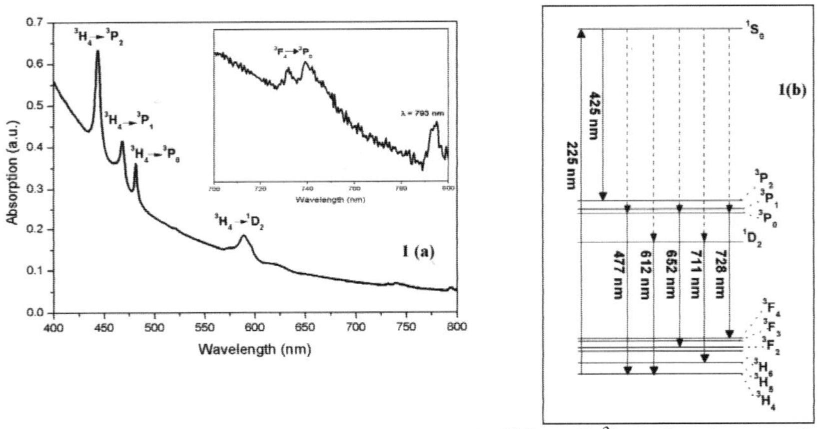

Figure 1. (a) Absorption spectrum of SiO_2: Pr^{3+} sol-gel monolith, (b) corresponding diagram levels.

Figure 2a, shows a characteristic RT emission spectrum of SiO_2:Pr^{3+} obtained to wavelength excitation at λ_{exc}= 225 nm and λ_{exc} = 240 nm, respectively. In the same figure the characteristic absorption spectrum of the same sample is shown. It consists of several sharp lines with transitions from the state $^3H_4$ to the excited state stark components of the $4f^2$ configuration of the Pr^{3+} ion. The intensity of the spectra was adjusted in order to compare their shapes. In figure 2b, we present a comparison of both emission spectra and it is observe that there is a missing line at 425 nm; it is important to see that the wavelength excitation at 225 nm is a level energy corresponding to $^1S_0 \rightarrow {}^3F_4$ absorption transition from 4f5d configuration of Pr^{3+}; whereas exciting at 240 we can observe a hole at the same wavelength. The reabsorption band in the intrinsic luminescence spectrum just appeared at the wavelengths where there is a transition of Pr^{3+}, the $^1S_0 \rightarrow {}^3P_2$, at 425 nm proximately. It is observed because an effective radiative transfer between the silicates glasses host and the praseodymium ions occurred in our samples.

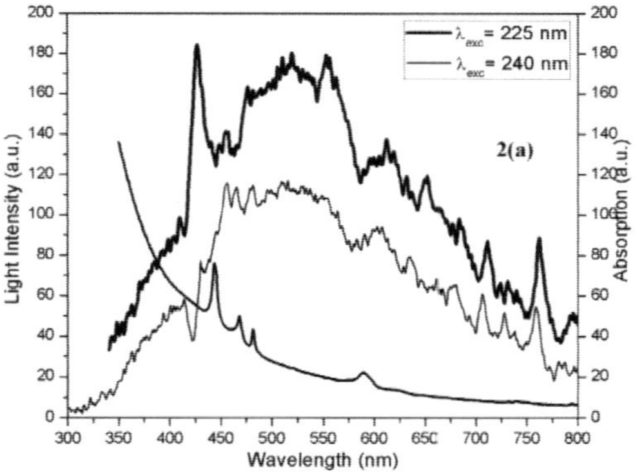

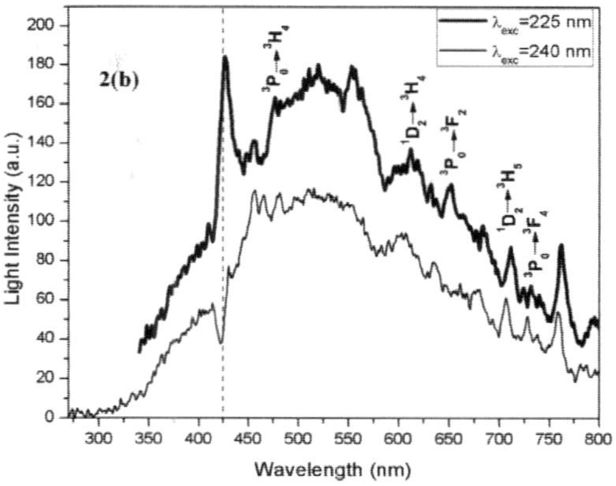

Figure 2. (a) Absorption and luminescence spectra measured to wavelength excitation 240 nm and 225 nm, respectively. (b) Identification of red and NIR transitions in the intrinsic emission of SiO_2: Pr^{3+}.

To account for the properties of the excited states it is applied a Gaussian fit deconvolution method to the optical absorption spectrum and it was evaluated the areas and the rate areas of each transition observed in absorption spectrum, see figure 3 and table 1. These were compared with reported data and it is observed that the $^3H_4$ - $^3P_2$ transition is highest. Figure 4, shows the composition of our sample that consist of 16.11 ± 5.4 of C, 71.15 ± 3.6 of O, 12.49 ± 1.8 of Si, 0.20 ± 0.03 of Cl and 0.05% ± 0.02 of Pr, all of them in atomic per cent. A homogeneous distribution of Pr ions in SiO_2 is presented in figure 5. It is relevant to know the dopant ions distribution to avoid the concentration quenching effect due to migration excitation energy among Pr^{3+} ions. Here we have enough low concentration of Pr^{3+} ions to consider it.

Table 1. Characteristics transitions of Pr^{3+} incorporated to SiO_2 glass with area under each absorption band, respectively.

Transitions	Experimental Wavelength (nm)	Experimental Wave Number (cm^{-1})	Reported Wavelength (nm)	Reported Wave Number (cm^{-1})	Area	% Area
$^3H_4 \rightarrow ^3P_2$	444	22523	444	22520	737.05	28.30
$^3H_4 \rightarrow ^1I_6$	449.5	22247	450	22211	126.43	4.85
$^3H_4 \rightarrow ^3P_1$	468	21368	469.4	21330	211.13	8.11
$^3H_4 \rightarrow ^3P_0$	481.5	20768	481.9	20750	109.78	4.21
$^3H_4 \rightarrow ^1D_2$	589.5	16964	593.8	16840	211.01	8.10

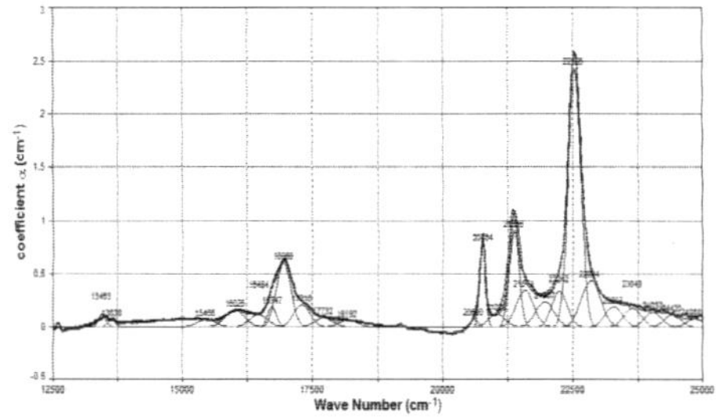

Figure 3. Gaussian fit of absorption spectra to account the individual intensity of optical transitions of the Pr^{3+}.

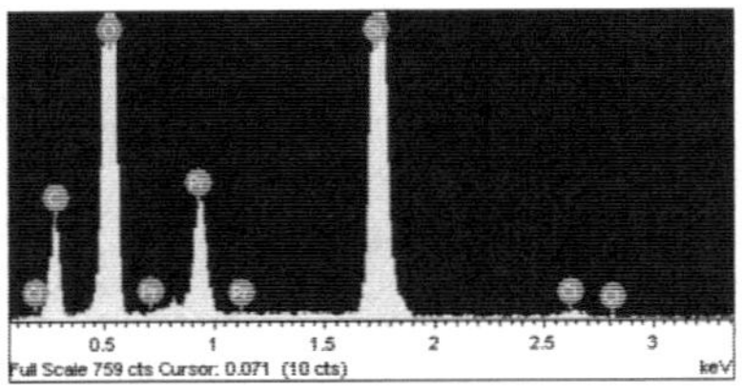

Figure 4. Electron by Dispersion Scattering X-Ray technique about the composition of SiO_2:Pr^{3+} samples.

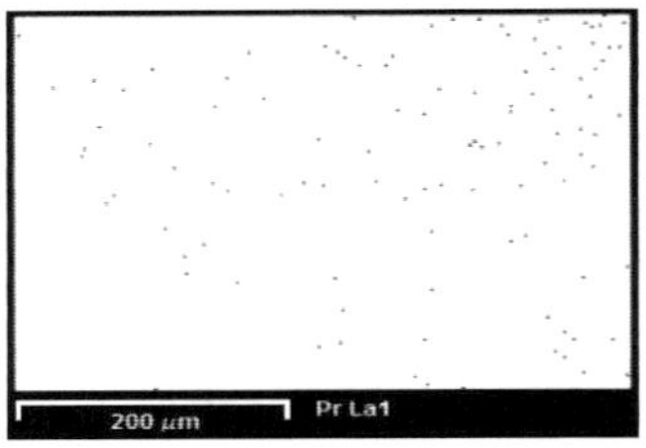

Figure 5. Photography of the Pr^{3+} ion distribution obtained by EDS spectroscopy.

4. Conclusions

Red and NIR emission at 612 nm, 652 nm, 711 and 728 nm, that corresponding to $^3P_0 - {}^3H_4$, $^3F_2$ and $^3F_4$ and from $^1D_2 \rightarrow {}^3H_4$, $^3H_5$, respectively, from Pr^{3+} ions in SiO_2 was analysed. Reabsorption line at 423 of the Pr^{3+} ions in the emission host is observed when we exciting at 240 nm but when λ_{exc} = 225 nm it is observed a peak at 425 nm, associated to the transition $^1S_0 \rightarrow {}^3P_2$ at 425 nm. It is because an effective radiative transfer between the silicate glasses host and the praseodymium ions is observed and discussed. The possible mechanism of energy transfer from host SiO_2 to Pr^{3+} was discussed.

Acknowledgments

Thanks to A M Arrieta Castañeda for the EDS data of the Central Laboratory at the Universidad Autónoma Metropolitana, Iztapalapa.

References

[1] Mhlongo G H, Ntwaeaborwa O M, Swart H C, Kroon R E, Solarz P, Ryba-Romanowski W and Hillie K T 2011 *J. Phys. Chem. C* **115** 17625-17632
[2] Ryba-Romanowski W, Dominiak-Dzik G, Solarz P, Klimesz B, Zelechower M J 2004 *Non-Cryst. Solids*, **348&346** 391-395
[3] van der Kolk E, Dorenbos P, Vink A P, Perego R C and van Eijk C W E 2001 *Phys. Rev. B* **64** 195129
[4] Fu Y, Zhang G, Qi Z, Wu W and Shi C 2007 *J. Lumin.* **124** 370–374
[5] Carnall W T, Fields P R and Rajnak K 1968 *J. Phys. Chem* **49** 4424-4442
[6] Juestel T, Mayr W, Schmidt P J, Tuning the 4f1 - 5d-4f2 UV Emission of Pr3+ Philips Research Laboratories, Aachen, Germany

Journal of Physics: Conference Series **582** (2015) 012040 doi:10.1088/1742-6596/582/1/012040

Effects of psychological stress test on the cardiac response of public safety workers: alternative parameters to autonomic balance

M R Huerta-Franco[1], F M Vargas-Luna[2], I Delgadillo-Holtfort[2]

[1]Department of Applied Sciences to Work, Health Science Division, University of Guanajuato,
Aquiles Serdán # 924, Col. Obregón, 37320, León, Guanajuato, Mexico
[2]Department of Physical Engineering, Science and Engineering Division, University of Guanajuato,
Loma del Bosque 103, Col Lomas del Campestre León, Guanajuato 37150, Mexico

E-mail: huertafranco@hotmail.com

Abstract. It is well known that public safety workers (PSW) face many stressful situations that yield them as high-risk population for suffering chronic stress diseases. In this multidisciplinary research the cardiac response to induced psychological stress by a short duration Stroop test was evaluated in 20 female and 19 male PSW, in order to compare traditionally used cardiac response parameters with alternative ones. Electrocardiograms have been recorded using the Eindhoven electrodes configuration for 1 min before, 3 min during and 1 min after the test. Signals analysis has been performed for the heart rate and the power spectra of its variability and of the variability of the amplitude of the R-wave, i.e. the highest peak of the electrocardiographic signal periodic sequence. The results demonstrated that the traditional autonomic balance index shows no significant differences between stages. In contrast, the median of the area of the power spectrum of the R-wave amplitude variability in the frequency region dominated by the autonomous nervous system (0.04-to-0.4 Hz) is the more sensitive parameter. Moreover, this parameter allows to identify gender differences consistent with those encountered in other studies.

1. Introduction

Electrocardiograms are the most frequently used cardiac electric signals because of their noninvasive character and the information they can provide. The characteristic shape of an electrocardiogram (ECG) consists of a periodic sequence of a succession of alternating local maxima and minima denoted as PQRST complex that are presented in each heartbeat. The central peak maximum, usually the highest peak of the complex when using the Eindhoven electrodes configuration, is known as R-wave. This complex is due to the different phases of depolarization, repolarization and electrical recovery in the atria and ventricles of the heart. However, due to the characteristics of the measurement, the respiratory function also influences the ECG-signal [1-3].

Among the parameters that can be obtained from an ECG, the heart rate (HR) or cardiac frequency is one of the most important. It is the reciprocal of the time intervals among successive R peaks, also known as RR intervals. In an individual there is a natural variation in the RR intervals and therefore in the HR. Such variability is known as heart rate variability (HRV) [2, 3]. Another parameter, among others, that is usually extracted from the ECG, although strongly influenced by the thoracic movements during respiration, is the height of the R-wave with respect to the baseline of the signal. This parameter is known as R-wave amplitude, and is related to changes on the volume of the left ventricle [4, 5].

The ECG signals are mediated by the autonomic nervous system (ANS) through the sympathetic (SNS) and parasympathetic (PSNS) nervous system [6]. Thus, stimuli that influence the ANS of an individual affect its electrocardiographic signal. The differences can be interpreted with respect to the ECG baseline as the individual's response to such stimulus. As a result, the ECG can be used to assess the level of stress experienced by a subject. To this aim, the most commonly used analyses employ the HRV and to a lesser extent, any other obtainable ECG parameters such as the amplitude of the R-wave or the QT interval (time interval between the peaks Q and T within a given PQRST complex) [7,8].

In the assessment of stress by HRV, the power spectrum of the variation of the RR intervals of the ECG, obtained by the fast Fourier transform (FFT) is firstly determined. Then the two areas under the curve of this spectrum within the so-called "low" and "high" frequency ranges, defined according to international standards of HRV analysis, are considered [9]. The low frequency (LF) region ranges from 0.04 to 0.15 Hz, while the high frequency (HF) interval goes from 0.15 to 0.4 Hz.

It has been determined that the LF range is associated to the SNS activity, whereas within the HF region both the SNS and PSNS are involved, and consequently the whole LF + HF region is dominated by the ANS [9]. The ratio of the LF to the HF areas define what is known as (autonomic) balance index (BI) of the ANS, which is influenced for states of anxiety and stress [6,10,11]. Another common way to observe the ANS behavior is through the so-called Poincare plot of the HR, which is a graph that reflects the dispersion of the data. In this case, the set of segments RR (in pairs) is brought into a two-dimensional plot, in which the abscissa of point n is given by the interval R_n-R_{n+1} (or the corresponding HR(n)) and its ordinate by the following R_{n+1}-R_{n+2} interval (or HR(n+1)), so that a delay map is obtained. In the plot a roughly elliptical shaped group of data is usually obtained, so that up to a certain percentage the data can be enclosed in an ellipse. The longitudinal diameter or major axis (with positive slope) of the ellipse describes the deviation of the long-term HR, whereas the cross-sectional diameter or minor axis (negative slope) characterizes the changes in the HR. With help of both values, it is the possible to quantify spontaneous and long-term changes of the HRV [12,13]. HRV has been used in some studies related with psychological stress in wich the authors use the usual parameters as well as alternative methodologies of measurement [14].

In this investigation we measured and evaluated the cardiac response of a sample of male and female public safety workers (PSW) subjected to induced psychological stress by a short duration Stroop test [15]. PSW face many stressful situations that yield them as high-risk population for suffering chronic stress diseases. Objectives of the study were to determine whether the traditionally used parameters for stress-psychological assessment through the cardiac response were sensitive enough for this test and sample, and to compare them with alternative parameters. Results are additionally discussed in terms of differences between the groups of male and female subjects.

2. Methodology

This study was approved by the Department of Public Safety (DPS) of Leon, Guanajuato, and for the Division of Health Sciences Campus León, from the University of Guanajuato, Mexico. For purposes of comparison the same methodology applied in previous studies [16] has been used here. Written informed consent was obtained from all volunteers, and the study was conducted according to the guidelines outlined in the Declaration of Helsinki World Medical Organization, 1996 [17].

Journal of Physics: Conference Series **582** (2015) 012040 doi:10.1088/1742-6596/582/1/012040

The procedures were explained to the volunteers and the evaluations were done during the morning (from 8:00 to 14:00) at the DPS of Leon, Guanajuato, Mexico.

A group of 20 women and 19 men from 18 to 40 years of age was studied in an experimental clinical trial. The sample of volunteers was a randomly selected from a universe of 1000 polices of the DPS of Leon, Guanajuato, Mexico. Workers belong to a medium low socioeconomic level which was determined as in our previous study [16]. Inclusion criteria considered only subjects who had at least nine years of schooling level, without cardiovascular, neither endocrine or psychiatric illnesses that could affect the variables studied, and also with non-antecedents of use of any drug or medication which may have effects in the central nervous system or at the cardiovascular level. Volunteers were divided in two groups according to gender. Before starting the experiment, clinical assessments were conducted for all subjects.

ECG signals from each participant were recorded at three different stages in the study: before (initial stage: 1 min), during (test stage: 3 min), and after (final stage: 1 min) the Stroop test. During the Stroop test words in colour are projected upon a screen at the rate of one word per second. The subject was asked to speak up the colour of the word. During the first 40 s, words of several objects were projected on a scream, followed by one minute of words that name colours. The written colour sometimes corresponds to the colour of the word and sometimes does not. Finally, again words of several objects, excluding names of colours are shown [18,19]. During the whole measurement, the subjects rest in a supine position having a light slope in order to let them comfortably read the projected words on the screen. The electrodes are positioned in the Eindhoven configuration and the ECG measurements performed with a BIOPAC MP150 system and corresponding ECG module.

From the recorded ECG signals the R-wave amplitudes have been considered additionally to the RR time intervals, from HR and the HRV have been determined. The usual autonomic BI has been obtained from the HRV power spectrum (BI of the HRV). Poincare parameters for the HR have also been considered for each subject and experimental stage using 95 % data inclusion for the Poincare ellipse. Additionally, the HRV-related alternative parameters of relative area LF to the total area LF + HF, and the median of this total area have been defined for both the HRV and the power spectrum of the variability of the R-wave amplitude or simply R-wave variability (RWV). For the RWV an alternative BI (BI of the RWV) has been analogously defined as the usual BI, as the rate between the corresponding LF and HF areas.

For the complete data set, differences among stages were tested using dependent t-test. Significance was considered when p was < 0.05. Statistical analysis was performed with STATISTICA, Stat-Soft, Inc. Tulsa OK, version 7.0 and all plots were generated with Origin Pro 8.0.

3. Results and discussion

Figure 1 shows typical results for one subject during the three experimental stages. In figure 1a the plot of the HR can be seen. Here, the high HRV before and after the Stroop test is illustrated. For the test stage, the HR typically rises, whereas the HRV diminishes for most the volunteers. This indicates that the Stroop test does work as stressor even for SSP, as other studies confirm [15]. The diminishing of the HR and the slight increase of its variability during the last part of the test stage apparently come from a partial relaxation of the subject. In the Poincare plot shown in figure 1b, as expected, the data are distributed in three regions of roughly elliptical shape for each stage. The plot shows that in comparison to the data obtained for the initial stage (black circles), the points corresponding to the test stage (white squares) not only present a shift to higher HR values but the ellipse that can enclose them is narrower. For the final stage, the dots of the Poincare plot return to lower HR values. These results are consistent with the results observed in figure 1a.

In figure 2 the results for the HRV analysis for the groups of men and women for all

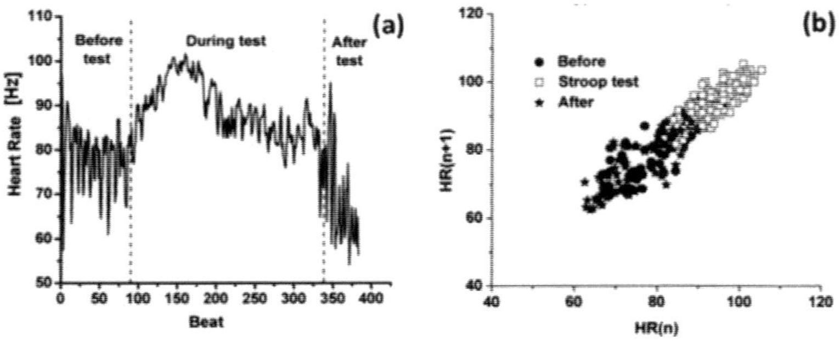

Figure 1. Typical results for one subject for (a) the HR, and (b) the HR delay map (Poincare plot) for all experimental stages.

experimental stages are shown. There are not any statistically significant (SS) differences neither between the different stages nor between the gender groups for neither the BI, relative area LF / (LF + HF) nor median of total area LF + HF.

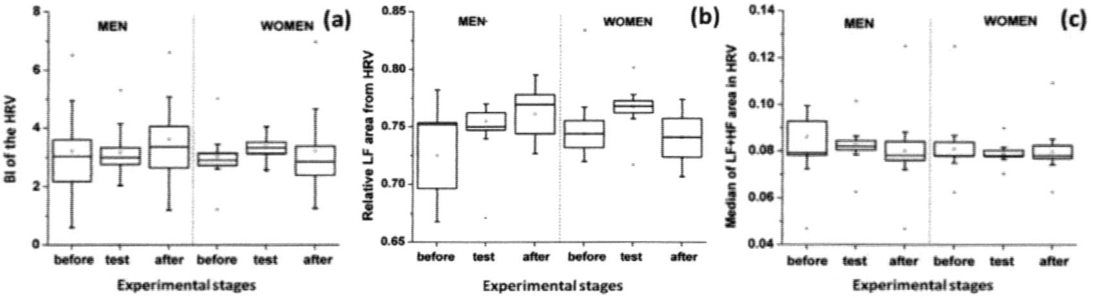

Figure 2. Results obtained for the HRV analysis for men and women for all experimental stages. **(a)** Autonomic balance index. **(b)** Relative area LF / (LF + HF). **(c)** Median of total area LF + HF.

For the BI (figure 2a) however, for both the genders a slight tendency to increase between the initial and the test stages is observed This tendency is more noticeable and with less statistical dispersion for the women than for the men. However, whereas this rising tendency holds for men during the final stage, for the women a diminishing tendency is observed after the Stroop test. As for the relative area LF to LF + HF (figure 2b), the same tendencies described for the BI of the HRV are also here observed although in a more emphasized way.

The increases observed in the BI and in the relative area LF to LF + HF between the initial stage and the test stage indicate that, according to [5] and [15], the Stroop test generates a response of the sympathetic nervous system of the subjects. As a consequence, the medians of the total areas LF + HF shift to the LF region, where the SNS dominates, as is shown in the box plot of figure 2c. Here, the tendencies between stages agree with those expected in presence of stress and do not indicate any gender difference besides the different intensity of the changes between stages and the less dispersion obtained for the group of women. However, like for the former parameters extracted from the HRV, lack of significance is observed. A likely explanation of this is that some partial relaxation mechanism sets on during the Stroop test, as suggested by the cardiac frequency diminishing observed in figure 1 during this test stage.

The group results of the Poincare plots analysis are summarized in figure 3 as box plots of the Poincare ellipses semi-axes obtained for the three experimental stages and both the genders. In contrast to the parameters obtained from the HRV analysis the Poincare parameters do show differences that are statistically significant or at least in the limit of significance. For the semi-major axis, more directly associated with changes in the HR, the SS changes take place between the initial and both the test and final stages for both genders. For the semi-minor axis, Y, directly associated with changes in the HRV, the most important changes occur between the final and the previous stages. These changes are SS for men and *quasi*-SS for women. Contrast in the tendencies of the changes between the different stages between men and women are to be noticed. These partially SS results seem to point out that whether men relax after the test, women do not.

The results for the RWV analysis for the groups of men and women for all experimental stages are shown in figure 4. Similarly as for the Poincare parameters, and in contrast to the results for the HRV analysis, these parameters present SS changes. Among all parameters analyzed in this work, the median of the total area LF + HF obtained from the RWV (figure 4c) shows the largest number of SS differences between stages.

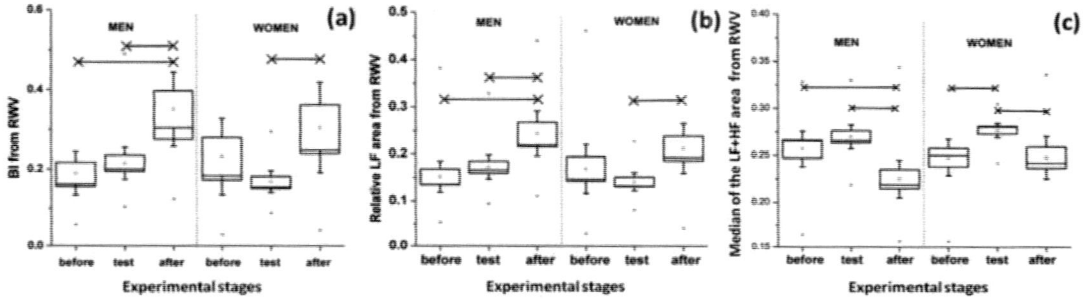

Figure 4. Results obtained for the RWV analysis for men and women for all experimental stages. **(a)** Analogous balance index. **(b)** Relative area LF / (LF + HF). **(c)** Median of total area LF + HF. The horizontal solid (dot-) lines indicate SS (or *quasi*-SS) differences between stages marked with large crosses.

The SS differences shown by the BI and the relative area LF to LF + HF of the RWV (figures 4a and 4b) are referred to the final stage for both genders. This indicates that since the R-wave amplitude is influenced by respiration, a larger relaxation is obtained after the Stroop test than before or during the test. It is however very remarkable that for the median of the total area LF + HF (figure 4c) this does not happen for the women group. For this case, the referent for SS changes is the test stage.

It is important to mention that there are several studies that indicate the existence of gender differences in the response to stress [20,21]. Nevertheless, for the short test presented here, it is found that gender response difference is not clearly recognizable either from the BI nor from any of the other HRV considered parameters. Gender differences can be seen from the Poincare analysis, which only gives information of non-parametric values (ranges). From the median of the total area

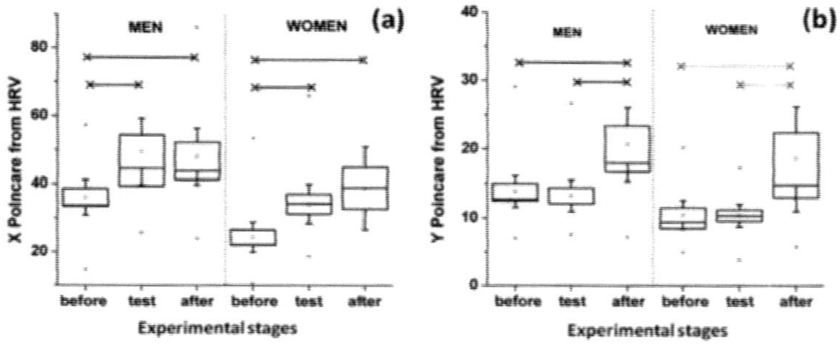

Figure 3. (a) Semi-major and **(b)** semi-minor axes of the HR Poincare analysis for both genders and all experimental stages. The horizontal solid (dot-) lines indicate SS (or *quasi*-SS) differences between stages marked with large crosses.

Journal of Physics: Conference Series **582** (2015) 012040 doi:10.1088/1742-6596/582/1/012040

LF + HF of the R-wave variability it is evidenced that the response to acute stress, as the induced by Stroop test, has similar behaviors in both genders (figure 4c), but for the women group, the differences are higher for women than for men, which is again in agreement with published results [20-23].

4. Conclusion

Here we confirm that the Stroop test is an appropriate procedure to evaluate the psychological stress response through analysis of the ECG signals. To this aim however, appropriate ECG signals parameters have to be considered. We found that neither the HRV, nor the BI, nor the here proposed HRV-related alternative parameters present SS changes during the stages of the test. SS gender differences are also not shown by any of these parameters. In contrast, the RWV and the Poincare plots do identify SS changes due to the Stroop test and allow to observe gender differences. From all the analyzed parameters, the median of the total LF + HF area of the RWV is the more sensitive. It also permits to identify gender differences consistent with those encountered in other studies.

References

[1] Klabunde RE 2004 *Electrical activity of the heart: Cardiovascular physiology concepts.* Lippincott Williams & Wilkins. ISBN 0-7817-5030-X.

[2] Hobbie RH 1996 *Intermediate Physics for medicine and biology.* 3a Ed, Biological Physics Series.

[3] Valdés-Pérez FE, Valdés-Villarrubia F, et al. 2006 Nuevo sistema para adquirir el electrocardiograma y el Fonocardiograma. Conferencia Internacional FIE'06, Santiago de Cuba, Cuba.

[4] Brody DA 1956 A theoretical analysis of intracavity blood mass influenced on the heart-lead relationship. *Circ. Res.* **4** 731-8.

[5] Lorne E, Mahjoub Y, Guinot PG, et al. 2012 Respiratory variations of R wave amplitude in Lead II are correlated with stroke volume variations evaluated by transesophageal Doppler echocardiography. *J Cardiothor Vasc An.* **26** 381-6.

[6] Yamamoto S, Iwamoto M, Inoue M, Harada N 2007 Evaluation of the effect of heat exposure on the autonomic nervous system by heart rate variability and urinary catecholamines. *J Occup Health.* **49** 199-204.

[7] Gábor Andrássy, M.D. 2007 The effect of different stressors on the QT interval and the T wave, Doctoral Thesis, Semmelweis University Doctoral School of Basic Medicine, Budapest.

[8] Doi Y, Yoshida T, Hiroki T, Arakawa K 1983 Effect of mental stress on R wave amplitude of the electrocardiogram in young healthy male subjects, *Jpn Heart J.* **24** 189-98.

[9] Task Force of the European Society of Cardiology the North American Society of Pacing Electrophysiology. Heart rate variability. Standards of measurement, physiological interpretation, and clinical use. 1996 Circulation **93** 1043-65 and Eur Heart J.1996 **17** 354–81.

[10] Langelotz C, Scharfenberg M, Haase O, Schwenk W 2008 Stress and Heart Rate Variability in Surgeons during a 24-Hour Shift. *Arch Surg.* **143** 751-5.

[11] Ruediger H, Seibt R, Scheuch K, Krause M, Alam S 2004 Sympathetic and parasympathetic activation in heart rate variability in male hypertensive patients under mental stress. *J Hum Hypertens.* **18** 307-15.

[12] Kamen PW, Krum H, Tonkin AM 1996 Poincaré plot of heart rate variability allows quantitative display of parasympathetic nervous activity in humans. *Clin Sci Lond.* **91** 201-8.

[13] Brennan M, Palaniswami M, Kamen P 2001 Do existing measures of Poincare plot geometry reflect nonlinear features of heart rate variability? *IEEE Trans Biomed Eng.* **48** 1342-7.

[14] Melillo P, Bracale M, Pecchia L 2011 Nonlinear Heart Rate Variability features for real-life stress detection. Case study: students under stress due to university examination. *Bio Medical Engineering On Line* **10** 96
[15] Hoshikawa Y, Yamamoto Y 1997 Effects of Stroop color-word conflict test on the autonomic nervous system responses. *Am J Physiol.* **272** H1113-21.
[16] Vargas-Luna M, Huerta-Franco MR, Montes-Frausto JB 2012 Evaluation of the Cardiac Response to Psychological Stress by Short-Term ECG Recordings: Heart Rate Variability and Detrended Fluctuation Analysis. World Congress on Medical Physics and Biomedical Engineering *IFMBE Proceedings 2013* **39** 333-5.
[17] Declaration of Helsinki 1964 *Br Med J.* **313** 1448-9.
[18] Delaney JP, Brodie DA 2000 Effects of short term psychological stress on the time and frequency domains of heart-rate variability. *Perceptual & Monitor Skills* **91** 515-24.
[19] Lee MH, Yang G, Lee HK, Bang S 2004 Development stress monitoring system based on personal digital assistant (PDA). *Conf Proc IEEE Eng Med Biol Soc.* **4** 2364-7.
[20] Garcia S, Mariscal MA, Garcia J, Ritzel DO 2012 Influence of task demands on occupational stress: Gender differences. *J Safety Res.* **43** 365-74.
[21] Matud MP 2004 Gender differences in stress and coping styles. *Pers Indiv Differ.* **37** 1401-15.
[22] Stroud LR, Salovery P, Epel ES 2002 Sex differences in stress response: social rejection versus achievement stress. *Biol Psychiatry* **52** 318-27.
[23] Verma R, Balhara YPS, Gupta CS 2011 Gender differences in stress response: Role of developmental and biological determinants. *Ind. Psychiatry J.* **20** 4-10.

Acknowledgments
The authors thank the DAIP of the University of Guanajuato for financial support and the authorities of the Department of Public Safety of Leon, Guanajuato.

Journal of Physics: Conference Series **582** (2015) 012041 doi:10.1088/1742-6596/582/1/012041

High precision speed measurement by using interferometric techniques

M A Rodríguez Ávila, R Ochoa Valiente, L A García Trujillo
Facultad de Ciencias Físico Matemáticas. Universidad Autónoma de Coahuila. Saltillo, Coahuila de Zaragoza, México

Email: raul.ochoa.valiente@uadec.edu.mx,

Abstract: In this work we present the experimental realization of speed measurement by the use of a two wave interferometer and digital signal processing techniques. We built an automated Michelson interferometer and using an He-Ne laser and with the use of the Fast Fourier Transform (FFT) and computer algorithms we derived a method for finding the speed of displacement. We report uncertainties in the order of 2 -3 μm/s. with the use of this procedure. This brings the potential of another physical variable measurement like distance or pressure by this indirect measurement method. This approach is compared with an ultrasonic Logger Pro ® speed measurement system, and the results are compared between systems.

1. Introduction

High precision measurement techniques are a very important subject in development of Physics. It impacts also in the applied, industrial and commercial fields around the world. It is required to have standards that allow keeping congruent measurements in different parts of the world. With this idea in mind, many different techniques and devices are used and designed in order to achieve better measurements for specific purposes [1],[2]. In this work we use a Michelson interferometer and a a data acquisition system to store and process the interference signal in order to measure the velocity of a device which moves on a linear path. This device is coupled to an arm of the interferometer. By the use of digital signal processing algorithms we find indirectly the instantaneous velocity and acceleration of the linear displacement by the analysis of the frequency of the acquired signal. In the next section we describe the theory that helps to sustain our hypothesis, followed by the experimental setup, the results which includes the error analysis.

2. Theorethical Background

The normalized intensity equation at an interferometer photodetector using single-mode light with wavelength λ is [4]:

$$I = \frac{I_0}{2} + \frac{I_0}{2}\cos(k\ OPD). \tag{1}$$

Previous expression is also known as the interference equation, where the propagation number $k = 2\pi/\lambda$, and *OPD* is the optical path difference. I_0 refers to the sum of the intensity of each arm of interferometer. Michelson interferometer experimental setup diagram is presented in figure 1.

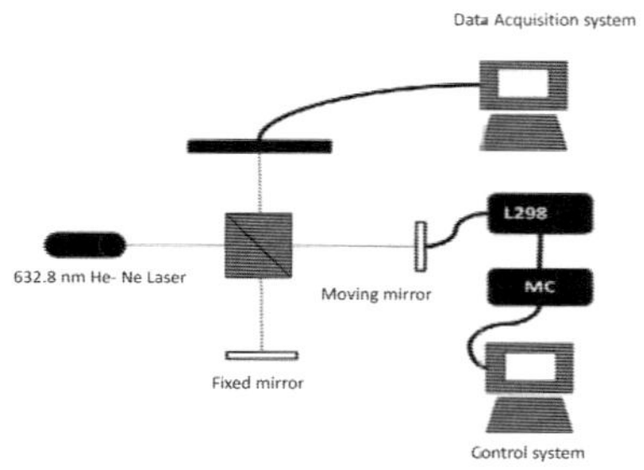

(a)

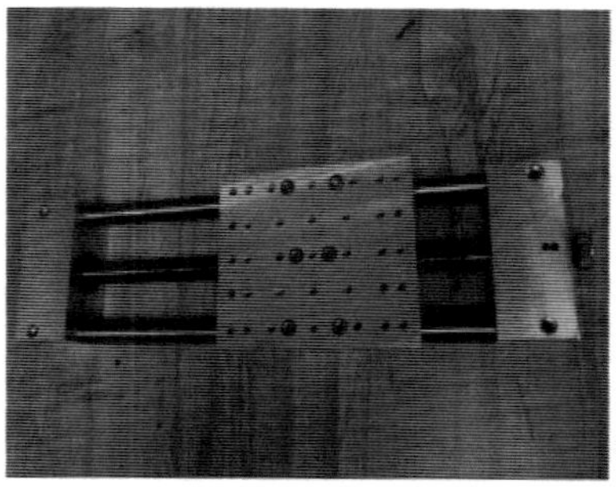

(b)

Figure 1. Block diagram of experimental setup using a Michelson interferometer (a) Electromechanicad displacement device used (b).

This interferometer provides an optical path difference $OPD = 2(d_1 - d_2)$, where d_1 and d_2 are the respective lengths of each arm of the interferometer. The previous expression (1) represents the interaction of two beams. The second part of the equation with the cosine function is also known as the interference term. When *OPD* is an even multiple of $\lambda/2$, the intensity I is maximum, while being odd multiple of $\lambda/4$, intensity is minimum.

By moving an arm of the interferometer with constant speed v, the interference terms becomes:

$$\cos\left(k\ OPd(t)\right) = \cos\left(\frac{2\pi}{\lambda}2\left(d_1 - d_2 + vt\right)\right) = \cos\left(\frac{2\pi(2vt)}{\lambda} + \phi_0\right), \tag{2}$$

where ϕ_0 is considered an initial phase. By considering the following analogy:

$$\cos\left(2\pi f\, t + \phi\right) = \cos\left(\frac{2\pi(2vt)}{\lambda} + \phi_0\right). \tag{3}$$

Term to term, it is possible to see the following relationship between frequency and speed:

$$f = 2v/\lambda \tag{4}$$

The main objective is to take several measurements of an interference signal. With the aid of digital signal processing techniques, the acquired data will be filtered and processed with fast Fourier transform (FFT) [5],[6], and the data collected from those processes will be error analyzed [7],[8].

All the data samples are processed thru a window of 64 samples using a computer algorithm with FFT. By this way we compute the speed and record it, sample by sample. The process is repeated and the operating voltage of the displacement device is changed to achieve a different speed. With the recorded data, we use an error analysis method [8] to verify the uncertainty of our measures. This process has the following steps.

- Compute the probability distribution function of the multiple speeds using histograms.

Journal of Physics: Conference Series **582** (2015) 012041 doi:10.1088/1742-6596/582/1/012041

- Calculate the i-th probability by computing the following quotient

$$P_i = N_i / N_t \,, \tag{5}$$

where N_i is the number of speeds that belong to the i-th bin and N_t is the total number of occurrences.

- Calculate the expected speed $<v>$ for each measure with the following expression

$$<v> = \sum_{i=1}^{N} v_i P_i \tag{6}$$

with v_i representing all possible speed values during the experiment.

- With the expected value, standard deviation is computed by

$$\sigma_x = \left(\sum_{i=1}^{n} P_i (v_i - <v>)^2 \right)^{1/2} \tag{7}$$

and the associated uncertainty of measurement is calculated with the following expression.

$$\sigma_{\bar{x}} = \sigma_x / \sqrt{N_t} \,. \tag{8}$$

In the next section the experimental setup and frequency analysis is presented.

3. Experimental Setup

For the electromechanical part, the displacement device is a motor driven worm gear where a mirror is attached. It has an small step size about 0.5 mm/rev. The motor is controlled with an Arduino microcontroller with a pulse width modulation driver (L298 IC). Coherent light source is provided by an He-Ne laser with $\lambda = 632.8$ nm as specified by manufacturer.

For the data acquisition system a photodetector (Thorlabs FDS100) was used with a soundcard of a Netbook. The control system used a different network in order to reduce the amount of electrical noise. Figures 2 and 3 present an example of signal recorded with sampling frequency $F_s = 44.1$ kSamples/s. For different speeds the control system decreased motor voltage in steps of 0.5 volts, starting from 12 V to 9.5 V.

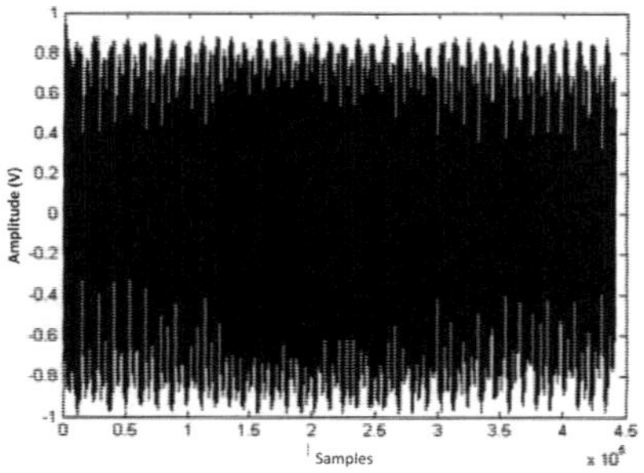

Figure 2. Example of acquired data in one second Voltage used for the motor is 12 V.

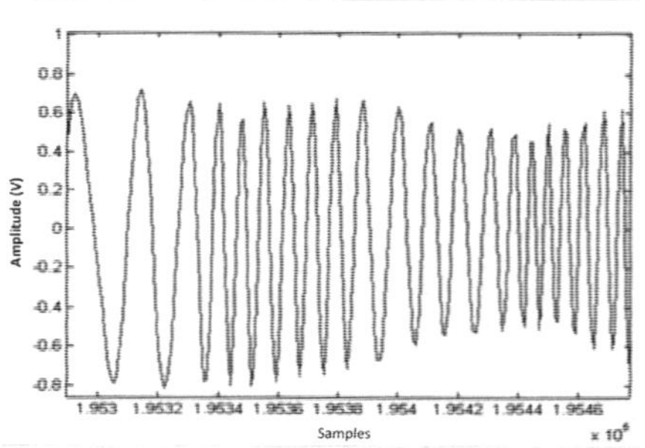

Figure 3. Zoom of a small amount of time of the signal. It can be shown that frequency is not constant because of different speed of the displacement device.

To calculate speed from the recorded data, the following algorithm is presented:

- In an empty data window with N blank spaces is generated.
- A single data (from left to right) is introduced in the last position of the window. A Hamming window [6] is used to multiply the window point to point to improve frequency response.
- A FFT with M points is computed and the maximum amplitude in the magnitude spectrum is taken. Corresponding frequency at this point is located with a resolution of $Fs/2\,M$.
- Speed is computed from expression in (4) and is recorded in another data set.
- Data in window is shifted one position to the left and the procedure is repeated from the second step of this algorithm.

Figures 4 shows an example of this procedure using N = 64 and M = 16384. Figure 5 shows calculated speed and acceleration computed with this method.

Simultaneous to this experiment, there was the need to compare with another measurement device. This device was an ultrasonic speed measurement device Logger Pro ®.

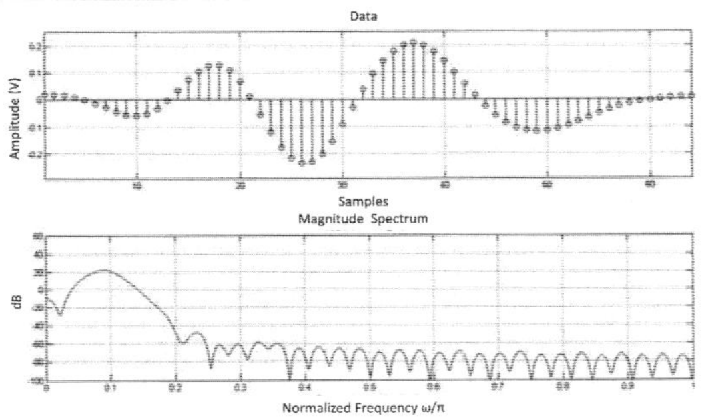

Figure 4. Example of frequency analysis used to measure the speed of displacement device.

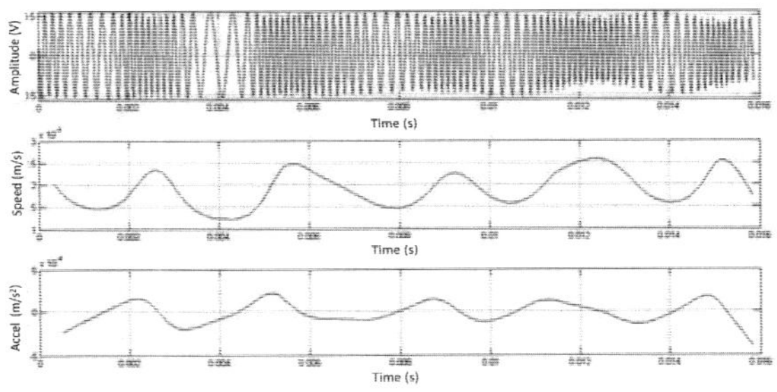

Figure 5. Speed and acceleration measured using this approach. Interference fringes are shown to verify compression and expansion in function of speed increase or decrease respectively.

4. Results

These experiments were error analyzed using expressions (5) to (8) and the corresponding results (averaged) of the speed with associated uncertainty for proposed and ultrasonic method are presented in table 1.

Table 1. Speed measurement using frecuency and ultrasonic method.

Volts	**Speed (*mm/s*)**	**Logger Pro (*mm/s*)**
12.0	2.150±0.002	2.10±0.10
11.5	2.050±0.002	2.06±0.08
11.0	2.000±0.002	1.97±0.08
10.5	1.900±0.002	1.92±0.10
10.0	1.835±0.003	1.90±0.12
9.5	1.770±0.003	1.80±0.11

The results presented shows three orders of magnitude in associated uncertainty. It can be shown that proposed method has three orders of magnitude of improvement compared to ultrasonic method.

5. Conclusions

This work presents an approach of measuring speed using frequency analysis. The method uses dynamic frequency analysis of an interference pattern from a Michelson interferometer attached to a displacement device. Error analysis shows that precision of this method is better than conventional methods using ultrasound. This approach needs to be improved and the potential of use includes the measurement of other physical variables like distance, pressure, optical properties, refraction index or temperature variations. More work is in progress to achieve a better measure system and test with other optical sources.

References

[1]. Demarest. Frank C , 1988. "High-resolution, high-speed, low data age uncertainty, heterodyne displacement measuring interferometer electronics", *Meas Sci and Tech* Vol 9 Number 7 pp. 1024, IOP Publishing

[2]. Scalise L, Yanguang Yu ; Giuliani, G ; Plantier, Guy ; Bosch, T. 2004. "Self-mixing laser diode velocimetry: application to vibration and velocity measurement", IEEE Trans on Instr and Meas Vol 53, Issue: 1, pp. 223 -232,

[3]. Yanguang Yu ; Wollongong; Jiangtao Xi ; Chicharo, JF 2007 "Improving the Performance in an Optical feedback Self-mixing Interferometry System using Digital Signal Pre-processing", IEEE Int Symp on Intell Sign Proc,. WISP 2007. pp. 1-6.

[4]. Hetch, E., 2000 "Óptics"; Ed. Addison Wesley.

[5]. Rao KR, Kin DN, Hwang J J; 2010. "Fast Fourier Transform: Algorithms and Applications"; Ed. Springer

[6]. Mitra S K; 2001 "Digital Signal Processing: A computer-based approach"; 2nd Edition; McGraw-Hill;.

[7]. Quarteroni A; 2006 "Scientific Computing with MATLAB and Octave", 2nd Edition; Springer;.

[8]. Taylor J R; 1997 "An Introduction to Error Analysis", 2ª Edition; University Science Books.

Acknowledgements

Authors thank the Facultad de Ciencias Físico Matemáticas of Universidad Autónoma de Coahuila for the support to this work.

Journal of Physics: Conference Series **582** (2015) 012042 doi:10.1088/1742-6596/582/1/012042

Using the Finite Elements Method (FEM) for Nanotechnology Education. A rectangular cantilever as a mass sensor

H. Aya Baquero
Facultad de Ingeniería
Universidad Distrital Francisco José de Caldas
Bogotá, Colombia

E-mail: haya@udistrital.edu.co

Abstract. The Finite Element Method FEM can be used in the context of physics engineering education, particularly in nanotechnology training. Cantilevers and cantilevers arrays have been implemented as sensors within lots of applications. In the present paper, FEM was used to assess validity of basic models where cantilevers are used as mass sensors. Resonance frequency of a cantilever transversal vibration was found; this was a silicon one-side clamped cantilever. A number of minor mass elements Δm was added on the cantilever's free side. Then in each case, a new resonance frequency was found; this led to obtain the Δm values from shifts of resonance frequencies. Finally, those values were compared with CAD model values.

1. Introduction

The Finite Element Method (FEM) is a numerical method, intended to solve mathematics, physics and engineering problems (structural analysis, heat transfer, fluids, mass transport, electromagnetism, and more). Nowadays, FEM is applied in several areas such as biomechanics, aeronautics, geology, electronics, and micro technology. Basically, FEM is a method that allows reaching a numerical approximation of differential equations solutions and partial derivatives solutions [1]. It is known that except for special cases, reaching an analytical solution to those equations is impossible, and the only remaining possibility is to get estimated answers. FEM is maybe used as an educational tool in Micro and Nano technologies. Mostly, it allows checking the validity of cantilevers as mass sensors [2,3]. These micromechanical sensors can be operated in both static and dynamic mode. In the static mode, bending of the cantilever occurs due to changes in surface stress. In the dynamic mode, resonance frequency shift occurs due to changes in mass or/and spring constant of the cantilever. In the last decade, researchers have proved its discrimination sensitivity of DNA mismatches [4,5] in static mode. In dynamic mode cantilevers have been used in many kinds of applications such as sensors to detect the mass of a single virus particle [6,7,8]. FEM has been used as an active virtual learning tool that allows students to explore a range of concepts and detect techniques in cantilevers eliminating the use of tests on real devices.

2. Theory

The resonance frequency of a cantilever sensor vibrating in flexural mode in vacuum is stated as [7]

$$f_0 = \frac{1}{2\pi}\sqrt{\frac{k}{m}}\ , \tag{1}$$

where m is the cantilever mass and k is the spring constant, which for a rectangular cantilever of length l, thickness d and width w is calculated as

$$k = \frac{E\,w\,d^3}{4\,l^3}\ , \tag{2}$$

where E is the Young's modulus.

In contrast, mass changes can be accurately determined by tracking the resonance frequency of the cantilever during mass adsorption or desorption. The cantilever elastic properties remain unchanged during the molecule adsorption/desorption process and damping effects are minor. This operation mode is called dynamic mode. The new resonance frequency may be stated as

$$f_1 = \frac{1}{2\pi}\sqrt{\frac{k}{m+\Delta m}}\ , \tag{3}$$

where Δm is the mass change due to adsorption.

Mass change on a rectangular cantilever is calculated according to

$$\Delta m = \frac{k}{4\pi^2}\left(\frac{1}{f_1{}^2} - \frac{1}{f_0{}^2}\right) \tag{4}$$

Frequency f_1 may be written in terms of f_0, as follows

$$f_1 = \frac{f_0}{\sqrt{1+\frac{\Delta m}{m}}} \tag{5}$$

A first-order Taylor series taking into account $f_1 = f_0 + \Delta f$ results in:

$$f_1 = f_0\left(1 - \frac{\Delta m}{2m}\right), \tag{6}$$

then

$$\sigma = \frac{\Delta m}{\Delta f} = \frac{2m}{f_0}\ , \tag{7}$$

where the term σ represents mass change sensitivity of cantilever and Δf is the frequency change.

3. Finite Element Method to measure the mass change

A silicon micro cantilever CAD model was created (Young's Module: E=131 Gpa, density: ρ=2330 Kg/m^3) with the following dimensions: length l=600 μm, wide: w=50 mμ and height: d=18 μm. See figure 1.

Figure 1. CAD model of silicon micro cantilever with l=600 μm, w=50 mμ and d=18 μm

By FEM, resonance frequency f0 of the first mode of transverse vibration of the micro cantilever clamped at one end was obtained as shown in figure 2. Being the micro cantilever resonant frequency f0=60794.47 Hz and its mass m=1.258 μg, then the cantilever mass change sensitivity is σ=0.041 ng/Hz. A number of CAD models were created; each model has its own Δm value, see figure 3. Then, for each case, a new resonant frequency value f1 was obtained by FEM and Δm* values were taken out from (4), and finally, error was evaluated. Results are featured in table 1.

Figure 2. FEM resonance frequency of silicon micro cantilever

Free-software was used to do this work; this means that some limitations appeared mostly in the meshing step. The reason for this limitation was the restriction of node number and elements. However, the data given in table 1 show satisfactory results. Using FEM commercial packages (ANSYS, COMSOL, etc.) better performance micro sensors with greater geometric complexity can be analyzed and designed.

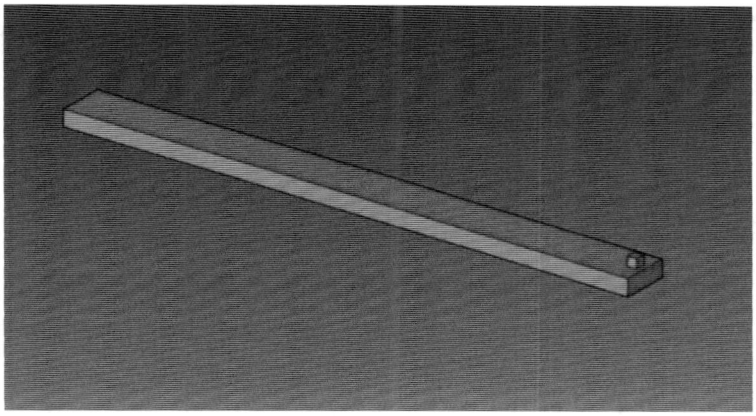

Figure 3. CAD model of micro cantilever with added mass Δm.

Table 1 Results obtained from FEM implementation.

Δm [ng]	f_0 [Hz]	f_1 [Hz]	Δm^* [ng]	$\left\lvert\frac{\Delta m^* - \Delta m}{\Delta m}\right\rvert 100$
2.097	60794.47	60589.14	2.055	2.03
4.194	60794.47	60388.93	4.079	2.73
6.291	60794.47	60185.49	6.013	2.13

8.388	60794.47	59985.66	7.493	2.02
10.485	60794.47	59789.41	10.262	2.12

Δm^* calculated using FEM and (4).

4. Conclusions

Results show a high coincidence between the theoretical model and the virtual solution to the problem. The comparison of mass values Δm between CAD model and FEM simulations, using the expression (4), shows an error close to 2%. Using FEM in the design and application of MEMS sensors has great potential of future development of new learning strategies of new technologies. 3D CAD modeling FreeCAD 0.13 and FEM LISA 8.0.0 free-software were used. H. A. acknowledges partial financial support from Facultad de Ingeniería at Universidad Distrital through Proyecto Curricular de Ingeniería Electrónica.

5. References

[1] Aya B H, Cano M R and Zhevandrov B P 2011 *Ingeniería e investigación*, **31** 1 7-15
[2] Gimzewski J K, Gerber Ch, Meyer E and Schlittler R R 1994 *Chem. Phys. Lett.* **217** 589
[3] Thundat T, Warmack R J, Chen G Y and Allison D P 1994 *Appl. Phys. Lett.* **64** 2894
[4] Fritz J, Baller M K, Lang H P, Rothuizen H, Vettiger P, Meyer E, Güntherodt J G, Gerber Ch and Gimzewski J K 2000 *Science* **288** 316
[5] Hansen K M, Hai-Feng Ji, Wu G, Datar R, Cote R, Majumdar A and Thundat T 2001 *Anal. Chem.* **73** 1567-1571
[6] Gupta A, Akin D and Bashir R 2004 *Appl. Phys. Lett.* **84** 1976
[7] Chen G Y, Warmack R J, Thundat T, Allison D P and Huang A 1994 *Rev. Sci. Instrum.* **65** 2532
[8] Subhashini S and Vimala Juliet A 2013 *Computer Networks & Communications (NetCom). Lecture Notes in Electrical Engineering* **131** 75-80.

Journal of Physics: Conference Series **582** (2015) 012043 doi:10.1088/1742-6596/582/1/012043

Low Coherence Optic Source Characterization

C E Flores-Domínguez, R Ochoa-Valiente, L A García-Trujillo
Facultad de Ciencias Físico Matemáticas. Universidad Autónoma de Coahuila. Saltillo, Coahuila de Zaragoza, México.

Email: raul.ochoa.valiente@uadec.edu.mx

Abstract: This work presents the results of characterization of the coherence length of an optic source using interferometric techniques and digital signal processing. Optic sources are not ideal because of random behavior in the emission process and spectral dispersion. Optical coherence is the ability of light to generate interference, either temporal or spatial. In time domain, coherence is expressed by the autocorrelation function. In case of monochromatic laser, it has larger coherence length, in the order of tenths to hundredth of meters, rather than a superluminiscent diode (SLD), which is shorter, in the orders of millimeters. This work presents a method for measuring coherence length using an automated Michelson interferometer and a SLD with central wavelength λ_0 = 1302.4 nm and acquisition system.by means of a soundcard in a personal computer.

1. Introduction

The term optical coherence refers to the property of light of keeping the same behavior at different times or different places. The first is referred as temporal coherence while the second refers itself as spatial coherence. One of the properties of coherence is the ability to produce interference fringes, in the case of temporal coherence this only ocurrs while temporal delay τ is less than or equal the coherence time τ_c. For example, monochromatic light has long coherence time rather than wider spectrum light. For further examination of this, it is possible to refer to [1].

2. Theoretical background

To verify the coherence of an optical source, the autocorrelation function $G(\tau)$ is expressed as:

$$G(\tau)=\langle U^*(t)U(t+\tau)\rangle=\lim_{T\to\infty}\frac{1}{T}\int_{-T}^{T}\left(U^*(t)U(t+\tau)\right)dt, \tag{1}$$

where $U(t)$ represents a complex wave function and T a time period where the function is measured. This previous expression represents the intensity of a two wave function from the same source while having a time delay between them. Power spectral density $S(\nu)$ is defined as the Fourier transform of autocorrelation function. The inverse Fourier transform, if the power spectral density is known a priori helps to calculate the autocorrelation function.

$$S(\nu)=\int_{-\infty}^{\infty} G(\tau)e^{-j2\pi\nu\tau}d\tau,\quad G(\tau)=\frac{1}{2\pi}\int_{-\infty}^{\infty} S(\nu)e^{j2\pi\nu\tau}d\nu. \tag{2}$$

This relationship is known as Wiener -Kinchine theorem [1]. This means that is possible to know the power spectral density $S(\nu)$ of a source by means of the autocorrelation function or, if $S(\nu)$ is known a priori, it is possible to compute the autocorrelation function $G(\tau)$.

For the characterization of optical sources, one important parameter is the coherence length. This is computed by

$$\tau_c=\int_{-\infty}^{\infty}|g(\tau)|^2 d\tau \tag{3}$$

where $g(\tau)=G(\tau)/G(0)$, also known as the normalized correlation function. Coherence length is computed as:

$$l_c=\tau_c c \tag{4}$$

All previous expressions are referred to ideal optical sources whose mathematical model is possible to analyze with, such as Gaussian or quasi monochromatic sources. Figure 1 presents the emission spectrum of a sample superluminiscent diode (SLD). From this image it is possible to see that this is not an ideal Gaussian spectrum, but it is possible to approximate it to. Also, specifications vary slightly from each device fabricated. Because of this reason, the motivation for this work is to characterize a SLD with the tools in hand, as an interferometer.

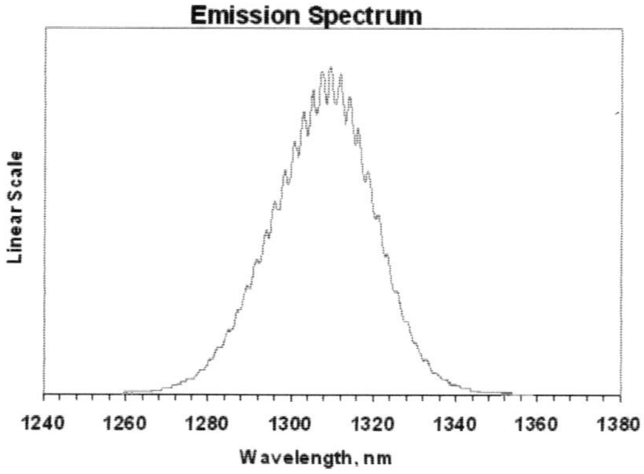

Specifications from sample datasheet:
Center wavelength: λ_0 =1314 nm.
Spectral width @ 1 nm resol: Δ_λ = 30 nm
Operating current 247 mA.

Figure 2. Emission Spectrum of a sample superluminiscent diode QSDM-1300-9 Qphotonics (Image from specs at http://www.qphotonics.com/Superluminescent-diode-10mW-1300nm.html)

This work presents one experiment to measure the coherence length of a QSDM-1300-9 SLD using an automated Michelson interferometer as autocorrelator[2] with a moving mirror at constant speed. The use of digital signal processing techniques allows to measure the coherence length of the source. The experimental setup is presented in the next section

3. Experimental setup

Figure 2 presents the experimental setup used to measure the coherence length of a SLD. The moving mirror moves at constant speed v = 1.1 μm/s. The photodetector used is a Thorlabs FGA20

Journal of Physics: Conference Series **582** (2015) 012043 doi:10.1088/1742-6596/582/1/012043

and the data was acquired with the soundcard of a PC computer using sampling frequency F_s = 44.1 kSamples/s ($T = 1/F_s$ = 22.68μs/sample).
The SLD used was Qphotonics QSDM-1300-9 with central wavelength λ_0 = 1302.4 nm with Δ_λ = 34.96 nm operating at 247 mA. These values are specific for the diode used in the experiment.

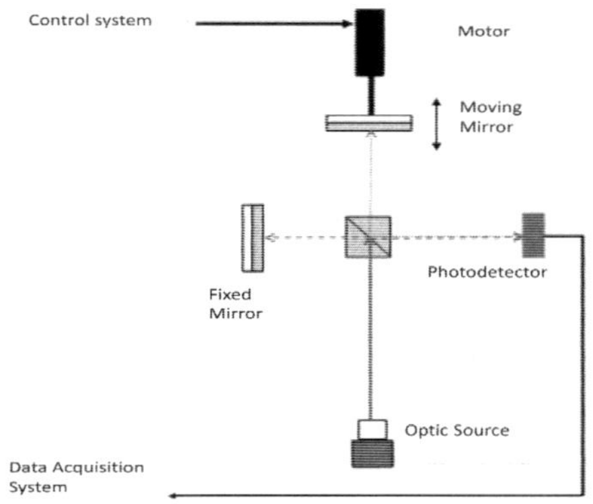

Figure 2. Automated Michelson interferometer used to measure the coherence length of an optical source.

Aproximating the data to a gaussian power spectral density:

$$S(\lambda) = P_0 e^{-\frac{(\lambda-\lambda_0)^2}{2\sigma_\lambda^2}} \tag{5}$$

where $\sigma_\lambda = \Delta_\lambda / 6 = 5.827\,\text{nm}$. To change previous expression to the frequency domain it is necessary to see the variations [1],[3]:

$$\Delta\lambda = \Delta\left(\frac{c}{\nu}\right) = -\frac{c}{\nu^2}\Delta\nu = -\frac{c}{\left(\frac{c}{\lambda}\right)^2}\Delta\nu = -\frac{\lambda^2}{c}\Delta\nu . \tag{6}$$

The minus sign indicates that this relationship is inverse, so, it is possible to take absolute value to change between domains. Around $\lambda = \lambda_0$ the previous expression changes to $\Delta\lambda = -\left(\lambda_0^2/c\right)\Delta\nu$.

Expression (5) in the frequency domain is

$$S(\nu) = P_\nu e^{-\frac{(\nu-\nu_0)^2}{2\sigma_\nu^2}} , \tag{7}$$

where $P_\nu = \left(c/\lambda_0^2\right)P_\lambda$, $\sigma_\nu = \left(c/\lambda_0^2\right)\sigma_\lambda$ = 1.031 THz, $\nu_0 = c/\lambda_0$ = 230.3 THz.

By the use of (2), the autocorrelation function can be expressed as:

$$G(\tau) = \frac{P_\nu \sigma_\nu}{\sqrt{2\pi}} e^{-4\pi^2 \frac{\tau^2\sigma_\nu^2}{2}} e^{j2\pi\nu_0\tau} . \tag{8}$$

This function shows a gaussian function in time domain that multiplies an harmonic function with frequency ν_0.

The coherence length can be computed by (3) and it is equal to[3]:

$$\tau_c = \left(\frac{1}{4\pi\sigma_v^2}\right)^{\frac{1}{2}} = 273.6fs \,. \tag{9}$$

Then, the theoretical coherence length using (4) is l_c = 82.08 μm.

By using the experimental setup we took several measures. Figure 3 presents an example of an interferogram for the SLD.

In order to measure the coherence length it is necessary to determine it from the measurements. The following procedure with the help of digital signal processing [4]-**¡Error! No se encuentra el origen de la referencia.**, was used to achieve this.

- Signal was filtered with high-pass filter; helping with this that any DC component was diminished.
- Signal was processed to obtain its absolute value.
- Signal was low-pass filtered; the objective is to get the envelope of the signal and then it was normalized. This shows a gaussian shape function that covers the coherence length.
- Using previous data, the position of the two closest values to 0.5 were taken. The number of samples is multiplied by the period T to get the distance time. This time distance is considered as full width at half maximum Δ_{FWHM} for the new gaussian-shape function.
- From this value, its corresponding standard deviation σ is computed by

$$\sigma = 0.4247\Delta_{FWHM} \tag{10}$$

- The coherence time using the interferometer τ^* is computed as

$$\tau^* = 6\sigma \tag{11}$$

- Finally, the coherence length of the optical source is calculated with the help of the speed of the interferometer displacement arm by

$$l_c = \tau^* v \tag{12}$$

Figure 4 shows the example of the envelope obtained using this approach.

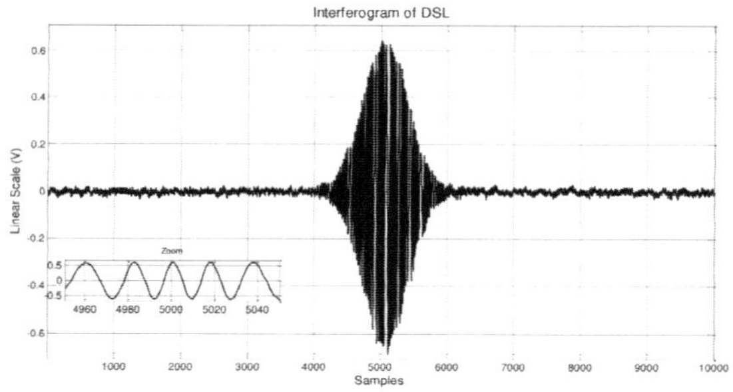

Figure 3. Example of an interferogram using Michelson interferometer

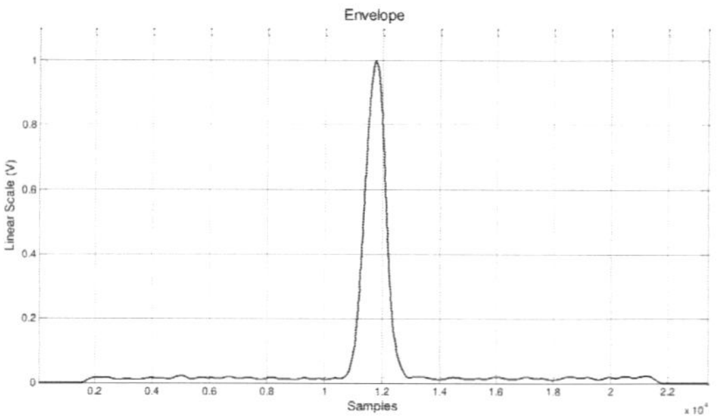

Figure 4. Normalized envelope of interferogram.

4. Results

For this experiment, 15 measures were made, and the results are presented in table 1. For these measurements, the coherence length l_c found is 50.2 μm. This differs from the theoretical value of 82.08 μm. Statistically, the deviation coefficient is acceptable but it has to be improved.

5. Conclusions

A different approach for measuring the coherence length of a SLD is presented. This approach tries to fit measures obtained by the use of a Michelson interferometer used as an autocorrelator device. This gives the opportunity of a less expensive method rather than an expensive optical analyzer. The coherence length can be explained with the fact that current SLD power spectral density is not an ideal gaussian function, but it approximates to that. Also, it presents ripples that modifies the right autocorrelation function. Improvements in electromechanical devices have to be done to get better measurements. Further work will be in the direction of calculating l_c for different optical sources with other spectral densities.

Table 1. Measurements of coherence length.

# Meas	Δ_{FWHM} (s)	σ (s)	τ^* (s)	l_c (m)
1	1.74E-02	7.39E-03	4.43E-02	4.88E-05
2	1.88E-02	7.98E-03	4.79E-02	5.27E-05
3	1.74E-02	7.39E-03	4.43E-02	4.88E-05
4	1.86E-02	7.90E-03	4.74E-02	5.21E-05
5	1.91E-02	8.11E-03	4.87E-02	5.35E-05
6	2.00E-02	8.49E-03	5.10E-02	5.61E-05
7	1.80E-02	7.64E-03	4.59E-02	5.04E-05
8	1.62E-02	6.88E-03	4.13E-02	4.54E-05
9	1.73E-02	7.35E-03	4.41E-02	4.85E-05
10	1.77E-02	7.52E-03	4.51E-02	4.96E-05
11	1.87E-02	7.94E-03	4.76E-02	5.24E-05
12	1.59E-02	6.75E-03	4.05E-02	4.46E-05
13	1.76E-02	7.47E-03	4.48E-02	4.93E-05
14	1.85E-02	7.86E-03	4.71E-02	5.19E-05
15	1.76E-02	7.47E-03	4.48E-02	4.93E-05
		MEDIA		**5.02E-05**
		STD. DEV.		**3.0079E-06**
		DEV COEF (%)		**5.99%**

References

[1]. Saleh B E A, Carl Tech M, 1991 "*Fundamentals of Photonics*", (Wiley & Sons).

[2]. Hetch, E., 2000 "*Óptics*"; (Addison Wesley).

[3]. Ochoa-Valiente R, 2010 "*Detección y medición de campos eléctricos utilizando retardadores electroópticos en tecnología óptica integrada*". (Instituto Nacional de Astrofísica, Óptica y Electronica INAOE).

[4]. Rao K R, Kin D N, Hwang J J; 2010. "*Fast Fourier Transform: Algorithms and Applications*"; (Springer).

[5]. Mitra S K; 2001 "*Digital Signal Processing: A computer-based approach*"; (McGraw-Hill).

[6]. Quarteroni A; 2006 "*Scientific Computing with MATLAB and Octave*", (Springer).

Acknowledgements

Authors thank the Facultad de Ciencias Físico Matemáticas of Universidad Autónoma de Coahuila for the support to this work.

Vanishing condition for the heat flux and slow evolution of a spherically fluid distribution

M Romero-Muñoz[1], **L Dagdug**[1], **G Chacón-Acosta**[2]
[1]Physics Department, Universidad Autónoma Metropolitana-Iztapalapa, San Rafael Atlixco 186, México D.F. 09340, México
[2]Applied Mathematics and Systems Department, Universidad Autónoma Metropolitana-Cuajimalpa, A. Vasco de Quiroga 4871, México D.F. 05348, México

E-mail: gchacon@correo.cua.uam..mx

Abstract. Recently, it has been found conditions for the heat flux where its introduction into the energy-momentum tensor, given his no mechanical nature, has no problems [1]. This has been achieved by checking the validity of the second law of thermodynamics in a fluid that is boosted by a Lorentz transformation of a non-commoving frame [2]. It is known that the condition, that turns out to be the null energy condition, involves the presence of a set of reference frames where Landau-Lifshitz frame is one of them. Moreover the entropy production remains positive, implying that there are no problems or issues that violate the second law of thermodynamics. In the present contribution we review the above condition, and apply it for a self-gravitating relativistic fluid in a spherically symmetric distribution in pure local coordinates. We find that our condition influences Schwarzschild fields, relating the condition of slow evolution with thermal quantities of the fluid.

1. Introduction

The Heat flow and its constitutive equations in relativistic systems have been the subject of several discussions. In Eckart's formalism [3] the constitutive equation of heat flow contains an acceleration term that seems to be the major source of general instabilities in the fluctuations which implies that water at room temperature becomes unstable [4], also it implies the nonexistence of the Rayleigh-Brillouin spectrum [5,6], and also the disappearance of the gravitational Jeans instability responsible for structure formation [7]. There have been proposed different constitutive equations that attempt to solve this problem [4,6,8]. Among them, one perspective comes from the relativistic kinetic theory [9], where a pressure gradient appears naturally instead of the acceleration term, and also turns the theory stable [6,8].

In the mid 80's the relativistic extension of the Meixner Prigogine formalism was made [10]. In this theory the heat flux is not included in the energy-momentum tensor [11], it appears as an independent 4-vector and as a consequence such instabilities does not appear. It was argued that the single reason to include the heat flux in the energy momentum tensor is to fulfil the energy conservation [1]. It was claimed that because the heat is a non-mechanical form of energy, and has properties of momentum and inertia; it couldn't be included into energy-momentum tensor.

Given such objections, the question whether to introduce the heat flux in the energy momentum tensor is posted. For its transformation properties [12], and also from the kinetic theoretical point of view the heat flux is part of the energy tensor [9]. In previous work [2], we address this issue from a

rather simple view performing Lorentz transformations. We founded that in the commoving frame the fluid has non zero heat flux while in the moving one it does not. It can be thought that the frame with no heat flux is just the Landau-Lifshitz frame [13], the so-called frame *that moves with the heat*. However, in [2] we obtain the condition which heat flux needs to satisfy, namely, just imposing to be smaller than a fixed amount, which is not negligible.
In this contribution we study the vanishing condition for the heat flux for an spherically symmetric self-gravitating fluid distribution. We also consider the impact of this condition on Schwarszchild field. It is noteworthy that the restriction reduces to well known conditions for the slow evolution regime.
In next sections we review the vanishing heat flux condition for the flat case and for the spherically symmetric case, sections 2 and 3 respectively. In the last one we summarize our conclusions.

2. Vanishing heat flux in a moving fram

From the point of view of kinetic theory, the energy-momentum tensor is identified as the second statistical moment of the distribution function and the heat flow q is part of the energy-momentum tensor [9].
Let us consider a dissipative fluid with a heat flux in x direction in the commoving frame, and suppose it is a component of the energy-momentum tensor. If we calculate a Lorentz transformation in x direction, we notice that the component $\bar{T}^{01}$, corresponding to the heat flux in that frame, can be split into two parts; a convective and non-convective flux [2]

$$\bar{q}_x = -\beta\gamma^2(ne + p) + q_x\gamma^2(1 + \beta^2), \tag{1}$$

where $\beta = \frac{v}{c}$, γ Lorentz factor, ne energy density, p pressure and q heat flux.

If we assume the flow is zero in the mobile system, the fluid in the commoving frame should satisfy the following condition [2]

$$q_x < \frac{1}{2}(ne + p). \tag{2}$$

To make contact with the Landau-Lifshitz scheme form the Eckart's frame, is possible to define a Lorentz frame only if $(q_x/nh_E)^2 \approx 0$, this condition is contained in Eq. (2) when $(q_x/nh_E)^2 < 1/4$, and where $h_E = e + p_x/n$ [9]. Notice that these imply that there is more than one reference frame where the heat flux vanishes [2]. The condition (2) is simple and can be obtained from another restriction, namely, the null energy condition. It states that the energy density of any matter distribution measured by any null observer must be non-negative [14]. Taking the heat flux in x direction and choose a simple null vector, we obtain the same condition given in (2). The difference is that we find (2) using transformations between frames [2]. In order to verify the validity of the second law on a fluid under condition (2), the entropy production is calculated in the Eckart frame [3], and it turns to be always positive and satisfies the second law of thermodynamics [2].

3. Heat flux in a spherically symmetric fluid distribution

Let us consider an spherically symmetric distribution of a fluid, which we assume to be locally anisotropic, undergoing heat flow dissipation limited by a spherical surface [15]. The line element is given in spherically symmetric by

$$ds^2 = e^\nu\, dt^2 - e^\lambda\, dr^2 - r^2\,(d\theta^2 + sin^2\,\theta d\phi^2)\,, \tag{3}$$

where $\nu(t,r)$ and $\lambda(t,r)$ are central functions. This metric need to satisfy Einstein field equations [10]

$$G_\mu^\nu = -8\pi T_\mu^\nu. \tag{4}$$

In the case of heat flow we are interested in the T_{01} component given by [15]

$$-8\pi T_{01} = -\frac{\dot{\lambda}}{r}, \tag{5}$$

where dots denotes partial differentiation with respect to t. Following Bondi let us introduce purely local Minkowski coordinates as follows

$$d\tau = e^{\nu/2}dt, \quad dx = e^{\lambda/2}dr, \quad dy = rd\theta, \quad dz = rsin\theta d\phi. \tag{6}$$

Relative to this coordinates the energy momentum tensor component is scaled as $e^{-(\nu+\lambda)/2}$. We suppose that, when viewed by an observer moving relative to these coordinates with velocity v_r in the radial direction, the anisotropic fluid is described by a tensor with energy density ne, radial pressure p_r, tangential pressure $p_\perp$ and radial heat flux q_r, which has the same form of the tensor we have considered so far. Thus, by a Lorentz transformation the (01)-component is

$$\bar{T}_{01} = -\frac{(ne+p_r)\beta e^{(\nu+\lambda)/2}}{1-\beta^2} + q_r e^{(\nu+\lambda)/2}\frac{(1+\beta^2)}{(1-\beta^2)}. \tag{7}$$

Let us notice the similarity between eqs. (7) and (1). Indeed, if we impose that convective and non-convective flows are canceled $\bar{T}_{01} = 0$, we obtain the same condition $q_r < \frac{1}{2}(ne+p)$, because the Swcharszchild fields are simplified. By substituting previous equation in (5) we obtain an extra condition that the field $\lambda(t,r)$ needs to fulfill,

$$\dot{\lambda} < 4\pi r(ne+p). \tag{8}$$

Unexpectedly, because of the positivity of the energy density, we can relate above condition with the so-called slowly evolving approximation [16,17]. A slowly evolving system is always close to hydrostatic equilibrium. This approximation demands, among others conditions, that $\dot{\lambda}^2 \approx 0$, in order to start with a system very close to the hydrostatic equilibrium. Thus, above expression becomes $\dot{\lambda}^2 < 16\pi^2 r^2(ne+p)^2$, and then, to find a Lorentz frame where vanishing heat flux, it must also satisfy the slow evolving condition [16,17].

4. Conclusions

We study the heat flow behaviour contained in the energy-momentum tensor under Lorentz transformations finding there is a total heat flow that contains a convective and non-convective part. When the heat flux is in the same direction as the Lorentz transformation was showed that condition (2) is satisfied, $q_x < \frac{1}{2}(ne+p)$. This condition can also be found through the null energy condition and was shown that do not violate the second law of thermodynamics [2]. With condition (2) is possible to find the reference systems where the heat flux vanishes, besides the so-called Landau-Lifshitz frame. For a spherically symmetric distributions of self-gravitating fluid, undergoing heat flow dissipation we find that relation (2) is satisfied by the radial component of the heat flux. When we introduce such conditions into the Schwarzschild field it turns out that $\dot{\lambda} < 4\pi r(ne+p)$. This condition can be related with the slow evolving condition that is the basis of the description of the life of most stars. This is because most of the relevant processes that occur in stars interiors happen on much longer time scale than the hydrostatic one [16,17]. This work is in the same direction as [7], in the sense of seeking conditions where relativistic heat flow is relevant and to what extent.

Acknowledgements
We thank the partial support from CONACyT grant No. 267825 (MRM), CONACyT grant under the project No. 176452 (LD).

References

[1] L.S. García Colín and A. Sandoval-Villalbazo, 2006 *J. Non-Equlibrium Thermodynamics* **31,** 11
[2] M. Romero-Muñoz,L. Dagdug, G. Chacón-Acosta, Vanishing condition for the heat flux of a relativistic fluid in a moving frame 2014, *Journal of Conference Series*
[3] C. Eckart, 1940 *Phys. Rev.* **58,** 919
[4] W. A. Hiscock, L. Lindblom, 1985 *Phys. Rev. D* **31,** 725
[5] A. L. García-Perciante, L.S. García-Colín y A. Sandoval-Villalbazo, 2009 *Phys. Rev. E* **79,** 066310
[6] A. Sandoval-Villalbazo, A. L. García-Perciante, L.S. García Colín, 2009 *J. Non-Equlibrium Thermodynamics* **388,** 3765
[7] J. H. Mondragon-Suarez, A. Sandoval-Villalbazo, A. L. García-Perciante, 2013 *General Relativity and Gravitation* **45**
[8] A. L. García-Perciante, L.S. García Colín, A. Sandoval-Villalbazo, 2008 *General Relativity and Gravitation* **41**
[9] C. Cercignani y G.M. Kremer, 2002 *The Relativity Boltzmann Equation Theory and Applications,* Birkhäuser Verlag, Basel, Boston, Berlin
[10] S.R. de Groot and P. Mazur, 1984 *Non-Equilibrium Thermodynamics* (Dover publications, Toronto, Ontario)
[11] L.S. García-Colín and A. Sandoval-Villalbazo, 1996 *Physica A* **234,** 358-370
[12] B. Schutz, 1985 *A first coure in generl relativity,* 2nd edition, Cambridge University Press
[13] C. Cercignani, 1975 *Theory and application of the Boltzman equation,* Instituto di Matematica, Politecnico di Milano, Italy
[14] R. M. Wald, 1984 *General Relativity,* University of Chicago Press
[15] L. Herrera, W. Barreto, A. Di Prisco and N.O. Santos, 2002 *Physical Review D* **65,** 104004
[16] L. Herrera, A. Di Prisco, J.L. Hernández-Pastora, J. Martínez, 1997 *Class. Quantum Grav.* **14,** 2239-172247
[15] L. Herrera and N.O. Santos, 1995 *Gen. Relativ. Grav.* **27,** 1071

Journal of Physics: Conference Series **582** (2015) 012045 doi:10.1088/1742-6596/582/1/012045

Erbium Doped Fiber Sensor for Ammonia Detection into Water

G G Pérez-Sánchez, E F Pinzón-Escobar, G E Sandoval-Romero, J A Álvarez Chávez

Av. 16 de Septiembre # 54, C.P. 55700, Col. Cabecera Municipal, Coacalco de Berriozábal, Estado de México

E-mail: i.m.g.r.e.t.h.e.l.1@gmail.com

Abstract. The water pollution is a health problem around the world. One of the most common pollutants in water is the Ammonia. Some sensors have been developed for Ammoniac detection even though most of them are not in real time and could be expensive. In this work an Erbium-doped fibre-sensor for Ammonia detection with a mechanical long period fibre grating and a taper in cascade for improving the sensitivity; the detection bandwidth is in the 1550 nm region from 1460 nm to 1640 nm. Output optical spectrum for Ammonia concentrations from 1ml to 5 ml is shown

1. Introduction

In some under-developed countries water treatment is not sufficient, and in most cases the water after domestic or industrial usage travels down the streets were time ago river existed. One of the most common pollutants in water is ammonia.

Ammonia pollution in water is a serious health problem. Exposure to ammonia at environmental concentrations is unlikely to have adverse effects on health. However, exposure to high concentrations following an accidental release or in occupational settings could cause irritation of the eyes, nose and throat as well as burning the skin on direct contact.

Some methods for pollution detection in water have been developed on which luminescence methods, nanotechnology, chemical, spectroscopy are included [1-5].

In this work, we propose an easy to implement sensor for ammoniac detection in real time with no aggressive methods. The sensor is implemented with an Erbium superluminiscent source along with one of the most used fibre components: a LPFG. It consists of a periodic perturbation of the refractive index of the fibre core that couples energy between two co-propagating modes, typically the fundamental mode and a cladding mode. LPFGs formed by press-induced methods have generated great interest due to the simplicity in their process of fabrication. In these grating the fibre is subject to periodical stress, which results in alternated regions under compression and stretching that modulate the refractive index via photo-elastic effect and micro bending. For the testing probe a tapered region is proposed for detecting the refractive index change when different concentrations are in contact with the fibre.

2. Er ASE theoretical model

We used the model in [6].Whose solution allows describing the evolution of pump and signaling powers along the fibre:

$$\frac{dP_P(z)}{dz} = -\gamma_P(z)P_P(z) \quad (1)$$

$$\frac{dP_S^{\pm}(z,\lambda_i)}{dz} = \pm\left\{G_e(z,\lambda_i)\left[P_S^{\pm}(z,\lambda_i)+P_0\right]-G_a(z,\lambda_i)P_S^{\pm}(z,\lambda_i)\right\} \quad (2)$$

Where $P_P(z)$ is the pump power propagating in z direction parallel to the doped optical fibre axis and $P_S^{\pm}(z,\lambda_i)$ is the output power in forward and backward direction, $\gamma_P(z)$ is the absorption coefficient, $G_e(z,\lambda_i)$ is the amplification of spontaneous emission, $G_a(z,\lambda_i)$ is the absorption coefficient of spontaneous emission and P_0 represents an equivalent input noise power:

$$P_0 = 2h\nu_s\Delta s \quad (3)$$

$$\Delta s = \left(\frac{c}{\lambda_s^2}\right)\Delta\lambda_s \quad (4)$$

When $P_P(z) \gg P_P^{th}$ we can consider $\gamma_P(z)$ as a constant [7]. P_P^{th} is the threshold power

$$P_P^{th} = \pi a^2 \frac{h\nu_P}{\sigma_P\tau} \quad (5)$$

From [7] we consider eq. (1) as:

$$\frac{dP_P(z)}{dz} = -N_T\pi a^2\frac{h\nu_P}{\tau} \quad (6)$$

Solving (6) with boundary conditions $P_P(0) = P_{in}$, where P_{in} is the initial pump power, then:

$$P_P(z) = P_{in} - \left(N_T\pi a^2\frac{h\nu_P}{\tau}\right)z \quad (7)$$

And from eq. (2) we obtain:

$$P_S^{+}(z,\lambda_i) = \left(\frac{G_e(z,\lambda_i)}{G_b(z,\lambda_i)}P_0e^{zG_b(z,\lambda_i)} - \frac{G_e(z,\lambda_i)}{G_b(z,\lambda_i)}\right) \quad (8)$$

$$P_S^{-}(z,\lambda_i) = \left(\frac{G_e(z,\lambda_i)}{G_b(z,\lambda_i)}P_0e^{(L-z)G_b(z,\lambda_i)} - \frac{G_e(z,\lambda_i)}{G_b(z,\lambda_i)}\right) \quad (9)$$

$$G_b(z,\lambda_i) = G_e(z,\lambda_i) - G_a(z,\lambda_i) \quad (10)$$

3. Long Period Fibre Grating working principle

A LPFG can couple light between the fundamental guided mode and forward-propagating cladding modes inside the fibre. As a result, several resonant modes are manifested as loss notches in the corresponding transmission spectrum. The resonant wavelength of the mth order mode λ_m is defined by the phase matching condition: $\lambda_m = (n_{eff1} - n^m_{eff2})\Lambda$; where Λ is the grating period n_{eff1} and n^m_{eff2} are the effective indexes of the LP01 fundamental mode and the mth cladding mode, respectively [8-10]. The coupled-mode equations are:

$$\frac{dA}{dz} = \kappa B e^{i\Gamma z} \tag{11}$$

$$\frac{dB}{dz} = \kappa A e^{-i\Gamma z} \tag{12}$$

$$\Gamma = \beta_1 - \beta_2 - \kappa \tag{13}$$

κ is the coupling coefficient and the LPFG period is defined as $\Lambda = \frac{2\pi}{\beta_1 - \beta_2}$ where $\beta_1 = \frac{2\pi n_{eff1}}{\lambda_0}$ and $\beta_2 = \frac{2\pi n_{eff2}}{\lambda_0}$ are propagation constants of the core and the cladding respectively.

4. Experimental results

For the experiment the sensor array consists of a 980 nm pump power laser at 350 mW connected to an Erbium-doped fiber segment; at the end of the fiber, another SMF-28 fibre segment is connected, a long period fibre grating is allocated on to the fiber with 147 N force; at the end segment a taper fibre section is connected as a testing probe. The output power spectrum is observed in an optical spectrum analyzer.

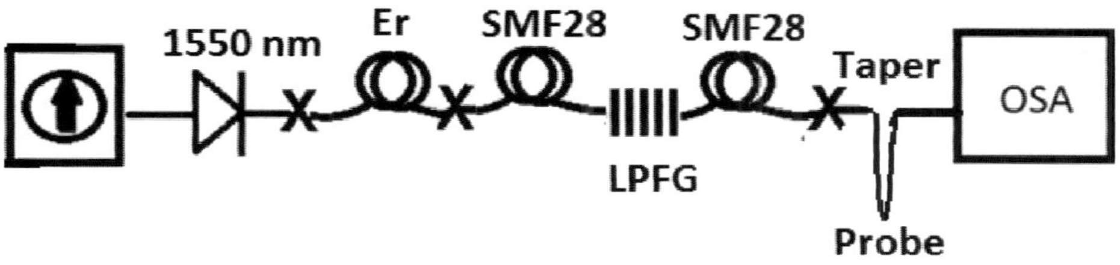

Figure 1 Experimental sensor setup.

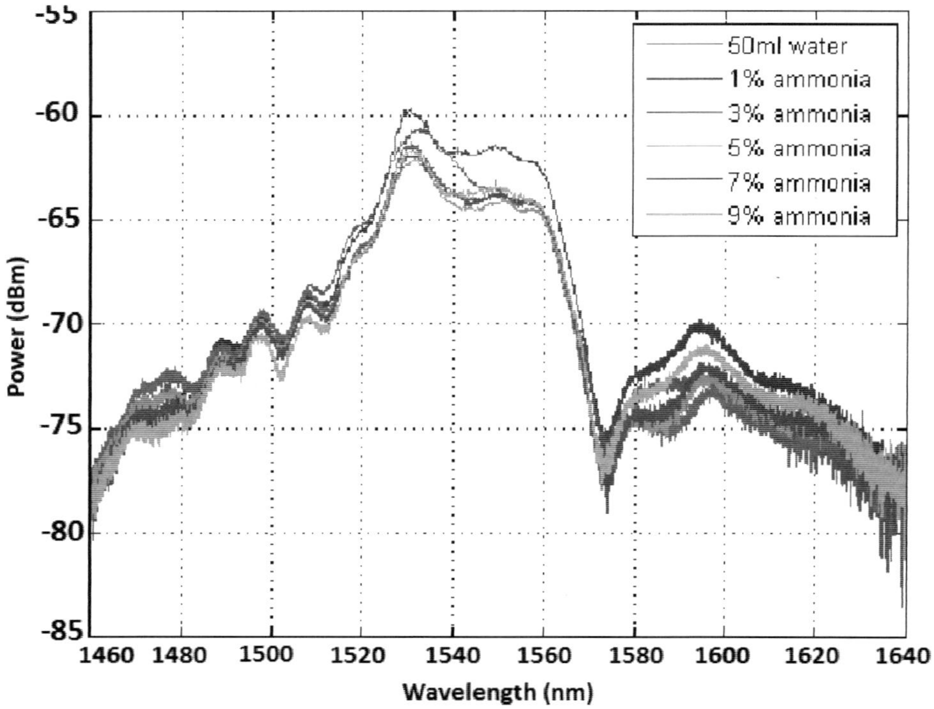

Figure 2 Power-wavelength spectrum of ammonia fiber sensor.

The testing probe was introduced in deionized water and the output spectrum was captured. Afterwards, ammoniac percentages from 1 % to 9 % were added and measured. In figure 2 we can observe the output power responses for different ammonia concentrations.

From figure 2 from 1460 nm to 1520 nm it is possible to observe how the output power decreases when the ammoniac concentration increases, also it is possible to observe at 1505 nm that a filtered band is deeper for higher ammonia concentrations.

We have to take into account that the fibre tapered probe detects the changes of refractive index when it is exposed to certain substances. Clearly, the erbium doping increases the sensitive at around 1550nm.

5. Conclusions

An erbium-doped fibre sensor for ammonia detection was proposed with encouraging experimental results which were included. The present work shows a broad spectrum from 1460nm to 1640 nm, even though the higher sensitivity is centred at 1505 nm with 2.5 dBm extinction ratio. Future work is to process de optical results with an opto-electronic device for obtaining a portable sensor.

References

[1] Slooff W, de Zwart D, Marquenie J M 1983 Detection limits of a biological monitoring system for chemical water pollution based on musel activity *B. Environ. Contam. Toxicol* **30** 400-405

[2] Nasu Y, Kugimoto M 1091 Lemnac (Duckweed) as an indicator of water pollution in the sensitivity of Lemna paucicostatato heavy metals *Archives of Environ. Contam. and Toxicol.* **90**159-169

[3] Rosiev P, Buich A S 2011 State of the art molecular markers for fecal pollution source tracking water", *Appl. Mocrobiol. Biotechnol.* **84** 1341-1355

[4] Marchetti D M 2013 Water Pollution Detection, *4th Imeko TC19 Symp. on Environment. Ins. and Meas. Protecting Environment, Climate Changes and Pollution Control (Lecce, Italy)*

[5] Moo Yow Chong, Jafri, M.Z.M., Lim Hwee San, Ton Chin Ho 2012 Detection of nitrate ions in water by optical fiber *Computer and Communication Engineering (ICCCE) IEEE,* 101109/ICCCE

[6] Pinzón-Escobar E, Ramírez-Ibarra A, Moreno-Hernández J C, Sandoval-Romero G E 2010 Experimental Results of the Superluminescent Fiber Laser Sources for Fiber Optic Sensors" *Proc. SPIE 7839, (WSOF-2)*

[7] Sandoval-Romero G E, Nikolaev V A 2003 Límite de detección de un giroscopio de fibra óptica usando una fuente de radiacion superluminiscente *Revista Mexicana de Física* **49** 2 155-165

[8] Krcmarík D, Slavík R, Karásek M, and Kulishov M 2009 Theoretical and Experimental Analysis of Long-Period Fiber Gratings Made Directly Into Er-Doped Active Fibers *J. Lightwave Technol.*, **27** 13

[9] Li L, Xia L, Xie Z, and Liu D 2012 All-fiber Mach-Zehnder interferometers for sensing applications *Optics Express* **20** 10

[10] Cardenas-Sevilla G A, Monzon-Hernandez D, Torres-Gomez I, Martınez-Rios A 2012 Tapered Mach–Zehnder interferometer based on two mechanically induced long period gratings as refractive index sensor *Opt. & Laser Technol.*, **44**, 1516-1520

Journal of Physics: Conference Series **582** (2015) 012046 doi:10.1088/1742-6596/582/1/012046

Electrosynthesis and characterization of $Hg_{1-x}Cd_xSe$ films

M Sánchez, F A González-Camacho, D A Miranda, A M Meléndez

CIMBIOS, CMN, Universidad Industrial de Santander, Cra 27 Cll 9, Bucaramanga, Colombia.

E-mail: dalemir@uis.edu.co

Abstract. An electrodeposition process from an aqueous bath of $CdSO_4$, SeO_2 and Hg_2Cl_2 at pH 2.3 was performed to synthesize two $Hg_{1-x}Cd_xSe$ alloys with $x \geq 0.8$ onto TiO_2. In order to establish an intimate contact between $Hg_{1-x}Cd_xSe$ and the TiO_2 substrate, electrodeposition of $Hg_{1-x}Cd_xSe$ films was investigated. Reduction processes of H_2SeO_3 and Cd^{2+} to produce CdSe and $Hg_{1-x}Cd_xSe$ were studied by cyclic voltammetry. Initially, deposition of selenium allows codeposition of Cd^{2+} and Hg^{2+}. $Hg_{1-x}Cd_xSe$ film was obtained by holding the electrode at a constant potential. The characterization of the films was performed by GIXRD, FESEM, and the chemical composition of the film was obtained by energy-dispersive X-ray spectroscopy, photoluminescence and Raman spectroscopy.

1. Introduction

In recent years, pseudobinary II–VI chalcogenide semiconductors have received a lot of attention because of their potential applications in optoelectronic devices, solar cells and photocatalysts [1]. $Hg_{1-x}Cd_xSe$ is a ternary semiconductor/semimetal chalcogenide which can be tuned across the infrared spectrum by controlling the Hg and Cd molar fractions. These materials can be synthesized by numerous preparation methods in the form of thin polycrystalline films. In particular, cathodic electrodeposition has several advantages over vapor phase techniques because it is a potentially low-cost, non-vacuum, high-rate deposition process that can easily deposit uniform film on different conducting substrates at low temperature [1]. Himei and Muto [2] reported the preparation of $Hg_{1-x}Cd_xSe$ films onto ITO substrate by varying the Hg content, whereas Singh and Tanveer [3,4] published the electrochemical formation of $Hg_{1-x}Cd_xSe$ films on titanium foils. Although, $Hg_{1-x}Cd_xSe$ films exhibited photoresponse [3,4], composition of the electrodeposits was not determined. In order to study chemical composition of $Hg_{1-x}Cd_xSe$ alloys, in this study several measurement were performed: FESEM-EDS, photoluminescence and Raman spectroscopy. In addition, to the best of our knowledge, the mechanism of $Hg_{1-x}Cd_xSe$ electrodeposition from aqueous solutions containing Hg(II), Cd(II) and Se(IV) have not been investigated. In an effort to further elucidate this question, we have performed cyclic voltammetric studies.

2. Materials and methods

1. *Electrosynthesis*

The electrodes employed were made from a 99.99 % purity Alfa-Aesar Ti rod embedded in Teflon, and with a 0.07 cm^2 circular area left uncovered. Prior to the electrochemical experiments Ti was polished with a series of silicon carbide emery papers, subsequently obtaining a mirror-like finish with 50 nm alumina powder and washed successively with acetone, distilled water and etched in 10% (v/v) HF. The experiments were carried out using a conventional three-electrode cell with a graphite rod as counter electrode and a reference electrode Ag/AgCl/3 M KCl (0.210 V vs NHE). All potentials in this work are given with respect to this reference electrode. Electrochemical experiments were performed with a potentiostat Autolab PGSTAT 302N (Eco Chimie, Utrecht, The Netherlands) and NOVA software. Cyclic voltammetry studies were recorded in negative direction at 100 mVs^{-1}. Before electrodeposition, nitrogen was purged through the bath for 20 min to remove oxygen from the solution. The solutions were prepared from Millipore Milli-Q grade water and analytical-grade reagents (99% purity, Merck). The pH of solution was adjusted at 2.3 with H_2SO_4. Depositions were carried out from the aqueous solutions containing 50 mM $CdSO_4$, 10 mM SeO_2, 1 mM and 10 mM Hg_2Cl_2 The films used in the characterization were grown on Ti foils by imposing a potential of -0.60 V during 1800s.

2.1 Crystalline structure

Grazing incidence X-ray diffraction measurements were made using a Bruker D8-Discover diffractometer working at 40 kV with the Cu Kα radiation in a glancing geometry and incidence angle fixed at 2°.

2.2 Morphology and chemical analysis

The morphology of $Hg_{1-x}Cd_xSe$ films was investigated by field emission scanning electron microscopy FESEM, QUANTA FEG 650.

2.3 Composition

Energy dispersive spectroscopy (EDS) was used to study atomic composition with a FESEM, QUANTA FEG 650 scanning electron microscopy, Noran detector and analysis system TN-5400. Furthermore, photoluminescence and Raman were measured at 291K with a LabRam Evolution Horiba spectroscope. A laser wavelength of 532nm and 785nm were used to obtain photoluminescence and Raman spectra, respectively. Gaussian peaks models were implemented to analyse photoluminescence and Raman spectra and Summers-Broerman relation for band gap [5] and Miranda et. al. model for phonons in $Hg_{1-x}Cd_xSe$ alloys [6] were implemented to achieved alloy composition, respectively.

3. Results and discussion

In order to investigate the mechanism of $Hg_{1-x}Cd_xSe$ electrodeposition onto TiO_2 substrate, the electrochemical behavior of Ti electrode in solutions of individual components of the electrolyte bath was studied. In addition, cyclic voltammetry was used to study the electrochemical reactions in solution containing 50 mM $CdSO_4$, 10 mM SeO_2, and 1 mM Hg_2Cl_2 at pH 2.3. All voltammetry curves were initiated in the negative-going direction. The background current of TiO_2/H_2SO_4 interface is observed in figure 1 (aI); it is very small compared to those currents obtained from the Cd(II) and Se(IV) within the potential window studied. Typical cyclic voltammograms for 50 mM $CdSO_4$ is displays in figure 1 (aII). In the forward scan, a visible silver gray film appear on the surface of working electrode at a potential about -0.70 V, confirming that the increase of current is due to the cadmium reduction. In the backward scan, the anodic peak corresponding to oxidation of metal Cd is about -0.5 V. The currents on the inverse scan are higher than those on the forward scan, thereby appearing on reverse scan a crossover potential. This behavior is described in the literature as a nucleation-loop phenomenon [7] and it has provided direct experimental evidence to diagnose a process nucleation-controlled. Hence, reduction of cadmium ions occurs more easily on the cadmium surface than that on the TiO_2 surface.

A cathodic peak with onset about -0.4 V can be observed in figure 1 (aIII) where the wide cathodic peak corresponds to the irreversible reactions (1) and (2) [8].

$$H_2 Se\ O_3 + 4H^+ + 4\ e = Se + 3H_2O \quad (1)$$

$$Se + 2H^+ + 2\ e = H_2Se + 3H_2O \quad (2)$$

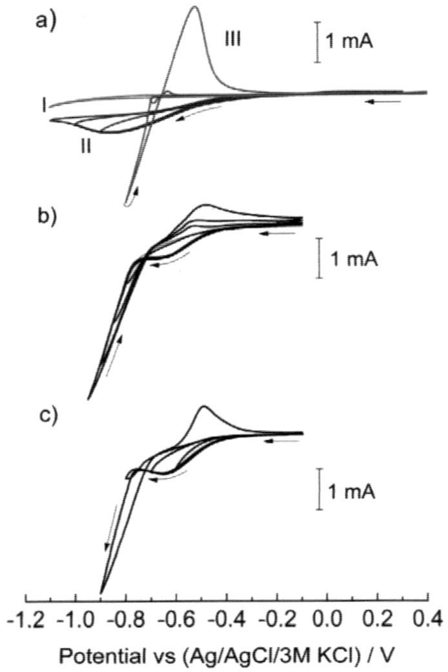

Figure 1. Typical cyclic voltammograms obtained on TiO_2 for aI) 0.1 M H_2SO_4, aII) 50 mM Cd^{2+}, aIII) 10 mM SeO_2, b) 50 mM Cd^{2+} + 10 mM SeO_2 and c) 50 mM Cd^{2+} + 10 mM SeO_2 + 1 mM Hg^{2+}. Different inversion potentials values were applied to the substrate. Scan rate v = 0.1 Vs^{-1}.

The electrodeposition of $Hg_{1-x}Cd_xSe$ is very similar to CdSe, figure 1 (b) and (c), as such streamlining of mercury in the solution is very low. In this deposition process when a suitable potential is applied between the electrodes, the ions undergo the discharge process and the deposits of $Hg_{1-x}Cd_xSe$ is formed. The potential range reported in the literature for obtaining the chalcogenide compounds on TiO_2 is about of -0.62 V, hence, Se must be initially deposited on TiO_2, thereby Hg and Cd are codeposited to form $Hg_{1-x}Cd_xSe$ alloy.

The identification of Hg1-xCdxSe was confirmed with XRD measurements, figure 2. Diffraction peaks exhibit the characteristic reflections (111), (220), (311) and (400) of Cd0.733Hg0.267Se phase (PDF 000-48-1450). The peaks ratios shown the deposited film has a preferential orientation in (111) direction in contrast with non-oriented alloys in which not preferences in an orientation are obtained. [9].

The surface morphology of the films is shown in figure 3. The deposits are compact and exhibit cauliflower morphology. The quantitative analysis of the film was carried out by EDS, to determine the stoichiometry of the films. The elemental analysis was carried out only for Cd and Se, the average atomic percentage; it was found that the film deposited is Hg0.1Cd0.9Se.

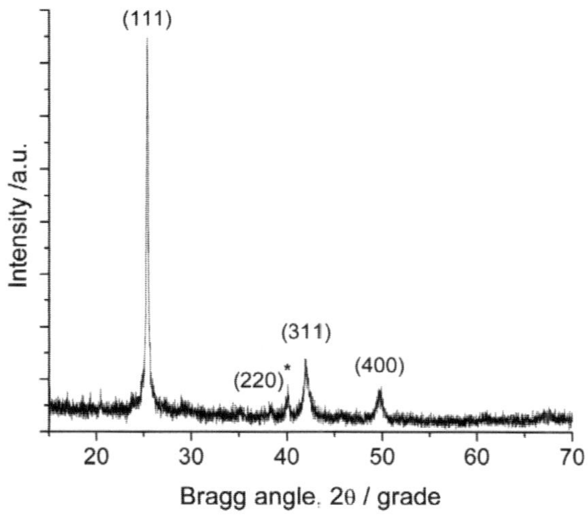

Figure 2. X-ray powder diffraction profile of electrodeposited $Hg_{1-x}Cd_xSe$ film from 50 mM Cd^{2+} + 10 mM SeO_2 + 1 mM Hg^{2+}. Asterisk (*) indicate peak from the Ti support.

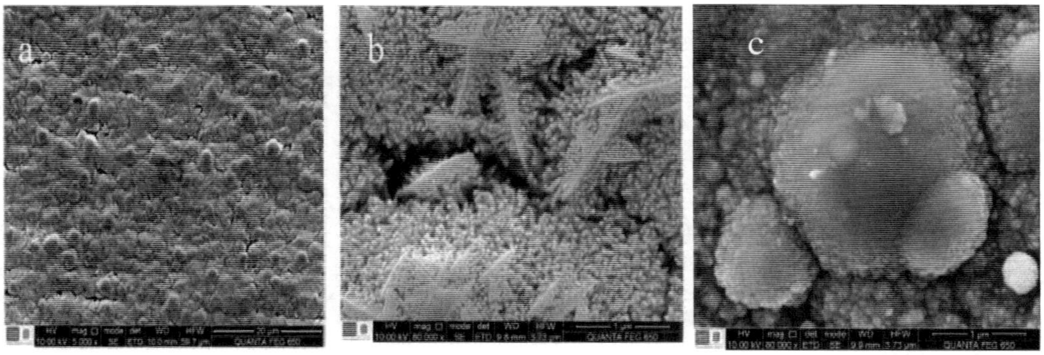

Figure 3. Scanning electron micrograph of electrodeposited $Hg_{0.1}Cd_{0.9}Se$ (a) 5000X, (b) 80000X and (c)$Hg_{0.2}Cd_{0.8}Se$, 80000X.

Photoluminescence and Raman spectra are shown in figures 4 and 5, respectively. Raman spectra were modeled by three Gaussian peaks at 150 cm^{-1}, 199 cm^{-1} and 204 cm^{-1}. Miranda et. al. model [6] applied to Raman peaks suggest the electrodeposited alloy are conformed by two phases: $Hg_{0.1}Cd_{0.9}Se$ and CdSe. Otherwise, photoluminescence spectra were modeled by fourth Gaussian peaks at 1.3 eV, 1.4 eV, 1.5 eV and 1.8 eV that can be associated to band gap of $Hg_{0.1}Cd_{0.9}Se$ alloy, 1.4 eV [5] and impurity and band to band transitions. The suggestion of impurity transition is based in the presence of low energetic peak and the expectation of impurities in the material due to fabrication process. The band to band transition is proposed to explain the more energetic peak that cannot be associated with band gap, i.e. 1.5 eV and 1.8 eV.

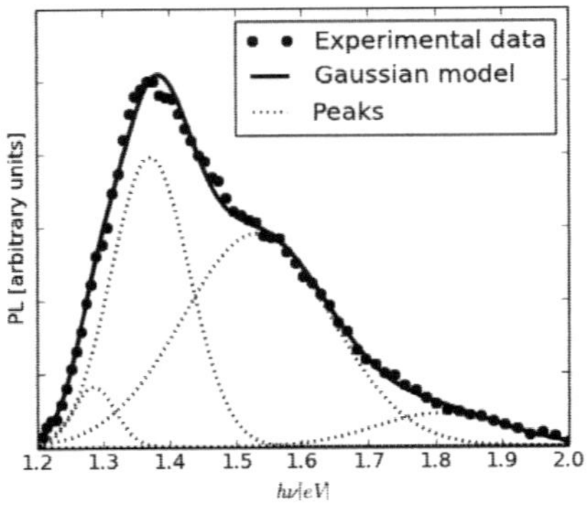

Figure 4. Photoluminescence spectra of electrodeposited alloy.

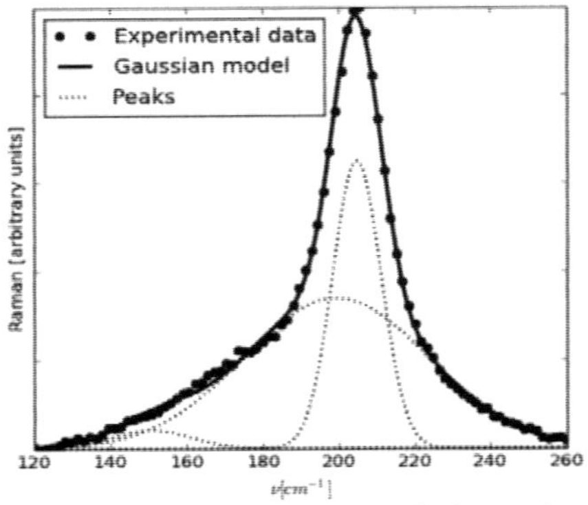

Figure 5. Raman spectra of electrodeposited alloy.

4. Conclusions

$Hg_{1-x}Cd_xSe$ thin films have been successfully electro-synthesized. Cyclic voltammetry studies revealed that deposition of selenium allows codeposition of Cd^{2+} and Hg^{2+}, furthermore, the reduction processes of H_2SeO_3 and Cd^{2+} allows deposition of CdSe and $Hg_{1-x}Cd_xSe$. Electron microscopy, DRX, EDS, photoluminescence and Raman studies show that electrochemically films are globular and polycrystalline with two dominant phases: $Hg_{0.1}Cd_{0.9}Se$ and CdSe.

Acknowledgments

We gratefully acknowledge the financial support of the Universidad Industrial de Santander, project number 5703. Furthermore, we appreciate the discussions, support and recommendations of professors José Antonio Henao, Rafael Cabanzo and Mario Macias.

References

[1] Bouroushian M 2010 *Electrochemistry of Metal Chalcogenides* (Berlin: Springer) Chapter 3 77–106.
[2] Himei N and Muto J 2000 *J. Mater. Scien.: Mater. In Electr.* **11** 145
[3] Singh K and Tanver R 1993 *J. Of Mater. Letters.* **12** 737
[4] Sing K and Tanverr R 1993 *J. Mater. Chem.* **3** 1295-98
[5] Summers C and Broerman J. 1980 *Phys. Rev. B* **21** 559-73
[6] Miranda D, Lópes S and Romero A. 2013 *AIP Conference Proceedings* **1566** 95-97
[7] Fletcher S and Halliday C 1983 *J. Electroanal. Chem.* **159** 267
[8] Natarajan C, Sharon M, Clément L and Neumann M 1994 *Thin Solid Films* **237** 118
[9] Miranda D, López S, Prias J and de los Rios M 2011 *AIP Conference Proceedings* **1416** 77

Journal of Physics: Conference Series **582** (2015) 012047 doi:10.1088/1742-6596/582/1/012047

CMOS sensor as charged particles and ionizing radiation detector

E Cruz-Zaragoza[1] and I Piña López[2]
[1] Unidad de Irradiación y Seguridad Radiológica, Instituto de Ciencias Nucleares, Universidad Nacional Autónoma de México, A.P. 70-543, 04510 México D.F, México.
[2]Facultad de Ciencias, Universidad Nacional Autónoma de México, Circuito Exterior C.U., 04510 México D.F., México.

E-mail: ecruz@nucleares.unam.mx

Abstract. This paper reports results of CMOS sensor suitable for use as charged particles and ionizing radiation detector. The CMOS sensor with 640x480 pixels area has been integrated into an electronic circuit for detection of ionizing radiation and it was exposed to alpha particle (Am-241, U_{nat}), beta (Sr-90), and gamma photons (Cs-137). Results show after long period of time (168 h) irradiation the sensor had not loss of functionality and also the energy of the charge particles and photons were very well obtained.

1. Introduction

Recently, the CMOS active pixel sensors have has been tested for several applications and thus have increased the demand for improving the performance thereof [1]. This type of sensors has an APS (Active Pixel Sensor) architecture, which is formed by an integrated circuit with a matrix of pixels where each pixel itself contains a photodetector [2,3]. At present, the pixel size is less than 2 micrometers which is possible thanks to microelectronics. In addition to imaging applications, new perspectives in various research areas could be opened by a different type of devices such as CMOS sensors. As an example, these devices could be used for detection of ionizing radiation with the possibility of a submicrometer measurement. In this work we have investigated the possibility of using a CMOS sensor with an area of 640x480 pixels, which previously has been integrated into an electronic circuit (figure 1) for detection of ionizing radiation and charged particles testing from radioactive sources, such as: Am-241, Natural Uranium (U-234, U-235 and U-238), Sr-90, and Cs-137.

1.1 CMOS sensor operation.

The operation of the CMOS sensor is characterized by its APS architecture, which is sensitive to the incidence of visible or infrared light, this way the polarized inversely occurs a certain circulation of current when pixel is excite by light generates a current power in each of the pixels, which varies with the intensity of light received.

In turn is incorporated an amplifier for electrical signal at each photo site where the device must send the electrical signal produced by each photosite sensor out of it and amplify on the PC.

Figure 1. The CMOS sensor integrated in the electronic circuit and attached to lap top.

1.2 Signal generation for the sensor CMOS.
According to the linear energy transfer (LET) [4], the load generated from several heavy ions is usually assumed to be proportional. Indeed for in situ particles in the surface area of the sensor, linear energy transfer remains constant, where the concept has worked well for older devices, and continues to have a reasonable approach to alpha particles. In this work the system has been exposed to alpha particles by Am-241 and natural uranium. Considering

$$\Delta F = -\nabla \Delta E$$

where ΔE is the energy loss of the charged particle due to electronic collisions while across a distance dx, excluding all secondary electrons with kinetic energies larger than E_0 from the count.
For beta radiation, electrons will have an electric current proportional to the decay energy, which is produced in each of the pixels, provided by the Sr-90 and Cs-137,

$$E_{e-} = I$$

Finally for gamma radiation, whose photon will impact in the surface of the sensor that is exciting each pixel sensor according to the photoelectric effect through the Cs-137,

$$hf = \phi + E_k$$

In this way, the energy that is incident on each of the pixels is directly proportional to the radiation from each radiation source.

2. Experimental part

2.1 System Calibration
The radioactive source of Americium-241 was characterized by constant emit alpha radiation, which has been used for the calibration of various sensors. From a sample of 0.2g and by mass-energy relation was calibrated system:

$$E_{Am-241} = (1.00495275 \times 10^{-29}\ \text{kg})(2.997 \times 10^{8}\ m/s)^2$$
$$E_{Am-241} = 5.63451604 MeV$$

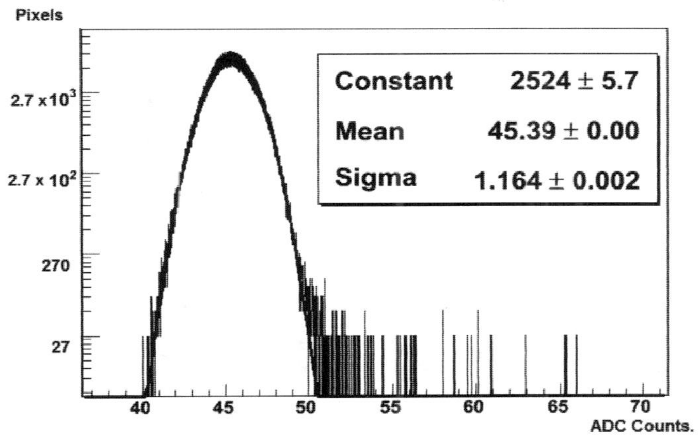

Figure 2. Distribution of digitization in the CMOS sensor pixel represented by a Gaussian.

2.2 Electrical response from CMOS sensor
The response of the sensor in the absence of radiation has been studied, for each of the pixels is taken the information from them; where the distribution was possible represent a Gaussian as shown in figure 2.

This means that the sensor could withstand the variation of current produced by different radioactive sources, which shows that for small variations of electric currents, the critical detection point will be through 2.7 x 10^3 photosites found in the pixels of the sensor and to larger variations will only need 27 to 270 photosites due to digitization of the signal.

3. Results

3.1 Sensor Response to Alpha radiations
The sensor was exposed to ionizing radiation at the Nuclear Sciences Institute (UNAM) of México. Figure 3 shows the impact of ^{4}He atoms on the sensor surface, and the total energy of the source (Am-241) was about 5.0 MeV where the peaks represent the total energy.

For the composition of Natural Uranium, was obtained a maximum energy of 4.0 MeV and minimum around 2.5 MeV as shown in figure 4. That behavior is due to the composition of the radioactive source which is composed of ^{234}U, ^{235}U, and ^{238}U. One influence is the types of nuclear decays occurring in it due to fluctuation of the energy that arrived to CMOS sensor.

3.2 Sensor response to Beta and Gamma radiations.
Sensor response to beta radiation was obtained by a strontium-90 source, as shown in figure 5, this source had magnitude of about 0.8 MeV of energy.

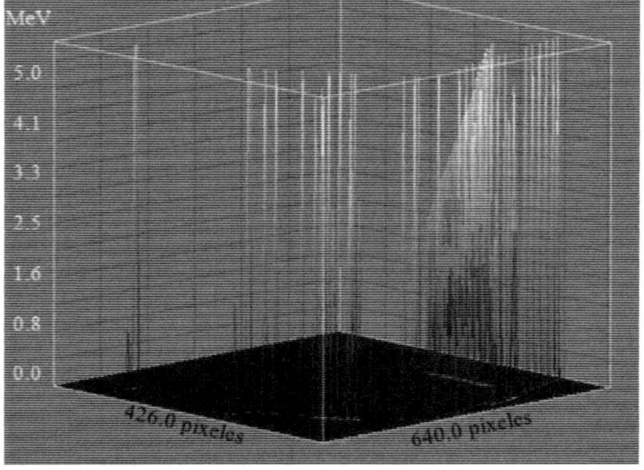

Figure 3. Alpha radiation detection from Am-241 on the sensor area.

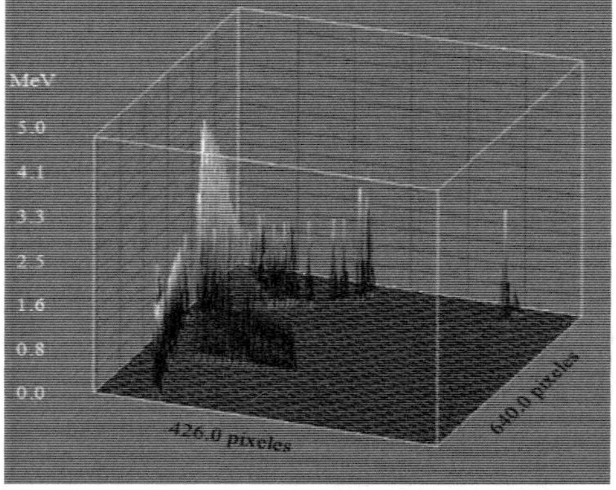

Figure 4. Natural Uranium radiations on the sensor area.

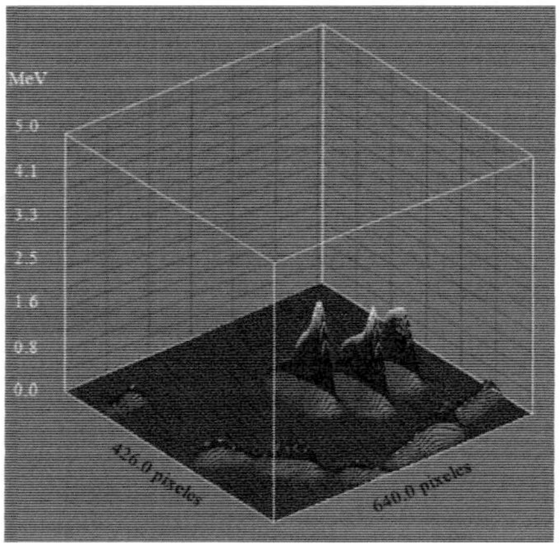

Figure 5. Beta radiation energy from Sr-90 on the sensor area.

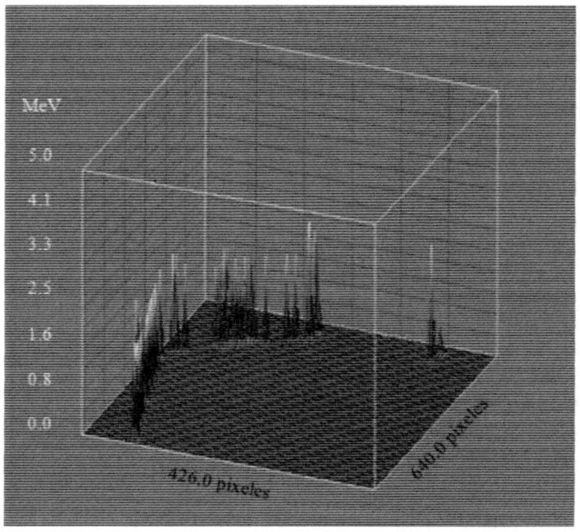

Figure 6. Photons from Cs-137 incidents on the sensor area.

Finally, the Cesium 137 source was used for testing CMOS sensor. Figure 6 shows beta and gamma radiation energy incidents on the sensor. It is important to mention that the gamma rays do not affect the functionality of the sensor and the energy recorded was 1.6 MeV.

3.3 Activity frequency.

Frequency of activity (hits per second) of the sources was carried out during 30 s irradiation and in this period of time the impacts or hits on the sensor were recorded (table 1). This readout was repeated for five times for each source.

Table 1. Activity frequency.

Sources	Hits/s
Am-241	1.295×10^4
U nat	1.244×10^6
Sr-90	5.12×10^9
Cs-137	3.15×10^{11}

4. Discussion and conclusion

The CMOS sensor response was appropriated in each variations of energy from the radioactive sources. The graph in the figures shows the area and positions of the sensor surface where the impacted particles and gamma rays.The higher peaks show each energy type from radiation source. The frequency of activity of the sources was very well determined. Due to the composition within the pixels by the photosites, exposure of the sources was in the absence of external light in order to avoid the outer photons excite each of the photosites. When the CMOS sensor was tested under continuous irradiation for a long period of 168 hours (1 week) using the largest energy source (Am-241), the sensor had not loss of functionality.

Acknowledgments

Authors would like to acknowledge the technical assistance from Benjamin Leal Acevedo (ICN UNAM) for irradiation sensor.

References

[1] Servoli L *et al* 2007 *Nucl. Instr. Meth.* **A 581** 335
[2] Abe K *et al* 1997 *Nucl. Instr. Meth.* **A 400** 287
[3] Kleinfelder S et al 2003 *Proc. SPIE 4784 Hard X-Ray and Gamma-Ray Detectors and Applications IV (Seatle, WA)* p 208 doi:10.1117/12.450826
[4] Attix F A *Introduction to Radiological Physics and Radiation Dosimetry* (John Wiley & Sons Inc., New York, 1986)

Modulation of radical pairs dynamics immersed in an ELF-EMF: The effect on hepatocarcinogenesis

G O López-Riquelme[1], E López-Sandoval[2], E Vera-Aguilar[3] and J J Godina-Nava[1]

[1]Departamento de Física, CINVESTAV, IPN, Ap. Post. 14-740, México, D.F. 07000
[2]Centro Brasileiro de Pesquisas Físicas, Rua Dr. Xavier Sigaud, 150, CEP 22290-180, Rio de Janeiro, Brazil
[3] Departamento de Farmacología, CINVESTAV, IPN, Ap. Post. 14-740, México, D.F. 07000

Email: jj@fis.cinvestav.mx

Abstract. The most suitable mechanism of action of electromagnetic fields (EMF) on biological systems is the effect on the radical pair (RP) recombination through the Zeeman effect and hyperfine interaction, which changes the rate of reactions or the product distribution. Enzyme reactions with RP intermediates can be altered by EMF, like those catalyzed by cytochrome P450 enzymes (CYP450), a heme-thiolate family protein that detoxifies xenobiotics and involved in chemical carcinogenesis. CYP450 activate chemical carcinogens producing an enormous amount of free radicals, which damage the DNA resulting in the malignant transformation of cells. During the activation, CYP450 produce spin-correlated RP intermediates that can either go to recombination or to continue the catalytic process. As CYP450 are electron carrier proteins, it is possible that RP intermediates may be affected by EMF. It was previously found that periodic treatment with extremely low frequency electromagnetic fields (ELF-EMF) inhibits more than 50% the number and area of preneoplastic lesions in rats with chemically induced hepatocarcinogenesis through reduction of cell proliferation. In this work, we developed a quantum mechanical model based on RP mechanism in order to explain the experimental effects of ELF-EMF on the free radicals produced in the early stages of chemical carcinogenesis.

1. Introduction

Although mild, several studies have demonstrated the influence of extremely low-frequency electromagnetic fields (ELF-EMF; < 1200 Hz, C≈ 1 km) on biological processes [1] and chemical reactions [2, 3]. One of the most studied mechanism of action of ELF-EMF on chemical reactions, and, therefore, biological systems, is the radical pair mechanism (RPM). Radicals, molecules or atoms which contain unpaired electrons in its outer orbital, are short-lived intermediates in a wide diverse of reactions that are produced in pairs, by what they are called radical pairs (RP) [3]. RP are involved in many biological processes like oxidation and, due to this, in aging and diseases. Since radicals have unpaired electron spins they, therefore, can present one of two spin states, singlets (S) or triplets (T). RP are produced simultaneously in reactions by an electron transfer reaction in spin-correlated states. These unpaired electrons are very unstable and hazardous. Although its lifetime is in the order of nanoseconds, they collide with other molecules and atoms generating more free radicals, creating a chain reaction in which millions of molecules could be altered in a few nanoseconds.

The RPM is used to study the effect of EMF on the rate of the RP recombination dynamics, *i.e.*, the interconversion of the S and T states through the Zeeman effect and hyperfine interaction due to the associated magnetic momentum to each electron spin [4]. RPM effects of ELF-EMF only are significant with EMF of more than 1mT [2] where they have measurable effects on the kinetics of chemical reactions through their effect on the spin correlation rate of the unpaired electrons and consequently over the radicals lifetime [5]. However, since ELF-EMF are too weak, they do not produce any temperature increase in biological systems [6].

In the absence of EMF, there is not S-T interconversion. On the other hand, ELF-EMF can alter the course of reactions through the regulation of the interconversion from S to T spin states or its regeneration from the S spin state. When an EMF is applied, two of the three triplets levels, namely T_{+1} and T_{-1}, become progressively decoupled restricting S-T interconversion to the S and T_0 states. Thus, RP can either to recombine to form the original molecule or to separate into free radicals and create products in other chemical reactions [3]. Since radicals usually react and recombine when they are in the S state [3], the result is a change in the rate of reactions or in the product distribution [2, 3, 7]. The dynamics of the RP is generated by a magnetic interaction provoking a Zeeman splitting of the energy levels and a hyperfine coupling between an electron and nucleon. The RP oscillates as a combination of S and T states [8]. RPM is considered an excellent, and currently the best, theoretical model for understanding the effects that magnetic fields exert on chemical reactions [3, 9].

Enzymes with RP intermediates exhibit magnetic field-dependant parameters [2], and numerous enzymes with radical mechanisms are known such as cytochrome P450 (CYP450) [10]. CYP450 enzymes are a heme-thiolate monooxygenase family of proteins expressed in the liver and other tissues [11]. The main function of CYP450 is the oxidation of substrates [12]. CYP450 use the energy from an electron provided by a donor protein to initiate its catalytic process [13]. CYP450, which normally detoxify xenobiotic compounds, are important enzymes because they are involved in chemical carcinogenesis since this enzyme activates procarcinogens [14]. Chemical carcinogenesis is a multi-step process characterized by the electrophile attack to DNA, which produces mutations, resulting in the malignant transformation of cells [15]. Although most chemical carcinogens tend to be stable (procarcinogens), once they are incorporated into the organism, suffer an activation in the liver by CYP450 [14] converting the procarcinogen into a toxic compound. During the oxygenation of the substrates by CYP450, bursts of electrophiles near the site of oxidation are generated [16] inducing oxidative stress (OS) [17, 18, 19]. The relevance of CYP450 in carcinogenesis has made this enzyme the target of chemoprevention [20].

CYP450 are electron carrier proteins that produce spin-correlated RP intermediates [2] susceptible of being modulated by ELF-EMF. Therefore, since ELF-EMF modifies the spin orientation and, therefore, the interconversion rate between S and T spin states of RP populations, it is plausible to expect that the formation of products derived from the metabolization of chemical carcinogens, *i.e.*, free radicals, might be modified.

2. Problem statement

In a previous work, our group experimentally analyzed the effect of ELF-EMF on chemically induced hepatocarcinogenesis employing the modified resistant hepatocyte model (MRHM) in Fisher-344 rats [21]. The main chemical carcinogen used in the MRHM, diethylnitrosamine (DEN) is a well-known potent hepatocarcinogenic agent [22] present in many products for human consumption [23]. DEN generates free radicals causing necrosis, hyperplasia, and tumors [24]. Besides, DEN produces single strand breaks in the DNA and also induces changes in enzymes

involved in the DNA repair [24]. The second chemical used in the MRHM is 2-acetilaminofluorine (2-AAF), which also induces tumorigenesis in the rat liver [25]. Both chemical carcinogens are activated by CYP450 producing an unregulated enormous wave of electrophiles that cause OS, leading to carcinogenesis initiation. After administration of carcinogens, partial hepatectomy promotes the development of preneoplastic lesions in the liver of rats [21]. The experimental results showed that periodic treatment with 4.5 mT - 120 Hz ELF-EMF (50 minutes each day) inhibits more than 50% the number and area of preneoplastic lesions in rats with chemically induced cancer through the reduction of cell proliferation [21]. Despite these findings, the molecular mechanisms responsible for these effects are still unclear. Nonetheless, RPM is an important mechanism that could explains part of the observed experimental effects of the ELF-EMF.
In this work, it was theoretically analyzed the experimental effects of ELF-EMF on RP intermediates and OS produced by CYP450 during procarcinogen activation. The objective was to establish a theoretical methodology to model the possible mechanism of the cytoprotective effect of ELF-EMF on the liver with induced carcinogenesis. The model uses the RP theory to study charge migration in the electron transfer reactions catalyzed by CYP450, involving free radicals dynamics interacting with the ELF-EMF. It is plausible to assume that through the RPM, the RP involved in free radicals generation could be modulated, as has been reported for other reactions [5, 26], and directed to a lower energetic states in such a way that the activation of oxidative products would be diminished and the electrophile damage reduced.

3. The model

Several attempts to model the effects of ELF-EMF on reaction kinetics have been developed [27]. Here, in order to model the dynamics of RP intermediates involved in OS produced by chemically induced hepatocarcinogenesis when an external ELF-EMF (consisting of both, static and pulsating fields) is applied, the RPM was used [3] because it involves the recombination of short-lived reactive free radicals with correlated electron spins [28]. A theoretical semi-classical (or semi-quantum) model was used to describe the possible modulation of the high amount of RP produced during CYP450 carcinogen activation by the ELF-EMF. Then, the spin Hamiltonian operator for the weakly coupled RP with an external EMF was used [3]:

$$\begin{aligned} \widehat{H}_{RP} &= \widehat{H}_{mag} + \widehat{H}_{ex}, \\ \widehat{H}_{mag} &= \mu_B \, g_a \, BS_{1z} + \alpha \vec{I} \cdot \vec{S_1} + \mu_B \, g_b \, BS_{2z}, \\ \widehat{H}_{ex} &= -J(\vec{r})\left(2\vec{S_1} \cdot \vec{S_2} + 1/2 \right), \end{aligned} \tag{1}$$

where $\widehat{H}_{mag}$ is the Hamiltonian for the ELF-EMF and $\widehat{H}_{ex}$ the exchange Hamiltonian operator term for the spin exchange interaction; $\mu_B g_a BS_{1z}$ and $\mu_B g_b BS_{2z}$ correspond to the Zeeman interaction, with g_a and g_b as isotropic values of the *a* and *b* radicals of the RP; and α is the hyperfine coupling constant with spin nucleus $\vec{I}$. The interaction energy depends on the distance by the term $J(\vec{r})$.

The chemical reactivity of free radicals is controlled by spin dynamics, which consists in the transformation of nonreactive T spin states into the reactive S state and *vice versa* [29]. Hyperfine coupling, the Zeeman effect and the differences in Larmor precession rates can alter the spin orientation [3]. Then, the recombination probability of S states of the RP through the EMF influence by the hyperfine interaction was studied to assess the spin states development during S-T inter-conversions. Since S-T interconversion can be influenced by an external EMF, the conservation of the angular momentum will favor one of the states, either S or T, over the other in one or another reaction pathways. Thus, the production rate of S-T populations of free radicals is modified due to differences in energy among eigenstates caused, in turn, by changes in the spin eigenstructure of the Hamiltonian of the system. Such effect is called spin selectivity of the RP recombination [30]. The hypothesis is that the periodic exposure to ELF-EMF will alter the progression of RP of free radicals generated by CYP450 catalysis of carcinogens in such a way that the electron transfer through S-T interactions is modified, with the result of a reduction of OS. The consequence is a reduction in the formation of preneoplastic lesions. Two main steps were followed to construct the model: 1) RP analytical solution to study the influence of the ELF-EMF on the S-T interconversion employing a system composed of two spins from electrons and one spin from one nucleus; 2) numerical solution by Lanczos method in order to study complex systems composed by many spins using the tridiagonal transformation matrix obtained from the Hamiltonian.

3.1. Modeling one nucleus and two electron spins

The first approximation was to solve analytically the three spin system composed by two electron spins and one nuclear spin modulated by the ELF-EMF. Interactions between RP intermediates and electron transfer were modelled employing the RPM (se equation 1). The recombination of the free radicals by pairs was studied through the magnetic influence of the hyperfine interaction including the damping of the two quantum states to analyze the spin states development during the S-T interconversion. The relaxation time of the enzymatic reaction was assumed to be greater than the lifetime of the RP. This feature provides the adequate boundary conditions that confer the system with the capability to sustain oscillatory behavior between the spin states due to the interaction between the two electrons and the nucleus. This was assured by allocating the elemental cell target in an average position where the OS was generated. The studies were performed in the steady state, and an isotropic hyperfine interaction was assumed as an essential mechanism in the time evolution of the spin states. In the recombination study, a simple kinetic model [27, 31, 32] was used in which the quantum modulation of RP by the EMF is involved. The purpose of this was to examine the behavior of the spin population in a steady state to evaluate the recombination probability (Fig 1). Since there has been included both, the static magnetic field and the pulsating, Timmel and Hore's theory was used [33]. As the system consisted of two electron spins and one nuclear spin modulated by hyperfine interaction, eight basic states were separated into two sets: 1) one with the energy equal to M/4 and degeneracy of 6, and the other with energy equal to -3M/4 and degeneracy of 2. For an EMF directed along the z-axis, the spin Hamiltonian operators are:

$$\hat{H}_0 = \alpha\overrightarrow{S_a}\cdot\vec{I} - \omega_{ELF}\ (S_{az} + S_{bz} + I_z) + \omega_S\ (S_{az} + S_{bz}),$$
$$\hat{H}_1 = \omega_t\ (S_{az} + S_{bz}).$$

The first term includes the static magnetic field $\omega_S(S_{az} + S_{bz})$ and the pulsating magnetic field $\omega_{ELF}\ (S_{az} + S_{bz} + I_z)$. S_{az}, S_{bz}, I_z are the spin operators and nuclear isospin.

3.2. Lanczos numerical approximation

RPM modelling allow us to understand the way in which the RPM works and to know what is the main effect of the ELF-EMF for the transition between S and T_0 spin state. For more complex systems composed by many pairs of electrons and nucleus and other transitions in the system, it is necessary to use numerical methods in order to simplify calculations. The Lanczos method with the Stochastic Liouville equation were used as a way to evaluate eigenvalues and eigenvectors of a matrix, which represents their typical dynamical evolution. With the Lanczos method [4, 34, 35] the matrix of the system was transformed to a tridiagonal form, which simplifies the finding of eigenvalues and eigenvectors. In this way, a vector space tailored can be created to contain the dynamical quantities needed to describe the time evolution of relevant parameters. The advantage of this method is that it allows managing smaller dimension of the vector space spanned than the original dimension of the system, eliminating redundant quantities as in the case of degenerate energy eigenvalues. The evolution equation is:

$$\frac{\partial}{\partial t}\langle a_k|\hat{\rho}|a_l\rangle = -\sum_{j,m} L(kl,mj)\langle a_m|\hat{\rho}|a_j\rangle,$$

where $L(kl,mj)$ is the matrix of the transition operator in the super-operator basis, $\hat{\rho}$ the spin population operator.

Journal of Physics: Conference Series **582** (2015) 012048 doi:10.1088/1742-6596/582/1/012048

4. Results and discussion

In this work, matrix methods were applied to the quantum model that describes the effects of EMF on the RPM in order to determine the evolution of the spin system proposed. The eigenstates and eigenvalues of the system were calculated and used to perform an expansion of an arbitrary state. To know the possibilities of the method, the cases in which results can be calculated analytically were determined. These results allowed to study the magnetic effects that appear as a result of the interplay between the spins of electrons and the spin of the nucleus.

We analyzed the dynamics of the RP (governed by rules of spin coupling and the conservation of angular momentum) generated from two weakly coupled spins when CYP450 enzymatic reaction is carried out. A cyclic kinetics was used to examine the EMF influence on the spin-states inter-conversions. Such kinetic processes were introduced using them as an intermediate oscillatory process to account for an energy conversion step due to the spin selectivity. In the calculations, it was considered an elemental target cell exposed to the OS generated by the DEN activation and submitted to a periodic external EMF. Figure 1 shows the densities of singlet populations $\rho_1(t)$, triplet populations $\rho_4(t)$, and both $|\rho_2(t)|^2$ as a function of ELF-EMF (B) and time (t). It can be seen that, when ELF-EMF is turned off (B = 0), there are no oscillations. By contrary, oscillations increase quickly up to a maximum value when ELF-EMF intensity increases, modulating the behavior of spin populations. The resultant combined effect in the recombination process was 40% from the static EMF plus 25% from the pulsating EMF. Thus, since there are less active S spins, the number of initiated cells diminish, and, therefore, the preneoplastic lesion formation, which is the main cytoprotective effect of the EMF found experimentally.

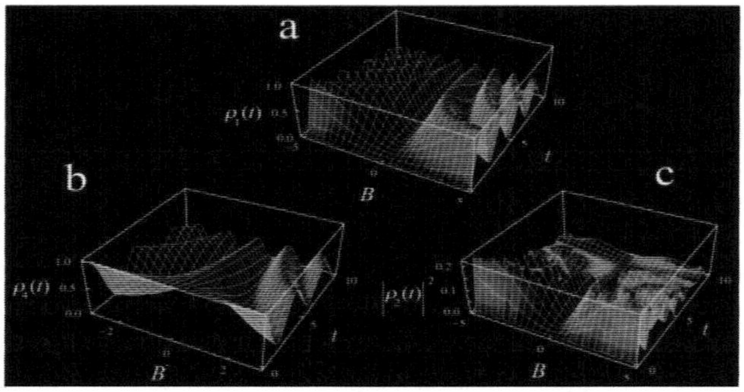

Figure 1. Time evolution of spin populations submitted to ELF-EMF for the components of the $\rho(t)$ for the initial state $\rho(0)$=(0,0,0,1). Plots for density of spin populations for a) singlets; b) triplets, and c) for S and T spin populations combined. B: EMF; t: time (see text).

The mathematical model presented in this work is a simple approximation that contributes to the understanding of chemical carcinogenic process based on the behavior of charged particles, which are the direct responsible of cell damage. Evidently, the scope of this approach is simplistic since biological systems are highly complex and indeed exist mechanisms that were not taken into account. For this reason, it is necessary to improve and complexify the approximations developed in the present study. As far as we know, this is the first study in which RPM is used to explain and model processes related with enzymatic RP intermediaries involved in OS and carcinogenesis. The results of this work could constitute the basis for the design of strategies or clinical applications of ELF-EMF as co-adjutant in the treatment of several diseases related with oxidative damage, including cancer.

Acknowledgments

This work was supported by grant 132432 to JJGN and GOLR from CONACYT, México. We thank Gabino Torres for his helpful advice concerning the Lanczos methods.

Journal of Physics: Conference Series **582** (2015) 012048 doi:10.1088/1742-6596/582/1/012048

References

[1] Paunesku T and Woloschak G E 2007 Effects of Radiofrequency and Extremely Low-Frequency Electromagnetic Field Radiation on Cells of the Immune System in *Biological and medical aspects of electromagnetic fields* Ed Barnes F S and Greenbaum B (Florida: Taylor & Francis) pp 1-20

[2] Grissom C B 1995 *Chem Rev* **95** 3

[3] Hayashi H 2004 *Introduction to dynamic spin chemestry Magnetic Field Effects on Chemical and Biochemical Reactions* (New Jersey: World Scientific) p 254

[4] Schulten K, Swenberg C E and Weller A 1978 *Z Phys Chem* , **111** 1-5

[5] Lacy-Hulbert A, Metcalfe J C and Hesketh R 1998 *FASEB Journal* **12** 395

[6] Adair R K 1991 *Phys Rev A* **43** 1039-1048

[7] Adey W R Electromagnetics in Biology and Medicine in *Modern Radio Science* Ed Matsumoto H. (Oxford) pp 231-249

[8] Neufeld A A and Pedersen J B 1998 *J. Chem. Phys.* **109** 8743

[9] Rodgers C T and Hore P J 2009 *PNAS* **106** 353

[10] Stubbe J A 1989 *Ann. Rev. Biochem.* **58** 257

[11] Guengerich F P 2003 *Mol interventions* **3** 194

[12] Bernhardt R 2006 *J Biotech* **124** 128

[13] Guengerich F P and Isin E M 2008 *Acta Chim. Slov.* **55** 7

[14] Nebert D W and Dalton T P 2006 *Nature Reviews Cancer* **6** 947

[15] Bertram J S 2000 *Mol. Aspects Med.* **21** 167

[16] Klaunig J E and Kamendulis L M 2004 *Annu. Rev. Pharmacol. Toxicol.* **44** 239

[17] Koop D R 1992 *FASEB* **6** 724

[18] Parke D V 1994 *Environ. Health Persp.* **102** 852.

[19] Guengerich F P 2008 *Chem. Res. Toxicol.* **21** 70

[20] Yang C S, Smith T J and Hong J Y 1994 *Cancer Res.* **54**1982s

[21] Jiménez-García M N, Arellanes-Robledo J, Aparicio-Bautista D I, Rodríguez-Segura M Á, Villa-Treviño S, and Godina-Nava J J 2010 *BMC Cancer* **10** 159

[22] Lawley P D 1990 N-nitroso compounds *Chemical Carcinogenesis and Mutagenesis I* (Springer Berlin Heidelberg) pp 409-469

[23] Mirvish S S 1995 *Cancer Lett.* **93** 17

[24] Malik S, Bhatnagar S, Chaudhary N, Katare D P and Jain S K 2012 *Protoplasma* **250** 175

[25] Tiwawech D, Hasegawa R, Kurata Y, Tatematsu M, Shibata MA, Thamavit W, Ito N 1991 *Carcinogenesis* **12** 985

[26] Hamilton C A, Hewitt J P, McLauchlan K A and Steiner U E 1988 *Mol. Phys.* **65** 423

[27] Eichwald C and Walleczek J 1996 *Biophys J* **71** 623

[28] Eveson RW, Timmel C R, Brocklehurst B, Hore P J, McLauchlan K A 2000 *Int. J. Rad. Biol.* **76** 1509

[29] Buchachenko A L and Berdinsky V L 1998 *Chem. Phys. Lett.* **298** 284

[30] Hansen M J and Pedersen J 2002 *Chem. Phys. Lett.* **361** 219

[31] Michel-Beyerle M E, Haberkorn R, Bube W, Steffens E, Schroder H, Neusser H J, Schlag E W and Seidlitz H 1976 *Chem. Phys.* **17** 139

[32] Pedersen J B, Mojaza M, and Lukzen N N 2010 *Chem. Phys. Lett.* **496** 212

[33] Timmel C R and Hore P J 1996 *Chem. Phys. Lett.* **257** 400

[34] Wassam Jr, W A and Torres-Vega G 1987 *Chem. Phys. Lett.* **134** 355

Journal of Physics: Conference Series **582** (2015) 012049 doi:10.1088/1742-6596/582/1/012049

Fractal dimension and turbulence in Giant HII Regions

H. E. Caicedo-Ortiz[1], E. Santiago-Cortes[1,2], J. López-Bonilla[3], H.O. Castañeda[4]

[1] Maestría en Ciencias de la Complejidad, Universidad Autónoma de la Ciudad de México, San Lorenzo 290, esquina Roberto Gayol, Col. del Valle Sur, Del. Benito Juárez, C.P. 03100, México D.F.

[2] Grupo de Ingeniería y Tecnologías Cuánticas, Corporación Universitaria Autónoma del Cauca, Calle 5 No 3-85, Popayán, Colombia

[3] ESIME-Zacatenco, Instituto Politécnico Nacional, Edif. 5, CP 07738, México D.F.

[4] ESFM, Instituto Politécnico Nacional, Edif. 9, 1er piso, CP 07738, México D.F.

E-mail: hecaicedo@gmail.com

Abstract. We have measured the fractal dimensions of the Giant HII Regions Hubble X and Hubble V in NGC6822 using images obtained with the Hubble's Wide Field Planetary Camera 2 (WFPC2). These measures are associated with the turbulence observed in these regions, which is quantified through the velocity dispersion of emission lines in the visible. Our results suggest low turbulence behaviour.

1. Introduction

In the last years, the concept of fractals has been used intensively in the study of the Universe. The fractal dimension is employed as tool to study, for example, the solar flares [1], solar granulation [2], the morphology of craters on planets [3], galaxy catalogs [4], the study of the distribution of HII regions in galaxies [5] and the fractal dimension of the interstellar medium [6].

In this paper, we compute the fractal dimension of the Giant HII Regions of the Hubble V and Hubble X and we propose that the result can be interpreted as an indicator of the degree of hydrodynamic turbulence within the regions.

2. Observations

For the analysis of the fractal structure, we used images obtained by the Hubble Heritage Team. Hubble X observations were made on September the 7th, 1997 using F656N (Hα), F502N ([O III]) and F487N (Hβ filters), with 2.6 hours of exposition time. On the other hand, Hubble V observations were performed on June the 18th in 1996 using F300W (U), F487N (H-β), F502N ([O III]), F547M (Strömgren y), F656N (Hα) and F658N ([N II]) filters and on August the 2nd in 1997 using the F555W (V) filter. The whole observations were processed using the WFPC tool of Hubble Special Telescope (HST). Figures 1 and 2 correspond to the Hubble X and Hubble V images used in this research.

Journal of Physics: Conference Series **582** (2015) 012049 doi:10.1088/1742-6596/582/1/012049

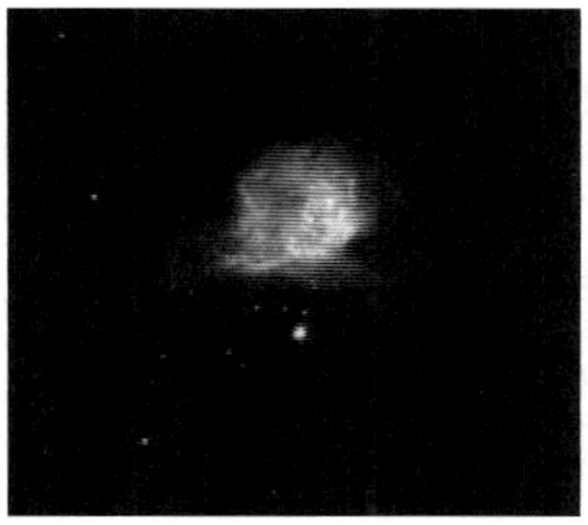

Figure 1. Hubble X. Hubble Heritage Team, NASA.

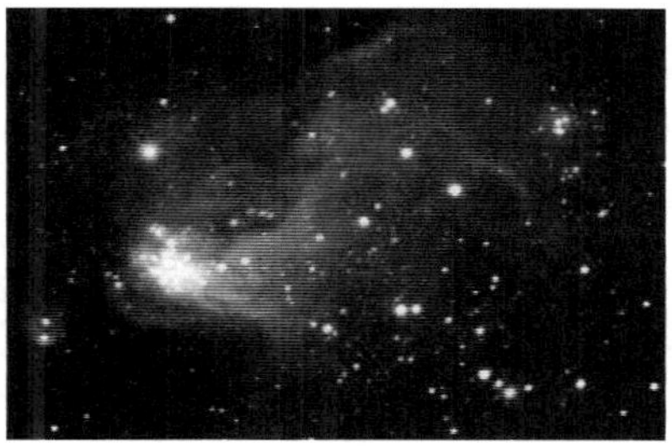

Figure 2. Hubble V. Hubble Heritage Team, NASA.

The spectra used in this study were obtained from data available at The Cambridge Astronomical Survey Unit (CASU) produced by the Isaac Newton Group of Telescopes located in the Canary Islands. The observations were carried out with the ISIS spectrograph of the William Herchel Observatory at the Roque de los Muchachos, using the technique of longslit spectroscopy, between August the 18th and 19th in 1992. The spectra for Hubble V (X) were obtained in 4 (8) different positions, all of them at position angle 90°, with a slit width of 1" and a separation of 2" between the centers of two consecutive slit positions. Two spectra were taken simultaneously for each position, one in the range between 6390 Å and 6849 Å (red arm) and one between 4665 Å and 5065 Å (blue arm), both with a dispersion approximately 0.4 Å /pixel. The slit length was 200", with a spatial sampling along the slit of 0.34"/pixel in the red arm and 0.36"/pixel in the blue arm. Data reduction was performed using the IRAF data reduction software, following standard procedures.

3. Fractal dimensión

The fractal structure is characterized by a parameter called fractal dimension, which is a generalization of the concept of Euclidean dimension. There are several methods to define the fractal dimension [7].

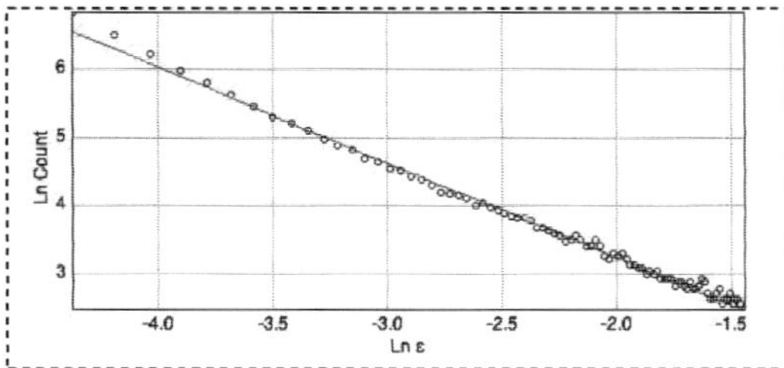

Figure 3. Fractal dimension of the perimeter of Hubble X.

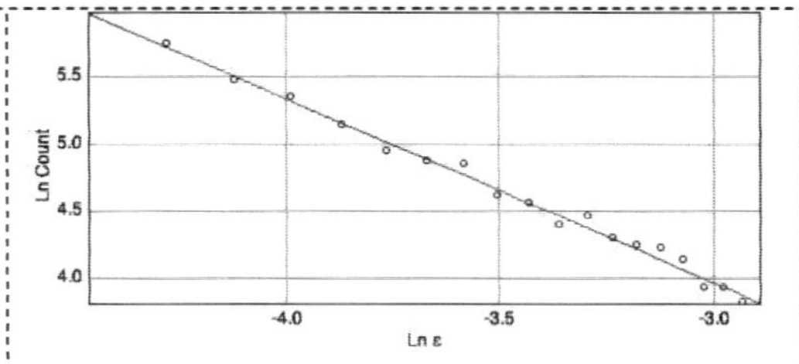

Figure 4. Fractal dimension of the perimeter of Hubble V.

The most popular algorithm to compute the fractal dimension is the Box Counting (D_B) [7,8]. In this method, the object is covered with a set of boxes with side length ε. The number of $N(\varepsilon)$ boxes is proportional to scale ε in the form $N(\varepsilon)\ \alpha\ 1/\varepsilon^{d}$, where the d exponent is the object dimension. The dimension is defined by:

$$D_B = \lim_{\varepsilon \to 0} \frac{\log N(\varepsilon)}{\log(1/\varepsilon)}. \tag{1}$$

This methodology can be applied in systematic way because of the simplicity in the creation of the box-covering and the computation is quickly performed, even for large images obtained with high resolution.

The fractal dimension of Hubble V and Hubble X was estimated using the automated dimension analysis plugin ***Fraclac*** of the images processing software ***ImageJ***. A set of 12 measures was taken. Log-Log graphs were plotted of the reciprocal of the side length of the square against the number of out-lines containing squares (Fig 3 and 4). The fractal dimension for each HII region is the average of these 12 measures being 1.4081 for Hubble X and 1.3818 for Hubble V.

4. Velocity dispersion

The traditional method used to determine turbulence in a gaseous nebula is by velocity dispersion (σ) obtained from the width of the emission lines. The width of the emission lines measured in the observations presents excess after we take into considerations the contributions of the thermal movement of gas and the instrumental width. To estimate the instrumental width associated to every emission line, we used the emission lines observed in the calibration spectrum associated to each spectrograph arm (red and blue) and for each observed object (one for Hubble V, another for Hubble X, for arm). The instrumental width at half maximum (FWHM) associated to each emission line, is the width at the average medium height of the two emission lines on the calibration spectrum that narrow from upper to lower.

Table 1. Velocity dispersion for Hubble X and Hubble V with integrated spectral.

Object	Hubble X	Hubble V	References
Hα	10 ± 2	10 ± 2	*
Hα	10.5 ± 0.2	10.3 ± 0.3	[9]
Hα	10.4 ± 0.9	9.5 ± 0.8	[10]
Hα	9.8	9.5	[11]

* This research

We have that the component of instrumental width-associated velocity is expressed by $\sigma_{inst} = \frac{0.6006\ FWHM_{inst}}{1.414}$. $FWHM_{inst}$ is the instrumental width at half height expressed in velocity units. The velocity-associated to width observed in the emission line is expressed by $\sigma_{inst} = \frac{0.6006\ FWHM_{obs}}{1.414}$, with $FWHM_{obs}$ as the observed width (in velocity units). The thermal component is sensitive to the temperature assumed in the gas and to the electron-emitted mass in the emission lines. $\sigma_{inst} = \frac{166.3\,t}{1.414\,m}$, where $t = T_e \times 10^{-4}$, T_e is the gas temperature, which in this case is 11500 K [12] and m is the atomic mass of the ion related to the emission line. The velocity dispersion is:

$$\sigma_{disp} = \left(\sigma_{obs}^2 - \sigma_{inst}^2 - \sigma_{term}^2\right)^{1/2}. \tag{2}$$

Table 1 contains the results of the velocity dispersion in Hα for Hubble X and Hubble V for the integrated spectra for this research and by other previous research.

5. Discussion

In the last years, the fractal theory has been applied at the study of fluid dynamics. These properties are present in fluids that show the transition from a laminar structure to a turbulent regime [13]. Sreenivasan and Meneveau [14] introduced the fractal theory into experimental fluid dynamics, and

Journal of Physics: Conference Series **582** (2015) 012049 doi:10.1088/1742-6596/582/1/012049

analyzed the fractal dimension of a low speed shear layer, jet, wake and boundary layer. Sreenivasan discussed the fractal and multifractal description and measurement of turbulent flow [15].

The value of the fractal dimension of 1.4081 for Hubble X and 1.3818 for Hubble V suggest a fractal behavior of these giant HIII regions. From the geometric point of view, the fractal dimension is a measure of the system irregularity. As the fractal dimension increases, the degree of irregularity of the curve is higher. Zhao YuXin et al [13], suggests that the transition from a laminar to a turbulent flow can be seeing as a process of breaking up of the streamline and contour. The streamline and contour of laminar flows are smooth curves. When the flow tends to be instable, they will curl and twist. Vortex of different scales will interact and transfer energy, and while the flow make the transition to turbulence, the streamline and contour will break up completely and it will show irregularity of the transitional mixing layer, which can be evaluated by the fractal dimension [13]. Table 2 presents the fractal dimension of different fractal objects, of Hubble X, Hubble V and another HIII regions for whose this property has been calculated. The few existent studies about fractal dimension made to HIII regions [16-19] only calculate it and suggest the possibility of associate the behavior of this HII regions with another fractal structures.

The Hubble X and Hubble V dimension fractal values suggest a behavior slightly irregular that can be interpreted as a low turbulence system. To check this hypothesis is necessary to determine the turbulence in Hubble X and Hubble V.

Table 2. Fractal dimension of different fractal objects and HII regions.

Object	D_B	Reference	Object	D_B	Reference
Vela nebula	1.2 ± 0.05	[16]	Paley I	1.677181	[17]
Eta Carinae nebula	1.37 ± 0.02	[16]	Rosette Nebulae (close-up)	1.78261	[17]
Hubble V	1.3818	*	Eagle Nebulae (close-up)	1.669622	[17]
Hubble V	1.4081	*	IC434 (col I)	1.811718	[18]
Julia set	1.679594	[17]	Helix Nebula (NGC 7293)	1.841059	[19]
Horsehead nebula	1.6965725	[17]			

* This research

As stated previously, the velocity dispersion is one of the methods used to measure turbulence in HII regions. The natural width is a diagnosis method of a turbulence field of velocity.

In order for HII regions to have a supersonic turbulence movement is necessary the continuous input of energy to replace the dissipated energy due to the viscosity of the gas and shocks. The first source of energy is radiation from ionizing stars. Another source that can produce supersonic movement of gases en HIII regions are the stellar wins O and B stars inject about $10^{-(5-6)}$ *Mo* per year with a velocity that exceeds the 1000 km/s. Considering that HIII regions are ionized by a large number of high mass stars ($10^2 - 10^3$), their stellar winds affect the gas cinematic and the kinetic energy associated to the wins could produce a considerable turbulence. A third mechanism associated to this kind of turbulent movement is gravity.

Giant HIII region are ionized by large number of massive stars. It is possible estimate to the total mass of regions if we use typical mass function for the ionizing clusters is expected ant the neutral and ionized gas and ionized gas is included, allowing to estimate masses of HIII regions higher than 10^6 *Mo* [20,21].

Our results for Hubble X and Hubble V show that the ionized hydrogen is described by subsonic behavior, with velocities lower than sound (of the order of 10 km/s). The dispersion velocity values are in the limit of supersonic-subsonic. Both HII regions exhibit a turbulent behavior that is located at the transition subsonic-supersonic, which is a very peculiar case for typical giant HII regions, because of for these types of astronomical objects; the nebular gas presents dispersion velocities on the average of 25 km/s.

The results obtained through the computation of the fractal dimension of Hubble X and Hubble V are compatible with the turbulent behavior of the ionized gas of these HII regions. The principal result of this research is to focus on the meaning of the fractal dimension in these Giant HII regions as an element that allows characterizing their turbulent properties. This interpretation gives a new tool and methodology for the future study of turbulence in other Giant HII Regions.

References

[1] Aschwanden M J and Aschwanden P D 2008 *AJ* **674** 530, Aschwanden M J and Stern R A and Güdel M 2008 *AJ* **672** 659
[2] Greimel R, Brandt P N, Guenther N and Mattig W 1990 *Vistas in Astronomy* **3** 413
[3] Ching D, Taylor G J, Mouginis-Mark P and Bruno B C 1993 *Lunar Planet Sci* **XXIV** 283
[4] Thanki S, Rhee G and Lepp S 2009 *AJ* **138** 941
[5] Sanchez N and Alfaro E J 2008 *ApJS* **178** 1
[6] Sanchez N, Alfaro E J and Pérez E 2009 *RevMexAA Conf. Ser.* **35** 76
[7] Peitgen H O, Jurgens H,and Saupe D 2004 *Chaos and Fractals - New Frontiers of Science* (New York: Springer-Verlag)
[8] Theiler J 1990 *J. Opt. Soc. Am. A* **7**(6) 1055
[9] Hippelein H H 1986 *A&A* **160** 374
[10] Melnick J, Moles M, Terlevich R and Garcia-Pelayo J M 1987 *MNRAS* **226** 849
[11] Roy J R, Arsenault R and Joncas G 1986 *ApJ* **300** 624
[12] Peimbert A, Peimbert M and Ruiz M T 2005 *ApJ* **634** 1056
[13] Zhao Y X, Yi S, Tian L F, He L and Cheng Z Y 2008 *Sci China Ser G-Phys Mech Astron* **51** 8 1134
[14] Sreenivasan K R and Meneveau C 1986 *J Fluid Mech* **173** 357
[15] Sreenivasan K R 1991 *Ann Rev Fluid Mech* **23** 539
[16] Blancher S and Perdang J 1990 *Vistas in Astronomy* **33** 393
[17] Datta S 2001 Fractal structure of molecular clouds Preprint astro-ph/0105036
[18] Datta S 2003 *A&A* **401** 193
[19] Datta S 2003 *Planetary Nebulae: Their Evolution and Role in the Universe* vol 209 (France: Astronomical Society of the Pacific) p 523
[20] Terlevich R and Melnick J 1981 *MNRAS* **195** 839
[21] Kennicutt R C Jr 1984 *ApJ* **287** 116

Journal of Physics: Conference Series **582** (2015) 012050 doi:10.1088/1742-6596/582/1/012050

Thermoeconomical analysis of a non-endoreversible Novikov power plant model under different regimes of performance

J C Pacheco-Páez[1], F Angulo-Brown[1] and M A Barranco-Jiménez[2]
[1]Departamento de Física, Escuela Superior de Física y Matemáticas del IPN, Edif. 9, 2° Piso, UP Zacatenco CP 07738, D.F., México.
[2]Departamento de Formación Básica, Escuela Superior de Cómputo del IPN, Av. Juan de Dios Bátiz s/n. UP Zacatenco, CP 07738, D. F., México.

E-mail: jcpacheco@esfm.ipn.mx

Abstract. In this work, we study the thermoeconomics of a non-endoreversible simplified thermal power plant model, the so-called Novikov engine. Our study is made by means of the maximization of objective functions defined by the quotient of the characteristic functions (power output, efficient power and ecological function) and the total costs considered in the performance of the power plant. In our study three different costs are considered: a capital cost that is proportional to the investment and, therefore, to the size of the plant, a fuel cost that is proportional to the fuel consumption and a cost associated to maintenance of the power plant; that is, proportional to the power output of the plant. It is shown that under ecological conditions the plant dramatically reduces the amount of heat rejected to the environment, and a loss of profits is translated in an usage of fuels that dramatically reduces the heat rejected towards the environment in comparison to that obtained by means of maximum power regime.

1. Introduction

In 1995 [1], Alexis De Vos introduced a thermoeconomical analysis of an irreversible heat engine model, the so-called Novikov plant model (see figure 1) [2], in terms of the maximization of an objective function defined by the ratio of the power output and the running costs of the plant exploitation. De Vos [1] assumed that the running costs of the plant consist of two parts: a capital cost that is proportional to the investment and, therefore, to the size of the plant and a fuel cost that is proportional to the fuel consumption. De Vos [1] showed how the optimal efficiency (in the economical sense) smoothly increases from the maximum-power point (Curzon-Ahlborn efficiency [3]) corresponding to energy sources where the investment is the preponderant cost up to the Carnot value (Carnot efficiency), that is, for energy sources where the fuel is the predominant cost when the heat fluxes in the Novikov model are given by a linear Newtonian heat transfer law, that is, $\eta_{MP} < \eta_{opt} < \eta_C$, where the subscripts *MP*, *opt* and C mean maximum power, optimal efficiency and Carnot respectively. De Vos [1] found for the Novikov's model that η_{opt} is given by:

$$\eta_{opt}(\tau, f) = 1 - \frac{f}{2}\tau - \frac{\sqrt{4(1-f)\tau + f^2\tau^2}}{2}, \tag{1}$$

where f is the fractional fuel cost (see table 1), which is defined as the ratio of the cost of the fuel consumption and the running costs of the power plant; $\tau = T_L/T_H$ with T_H and T_L the temperatures of the hot and cold thermal reservoirs respectively (see figure 1). Later, Barranco-Jiménez and Angulo-Brown [4, 5] also studied the Novikov engine model following the thermoeconomical approach used by De Vos, but by means of the so-called ecological optimization criterion. The ecological optimization criterion consist in maximizing the well known ecological function [6, 7] defined by $E = W - \in T_L\sigma$, where W is the plant's power output, $\in$ is a parameter depending of the heat transfer law [7] and σ is the total entropy production of the endoreversible power plant model. Barranco-Jiménez and Angulo-Brown [4, 5] also obtained a thermoeconomical efficiency η_{opt}^{E} between η_{MP} and η_C that drastically reduces the entropy production of the engine. Besides the optimal efficiency satisfy the following inequality $\eta_{MP} < \eta_{opt} < \eta_{opt}^{E} < \eta_C$.

Table 1. Relative fuel cost (in percent) for various energy sources

Fuel	$f(\%)$ [1]
Renewable	0
Uranium	25
Coal	35
Gas	50

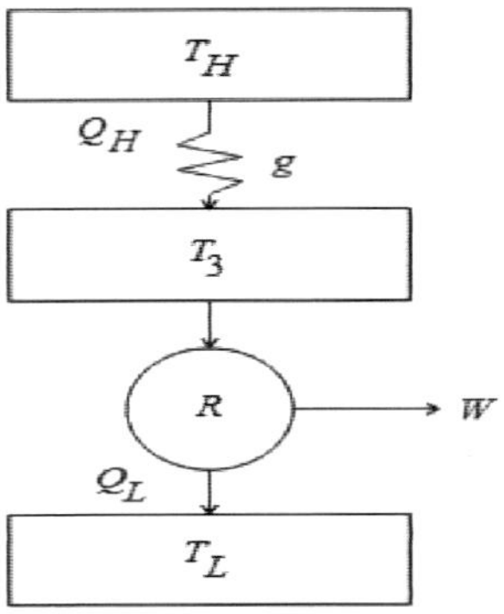

Figure 1. Novikov´s model for a thermal power plant.

In this work, we extend the thermoeconomical analysis following the De Vos's approach but maximizing three benefit functions defined by the ratio of the characteristic functions (Power Output [1], Efficient Power [8, 9] and Ecological Function [6]) and the total costs involved in the performance of the plants, that is: $F_{MP} = \frac{W}{C_t}$, $F_{EP} = \frac{\eta W}{C_t}$ and $F_{ME} = \frac{W - \varepsilon T_L\sigma}{C_t}$. Besides in the maximization of the benefit functions, the total costs C_t are taken as $C_t = aQ_{max} + bQ_H + cW$, where the proportionality constants a, b and c have units of *$/Joule ($:currency)*, the last term, is a cost associated to maintenance of the power plant that is proportional to the power output of the plant [10], and $Q_{max} = g(T_1 - T_2)$ is the maximum heat that can be extracted from the heat reservoir without supplying work [1].

2. Thermoeconomical optimization under different regimes of performance for a non-endoreversible Novikov plant model

Applying the first law of thermodynamics to Fig. 1, the power output is given by,

$$W = Q_H - Q_L, \tag{2}$$

where Q_H and Q_L are the heat transfer supplied by the hot source to working fluid and the heat transfer from the working fluid to cold source respectively. On the other hand, the internal efficiency of the engine is given by [5],

$$\eta = \frac{W}{Q_H} = 1 - \frac{\tau}{R}\frac{1}{\theta}, \tag{3}$$

where $\theta = T_3/T_1$ and the parameter $R = \Delta S_1/|\Delta S_2|$ is the non-endoreversibility parameter [11-13] (which characterizes the degree of internal irreversibility that comes from the Clausius inequality) ΔS_1 being the change of the internal entropy along the hot isothermal branch and ΔS_2 the entropy change corresponding to the cold isothermal compression. This parameter in principle is within the interval $0 < R \leq 1$ ($R = 1$ for the endoreversible limit, see [11]). If we consider that the heat transfer between the hot source and the working fluid obey a Newton heat transfer law, and applying the first and second laws of thermodynamics we can write the benefit functions defined in the previous section as [10],

$$aF_{MP} = \frac{\left(1 - \frac{\tau}{R}\frac{1}{\theta}\right)(1-\theta)}{(1-\tau) + \beta(1-\theta) + \gamma\left(1 - \frac{\tau}{R}\frac{1}{\theta}\right)(1-\theta)}, \tag{4}$$

$$aF_{EP} = \frac{\left(1 - \frac{\tau}{R}\frac{1}{\theta}\right)^2(1-\theta)}{(1-\tau) + \beta(1-\theta) + \gamma\left(1 - \frac{\tau}{R}\frac{1}{\theta}\right)(1-\theta)}, \tag{5}$$

$$aF_{ME} = \frac{\left[(1+\varepsilon)\left(1 - \frac{\tau}{R}\frac{1}{\theta}\right) - \varepsilon(1-\tau)\right](1-\theta)}{(1-\tau) + \beta(1-\theta) + \gamma\left(1 - \frac{\tau}{R}\frac{1}{\theta}\right)(1-\theta)}, \tag{6}$$

where $\beta = b/a$ *and* $\gamma = c/a$. Fig. 2 shows that exists an optimal efficiency value which depends on the parameter R and the optimum value of θ. We can also observe that the benefit diminishes as the internal irreversibilities (parameter R) increase. Besides, for the case of maximum power conditions, the optimum value of θ does not change the maximum of benefits when the maintenance costs (parameter γ, dashed lines) are present in the optimization. Therefore, taking the derivatives of aF_{MP}, aF_{PE} and aF_{ME} with respect to θ and setting them equal to zero we obtain the optimum working fluid temperature $(\theta^* = T_3^*/T_1)$ for each value that maximize equations (4), (5) and (6) respectively, and by using equation (3), the optimal efficiencies are obtained at maximum power output, efficient power and ecological function conditions respectively:

$$\eta_{MP}(\tau, f, R) = 1 - \frac{f}{2R}\tau - \frac{\sqrt{4(1-f)R\tau + f^2\tau^2}}{2R}, \tag{7}$$

$$\eta_{EP}(\tau, f, \gamma = 0, R) = 1 - \frac{(1+f)}{4R}\tau - \frac{\sqrt{8(1-f)R\tau + (1+f)^2\tau^2}}{4R}, \tag{8}$$

$$\eta_{ME}(\tau,f,\gamma,R)=1-\frac{2\tau(f-1)\left[R+(R-\gamma)\sqrt{\tau}\right]}{R\left[f\tau+2\gamma(1-f)\sqrt{\tau}+f\tau-\sqrt{\left(\tau+\sqrt{\tau}\right)\left(4(f-1)((f-1)\gamma-1)R\right)}+\Lambda(\tau,f,\gamma,R)\right]} \tag{9}$$

where $\Lambda(\tau,f,\gamma,R)=\left[4(R-\gamma)+f\left(f-4R+4\gamma(2-f)\right)\right]\sqrt{\tau}+f^2\tau$. In figure 3 we show the behavior of the optimal efficiencies (equations (7), (8) and (9)). Figure 3 also shows how the optimal efficiencies smoothly vary from the value $f=0$, corresponding to energy sources where the investment is the preponderant cost up to the Carnot value for $f=1$, that is, for energy sources where the fuel is the predominant cost, in an analogous way to De Vos-efficiency [1].

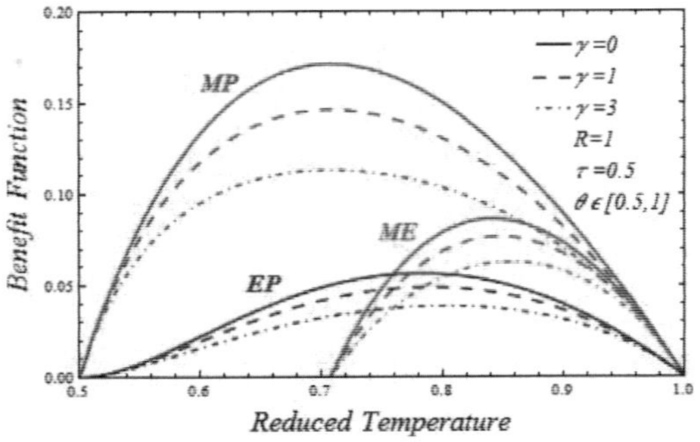

Figure 2. Comparison between the three thermoeconomic objective functions with respect to the reduced temperature θ.

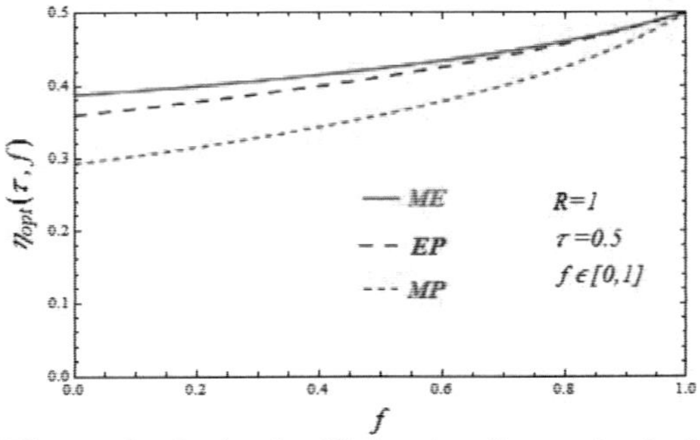

Figure 3. Optimal efficiencies for ecological regime, efficient power and maximum power output conditions.

3. Environmental impact

Applying the first and second laws of thermodynamics to the engine model shown in figure1, we obtain the expression for the heat rejected (Q_L) to the environment by the power plant given by [10],

$$Q_L\left(R,\tau,\theta^*\right)=gT_H\left[\frac{\tau}{R}\frac{1}{\theta^*}\right]\left(1-\theta^*\right). \tag{10}$$

From equation (10) we can calculate de heat rejected to the environment for each value of f and under different ways of operation of the power plant. In Fig. 4, it can be seen how the heat rejected under ecological function conditions is lower that the heat rejected under both maximum power output and maximum efficient power conditions. In Fig. 5 we show the ratio between the amounts of rejected heat considering the case when $\gamma\neq 0$ (red dashed line).

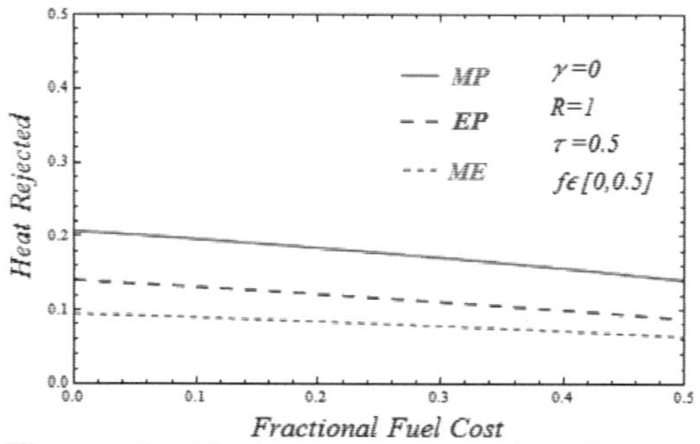

Figure 4. Maximum power rejected heat, efficient power rejected heat and ecological rejected heat for a linear heat transfer law.

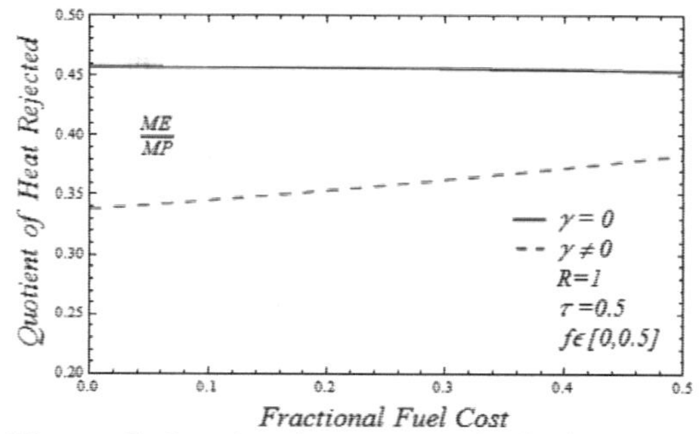

Figure 5. Quotient between ecological rejected heat and maximum power rejected heat in terms of fractional cost.

We also observe in figure 5 that for the case of $\gamma = 0$ (blue line), we obtain the result recently reported in [5] and previously reported by Velasco et al [14]. In addition, we can calculate the total entropy production and we also analyze the effect on the reduction of power output in terms of an internal irreversibility for the Novikov model for both ecological function and maximum power conditions [10].

3. Concluding Remarks

In this work, we have made a thermoeconomic study of a non-endoreversible simplified thermal power plant model (the so-called Novikov engine). This non-endoreversible case improves the results obtained by means of an endorreversible model due to the inclusion of the engine's internal dissipation through the lumped parameter *R*. In our study, we also take into account a cost associated to maintenance of the power plant; that is, proportional to the power output of the plant. In our study we consider different regimes of performance: Maximum Power Output, Maximum Efficient Power and Maximum Ecological Regime. We found that when the Novikov model maximizes the ecological function, it reduces the rejected heat to the environment down to about 55% (see figure 5) of the rejected heat in the case of a power plant model working under maximum power conditions. Besides, we analyze the effect on the reduction of power output in terms of an internal irreversibility.

Acknowledgments
This work was supported in part by CONACYT, COFAA and EDI-IPN-México.

[1] De Vos A 1995 Endoreversible thermoeconomics. *Energy. Convers. Magnagement.* **36** 1 1-5.
[2) Novikov I I 1957 The efficiency of atomic power stations (a review) *Atomimaya Energiya* **3** 409 (English translation 1958. *Nuclear Energy* **7** 125)
[3] Curzon F. and Ahlborn B 1975 Efficiency of a Carnot engine at maximum power output *Am. J. Phys.* **43** 22-24.
[4] Barranco-Jiménez M A and Angulo-Brown F 2007 Thermoeconomic optimization of a Novikov power plant model under maximum ecological conditions *J. of the Energy Inst.* 80:2(1-4):96-104, 2007.
[5] M. A. Barranco-Jiménez "Finite-time thermoeconomic optimization of a non endoreversible heat engine", Rev. Mex. Fis. 55 (3) 211-220 (2009).
[6] Angulo-Brown F 1991 An ecological optimization criterion for finite-time heat engines *Appl. Phys.* **69** 7465-7469

[7] Angulo-Brown F and Arias-Hernández L A 2001 *J. of Appl. Phys.* **89** 1520
[8] Yilmaz T 2006 A new performance criterion for heat engine: efficient power. *J. Energy Inst.* **79** 1 38-41.
[9] Arias-Hernández L A, Barranco-Jiménez M A and Angulo-Brown F 2009 Comparison between two ecological-type modes of performance for a simple energy converter", *J. of the Energy Institute* **82** 3 193-197
[10] Proceedin gs of 27th International conference, ECOS2014 2014 (Turku Finland) 15-19
[11] Chen J 1994 The maximum power output and maximum efficiency of an irreversible Carnot heat engine. *J. Phys. D: Appl. Phys* **27** 1144–1149.
[12] Öskaynak, S, Göktun, S, Yavuz, H 1994 *J. Phys. D: Appl. Phys.* **27** 1139–1143.
[13] Wu, C., Chen, L., Chen, J., eds., 1999 Recent Advances in Finite-Time Thermodynamics, Nova Science (New York)
[14] Velasco S, Roco J. M, Medina A, White J A and Calvo-Hernández A. 2000 *J. Phys. D.* **33** 355-359

Journal of Physics: Conference Series **582** (2015) 012051 doi:10.1088/1742-6596/582/1/012051

Unexplored Indoors method for pyranometers calibration traceable to SI

H A Castillo-Matadamas[1], J C Molina-Vazquez[1], R Quintero-Torres[2].

[1] Centro Nacional de Metrología, Dirección de óptica y radiometría, El Marques Querétaro, C.P. 76246.
[2] Centro de Física Aplicada y Tecnología Avanzada, UNAM campus Juriquilla, Boulevard Juriquilla 3001, Juriquilla Querétaro 76230, México.

E-mail: hcastill@cenam.mx

Abstract. A method to calibrate pyranometers with direct traceability to the International System of Units (SI) is presented, the method use an electrically calibrated pyroelectric detector (ECPR) as standard and offers numerous advantages over outdoors conventional calibration methods, such as reducing the uncertainty from the reference standard and the final uncertainty of the sensitivity coefficient of the calibrated pyranometer; the measurement uncertainty achieved with this method at normal irradiance is 2.1 % for a coverage factor k = 2 and could be reduce if one reduces the uncertainty level of the reference standard.

1. Introduction

According to various calibration standards reviewed in [1], traditionally the calibration of pyranometers with pyrheliometers as reference [2] and pyranometers with pyranometers as reference [3] have been carried out with instruments traceable to the radiometric scale of the world radiometric reference (WRR) [4], establishing the equivalence of the WRR scale and the SI radiometric scale by comparisons at the primary level [5] through cryogenic radiometer measurements as part of the SI scale and absolute cavity radiometers which are the reference of the WRR. However, direct measurement of pyranometers with conventional radiometric instruments of radiometric laboratory is possible with various inherent benefits from the indoors calibration and decreasing traceability chain to a primary standard and the improvement in the uncertainty of the experiment.

The actual indoor calibration systems of pyranometers, essentially make use of international standards recommendations [3] and the calibration is performed using pyranometers as reference standard and translation systems to ensure same source strength in the measurand and the reference. In these cases it is required a calibrated pyranometer which is referred to a pyrheliometer by outdoors calibration.

The main benefit of the indoors calibration of this kind of instruments is the reduction in time due to constant availability of the light source instead of the natural source and the control of all the related parameters involve in the measurement.

Journal of Physics: Conference Series **582** (2015) 012051 doi:10.1088/1742-6596/582/1/012051

2. The Pyranometers calibration method

In the definition of the measurand for the calibration of a pyranometer it is relevant to establish that the sensor of the instrument is a set of thermal detectors (thermopiles) arranged in series, the sensitivity to the intensity of the incident light is kept low and is not possible to measure with a laser beam smaller than the size of the area of the sensor due to high spatial dependence of the sensitivity or the uniformity in the response of the detector, therefore the average irradiance on the entire area of the detector is the magnitude to be compare to the standard and is expressed in average irradiance units during a calibration exercise.

The ECPR are radiometric reference standards, like pyranometers are based on thermal detector but pyroelectric principle instead of thermoelectric phenomena of a thermopile, the pyroelectric detector with proper conditioning is able to achieve high sensitivity through reduction systems noise of lock-in filters, and the is electrically calibrated for power measurements using coated sensor with high absorption efficiency of light [6]. The exposed area of an ECPR is defined by a metal mask that can be measured with low uncertainty for example with a non-contact dimensional measurement.

The measurement process using ECPR requires the use of a modulator of the input light to the sensor and the experimental setup use a chopper with 50 % average light exposure. The measurement scheme used is shown in Figure 1.

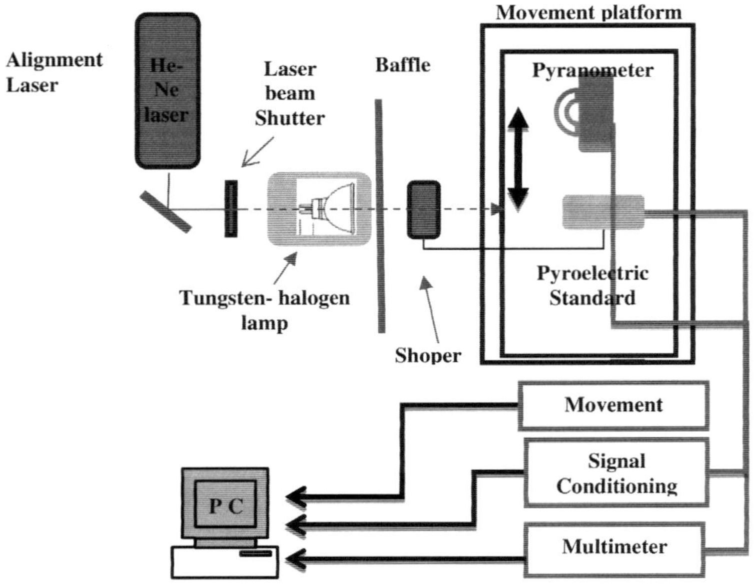

Figure 1. Experimental setup for a pyranometer calibration

The proposed method is based on the comparison of measurements of the same irradiance source on both sensors. If a source with flat front of intensity distribution without divergence is consider, one would expect than the irradiance of the source is constant in both sensors regardless of the size of the illuminated sensor area covered by the source. However using a conventional filament source as tungsten-halogen type and MR16 or MR11 diffuser, a radial symmetric intensity distribution is expected at certain distance with a cover size area defined by the normal distance of the source to sensor plane.

Due to the radial intensity distribution of real filament source, it is necessary to correct for the difference in size between the sensitive areas of the pyranometer and ECPR used as a standard and the correction will have no effect when both sensors have the same area or the beam divergence light source were negligible.

Journal of Physics: Conference Series **582** (2015) 012051 doi:10.1088/1742-6596/582/1/012051

For the propose measurement, the plane sensor of the pyranometer sensor must be aligned normal to the source and ideally coplanar to the plane of the reference standard sensor. In practice the alignment is achieved by aligning the reference base of the pyranometer and the plane of the ECPR with a flat mirror simulating sun direct exposure similar to outdoors solar tracker use. In this manner the calibration is carried out considering only direct radiation from the source and the same irradiance magnitude if both sensors have the same distance to the source.

2.1 Definition of the measurand

The pyranometer exposed to outdoor operation will have a response voltage V usually expressed as follows:

$$V = R(\theta)\left(I_{direct}\cos(\theta) + I_{diffuse}\right) \tag{1}$$

With θ as the solar zenith angle, I_{direct} the direct irradiance, $I_{diffuse}$ the diffuse irradiance collected by the dome of the pyranometer in the measured orientation and $R(\theta)$ the sensitivity coefficient as a function of the orientation angle of the tool relative to the sun with units of [V m2 / W]. However, in laboratory conditions, diffuse irradiance source is avoided and the plane of the pyranometer and the pattern should be normal to the maximum intensity of the source. So that the voltage response of the pyranometer can be defined according to the following expression:

$$V = \int_0^b R(\theta_0) I_{direct} F(r)dr = I_{direct} R(\theta_0)\left[\int_0^a F(r)dr + \int_a^b F(r)dr\right] \tag{2}$$

With I_{direct}, the maximum irradiance at the center point of the distribution. $F(r)$ is the normalized spatial distribution function of the intensity.

In this case the Pyroelectric detector area is circular with radius b larger than the ECPR area with detector radius a.

The response of the standard is similar and according to the following expression:

$$I_{standard} = \int_0^a I_{direct} F(r)dr \tag{3}$$

The expressions (2) and (3) do not include any correction for spectral response because both sensors have a flat response in the visible and near infrared.

From (2) and (3) the response of the pyranometer irradiance at normal incidence is:

$$R(\theta_0) = \frac{V}{I_{direct}\left[\int_0^a F(r)dr - \int_a^b F(r_a - r_b)dr\right]} = \frac{V}{I_{standard} - I_{direct}\int_a^b F(r_a - r_b)dr} \tag{4}$$

The spatial distribution function should be measured during the calibration process and its normalized value has slight variations with the change of the normal distance from the source to the plane of the detectors, such variations are due only to the uncertainty of the alignment process and the weak radial asymmetry of the source.

The determination of the irradiance on the plane of the detectors is based on the premise of having the same normal distance from the source to each of the planes of the detectors, which can be secured by the mechanical alignment of the edges of dome external reference of the pyranometer and known reference pyroelectric plane. The distance from source to detector plane is arbitrary and only limits the maximum possible intensity, which in the case of type 35W MR11 lamps at the evaluated distances reaches a maximum of 800 W/m2.

3. Results

With the method above described, the calibration for different distances from the source to the plane of the sensor of a pyranometer K & Z model CMP6 against ECPR pattern was performed. Both instruments were mounted and aligned in accordance with Figure 1.

The ECPR detector was centered with a laser of 3 mm in diameter and the laser was used to align the plane normal to the source sensor. Additionally the maximum intensity was located and verified that the irradiance distribution with reference to that point was symmetrical. The pyranometer was aligned in a plane normal to the source-detector line, using the reference base and the laser was focused into the same position, displacing the ECPR and centering the laser in the pyranometer using the mobile platform.

The alignment of the planes of the detectors was performed by reference to the outer edge of the pyranometer dome. The normal distance of the edge to the source was adjusted in the ECPR to ensure the same value of the source irradiance.

The measurement results for the different distances from the source are shown in Figure 3.

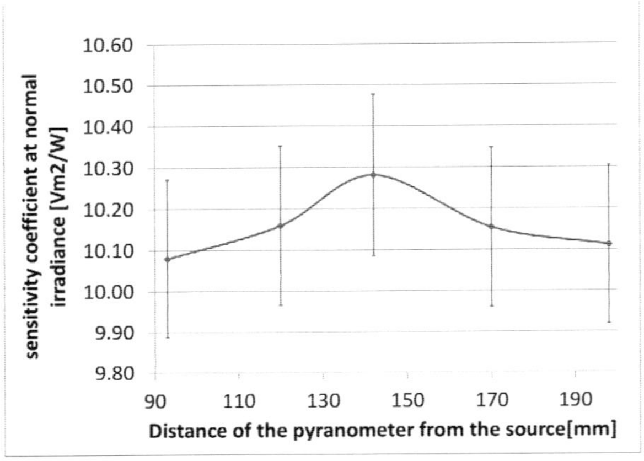

Figure 3. Calibrated values as a function of distance from the source. The measurement uncertainty is also shown.

The reproducibility achieved for a fixed length using the proposed method is better than 0.4 % and is the second contribution to the total uncertainty; the first is the calibration of the reference ECPR with an uncertainty less than 0.5 % both at a cover factor of k=1, the uncertainty of the standard includes the no linearity of the spectral response for the spectral emission of the lamp.

The combined uncertainty for the calibration values are shown in Table 1, calculated for a nominal irradiance of 500 W/m2 at the average distances of Figure 3, at lab temperature of 21 C.

The model used for the estimation of uncertainty is defined according to the expression 5 and is employed with the GUM methodology [7] to obtain the final estimate of the uncertainty indicated in expressions 6 and 7.

$$R_0 = \frac{V \cdot K_S}{I_{standard} - I_{direct} F_A} \tag{5}$$

$$\mu_{R_0}^2 = \sum \left[\frac{\partial R_0}{\partial x_i} \right]^2 \mu_{x_i}^2 + \mu^2{}_{rep} + \mu^2{}_{alin} \tag{6}$$

$$\mu_{R_0}^2 = \left[\frac{\mu_V K_s}{I_{standard} - I_{direct} F_A} \right]^2 + \left[\frac{\mu_K V}{I_{standard} - I_{direct} F_A} \right]^2 + \left[\frac{\mu_{I_{standard}} V K_s}{(I_{standard} - I_{direct} F_A)^2} \right]^2 + \left[\frac{\mu_{I_{direct}} V K_s F_A}{(I_{standard} - I_{direct} F_A)^2} \right]^2 + \left[\frac{\mu_{F_A} V K_s I_{direct}}{(I_{standard} - I_{direct} F_A)^2} \right]^2 + \mu_{rep}^2 + \mu^2{}_{alin} \tag{7}$$

Where $\mu_{R_0}^2$ is the value of the uncertainty of the different parameters in the expression 5, F_A is the exposure factor of the optical shopper during calibration, obtained by measuring the ratio of signal to shopper off operation and total openness, I_{direct} is the integral value of the radius b of the intensity distribution function of the source also measured during calibration.

The expression 7 is evaluated in Table 1, in it has been considered uncertainty and reproducibility of the method for further alignment sensor uncertainty, which was determined considering a maximum error of 1 mm in the determination of the distance from the source to the plane of the sensors.

The spectral distribution of the tungsten halogen source does not affect the measuring method and can be replaced by any other lamp that provides an intensity distribution with radial symmetry and the low possible divergence. A Si CCD spectrometer was used to measure the smooth curve of spectral distribution of several tungsten-halogen sources, the source used correspond to the more efficient commercial available source in the visible without dichroic filter.

Table 1. Budget of the uncertainties for the measurement results.

Uncertainty source	Standard uncertainty	Unit	Sens. Coef.	Coef. unit	Contribution	Degrees of fredom	Weight
Measurement of the Pyranometer voltage, V	7.00E-09	V	4.02E-03	m2/W	2.81E-11	200	0%
Exposition coeficient, Ks	1.00E-03		1.00E-05	Vm2/W	1.00E-08	100	1%
Irradiance calibration of the standard, Istandard	2.500	W/m2	4.03E-08	V(m2/W)^2	1.01E-07	50	89%
Nominal irradiance measure, (Idirecta)	1.000	W/m2	3.35E-10	V(m2/W)^2	3.35E-10	10	0%
Areas correction factor, (FA)	1.35E-08		1.01E-05	(Vm2/W)	1.37E-13	200	0%
Reproducibility of the measurements	3.42E-08	Vm2/W	1.00E+00		3.42E-08	200	10%
				u_c	0.107 μVm2/W		
				k	2.02		
				U	0.22 μVm2/W		
				U	2.1%		

3.1 Correction for difference in sensor areas

The area of the pyranometer sensor is determined by a laser beam of 3 mm in diameter aligned at normal incidence with respect to the base of the instrument using a flat mirror for such alignment; and by means of a sweep across the diameter of the detector, and the intensity distribution chart with respect to the displacement of the sensor.

According to the optical geometrical analysis from Figure 4, the calculation of the maximum beam refraction caused by the passage through both domes of the pyranometer will result a y_i variation in height of less than 0.03 % (0.06 % taking into account both ends) , this contribution to the uncertainty is negligible according to the values shown in the table 1 where the calculation of the uncertainty area correction factor, have included the uncertainty in estimating the diameter.

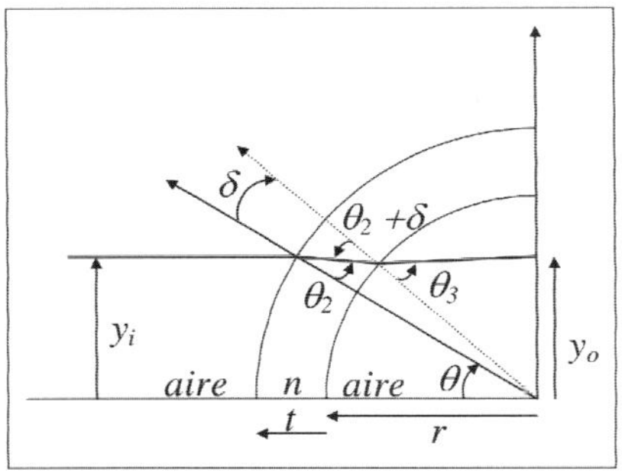

Figure 4. Diagram for the refraction analysis of laser beam through CMP6 pyranometer domes, for the analysis the beam enters horizontally (normal to the detector) with height y_i, considering a refractive index of glass n = 1.5. The beam leaves to the sensor at height y_0 which is only function of the income height.

The average coefficient of sensitivity obtained for different distances shown in Figure 3, is of 10.24 uVm2 / W with expanded uncertainty of ± 2.1 % (k = 2). A systematic error was identified related

Journal of Physics: Conference Series **582** (2015) 012051 doi:10.1088/1742-6596/582/1/012051

with the divergence of the illumination in the cases where the divergence of the source is greater than natural source, some test with the corrected divergence using lens, show than the error could be negligible but could be corrected due to the great reproducibility of the method.

3. Conclusions

The presented method for calibrating pyranometers using a pyroelectric radiometer traceable to cryogenic radiometer is novel and it could be contained in the recommendations of international standards, the results for the reproducibility of the method and traceability to the international system of units ensuring the reliability for the indoor calibration of pyranometers. Additionally, there are several advantages; the calibration do not required spectral correction or an ad hoc spectral distribution of the source, the method is significantly fast and simple compared to outdoors methods and provides the ability to obtain the response factor at normal irradiance for any instrument based on thermopile. The angular response of the instrument should be studied further with the appropriate rotational mount.

Systematic errors such as the position of the detectors plane respect to the plane of the reference ECPR occurs in all methods of laboratory, however its influence is negligible according to the estimated uncertainty if is taken the necessary attention during the measurement and the divergence of the sources is low. The error by the intensity distribution and the difference in area of the detector is appropriately adjusted when the diameter of the detector is known or measured using a laser beam. The uncertainty value obtained by the present method is lower than that obtained for some instruments at factory and is mainly limited by the calibration uncertainty of the standard used.

References

[1] Ossenbrink H. 2003 Calibration procedures—state of the art. *Proceedings of the 3rd WCPE (Osaka, Japan)* 2177–2181.
[2] Calibration of a pyranometer using a pyrheliometer 1993 *ISO 9846 Solar Energy*
[3] Calibration of field pyranometers by comparison to a reference pyranometer.1992 *ISO 9847 Solar Energy*
[4] IEC 60904-4 Reference solar devices 2009 *Procedures for establishing calibration the traceability*
[5] A Fehlmann et al. 2012 Fourth World Radiometric Reference to SI radiometric scale comparison and implications for on-orbit measurements of the total solar irradiance, *Metrologia* **49**
[6] Rs-5900 Electrically Calibrated Pyroelectric Radiometer Manual, Laser Probe inc.
[7] JCGM 100:2008 (GUM 1995 with minor corrections) 2008 *Evaluation of measurement data Guide to the expression of uncertainty in measurement.* BIPM. First Edit.

Journal of Physics: Conference Series **582** (2015) 012052 doi:10.1088/1742-6596/582/1/012052

Emulating a robotic manipulator arm with an hybrid motion-control system

G Aragón-González, A León-Galicia, M Noriega-Hernández, A Salazar-Hueta
Universidad Autónoma Metropolitana – Unidad Azcapotzalco. Programa de Desarrollo Profesional en Automatización. Av. San Pablo # 180. La Reynosa Tamaulipas. Azcapotzalco, 02200, D.F.

E-mail: alg@azc.uam.mx

Abstract. A motion control system with four and ½ degrees of freedom, designed to move small objects within a 0.25 m3 space, parallel to a horizontal table, with high speed and performance similar to a robotic manipulator arm was built. The machine employs several actuators and control devices. Its main characteristic is to incorporate a servomotor, steeper motors, electromechanical and fluid power actuators and diverse control resources. A group of actuators arranged on a spherical coordinates system is attached to the servomotor platform. A linear pneumatic actuator with an angular grip provides the radial extension and load clamping capacity. Seven inductive proximity sensors and one encoder provide feedback, for operating the actuators under closed loop conditions. Communication between the sensors and control devices is organized by a PLC. A touch screen allows governing the system remotely, easily and interactively, without knowing the specific programming language of each control component. The graphic environment on the touch screen guides the user to design and store control programs, establishing coordinated automatic routines for moving objects in space, simulation and implementation of industrial positioning or machining processes.

1. Introduction

Many engineering applications need to control the position, speed and acceleration of a tool or an object involved in some manufacturing process. It is frequent to choose a motion control solution based on a robotic manipulator arm, even when it is possible to use other resources to complete the task. But the motion control technologies evolve so quickly that many devices are rejected, under the false assumption that they are obsolete or incompatible [1].

The machine was conceived as training equipment, for engineering students learning how to assemble, communicate and program motion control systems. Even though its hybrid composition, the system allows operating five actuators under closed loop conditions and programing small objects positioning.

2. Power components

A group of actuators arranged on a spherical coordinates system is coupled to the platform of a linear servomotor. The conjugated motion of the actuators can translate a 40 N load, raising it over the table surface and moving it with high precision within a 0.25 m^3 space; figure 1.

Figure 1. Actuators arranged on a spherical coordinates system, over the servomotor platform.

The servomotor platform has 1 m horizontal trajectory, with 1 μm resolution and capacity to accelerate up to 5g. The φ and ψ angular displacements come from two hybrid biphasic unipolar stepper motors, attached to crown-worm gear 55:1 reductions. Both axes have 0.0327°/motor step resolution; figure 2. The stepper motors are SKC, NEMA 23 size, with 1.8° step angle and 55.6 Nm holding torque.

The φ vertical axe protrudes for an enclosure attached to the linear servomotor platform. The second stepper motor is linked to this axe and its worm gear reduction has the ψ axe oriented horizontally. A linear pneumatic actuator is coupled to this second axe and provides 0.1 m radial extension. The linear actuator rod has a pneumatic angular grip, to pick up and hold the load during translation; figure 3 [4].

3. Control components

The linear servomotor is controlled with a Parker Gemini GV6K digital servo drive. The Gemini GV6K offers connectivity to any ASCII-based serial device, including HMI panels and PLC ASCII modules. It has 8 digital inputs and 6 digital outputs onboard I/O capability, expandable up to 256 additional discrete digital I/O points.

Eight digital outputs are needed to feed the steeper motors coils through specific drives for each motor. Four more digital outputs are required for the solenoids, to actuate the two pneumatic directional control valves. To avoid the Gemini I/O expansion unit it was used an Arduino Uno microcontroller board, in order to increase and organize the output signal capability. The Arduino Uno is based on the ATmega328 microcontroller and has 14 digital input/output pins and 6 analog inputs.

The output signals from the Gemini GV6K are + 24 V, while the Arduino Uno receives + 5 V input signals. To solve this incompatibility a converter was designed, based on six optocouplers 4N35 (light emitting diodes optically coupled to photo-transistors). The + 5 V output signals to actuate the direc-tional control valves come out from the Arduino Uno; they energize the coils from

Figure 2. The crowns angular advance is 6.55° for each revolution in the stepper motor.

Figure 3. Actuators assembly attached to the servomotor platform.

four Sun Hold, model RAS-0510, SPDT compact relays. As each relay closes one solenoid receive 127 VCA to commute a valve; figure 4.

The servomotor has a linear encoder and one *home* and two *limit* sensors, communicated with the Gemini GV6K through the encoder/hall cable and the limit/home cable. Another four inductive proximity sensors and one encoder provide feedback for operating all the actuators under closed loop conditions; figure 5. Communication between the sensors and control devices is organized with a Mitsubishi PLC, model FX2n-64MR-ES/UL [5].

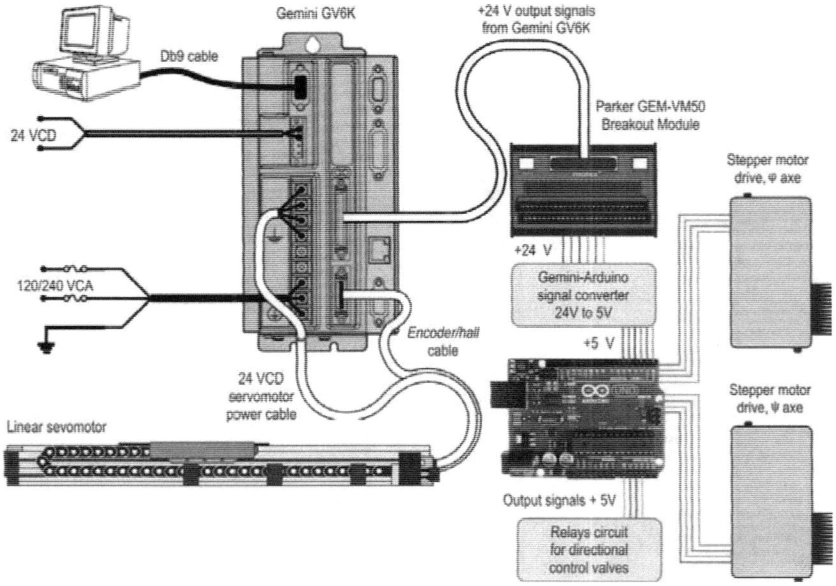

Figure 4. Control signals communication between three drives, the Arduino Uno and relays circuit.

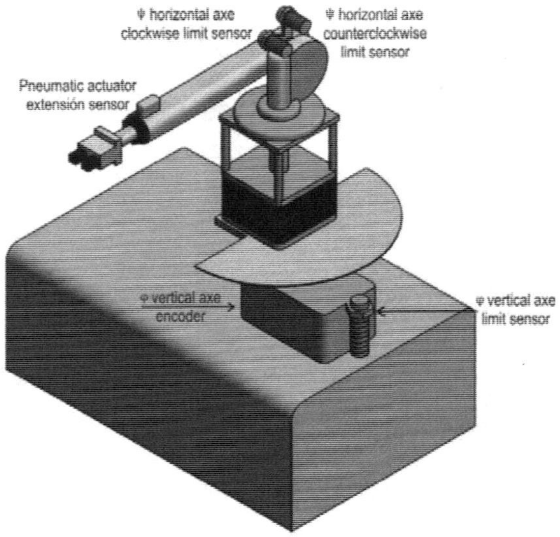

Figure 5. Limit sensors and encoder in the φ, ψ and r axes.

The two limit sensors for the ψ horizontal axe allow a maximum 300° angular displacement. Only one sensor is installed for the φ vertical axe, but it works with a half circle metallic surface and allows only 180° angular displacement (enough to move the load from the front to the rear of the servomotor and vice versa). The optical encoder in the φ axe is an Avago Technologies model HEDS-9700, with a 500 CPR code wheel.

The PLC receives the sensors signals and sends them off to the Gemini GV6K drive, through the Parker GEM-VM50 I/O breakout module; figure 6. The PLC also provides the 24 VCD sensors feeding. Two inductive sensors are PNP Lanbao model LR08BF02DPOY; another one is NPN Lanbao model LR08BF02DNOY. One 4N35 optocoupler was needed to conciliate the 5 V encoder output signal with the necessary 24 V PLC input signal [6].

The master control governing the drives and the PLC is a CTC-Parker power station, model PA10T-133. This touch screen power station is linked with the Gemini GV6K drive through a RS232 port; it allows real time tracing for the whole system and establish graphic communication with the users. Programs with motion control routines are stored in an external 32 Mb compact flash memory.

4. Conclusions

The motion control system performs very similar to a robotic manipulator arm, even though it is constructed with highly dissimilar actuators and diverse control resources, some of them far from being the most modern components. Its design intentionally shows how to overcome apparent incompatibility problems and take advantage of simple but efficient power and control solutions. Several routines have been designed to prove the flexibility and reliability of the system.

The touch screen is a key control component, because it communicates flawlessly the subordinated control drives. It also allows governing the system remotely, easily and interactively, without knowing the specific programming language of each control component.

The capacities of the system can be improved with the addition of free software control and programming platforms, like the Arduino or Raspberry microcontroller boards.

Journal of Physics: Conference Series **582** (2015) 012052 doi:10.1088/1742-6596/582/1/012052

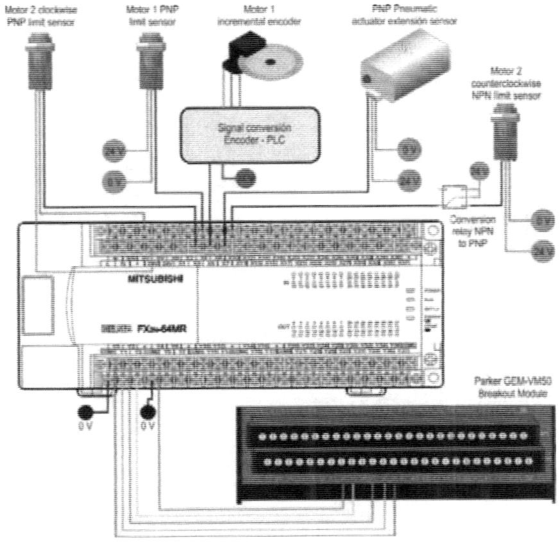

Figure 6. Sensors and encoder signals managed with the PLC.

References

[1] Moore P and Sheng Pu J 1996 Pneumatic Servo Actuator Technology *IEE Colloquium: Actuator Technology: Current practice and new developments.* (London UK)

[2] Manrique-Garay J et al 2012 Control de movimiento con un sistema de tres y medio grados de libertad. *XIII Con. Int. de Ing. Mec y Mecatrónica* (Hermosillo. México)

[3] Noriega-Hernández M and Salazar-Huerta A 2014 Integración de un banco de ensayos para control de movimiento. *Bachelor Thesis. UAM Azcapotzalco México*

[4] Aragón G, Canales P and León A 2014 *Introducción a la potencia fluida* Barcelona: Reverte

[5] Yoon Sang Kim and Hak-Man Kim 2013 Design of a New Virtual Interaction Based PLC Training Using Virtual Sensors and Actuators: System and Its Application. *International Journal of Distributed Sensor Networks* Volume 2013, Article ID 505920, 8 pages.

[6] Bishop R 2008 *Mechatronic Systems, Sensors, and Actuators: Fundamentals and Modeling* Boca Raton, Fl: CRC Press

Journal of Physics: Conference Series **582** (2015) 012053 doi:10.1088/1742-6596/582/1/012053

Effect of patterned coupled optical micro-cavities in two-dimensional Si-ZnO hybrid photonic structure

V M Carrillo-Vázquez and J G Murillo

Centro de Investigación en Materiales Avanzados S.C.
Miguel de Cervantes 120, Complejo Industrial Chihuahua, C. P. 31109, Chihuahua, Chih., México.

E-mail: jose.murillo@cimav.edu.mx

Abstract. The optical characterization of the Si-ZnO hybrid photonic device fabricated and studied in this work revealed its ability to selectively enhance the reflectance on specific wavelengths in the border of VIS-NIR range. This ability was attributed to the coupling of the embedded micro-cavities in the photonic crystal. The results found suggest the presence of a photonic band gap around the border of VIS-NIR range in the hybrid photonic structure studied.

1. Introduction

Photonic crystals (PhC) have demonstrated to be a good alternative to control the flow of light to develop new technology. Its interest have been growing incredibly every year since they were found as promising solutions for specific areas of study such like telecommunications [1-4] and light extraction on LEDs [5] for cite some of them. Two dimensional (2D) photonic crystals are of particular interest since many fabrication methods are currently available for this kind of structures like chemical assisted ion beam etching (CAIBE), reactive ion etching (RIE) and focused ion beam (FIB) milling [6, 7], among others. In this work a two-dimensional hybrid PhC was fabricated on a silicon (Si) substrate coated with a zinc-oxide (ZnO) thin film. The PhC was directly milled over the Si-ZnO hetero-structure using the FIB fabrication method. The photonic structure was described by circular air columns in square lattice and an array of 16 coupled quasi-circular micro-cavities embedded in the photonic crystal body. Reflectance measurements at normal incidence carried out on this photonic device revealed reflectance selectivity for wavelengths in the border of VIS-NIR ranges.

1. Sample fabrication

In order to fabricate the sample, a silicon substrate was coated with a ~240 nm thick ZnO thin film deposited by the Assisted Chemical Vapor Deposition (AACVD) method [9]. The hybrid PhC was milled directly on the Si-ZnO hetero-structure on a large area of 84 x 84 μm^2 using a Focused Ion Beam system. The photonic crystal milled showed a lattice constant a=1.05 μm and circular air columns of radius r=0.57a.The main parameters of the photonic crystal described above were achieved by setting a dose of 200 nC/μm^2 and 1000 pA of irradiating current during 15 seconds per single hole using the spot milling shape. A pattern of 16 quasi-circular micro-cavities of about 10 μm in diameter and a lattice constant of ~20 μm was embedded in the photonic crystal describing a secondary square lattice. The sample was designed including a pattern of micro-cavities in which

Journal of Physics: Conference Series **582** (2015) 012053 doi:10.1088/1742-6596/582/1/012053

each cavity has dimensions in the next order of magnitude compared to those of the regular square lattice photonic crystal. This was to identify whether these micro-cavities can be coupled between them to induce a photonic band gap in the PhC. Also, the square lattice pattern was selected with the aim in find a complete band gap when including the micro-cavities by measuring reflectance at normal incidence since this structure is well known to show partial band gaps for transversal electric (TE) or transversal magnetic (TM) modes [8]. Redeposition of eroded material was observed during the FIB milling process driving a rough surface on the photonic crystal area as result. Figure 1 shows the design of the complete photonic crystal structure and a single micro-cavity of the PhC device can be observed in the inset.

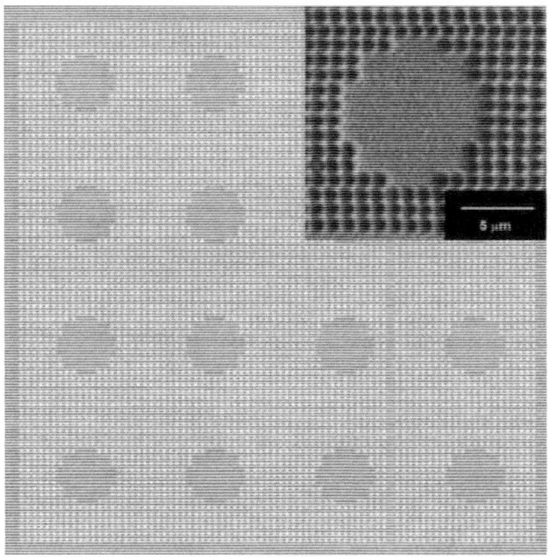

Figure 1. Design of the hybrid PhC with 16 patterned micro-cavities embedded in the PhC body. Inset: SEM micrograph of single optical micro-cavity.

2. Experimental details

Optical characterization of the hybrid photonic crystal was performed by measuring the optical reflectance at normal incidence using a similar experimental set up as used by Stumpf et al in Ref [10]. In order to measure reflectance on the photonic crystal two light sources as probe light were used: a He-Ne laser emitting at 633 nm and a multichannel fiber coupled laser source which emits 785, 852, 1310 and 1550 nm wavelengths. These wavelengths were used with the aim in identify the optical properties of the hybrid photonic hetero-structure in a discrete broad spectra on VIS-NIR ranges. The ZnO thin film was removed from the substrate surface once reflectance measurements were completed on the hybrid photonic crystal. Figure 2 shows a close view of the photonic structure after ZnO thin film removal where it can be observed the PhC topography that has been predicted in previous theoretical investigations for FIB fabrication method [11].

Figure 2. Micrograph of PhC on bare silicon after ZnO thin film removal in close view.

Journal of Physics: Conference Series **582** (2015) 012053 doi:10.1088/1742-6596/582/1/012053

In the optical characterization of the sample both polarizations of light TE and TM modes were considered. In fact, the results of optical reflectance obtained for both polarizations were very similar, showing only very small differences between them. This finding was expected due to the PhC symmetry which determines that the light at normal incidence observes the same pattern on the photonic crystal surface for either TE or TM modes. The optical characterization of the hetero-structure showed a high reflectance level on the hybrid photonic crystal area at the border of VIS-NIR range wavelengths and a low level for NIR wavelengths. Particularly, reflectance measurement carried out at the border of VIS-NIR range showed that level obtained for 852 nm was higher than level measured at 633, 785, 1310 and 1550 nm on the hybrid photonic crystal as depicted in Figure 3. The optical characterization was repeated on the PhC remaining on bare Si in order to determine the ZnO thin film contribution on the optical properties of the hybrid PhC. Figure 3 also shows the reflectance measurements of the PhC after the ZnO thin film removal. Here, it is noticeable that the reflectance magnitude was almost constant in the photonic structure on bare silicon for all the probe wavelengths 633, 785, 852, 1310 and 1550 nm. Furthermore, it was observed that the reflectance level of the photonic crystal on bare silicon only provides a uniform small gain compared to reflectance measured for uniform silicon substrate. Continuous dotted line in figure 3 represents the reflectance measured, using a Varian Cary 5000 spectrophotometer, on plain Si-ZnO substrate as a reference.

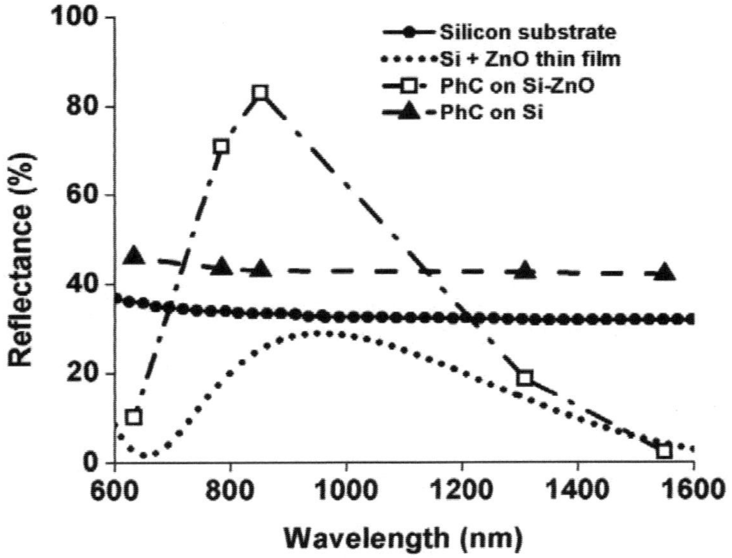

Figure 3. Reflectance measurement at 633, 785, 852, 1310 and 1550 nm wavelengths in PhC before and after ZnO thin film removal. Reflectance of silicon substrate and Si-ZnO hetero-structure are included for comparison purposes.

3. Discussion

It was observed a peak reflectance level at 852 nm probe wavelength in the hybrid PhC. Therefore, the hybrid photonic structure enhanced selectively wavelengths in the border of VIS-NIR ranges rather than NIR wavelengths. Moreover, reflectance measurements on the photonic crystal in bare silicon showed a constant reflectance level which was increased uniformly for all wavelengths

Journal of Physics: Conference Series **582** (2015) 012053 doi:10.1088/1742-6596/582/1/012053

probed in the border of VIS-NIR ranges compared to the reflectance measured for plain silicon substrate. This indicates that photonic crystal with embedded micro-cavities pattern fabricated on bare Si substrate did not contributed in the reflectance modulation as the hybrid Si-ZnO photonic crystal did. Hence, the role of ZnO thin film in the photonic hetero-structure was relevant for the optical behavior here obtained. It was noticeable that the whole hetero-structure system, i.e. the PhC with embedded micro-cavities milled on Si-ZnO, is coupled to selectively enhance the reflectance around 852 nm. This finding strongly suggests the presence of a photonic band gap around this wavelength in the hybrid photonic crystal studied in this work. Further experimentation has to be completed to confirm this statement sweeping in a continuous way a region around the border of VIS-NIR wavelengths.

4. Conclusions

It has been confirmed that the hybrid photonic crystal discussed in this work was able to promote the magnitude of reflectance at the border of VIS-NIR wavelengths range. This finding suggests the presence of a photonic band gap around the border of VIS-NIR ranges for this hybrid PhC. Furthermore, it is concluded that ZnO thin film coating on the photonic hetero-structure clearly improves the reflection in this silicon based photonic crystal promoting a photonic band gap appearance.

References

[1] Robinson S and Nakkeeran R 2012 *Optik* **123** 451– 457
[2] Yang D, Tian H and Ji Y 2011 *Sensor Actuat.* A **171** 146– 151
[3] Esmaieli A and Ghayour R 2012 *Photonics Nanostruct* **10** 131–139
[4] Rahmati A T and Granpayeh N 2011 *Optik* **122** 502–505
[5] Zhmakin A I 2011 *Phys. Rep.* **498** 189–241
[6] Krauss T F and De La Rue R M 1999 *Prog. Quant. Electron.* **23** 51-96
[7] Nellen P M, Callegari V and Bronnimann R 2006 *Microelectron. Eng.* **83** 1805–1808
[8] Hillebrand R, Hergert W and Harms W 2000 *Phys. Status Solidi* (b) **217**, 981
[9] Amézaga-Madrid P, Antúnez-Flores W, Ledezma-Sillas J E, Murillo-Ramírez J G, Solís-Canto O, Vega-Becerra O E, Martínez-Sánchez R and Miki-Yoshida M 2011 *J. Alloy Compd.* 509S, S490–S495
[10] Stumpf W C, Asano T, Kojima T, Fujita M, Tanaka Y and Noda S 2010 *Phys. Rev.* B **82** 075119
[11] Han J, Lee H, Min B and Lee S J 2010 *Microelectron. Eng.* **87** 1–9

Hydrogen production by a PEM electrolyser

G Aragón-González[1], A León-Galicia[1], R González-Huerta[2], J M Rivera Camacho[1] and M Uribe-Salazar[1]
[1] PDPA, UAM-Azcapotzalco, Av. San Pablo 180, Col. Reynosa, Azcapotzalco, 02200 México, D.F
[2]Laboratorio de electroquímica y corrosión, ESIQIE-IPN, Edificio 8 Unidad Profesional Adolfo López Mateos, Col. Lindavista, Gustavo A, Madero 07738 México D, F

E-mail: gag@correo.azc.uam.mx

Abstract. A PEM electrolyser for hydrogen production was evaluated. It was fed with water and a 400 mA, 3.5 V cc electrical power source. The electrolyser was built with two acrylic plates to form the anode and the cathode, two meshes to distribute the current, two seals, two gas diffusers and an assembly membrane-electrode. A small commercial neoprene sheet 1.7 mm thin was used to provide for the water deposit in order to avoid the machining of the structure. For the assembly of the proton interchange membrane a thin square 50 mm layer of Nafion 115 was used.

1. Introduction

The ever increasing world population demands more goods and services, and as a consequence an incremented necessity for energy supply. To find alternative sources of energy for fossil fuels, in order to diminishing the combustion gasses discharged to the atmosphere and the global warming that they generate, is the world most pressing problem. At present, one of the cleanest fuels is hydrogen and one with the highest probability to become the most used in a near future. Its versatility, low contamination and their quality of being renewable [1] made it very attractive for generalised use, in combination with fuel cells.

One of the most promising methods for the generation of hydrogen is water hydrolysis. This was achieved for the first time by William Nicholson and Sir Anthony Carlisle in 1800 [2]. In this process a water molecule is dissociated by the effect of an electrical current. That is, the action of an electrical current produces the gain or loss of electrons in direct proportion to the intensity of the current; which in time generates an oxidation-reduction reaction [3].

An electrolyser, the device that generates hydrogen, is an electrolytic cell that is fed by an electrical current from an external source to produce the electrochemical oxidation-reduction reaction [4-5]. The key aspects in the use of electrolysers are that it could be coupled with a renewable source of energy and thus diminish the emission of atmospheric contaminants [5].

The first electrolysers developed by the General Electric Co in 1966, were of the "Proton Exchange Membrane" (PEM) type; they used a solid polymer electrolyte [6]. When a potential difference is applied between the electrolyser's electrodes an oxidation reaction is produced. In the anode appear oxygen atoms, protons and electrons; whereas in the cathode a reduction reaction takes place, where the protons (H+) together with the electrons form molecules of gaseous hydrogen.

Journal of Physics: Conference Series **582** (2015) 012054 doi:10.1088/1742-6596/582/1/012054

The parts that constitute an electrolyser are: two plates that function as the anode and the cathode; two meshes that distribute the electrical current; two gas diffusers and the arrangement membrane electrodes (MEA) that is, the electrolyser main structure. The hydrogen produced in this electrolysers has great purity and therefore suitable for use in fuel cells.

A fuel cell is an electrochemical energy converter that transforms directly the chemical energy of a fuel into electric energy. The most important aspect of these devises is that generates clean direct current by the electro-oxidation of hydrogen (H_2) and the electro-reduction of oxygen (O_2) and only generates by products no contaminants (steam and heat).

2. Methodology

An electrolyser was built with two acrylic plates that serve as its supporting structure, the acrylic was chosen because is a light material, corrosion resistant an ease to machine. The plates were squares with 65 mm by side and a 10 mm thickness. These plates were provided with an exit for the gas and an admission for the water. To avoid the machining of the acrylic plates, the water reservoir was made with a sheet of commercial neoprene, 1.7 mm thick; Figure 1. The electrodes (anode and cathode) were built with an active area of 9.62 cm^2.

The current distribution meshes were built with stainless steel 1 mm thick and holes 0.5 mm diam. They have the function of holding the electrodes and to distribute the electrons over all the surface of the catalyser. The sealing is made of silicon 0.25 mm thick, avoiding the leaking and providing isolation between the anode and cathode plates. Both seals have a central recess, 9.62 sq cm. The gas diffusers are made with carbon fabric and are Teflon coated, their purpose is to improve the contact between the current distributing meshes and the catalytic surface. Both were cut as circles, 9.62 cm^2.

The core of the electrolyser is the Membrane Electrode Assembly (MEA). MEA is a set of 5 layers; at its centre the proton interchange membrane, which acts as the electrolyte, is located. The PEM separates the anode from the cathode structures and avoids the mixture of gases. Each electrode has a gas diffuser with a layer of electro catalyser located between the membrane and the diffuser.

Figure 1. PEM electrolyser construction

As the electrolyte a 50mm square membrane made with Nafion 115 and 127 μm thick was used. The membrane allows the passing of the ions of hydrogen that form in the anode to the cathode and thus generate the molecular hydrogen. To activate the membrane several baths of chemical solutions at 60°C were used. The preparation of the catalytic inks took as reference the reactive area of the electrolyser, 9.62 cm^2. The catalytic ink for the anode was made with 7.81 gr of IrO_2, 23.44 mg of RuO_2, 120 μl of liquid Nafion, 900 μl of ethylic alcohol and 200 μl of water. The catalytic ink for the cathode was prepared with 5 mg of Pt-Etek at 20% in weight, 100 μl of liquid Nafion, 700 μl of ethylic alcohol and 200 μl of water. To impregnate the inks over the Nafion membrane a sprayer

Journal of Physics: Conference Series **582** (2015) 012054 doi:10.1088/1742-6596/582/1/012054

technique with an aerograph was used. The sprayed area was the 9.62 cm^2 circle, previously marked.

The assembly of the MEA was made in a hydraulic press using the process of heat pressing. At the centre, the sprayed Nafion membrane with a gas diffuser was located and a 5 Ton/cm^2 pressure was applied at 120°C for 5 s; afterwards the pressure was reduced to 0.25 Ton/cm^2 , keeping it for 2 min to achieve a good integration. The machining of the acrylic plates was made in a vertical milling machine; on the exterior surfaces of the arrangement, four fast coupling stainless steel connectors where located to provide for the water and gasses admission and discharge; Figure 2.

•

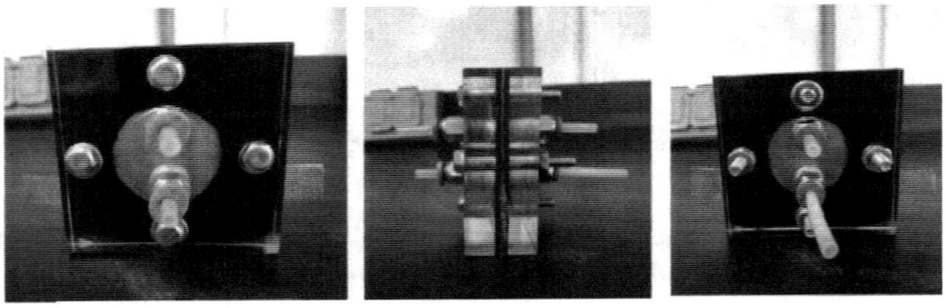

Figure 2. PEM Electrolyser final assembly.

3. Results and discussion

In order to correctly coupling of the electrolyser with the photovoltaic cells, the characteristic curve of each device is necessary. Then, the optimal performance point is determined so that the photovoltaic cell generate enough energy and the electrolyser can initiate the water electrolysis, and thus produce gaseous H_2 and O_2.

To characterise the photovoltaic panels three 250 W halogen lamps were located at 20 cm apart from then, so that solar radiation could be simulated. The system was connected to an arrangement of resistances where the open circuit voltage (VCA) and the short circuit current (CCC) theoretical values were adjusted. To obtain the panels real curve performance, resistances from 1 Ω to 200 Ω were applied. With this data graphs of voltage vs current were obtained.

The PEM electrolyser characteristic curve is obtained connecting it to an electrical power source, in this case the photovoltaic panels. The first test was made with an arrangement of three photovoltaic panels, serial connected, so that, a constant 100 mA (CCC) and 13.5 V (VCA) were obtained. A second test was made with the panels connected in parallel. In this case, 4.5 V (VCA) and 300 mA (CCC) were obtained. Both results were compared with the electrolyser characteristic curve; Figure 3.

The PEM electrolyser characteristic curve does not pass by any of the maximum power curve bends obtained from the photovoltaic modules. That is, the photovoltaic modules arrangement serial or parallel connections do not satisfy the electrolyser optimal energetic performance requirement. Therefore, the modules connection was modified to a serial-parallel arrangement. First, four arrangements were prepared, each one with two modules serial connected; then, these four arrangements were parallel connected, so that, 400 mA (CCC) and 3.5 V (VCA) were obtained.

In figure 4, the coupling of the photovoltaic serial-parallel system curve with the electrolyser performance curve is shown. In this new arrangement the electrolyser performance curve goes through the photovoltaic system maximum power curve bend.

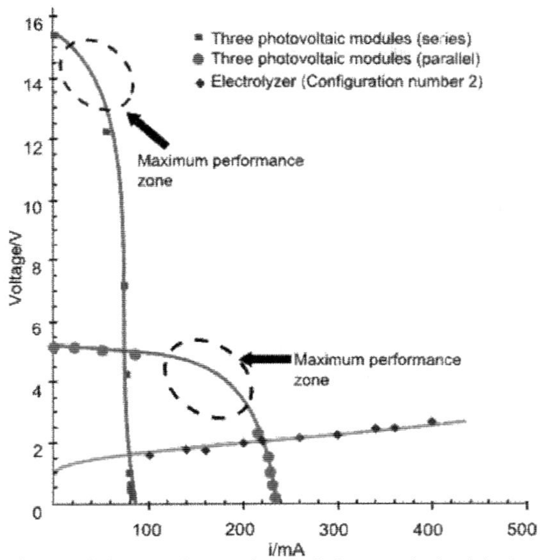

Figure 3. Photovoltaic modules, series and parallel, coupled with the stainless steel mesh electrolyser.

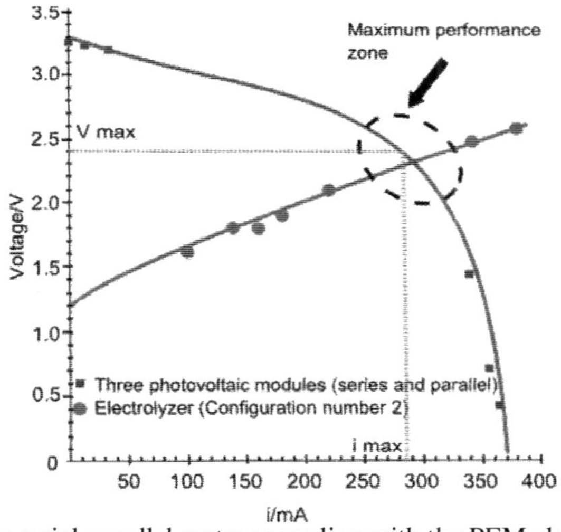

Figure 4. Photovoltaic serial-parallel system coupling with the PEM electrolyser performance curve.

4. Conclusion

The electrolyser design and the MEA assembly with the sprayed Nafion membrane gave satisfactory results in the hydrogen production, as well as to feed a low power fuel cell; all obtained from the electrical power of a photovoltaic system.

The use of a commercial neoprene sheet, in order, to form the water reservoir was successful and this in turn avoided the machining of the acrylic plates. The use of the acrylic plates to form the structure of the electrolyser was simple and at low cost.

The performance of photovoltaic cells alone and connected in serial and parallel arrays were obtained. It was found that a serial-parallel arrangement, giving a 400 mA short circuit current and a 3.5V open circuit voltage was satisfactory for the PEM electrolyser energy requirement.

5. References

[1] Aguado Alonso M, Alzueta Ibañez B, Garde Aranguren R. 2004 Altener.
[2] Chen J P., Chang S Y, Hung Y T 2007 *Physicochemical Treatment Processos, Handbook of Environmental Engineering* **5**
Lawrence K. Wang Yung-Tse Hung, Nazih K. Shammas *Handbook of Environmental Engineering* ISBN: 978-1-58829-860-7
[3] Yunez Cano A. 2013, Tesis, IPN-ESIQIE.
[4] Domínguez Reza M. A. 2010, Tesis, IPN-ESIQIE,
[5] Santillán Aragón G. 2012, Tesis, IPN-ESIQIE, p 8-18.
[6] Carmo M., Fritz D. L., Mergel J., Stolten D. 2013, *Int. J. Hydrogen Energy*, **38**, 4901-4934.
[7] Höök M, Tang X 2013 Depletion of fossil fuels and anthropogenic climate change, *Energ. And Police* **52** 797-809.

Journal of Physics: Conference Series **582** (2015) 012055 doi:10.1088/1742-6596/582/1/012055

Semiautomatic machine for turning inside out industrial leather gloves

G Aragón-González, M Cano-Blanco, A León-Galicia, L F Medrano-Sierra and J R Morales-Gómez

Programa de Desarrollo Profesional en Automatización (PDPA).
Universidad Autónoma Metropolitana - Azcapotzalco.
Av. San Pablo # 180. Col. Reynosa, Azcapotzalco, 02800, México, D.F.
Tel. and FAX: 5318-9057.

E-mail: gag@correo.azc.uam.mx

Abstract. The last step in the industrial leather gloves manufacturing is to turn the inside out so that the sewing be in the inside of the glove. This work presents the design and testing of a machine for that purpose. In order to quantify the relevant variables, testing was performed with a prototype glove. The employed devices and the testing proceeding were developed experimentally. The obtained information was used to build the turning inside out machine. This machine works with pneumatic power to carry the inside out turning by means of double effect lineal actuators. It has two independent work stations that could be operated simultaneously by two persons, one in each station or in single mode operating one station by one person. The turning inside out cycle is started by means of directional control valves operated with pedals. The velocity and developed force by the actuators is controlled with typical pneumatic resources. The geometrical dimensions of the machine are: 1.15 m length; 0.71m width and 2.15m high. Its approximated weight is 120 kg. The air consumption is 5.4 fps by each working station with 60 psig work pressure. The turning inside out operation is 40 s for each industrial leather glove.

1. Introduction.

An industrial glove manufacturing company located in Jalisco, Mexico, has been turning the inside out of the gloves they manufacture in a manual fashion. Even with top skilful workers this is a slow operation, which could turn out to be a bottleneck in the process. Therefore, a machine that could perform the turning inside out operation in less than 50 s, for all the variety of manufactured industrial gloves, was asked to the PDPA. Another design constraint was that the building of the machine be economically feasible and its construction be made with elements readily acquired in the Mexican market. This work describes the technological solution that was proposed to perform the turning inside out of the gloves in an automatic fashion.

The work started experimenting with a prototype driven with fluid power, in order to identify the steps involved in the process to turn the inside out of a glove by means of a machine. In this way the fundamental variables for the functioning of the machine were identified. After a long process of

Journal of Physics: Conference Series **582** (2015) 012055 doi:10.1088/1742-6596/582/1/012055

experimentation and modification of the prototype a simple technological solution was chosen. This solution is presented in the present work.

2. The manufacture of an industrial leather glove.

The factory can produce 300 different glove models, although the core production only span about 30 similar models as could be seen in the Figure 1. The manufacturing process is not much different between the models, therefore it could be said that a general method of manufacture exists [1].

2.1 Cut. The first step of the process is the cutting of all the pieces that form the glove model selected. The amount of pieces depends on the model and the thickness of the leather used is variable. The master cutter selects the material and decides upon the way of cutting, since the thickness is an important factor in the function of a particular section of the glove. The cutting is done with a hydraulic press to which the proper die-cutter is mounted.

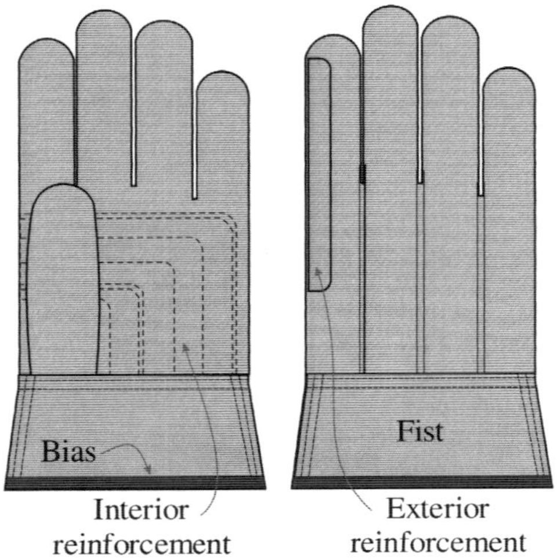

Figure 1. Typical industrial leather glove.

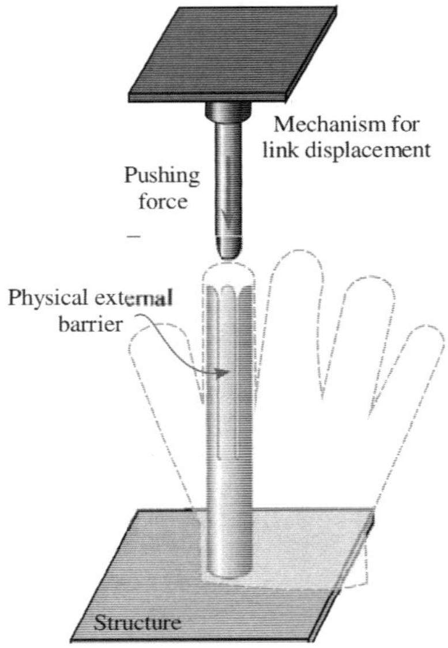

Figure 2. Basic concepts to invert the glove.

2.2 Sewn. The second step is carried after the pieces that form a complete glove is set. The sewing is realized by means of industrial sewing machines. The joining of the pieces is quit similar for all the models. It starts by sewing the palm with the interior reinforcement, then the fingers to the palm and then the fist is sewn. If the model requires a bias, this is sewn last.

2.3 Turn inside out. After the sewing is completed the process of turning the inside out is performed, since all the sewing is done in a way as to locate the seams in the inside of the finished glove. This operation is realized in a manual fashion, with the help of a cold rolled pipe upon which each finger of the glove is located, one by one; then, with the help of a thin bar, a force is applied as to induce the inside out of each finger. When the last finger is finished and before the bar is removed the palm is turned inside out by manually pulling the fist.

2.4 To press. After each glove is finished with the turning inside out operation it is convenient to press the glove. This is accomplished introducing an appropriate flat steel bar to each finger in a way as to accommodate the seams of every finger.

3. Develop of the solution.
After analysing the manual process, in order to identify the relevant variables and their quantification, the following result was obtained:
There are three relevant variables that should be controlled if an automatic solution is to be implemented:
a) The insertion into each finger of a physical barrier, which will guide the turning inside out of the leather that forms the glove.
b) The force required to push a solid cylindrical bar, so that the leather of the glove slides over the hollow flexible barrier. The leather of the glove is forced into the hollow barrier when the force is applied.
c) A mechanism that moves back and forth the link that applies the force, and a structure to support all of these elements; Figure 2.

Some machines have been developed earlier, to perform the turning inside out operation. But these machines were conceived to attend knitting, fabric or leather soft gloves manufacturing [2] [3] [4] [5]. All of them are unsuitable to work with stiff industrial leather gloves, or do not allow to be adjusted to receive many different glove models.

By experimenting with a physical model it was possible to reproduce and improve the turning inside out process. A power fluid system with the appropriate control elements was used. A design with two working stations was reached as is shown in the Figure 3. The geometrical arrangement to locate the mechanisms, the fixing systems, the actuators and the gloves were defined so that the gloves turning inside out were performed in a short time as possible.

The turning inside out operation was conceived thinking that one or two operators could work in two parallel stations. The turning inside out will be carried in three steps: a) in the first station all the central fingers will be turned; b) in the second station the thumb and the little finger will be turned and in a third station the cuff will be turned inside out.

3.1 Turning inside out mechanism. This device was built in two parts. Two or three support basis holding the flexible barriers are located over the guide plates and the whole assembly is on top of the workers table. (See Figure 3) In the upper part there is the second assembly: the pushing bars fixed to the actuators rods by means of a height and positioning sliding railing (see Figure 4). The flexible barriers are formed by a pair of slender bars with a ring attached to the upper end. The bars of the flexible barriers can slide vertically thru the bases that uphold them. Their height could be adjusted by mean of screws. In the upper part of the machine there is a rail that allows regulating the height and position of the pushing bars.

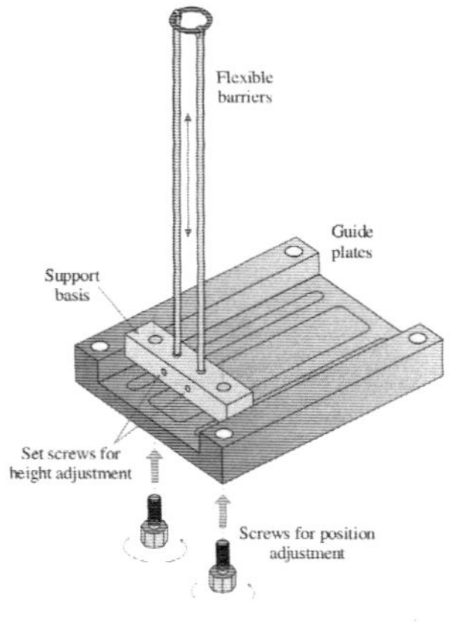

Figure 3. Guide plates with flexible roads.

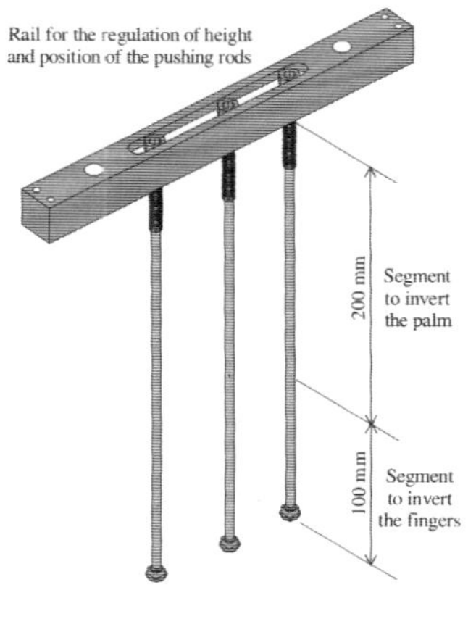

Figure 4. Rail with pushing bars.

Journal of Physics: Conference Series **582** (2015) 012055 doi:10.1088/1742-6596/582/1/012055

3.2 Machine configuration. The machine was built as a table with three horizontal levels. In the upper level, the pneumatic actuators are located; the working area is located at the medium level and in the third level the starting pedal. Figure 5, shows the arrangement of the three levels in relation with the body of the operator.

The shape of the table, the positioning of the workstations and the location of the column that supports the head of the machine, are shown in the Figure 6. The two work stations of the machine could be operated turning inside out two or three fingers of the glove according to what is needed. In each case, some arrangement should be made as to match the flexible barriers with the corresponding pushing rods.

3.3 Structure. The machine rests on three columns made with steel pipe. The central column is larger so that it could support the head of the machine. The head of the machine is built with a section of pipe to which two arms were welded. The arms support the pneumatic actuators and each one has a bar, which serves to reinforce the anchoring of the actuators.

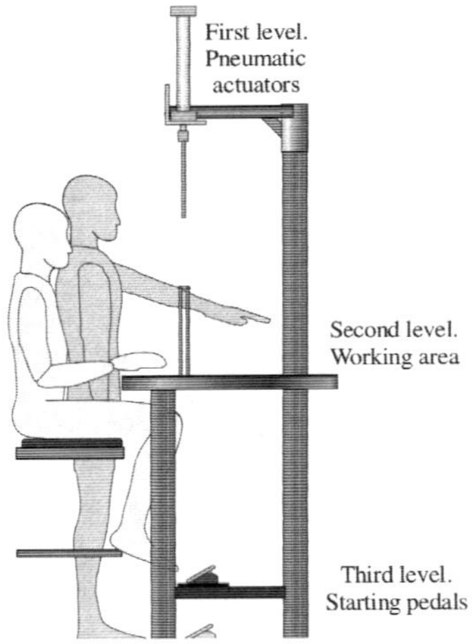

Figure 5. The machine is organized with three horizontal levels.

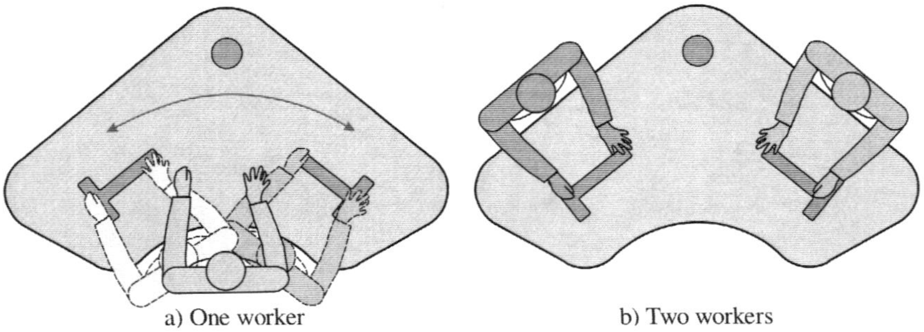

Figure 6. Work spaces distribution on the table.

3.4 Pneumatic power system. The force developed by the pushing bars is controlled with pressure regulators. The velocity at which the movements of the machine are carried is controlled with flow regulating valves. The machine is set in motion by means of a start pedal, which controls the inside out cycle. The buttons for controlling are located in places that are reached easily by the operator and their functioning is simple. This aspect was carefully considered mainly for the pressure

regulators and for the start pedal, since these elements are operated with more frequency than the others. The machine is equipped with two lineal actuators placed in parallel for each working station. Thus turning of the riel with the pushing rod is avoided. Moreover, the applied force is increased twice and the mounting of the tracks is carried in a simple way.
The advance and retiring velocity is fixed by the regulation of the air discharge, from the actuators ports to the atmosphere. In the control system limiting valves were used, which are controlled by the displacement of the rods. It is through these limiting valves that the change in pressure is ordered. The pressure during the piston displacement is diminished at first but increases at the end of the stroke. The operation with differentiated pressures produces a saving in the net consumption of compressed r. The air used in one turning inside out cycle is 5.4 scfm for each working station. The necessary force for turning inside out the three central fingers, for the most demanding glove, take 58 kgf.

4. Conclusions

The proposed arrangement used in the machine is simple but shows excellent performance to make the task. The process takes 40 s for the most rigid glove, without damaging it. The prototype shows many advantages that put their performance above the manual procedures used in the factory.

Among their advantageous characteristics are: its versatility to turn inside out a great variety of gloves b) the fluid power allows the easy regulation of the force and velocity of the linear actuators c) the machine is operated by one or two workers, d) energy consumption is reduced, e) their structure is light but strong, f) their fabrication cost is low and the maintenance requirements are simple, h) the risk of accidents during its operation is almost inexistent.

Every machine that works through an automatic cycle has gone by improvement steps, in order to obtain better performance. This machine is no exception. The use of other materials and the improvement of the process using the observed results will produce betterment to the presented design.

5. Reference

[1] Leather Products Development Institute 2005. *Leather gloves manufacturing technology.* (Pakistan)
[2] Crosby J E. 1909 *Machine for turning gloves, mittens, and the like*. US 978'434 A patent.
[3] Boldizzoni T G 1947 *Machine for turning and pressing gloves*. US 2'511'456 A patent
[4] Becker R E 1951 *Glove turning machine*. US 2'540'503 Patent.
Cripe O C 1952 *Glove turning machine*. US 2,601,504 Patent
[5] Travis H 1973 *Glove turning machine*. US 3'738'547 Patent

Optimization of an irreversible Stirling regenerative cycle

G Aragón-González*, M Cano-Blanco, A León-Galicia, J M Rivera-Camacho
PDPA, UAM-Azcapotzalco, Av. San Pablo 180, Col. Reynosa, Azcapotzalco, 02200 México, D.F

E-mail: gag@correo.azc.uam.mx,

Abstract. In this work a Stirling regenerative cycle with some irreversibilities is analyzed. The analyzed irreversibilities are located at the heat exchangers. They receive a finite amount of heat and heat leakage occurs between both reservoirs. Using this model, power and the efficiency at maximum power are obtained. Some optimal design parameters for the exchanger heat areas and thermal conductances are presented. The relation between the power, efficiency and the results obtained are shown graphically.

1. Introduction

At recent years the efficient use of energy has acquired a greater importance, one of the main points is the reuse of thermal energy wasted at many devices. All the mechanical devices like engines and turbines capable of produce electrical energy and power, waste important quantities of heat. Stirling engines are one of the few devices capable of reusing the wasted energy efficiently. Theoretical models [1-5] have been used to show how a regenerative engine can upgrade the overall efficiency of a thermal system. The engine model considered in this work has a regenerator coupled to improve the use of energy.

The Stirling engines can have several internal configurations, commonly known as alpha, beta and gamma. Throughout the years, these devices have been studied and their optimization at maximum power output has been developed [1]. Other approaches utilize some real design parameters as a non-ideal regenerator or dead volume to deal with its design analysis [2].

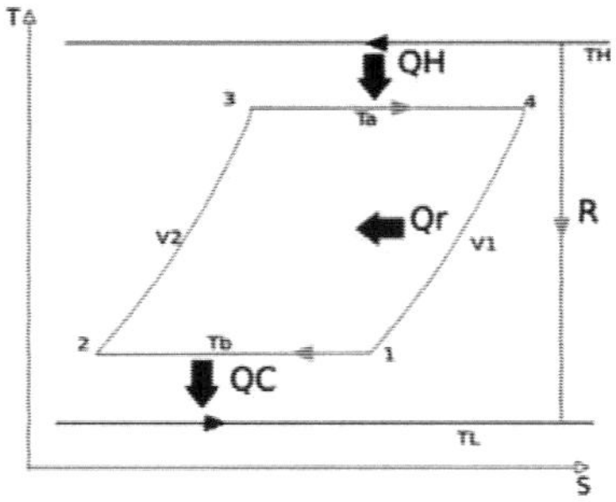

Figure 1. Stirling cycle with irreversibilities due to heat transfer between the hot and cold reservoirs and with ideal regenerator.

Journal of Physics: Conference Series **582** (2015) 012056 doi:10.1088/1742-6596/582/1/012056

Latest developed models have contributed with formal expressions for maximum power output conditions and maximum efficiency. The use of the expressions has contributed to optimize these engines [1-2] and [5-6]. Some of these ideas are developed here and some other irreversibilities are added. The irreversibilites included here are finite heat transfer between the hot and cold reservoirs and a heat leak between both of them. Adding up these design parameters, some useful expressions can be obtained at maximum power output and maximum efficiency. The obtained expressions relate power and efficiency of the engine with some real design parameters like the thermal conductance or the heat exchangers area. Expressions of this type have been done before using Carnot cycle models [3-5] and [7]. This method delivered relevant results including invariant optimal expressions for both heat exchangers [4].

2. Methodology

The modeled cycle in this power comprises two isothermal and two isochoric processes (Figure 1). A heat leak occurs at the engine and this energy cannot be used to perform power. Both heat exchangers can communicate a finite amount of heat. The engine uses an ideal regenerator to reuse his own thermal energy and improve his efficiency. According to the First Law of the Thermodynamics the power can be obtained by the heat flows incoming and leaving the engine:

$$\dot{P} = \dot{Q}_{2-1} + \dot{Q}_{3-2} + \dot{Q}_{4-3} + \dot{Q}_{1-4}. \quad (1)$$

$$\dot{Q}_H = \dot{Q}_{4-3}; \dot{Q}_L = \dot{Q}_{2-1}. \quad (2)$$

The heat flow from the hot heat exchanger is $\dot{Q}_{4-3}$, meanwhile the heat flow leaving the engine from the cold heat exchanger is $\dot{Q}_{4-3}$.

$$\dot{Q}_{1-2} = \beta(T_2 - T_L). \quad (3)$$

$$\dot{Q}_{3-4} = \alpha(T_H - T_1). \quad (4)$$

where α and β are the thermal conductances of the heat and cold heat exchangers. Thermal conductances are the heat transfer coefficient times the area of each heat exchanger [4]:

$$\beta = U_L A_L, \quad (5)$$

$$\alpha = U_H A_H. \quad (6)$$

The heat flow at each isochoric process is:

$$\dot{Q}_{2-3} = mC_V(T_a - T_b). \quad (7)$$

$$\dot{Q}_{4-1} = -mC_V(T_a - T_b). \quad (8)$$

Substituting all the heat flows at equation (1):

$$P = -\beta(T_b - T_L) + \alpha(T_H - T_a) + mC_V(T_a - T_b) - mC_V(T_a - T_b). \quad (9)$$

Journal of Physics: Conference Series **582** (2015) 012056 doi:10.1088/1742-6596/582/1/012056

but, both isochoric processes are equal, then:

$$P = \dot{Q}_H + \dot{Q}_L . \quad (10)$$

Because of there is not production of entropy inside the engine,

$$0 = \frac{\dot{Q}_L}{T_L} + \frac{\dot{Q}_H}{T_H} . \quad (11)$$

Eliminating $\dot{Q}_L$ and substituting into the equation (10):

$$P = \dot{Q}_H (1 - x) . \quad (12)$$

If $x = T_b / T_a$, the power can be expressed as:

$$P = \alpha (T_H - T_a)(1 - x) , \quad (13)$$

from the equation (11):

$$x = \frac{\dot{Q}_L}{\dot{Q}_H} . \quad (14)$$

Rewriting the heat flows at this equation:

$$x = \frac{\beta(T_b - T_L)}{\alpha(T_H - T_a)} , \quad (15)$$

but $T_b = xT_a$. So,

$$T_a = \frac{x\alpha T_H - \beta T_L}{x(\alpha + \beta)} . \quad (16)$$

Substituting (16) at (13) equation:

$$P = \alpha \left(T_H - \frac{x\alpha T_H - \beta T_L}{x(\alpha + \beta)} \right) . \quad (17)$$

Rewriting,

$$P = \frac{\alpha^2 \beta T_H \left(1 - \frac{\mu}{x} \right) (1 - x)}{\alpha + \beta} , \quad (18)$$

Journal of Physics: Conference Series **582** (2015) 012056 doi:10.1088/1742-6596/582/1/012056

where $\mu = T_L / T_H$. The efficiency is definite by:

$$\eta = P\left(\dot{Q}_E\right)^{-1}, \tag{19}$$

$$\dot{Q}_E = \dot{Q}_H + \dot{Q}_r + \dot{Q}_f, \tag{20}$$

where $\dot{Q}_r$ is the heat absorbed by the regenerator in the process 4-1. The same amount of heat is delivered in the process 2- 3. $\dot{Q}_f$ is the heat leak between both heat exchangers and Ω is the thermal conductivity per unit area of the regenerator. If the average to obtain the heat at the regenerator is applied:

$$\dot{Q}_r = \frac{\Omega}{2}\left(T_b - T_a\right), \tag{21}$$

and the heat leak,

$$\dot{Q}_f = k\left(T_H - T_L\right). \tag{22}$$

Substituting (2), (4), (21) and (22) in (19), the efficiency is obtained:

$$\eta = \frac{P}{T_a\left(-\alpha + \frac{\Omega}{2}(x-1)\right) + T_H\left(k(1-\mu)+\alpha\right)}. \tag{23}$$

If the heat exchangers between side hot is modeled by:

$$\alpha\left(T_H - T_a\right) = \dot{m} T_a C_V \ln \lambda . \tag{24}$$

The mass flow is $\dot{m}$ and the volume ratio is λ . Resolving T_a in (24) and substituting in (23):

$$\eta = \frac{\left(1-\frac{\mu}{x}\right)(1-x)\frac{\alpha\beta}{\alpha+\beta}}{\frac{\alpha}{g}\left(\Omega\left(\frac{1-x}{x}\right)-\alpha\right)+\left(k(1-\mu)+\alpha\right)}. \tag{25}$$

where $g = \dot{m} C_v \ln \gamma$. Factorizing α and with $G = 1/g$, finally is obtained:

$$\eta = \frac{\left(1-\frac{\mu}{x}\right)(1-x)\frac{\beta}{\alpha+\beta}}{G\left(\Omega\left(\frac{1-x}{x}\right)-\alpha\right)+\frac{\left(k(1-\mu)+\alpha\right)}{\alpha}}. \tag{26}$$

If $\Omega = cte$ and fixing the thermal conductances and areas of each heat exchanger (design rules [4,8]):

$$\alpha + \beta = \gamma, \tag{27}$$

$$A_H + A_L = A. \tag{28}$$

Let $\gamma = UA$, where U is the overall heat transfer coefficient and A the overall area of both heat exchangers. Now, equations (27-28) are parameterized of the following way:

$$\phi_1 = \frac{\alpha}{UA};1-\phi_1 = \frac{\beta}{UA}, \tag{29}$$

$$\phi_2 = \frac{\alpha}{U_H A};1-\phi_2 = \frac{\beta}{U_L A}. \tag{30}$$

Now, optimizing the power (equation (18)) with respect to x, ϕ_1 and ϕ_2. The same result for the areas and conductances are obtained as in [4].

$$\phi_1 = \frac{1}{2}, \tag{31}$$

$$\phi_2 = \sqrt{R}\left(1+\sqrt{R}\right)\left(R-1\right)^{-1}. \tag{32}$$

Moreover; $x_{opt} = \sqrt{\mu}$ and $R = U_L/U_H$. Substituting these values in (31), (32)

$$P_{\phi_1}^{\max} = \frac{1}{4}\left(1-\sqrt{\mu}\right)^2, \tag{33}$$

$$P_{\phi_2}^{\max} = R\left(-1+\sqrt{\mu}\right)^2\left(-1+\sqrt{R}\right)^{-2}. \tag{34}$$

Both expressions relate the maximum power at the engine with their heat exchangers. Now, substituting the same values at the efficiency:

$$\eta_{P_{\phi_1}} = -\frac{\left(-1+\sqrt{\mu}\right)^2\sqrt{\mu}}{8k\mu^{3/2} - 4\Omega G + 2\sqrt{\mu}\left(-2+G-4K+2\Omega G\right)}, \tag{35}$$

$$\eta_{P_{\phi_1}} = -\frac{R^{3/2}\left(1+\sqrt{R}\right)\left(1-\sqrt{\mu}\right)^2\sqrt{\mu}}{\left(1-\sqrt{R}\right)^2 k\left(1+\sqrt{\mu}\right)^2\left(1-\mu\right)\sqrt{\mu}+\sqrt{R}\left(-G\Omega-\sqrt{\mu}\left(-1+G\Omega\right)\right)}. \tag{36}$$

The equations (35-36) describe the behavior of the efficiency at maximum power output.

3. Results and Conclusions

Plotting efficiency at maximum power, the effects due to the regenerator are appreciated. The model contains more irreversibilities and if the regenerator efficiency can be improve, then the power and the efficiency increase (see figures 2 and 3, with $U_r A_r = C_r \varepsilon_r$; ε is la effectiveness of the regenerator). As consequence the model discussed in [7] is obtained, if $\Omega = 0$.

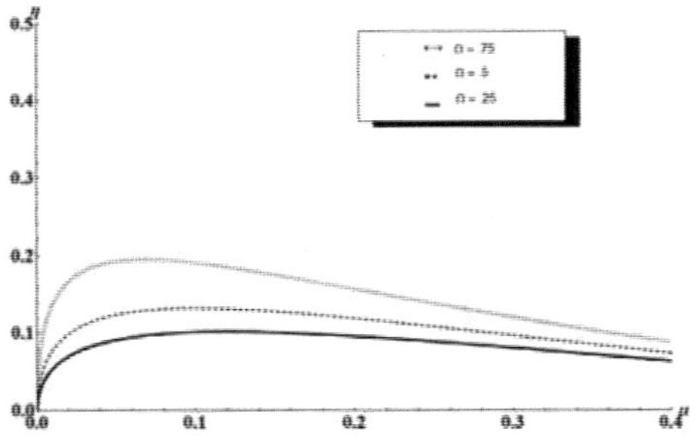

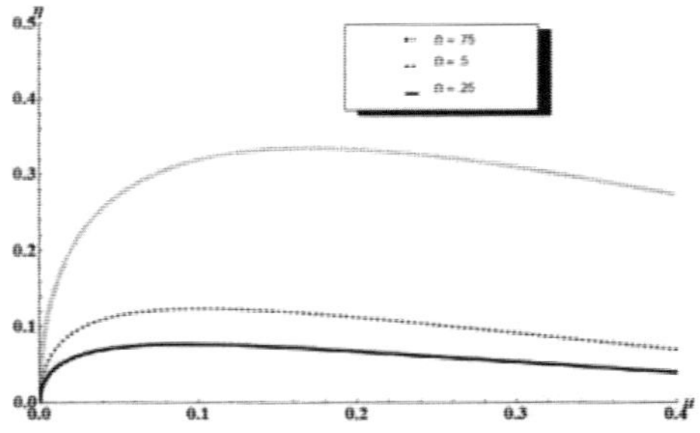

Figures 2 and 3: To the left is the behavior of the efficiency optimizing the heat area and to the right the behavior of the efficiency optimizing the heat conductances.

Solving equations (27) and (28) using the optimal values found at (31) and 32 can be obtained useful relationships for the heat exchangers areas.

$$A_H = \frac{1}{2} A \qquad (37)$$

$$A_L = \frac{1}{2} A \qquad (38)$$

Then if the total area is divided by half at every heat exchanger, best results for the performance of the engine are obtained. The power function has optimal values according to the design parameters chosen. Finally, both criteria are independent from the heat transfer law applied. This is relevant because independently of the law of heat transfer the same results can be obtained. Optimize the equation (26) for the efficiency is important to see if the equations (37-38) are satisfied. Further work is underway.

5. References

[1] Iskander T, Finite time thermodynamic evaluation of endoreversible Stirling heat engine at maximum power conditions. *Renewable and Sustainable Energy Reviews*, **16** (2012) 2234-2241.

[2] Puech P, Tishkova V . Thermodynamics analysis of a Stirling engine including regenerator dead volume. *Renewable Energy*, **36** (2011) 872-878.

[3] Chen J., The maximum power output and maximum efficiency of an irreversible Carnot heat engine, *J of Physics. D: Applied Physics*, **27** (1994) 1144.

Journal of Physics: Conference Series **582** (2015) 012056 doi:10.1088/1742-6596/582/1/012056

[4] Aragon Gonzáles G *et al*, The fundamental optimal relations of the allocation, cost and effectiviness of the heat exchangers of a Carnot like power plant. *Journal of physics A: Mathematical and theoretical*, **42** (2009) 425205.
[5] Gordon J. M., On optimizing maximum power heat engines, *Journal of Applied Physics*, **69** (1991) 1.
[6] Walker, G. et al, *In the Stirling Alternative Power Systems, Refrigerants and Heat Pumps*. Gordon and Breach Science Publishers, Switzerland, 1994.
[7] Organ, A. J., In *Thermodynamics and Gas Dynamics of Stirling Cycle Machine*. Cambridge University Press, **29** (1992).
[8] Feidt, M.C.a.M. The effect of the Overall Heat Transfer Coefficient Variation on the Optimal Distribution of Heat Transfer Surface Conductance or Area in a Stirling Engine. *Energy Conversion & Magnament*, **39** (1998) 1753-1761.

Journal of Physics: Conference Series **582** (2015) 012057 doi:10.1088/1742-6596/582/1/012057

Acoustic heart. Interpretation of Phonocardiograms by computer.

J Granados[1*], F Tavera[1], J M Velázquez[1], G López[2], R T Hernández[1] and A Morales[1]
[1]Universidad Autónoma Metropolitana Azcapotzalco, México
[2]Instituto de Educación Media Superior del Distrito Federal

Email: jalgras@yahoo.com

Abstract. In the field of Cardiology have been identified several heart pathologies associated with problems in valves and narrowing in veins. Each case is associated with a specific sound emitted by the heart, detected in cardiac auscultation. On the Phonocardiogram, sound is visualized as a peak in the wave. In the Optics Laboratory of the Universidad Autónoma Metropolitana - Azcapotzalco, we have developed a simulation of the Phonocardiograms of heart sounds associated with the main pathologies and a computer program of recognition of images that allows you to quickly identify the respective diseases. This is a novel way to analyze Phonocardiograms and the foundation for building a portable non-invasive cardiac diagnostic computerized analyzer system.

1. Introduction

The acoustics is an important branch of Physics for its applications in various fields of science. In particular, in medicine, where the audible signals emitted by the human body are fundamental to establish a diagnosis of a disease. They are detected and analyzed in the doctor's office, through the stethoscope, an acoustically defective instrument. However, in the case of cardiac pathologies, acoustic auscultation is the first indicator used to evaluate the performance of the heart. The correct interpretation of heart sounds is performed only by medical specialists in cardiology who have studied for many years and worked in hospitals and doctor's office a while until experienced. The Facultative identifies different heart sounds, or parts of them, concluding in a particular diagnosis, if your ears and your stethoscope are in optimal conditions, in what we consider as a subjective evaluation. For this reason, it is necessary to develop identification techniques of heart sounds totally objective as those based on a detector and a computerized Analyzer.

The so-called Phonocardiograms (PCG) are amplitude vs. time graphics of the sounds emitted by the heart, captured on the computer. To identify possible diseases present in the heart, it is necessary to correctly interpret the different peaks that appear on the PCG, and its location, since these indicate that an additional sound, that could be a signal of any abnormality, is present.

On the other hand, in clinics and hospitals, it is very important to have equipment to detect and measure physiological signals issued by the human body, which is functioning automatically and whose results support the medical diagnosis of people with heart diseases. With contemporary technologies, it is possible to build computational systems portable, economical, that collect and analyze Cardiologic data using this diagnostic method, which is applicable to patients of different ages and health conditions such as elderly people, young people, children and pregnant women, in towns away in where they are not present medical specialists.

Our proposal is to develop a new alternative, a technique that uses sound and digital images through recognition and image analysis software.

2. Antecedents

In several countries, at medical and academic institutions, the Phonocardiograms have been analyzed using various techniques and mathematical algorithms [1-4], somebody using the Fourier Transform, the Wavelet Transform and Gabor Transform [5-7]. Other authors [8-9] analyze time series between heartbeats that occurs in three groups: elderly healthy, young healthy subjects and patients with congestive heart failure and also Fractal and Multifractal analysis. Using the fractal dimension method proposed by Higuchi, find that the fractal organization is different for each group. They discuss their results in the context of the dynamics between heartbeats.

They also presented a daytime fractal analysis of time series of the interbeat of heart of young healthy subjects and patients with congestive heart failure [11]. Some differences between the groups using the calculation of some invariant exponents of scale and multifractal analysis of simulated sequences and its comparison with real data are described.

Multifractal analysis of the Phonocardiograms has recently been applied by Gavrovska et al, in Serbia, to develop a low-cost computer-aided diagnostic system, to detect pathologies of prolapsed Mitral valve in children [12].

3. The configuration of heart sounds

In this section, we will make a non-exhaustive description of the main sounds emitted by the heart and the physiological causes that produce them [5] and [13].

The cardiac cycle consists of two periods, systole and diastole. Diastole is filling with blood of the ventricles. Systole is the contraction of the ventricles of the heart to eject blood into the arteries pulmonary and aortic. In a normal heart on auscultation, sounds appears in rhythmical form, only two sounds called S1 and S2. S1 is the first sound that occurs at the beginning of the systolic phase of heart cycle, associated with the closure of the mitral and tricuspid valves. The sounds of the valves closing down are so close in time that will normally be heard as a single sound. Some illnesses change the direction of the blood flow resulting in that the two components are separate enough that they can hear, calling them as M1 and T1. S2 is the second sound of the heart, takes place at the end of the systolic and the diastolic phase of the heart cycle start. It is associated with the closure of the aortic and pulmonary valves and is usually audible as two "hits" together (split), which are labeled as A2 and P2 for aortic and pulmonary components respectively. Other sounds as S3 and S4, which already indicate any pathology can be heard. S3 occurs briefly before S2, and is related to early filling of the ventricle. It is present in about half of the normal children, appears in early diastole. When S4 is audible, it occurs briefly before S1 and is associated with a late diastolic filling. An audible S3 may be normal in young people under 35 years of age, while a S4 is always considered as abnormal.

Other components of the sound are puffs or murmurs, of longer duration. They can occur during systole, diastole or continue in both periods, these are caused by the turbulence in the flow of blood or vibrations in the tissues. Murmurs are type noise signals. Called innocent murmurs may occur in normal hearts. Those called organic, can reflect structural abnormalities [13].

May be heard another sound class, the clicks and snaps, always indicating abnormality. The most common is a systolic ejection click that occurs between S1 and S2, appears briefly after S1

Journal of Physics: Conference Series **582** (2015) 012057 doi:10.1088/1742-6596/582/1/012057

with the opening of the semilunar valves. When the mitral valve prolapses, a click appears usually half of systole. Snap is an abnormal heart sound associated with the opening of a valve. If it is present, it usually occurs with the opening of the mitral and tricuspid valves between S2 and S3, briefly after S2.
S4, clicks, snaps and organic murmurs, unlike innocent murmurs occur only under unusual circumstances.

3.1 Split sounds

Since the mitral and tricuspid valves are typically closed at one interval of less than 30 ms, S1 is perceived as one sound. The individual sounds due to the closure of individual components are usually not audible. However S2 is usually separable into two sounds (Split), just on the threshold of audibility. The width of the separation between A2 and P2 changes slightly with the inhalation and exhalation. There are several circumstances that may lead to an abnormal separation of S2, or an abnormal separation in S1. Elapsed time intervals between lock valves are, therefore, important data for the diagnosis.

3.2. Other related illnesses

Coarctation of the aorta is a congenital heart defect where there is a constriction in the aorta, restricting the flow of oxygen-rich blood from the heart to the body. The auscultatory results include one greater A2, a systolic murmur and possibly a systolic click.

Mitral valve prolapse (MVP) a defect of the heart where the leaflets of the mitral valve form an abnormal bulge up towards the left atrium, putting pressure on the connector cords and muscles. The valve may not close completely, causing regurgitation or insufficiency, reverse blood flow. The characteristic auscultation of MVP include a click half of systole and a blowing systolic half or retarded.

Pulmonary stenosis is a heart defect where a narrowing occurs at or near the pulmonary valve, therefore restricts the flow of blood from the right ventricle to the lungs. Auscultatory results typically include a systolic murmur of ejection and a soft, sometimes so soft P2, S2 shown only.

Ventricular septal defect (VSD) is a congenital heart defect in which there is a hole in the septum between the left ventricles and right (lower chambers) of the heart. This allows mixing oxygenated blood (left ventricle) and no oxygenated (right ventricle). VSD auscultatory results include a strong pansystolic murmur and a high P2. A breath in the middle of diastole may also be present.

4. Simulation of heart sounds

We have also built the part of simulation, in which through mathematical formulas plot a wave similar to the sound in the heart beat of different diseases mentioned in the previous section, but by space limitations only show that of a healthy person, figure 1, and those in which are present extra sounds between beats, figures 2 and 3, as it is the case in people with valve disease. Simulation corresponding to a healthy person is a graphic that only contains two peaks appearing periodically representing two clean heartbeats. By the other hand, Figure 4 shows a piece, 3 seconds, of a Phonocardiogram taken with an electronic stethoscope at 4000 samples for second, captured by LabVIEW software.

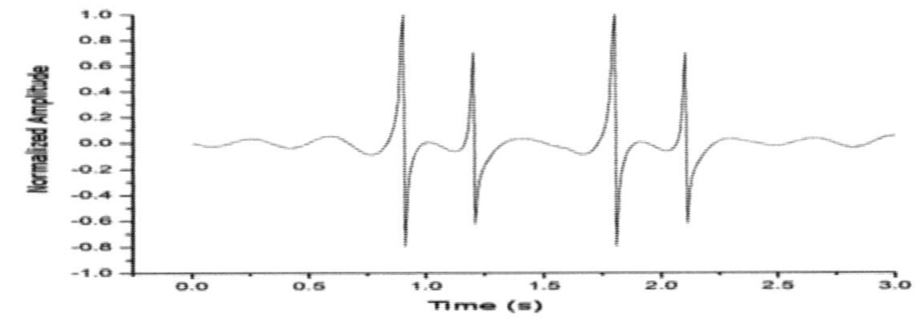

Figure 1. Simulation of two heartbeats S1, S2, from healthy heart.

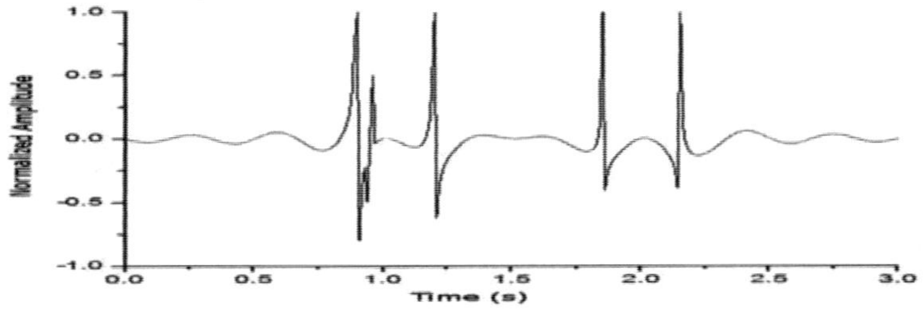

Figure 2. Simulation of two heartbeats with an intermediate noise between S1 and S2.

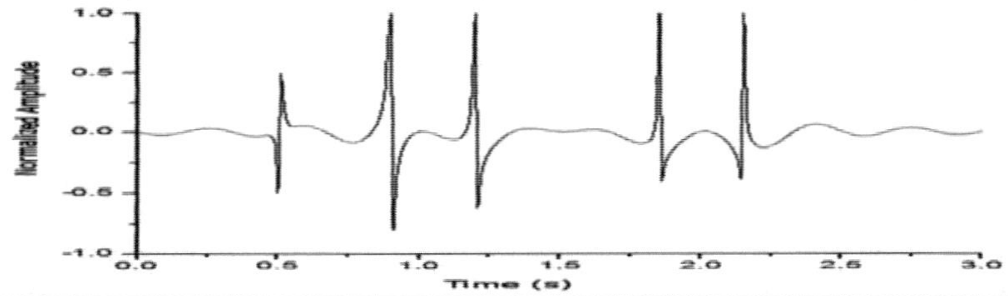

Figure 3. Simulation of two heartbeats with a sound before S1.

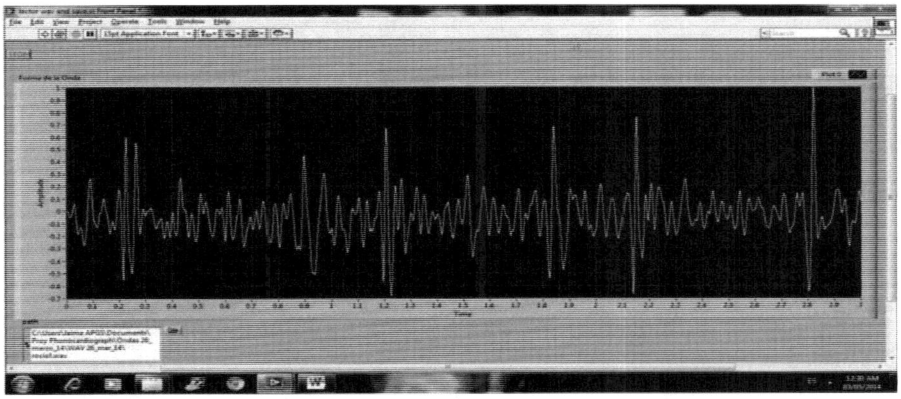

Figure 4. LabVIEW screen showing a Phonocardiogram of someone with heart disease.

5. Images recognition program

In order to identify automatically different pathologies present in the heart of a person which has taken a phonocardiogram, built a program of recognition of images, which is searching for sounds "abnormal spikes", its location in the respective image and measuring the time intervals between events. This methodology was applied in the first stage to images that simulate different most known diseases. For the analysis of graphs of the acquired sound was used a machine vision algorithm based on Ni vision builder using basic functions of recognition of patterns, contour wave analysis and measurement of distances.

Figure 5 shows the front panel of the computer program developed in LabView which acquired the image of a wave simulated with duration of 3 seconds. With artificial vision algorithm is measured the distance between two heartbeats, if the distance between systole and systole is constant means a heart without intermediate noises and heartbeat regular and a message in Dx as healthy, and a green light will be shown. If appear intermediate noises between peaks, DX will indicate the respective condition and the bright colour different from the green button will light.

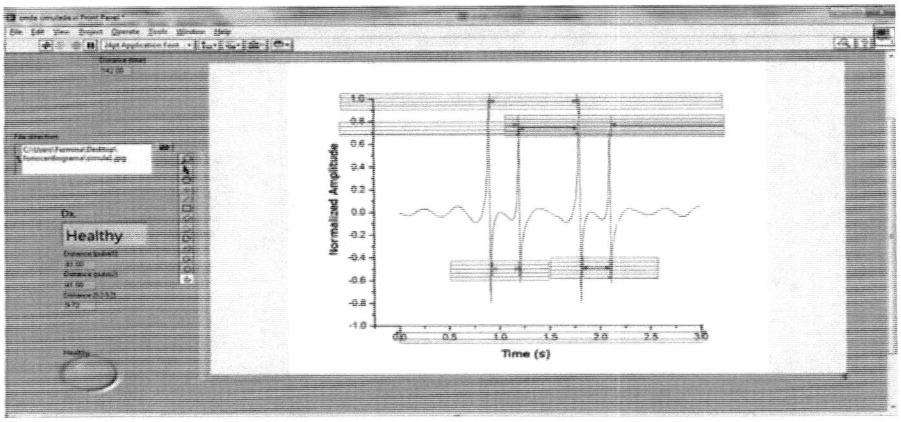

Figure 5. LabVIEW front panel of phonocardiogram analysis with machine vision.

6. Conclusions

Having as ultimate goal, the design and construction of a portable system of cardiac non-invasive diagnosis, in the Optics Laboratory of Universidad Autónoma Metropolitana - Azcapotzalco we have developed the stage of simulation of the Phonocardiograms of heart sounds associated with major diseases and built a computer program of recognition of images that allows you to quickly identify the respective diseases. This is a new and alternative test procedure. Our proposal is to develop a novel technique that uses sound and digital images through recognition and image analysis software.

References

[1] Delgado E, Castaño A M, Godino J and Castellanos G. 2007 *IFMBE Proceedings* **18**, 202-6.

[2] Quiceno A F, Delgado E, Acosta C and Castellanos G. 2007 *IFMBE Proceedings* **18** 162-6.

[3] Martínez-Alajarí J, López-Candel J and Ruiz-Merino R. 2006 *Computers in Cardiology* **33** 537–40.

[4] Martínez-Alajarín J. and Ruiz-Merino R. 2005 *Bioengineered and Bioinspired Systems II, edited by Ricardo A. Carmona, Gustavo Liñán-Cembrano, Proceedings of SPIE,* **5839**.

[5] Reed T, Reed N and Fritzson P. 2004 *Simulation Modelling Practice and Theory* ***12***, 129–46.

[6] Debbal S M, 2014 *Journal of Medical and Bioengineering 3, No. 1,* 5-11.

[7] Echeverry J D, López A F and López J F 2007 *Scientia et Technica* ***XII***, 139-43.

[8] Guzmán Vargas L and Angulo Brown F, 2003 *Physical Review E* **67**, 052901.

[9] Guzmán Vargas L, Diosdado M A and Angulo Brown F, 2005 *Physica* A **348** 304-6.

[10] Guzmán Vargas L, Calleja Quevedo E and Angulo Brown F, 2005 *Revista Mexicana de Física* **51** suplemento 2, 122-7.

[11] Guzmán Vargas L, Calleja Quevedo E and Angulo Brown F, 2003 *Fluctuation and Noise Letters* **3** No. 1 L83-9.

[12] A. Gavrovska, Zaji G, Reljin I, Reljin B. 2013 *Computational and Mathematical Methods in Medicine.* **2013**, 376152.

[13] Korytnicki L 1988 *Semiología Cardiológica* Librería Médica Uruguay.

[14] www.msd.com.mx/ msdmexico/hcp/cursos/ruidos/lesiones.html.

[15] Osorio V J S, Cuesta L L F and Gómez M F, 2007 *Revista de Ingeniería Biomédica* ***1*** 42-6.

[16] Granados S J, Hernández L R T, Velázquez A J M, Vargas C A and Tavera R F 2008 *Proceedings SOMI XXIII* Congreso de Instrumentación México.

[17] Granados S J, Rodríguez J L, Velázquez A J M, Fernández Ch J L, Jurado D T, Tavera R F and López G G 2010 Phonocardiograph signals acquisition and analysis system¨. *Proceedings of the 1st. International Congress on Instrumentation and Applied Sciences, Cancun, México.*

Journal of Physics: Conference Series **582** (2015) 012058 doi:10.1088/1742-6596/582/1/012058

Thermodynamic performance for a chemical reactions model

R E Gonzalez-Narvaez [1], N Sánchez-Salas [1], J C Chimal-Eguía [2]
[1] Escuela Superior de Física y Matemáticas, Instituto Politécnico Nacional, México.
[2] Centro de Investigación en Computación del IPN, Mexico.

E-mail: justicia.12@hotmail.com

Abstract. This paper presents the analysis efficiency of a chemical reaction model of four states, such that their activated states can occur at any point (fixed but arbitrary) of the transition from one state to another. This mechanism operates under a single heat reservoir temperature, unlike the internal combustion engines where there are two thermal sources. Different efficiencies are compared to this model, which operate at different optimum engine regimes. Thus, some analytical methods are used to give an approximate expression, facilitating the comparison between them. Finally, the result is compared with that obtained by other authors considered a general model of an isothermal molecular machine. Taking into account the above, the results seems to follow a similar behaviour for all the optimized engines, which resemble that observed in the case of heat engine efficiencies.

1. Introduction

In this work the thermodynamic properties of the steady-state model for a four-state chemical reaction is analyzed. Despite its simplicity, this model captures the essential features of various biochemical reactions, which in our consideration makes this work relevant in a biological context. The paper is organized as follows: Section 2, introduces the four-state model; section 3 shows a modified four-state model for chemical reactions. Finally, some results and concluding remarks are discussed in section 4.

2. The four-state model for chemical reactions

Consider the following model, which corresponds to a sequence of four states with reversible first-order reactions [2]

$$S_1 \underset{k_{-1}}{\overset{k_1}{\leftrightarrow}} S_2 \underset{k_{-2}}{\overset{k_2}{\leftrightarrow}} S_3 \underset{k_{-3}}{\overset{k3}{\leftrightarrow}} S_4 \quad . \qquad (1)$$

Let $S_i (i = 1,...,4)$ the i-th state of model reaction sequence, and let μ_i be the corresponding chemical potential. A condition for the reaction carried out is that $\mu_1 > \mu_2 > \mu_3 > \mu_4$. From the Arrhenius equation, the net velocities for the three reactions in equation (1) are [3]:

$$v_i = \xi_i (e^{-h_i/RT} - e^{-(h_i+\mu_i-\mu_{i+1})/RT}), \qquad (2)$$

where, i=1,2,3, R is the ideal gas constant and T is the temperature. Equation (2) obtains,

$$v_i = A_i(1-e^{-(\mu_i-\mu_{i+1})/RT}), \tag{3}$$

with $A_i = \xi_i \exp(-h_i/RT)$. For simplicity, it is assumed that $A_1 = A_3 = A$ and defined as $\alpha = A_2/A$. Consider the concentration of substrate and the final products remain constant, i. e., μ_1 and μ_4 are constants. On the other hand, μ_2 and μ_3 evolve in a free way since the system does not reach the steady state with $\mu_2 = \overline{\mu_2}$ and $\mu_3 = \overline{\mu_3}$. Then, if $\Delta = (\mu_1 - \mu_4)/RT$ and, remember that in the steady $v_1 = v_2 = v_3 = \bar{v}$ obtains, $(\mu_1 - \bar{\mu}_2)/RT = (\bar{\mu}_3 - \mu_4)/RT = \Delta(1-\eta)/2$.

Furthermore,

$$\alpha = \frac{1-e^{-\Delta(1-\eta)/2}}{1-e^{-\Delta\eta}} \tag{4}$$

Expression (4) shows η and α are related. In particular, $\alpha(\eta=1)=0$, $\alpha(\eta=1/3)=1$ and $\lim_{\eta\to 0}\alpha = \infty$. Finally, the rate in the steady state is given by $\bar{v} = A(1-e^{-\Delta(1-\eta)/2})$. It can be observed that $\bar{v}$ is a decreasing function of η, and $\bar{v}=0$ when $\eta = 1$. The maximum velocity of the reaction reaches $\eta = 0$, and is given by $\bar{v}_{\max} = A(1-e^{-\Delta/2})$.

2.1 Thermodynamics of the steady state

Consider the following chemical reaction $X \Leftrightarrow Y$, and suppose that the chemical potential of X and Y are μ_x and μ_Y, respectively. Thus, the change in molar free energy of the reaction is $\Delta G = \mu_Y - \mu_X$ [4]. If $\Delta G < 0$, the reaction could occur spontaneously, and the free energy of the system decreases $|\Delta G|$ units per mol in a single reaction. Moreover, this energy is also dissipated as heat. Then, the net velocity of the reaction is v and the rate of the consumption of chemical energy and dissipated as heat is: $|\Delta G| v = (\mu_x - \mu_y)v$ [4].

Now consider a sequence of reactions such that the substrates for a given reaction are products of the previous image, while products are substrates for the next reaction. From the viewpoint of thermodynamics, this whole sequence of reactions can be regarded as a single overall reaction where the change in the free energy is the sum of the free energy changes for the individual reactions [4]. Figure 1, shows the sequence of these three reactions. Then, from the above, the three reactions can be written as a single global reaction with $\Delta G_{Tot} = \mu_4 - \mu_1 = -RT\Delta$. On the other hand, the velocities of the reactions are all equal in the steady state ($v_1 = v_2 = v_3 = \bar{v}$), so the rate of energy consumption by all the system is,

$$\varepsilon_{in} = |\Delta G_{Tot}|\bar{v} = ART\Delta(1-e^{-(1-\eta)\Delta/2}). \tag{5}$$

Considering that the change of free energy in steady state for this reaction is $\Delta G_2 = \overline{\mu}_3 - \overline{\mu}_2 = -\eta \Delta RT$, it is possible to define the power output as,

$$P = |\Delta G_2| \overline{v} = ART\Delta\eta(1 - e^{-(1-\eta)\Delta/2}). \tag{6}$$

Observe that P is a concave function of η ; besides $P = 0$ in $\eta = 0$ and $\eta = 1$ and using the mean value theorem this function has a maximum value in a middle point of η .

In accordance with [5] the ecological function can be written as,

$$E = P - (\varepsilon_{in} - P) = ART\Delta(2\eta - 1)(1 - e^{-(1-\eta)\Delta/2}). \tag{7}$$

The definition of [7] is used since the analysis is on an isothermal system. The function E is concave and has roots in $\eta = 1/2$ and $\eta = 1$. This implies that using the mean value theorem, this function has a maximum for a given value of η in the range $(1/2,1)$. Finally the system's efficiency for energy conversion can be calculated as follows,

$$\eta = P/\varepsilon_{in} . \tag{8}$$

The definition of the parameter η is the maximum efficiency of the system. Moreover, this efficiency is completely determined by the parameter α (see equation (4)) which in turn depends on the chemical potential barrier (h_i) between states.

2.2 Thermodynamics characteristics of the steady state

Expression (6) defines the power output of a steady state as a function of the efficiency. This at maximum power output η_P can be calculated taking into account that the relationship $dP/d\eta = 0$ is satisfied. After performing some algebra, this relation can be simplified to the following transcendental equation:

$$e^{\Delta(1-\eta_P)/2} = 1 + \Delta\eta_P / 2. \tag{9}$$

Since it is not possible to obtain an explicit expression of the function in terms of Δ, proceed as follows: performing a Taylor series expansion for Δ around 0 on the left side of (9) until the second order. Then it is possible to obtain an analytic expression for η_P as a function of Δ, $\eta_P = \eta_P(\Delta)$, not presented in this text, instead write the Taylor series for this expression, the form is as follows [8],

$$\eta_P = \frac{1}{2} + \frac{1}{8}\cdot\frac{\Delta}{4} - \frac{1}{96}\cdot\frac{\Delta^2}{8} + O[\Delta]^3. \qquad (10)$$

If the same process is performed but now taking into account the ecological function (7), the relation which maximizes this function satisfies,

$$e^{\Delta(1-\eta_E)/2} = \left(1 - \Delta_4\right) + \Delta\eta_E/2, \qquad (11)$$

and its corresponding Taylor series is,

$$\eta_E = \frac{3}{4} + \frac{\Delta}{128} - \frac{1}{2}\cdot\frac{\Delta^2}{1024} + O[\Delta]^3. \qquad (12)$$

3. The modified four-state model for chemical reactions

Consider the following model, which corresponds to a sequence of four states with reversible first-order reactions as it shows in (1).

Let $S_i (i = 1,...,4)$ the i-th state of the sequence of model chemical reactions, and let G_i be the corresponding free energy of each state i. A prerequisite to carrying out the reaction is that $G_1 > G_2 > G_3 > G_4$, see Figure 1.

Now, suppose that some external agent or mechanism applied certain force in a steady fashion to transform substrates into products in each sub-reaction [6] and by considering the of Mass Action Law [7] obtains,

$$\frac{[S_{i+1}]_e}{[S_i]_e} = \exp\left[-\frac{G_{i+1} - G_i}{KT}\right] = \exp\left[-\frac{\Delta G_{i+1,i}}{KT}\right] = \frac{k_i}{k_{-i}} = K_{ei}, \qquad (13)$$

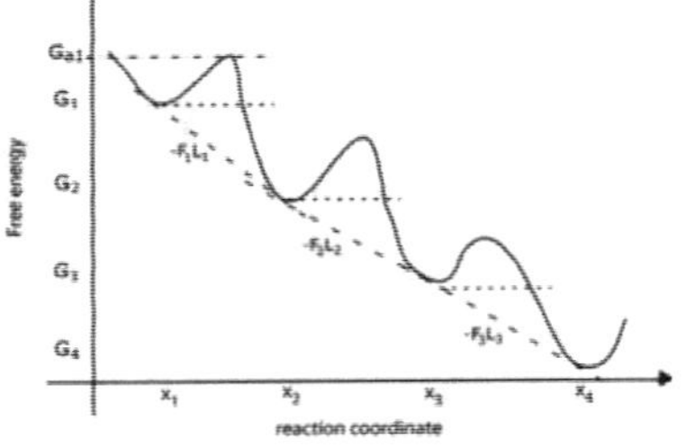

Figure 1. Behavior of the free energy for the above reaction takes place.

where $[S_i]_e$ is the chemical concentration of the i-th state, K is the Boltzmann constant and the subscript "e" denotes that the expression (13) holds only at equilibrium. i-states correspond to two minima of an appropriate free energy scheme. While physically, a scheme of free energy can actually be very complicated, thermal activation transitions between two states follows a preferential path connecting the two states through a saddle point. This is known as an activation state.

It is possible to project this path and introduce a one-dimensional reaction coordinate "x" with the corresponding free energy G (x). Thus, let $x_{i+1} - x_i = L_i$ and $x_{ai} = x_i + \lambda_i L_i$ the reaction coordinates of the minimum of free energy and the activated state, respectively [6]. To continue the analysis it is necessary that the substrates of sub-reactions are in equilibrium with the respective activated state. Besides, the activated states are equally likely to reach the substrates or products in each sub-reaction. Thus, from the Boltzmann law gives,

$$k_{ia} = \gamma_i \exp\left[-\frac{\Delta G_{ai,i}}{KT}\right] \ni \Delta G_{ai,i} = G_{ai} - \bar{G}_i \, . \tag{14}$$

Since the states $[S_i]$ and $[S_{i+1}]$ for each reaction are in equilibrium with the activated states, then, $k_i / k_{-i} = k_{ia} / k_{-ia} = K_{ei}$, substituting equation (14) in the former equation yields, $k_{-ia} = \gamma_i \exp\left[-\Delta G_{ai,i+1} / KT\right]$.

The net rate of each reaction is the forward rate (k_i) minus the reverse rate (k_{-i}) then $v_i = k_i - k_{-i} = k_{ia} - k_{-ia}$. The latter is because the change rates are constant. Now, consider the effect of the forces that produce the difference in free energy in the transition from one state to another, yields, $G_i(x_i) = G^0(x_i) - F_i x_i$, observe that the last equation is only valid for constant forces. Then, in v_i gives,

$$v_i = \gamma_i \exp\left[-\lambda_i \frac{\Delta G_{i+1,i}}{KT}\right]\left(1 - \exp\left[\frac{\Delta G_{i+1,i}}{KT}\right]\right). \tag{15}$$

Since there are four states, assume for simplicity that $\gamma_1 = \gamma_3 = \gamma$ and defining $\alpha = \gamma_2 / \gamma$. It is assumed that the concentration of the substrate and the product will remain constant i.e., G_1 and G_4 remain constant. Thus, the intermediate states, G_2 and G_3 evolve freely until the system reaches the steady state with $G_2 = \overline{G_2}$ and $G_3 = \overline{G_3}$, where, $\Delta = (G_1 - G_4) / RT$ y $\eta = (\overline{G}_2 - \overline{G}_3) / \Delta RT$. Now, recalling that, $v_1 = v_2 = v_3 = \overline{v}$ in the steady-state obtains $(G_1 - \overline{G}_2) / RT = (\overline{G}_3 - G_4) / RT = \Delta(1-\eta)/2$, while,

$$\alpha = \exp\left[\lambda_1 \frac{(1-\eta)\Delta}{2} - \lambda_2 \eta \Delta\right] \frac{1 - e^{-\Delta(1-\eta)/2}}{1 - e^{-\Delta\eta}} \, . \tag{16}$$

Finally, the velocity in the steady state is given by $\bar{v} = \gamma \exp[\lambda_1 \Delta(1-\eta)/2](1-e^{-\Delta(1-\eta)/2})$. The maximum of the last function is reached when $\eta = 0$ and this yields, $\bar{v}_{\max} = \gamma \exp[\lambda_1 \Delta/2](1-e^{-\Delta/2})$.

3.1 Thermodynamics characteristics of the steady state for the modified model

The entire sequence of reactions can be considered as a single overall reaction such that the free energy change is the sum of the free energy changes for individual responses. Then the rate of energy consumed by the entire system is,

$$\varepsilon_{in} = |\Delta G_{Tot}|\bar{v} = \gamma KT\Delta \exp[\lambda_1 \Delta(1-\eta)/2](1-e^{-(1-\eta)\Delta/2}). \quad (17)$$

Of the four states represented, only the second and third consume energy with a useful purpose, this is v to pass form state S_2 to state S_3, while the velocity is stationary, thus define the power output as,

$$P = |\Delta G_2|\bar{v} = \gamma KT\Delta\eta \exp[\lambda_1 \Delta(1-\eta)/2](1-e^{-(1-\eta)\Delta/2}). \quad (18)$$

Note that P is a concave function of η, besides $P = 0$ at $\eta = 0$ and $\eta = 1$, then a maximum at some point. Analogously, for the ecological function obtains,

$$E = P - (\varepsilon_{in} - P) = \gamma KT\Delta(2\eta - 1)\exp[\lambda_1 \Delta(1-\eta)/2](1-e^{-(1-\eta)\Delta/2}). \quad (19)$$

Equation (19) is also a concave function and has roots at $\eta = 1/2$ and $\eta = 1$, where it reaches a maximum at some point.

4. Results and Concluding remarks

Equation (18) defines the power output for the four-state modified model as a function of the efficiency η. From this expression it is possible to know the efficiency at maximum power η_P just by solving $dP/d\eta = 0$. A transcendental equation written as,

$$(2 + \Delta\eta_P(1-\lambda_1))e^{-\Delta(1-\eta_P)/2} = 2 - \lambda_1 \Delta\eta_P. \quad (20)$$

To obtain an explicit expression of the equation (20) in terms of Δ a Taylor expansion performs around 0 for the exponential on the left side of (20) to second order. It is possible to solve the efficiency at maximum power output depending on Δ. So the Taylor series for the efficiency is,

$$\eta_P = \frac{1}{2} + \frac{1}{16}(1-\lambda_1)\Delta - \frac{(1-\lambda_1)^3\Delta^3}{1024} + O[\Delta]^4 . \tag{21}$$

Taking into account the model proposed by Díaz-Hernández *et al.* [1], for this model the value of the parameter λ is equal to ½ [6] so this yields, $\eta_P(\lambda_1 = 1/2) = 1/2 + \Delta/32 - \Delta^3/8192 + O[\Delta]^4$, then, by comparing equation (10) this matches exactly in the linear term of Δ. The efficiency at maximum power is calculated for the extreme cases too: Eyring processes ($\lambda_1 = 0$) and Kramers ($\lambda_1 = 1$) [6] .

$$\eta_P(\lambda_1 = 0) = \frac{1}{2} + \frac{1}{16}\Delta - \frac{\Delta^3}{1024} + O[\Delta]^4 , \tag{22}$$

where $\eta_P(\lambda_1 = 1) = \frac{1}{2}$. Performing the same procedure with equation (19) as before, obtains the maximum efficiency for the ecological function which again is a transcendental equation of the form,

$$\left(4 + (2\eta_E - 1)(1-\lambda_1)\Delta\right)e^{\Delta(1-\eta_E)/2} = 4 + (2\eta_E - 1)\lambda_1\Delta . \tag{23}$$

In order to have an analytical expression for η_E, a Taylor series expansion around 0 for the exponential function of the left side of (23) was performed. It is possible to obtain η_E as a function of Δ. Thus, the Taylor series for the maximum ecological function is given by,

$$\eta_E = \frac{3}{4} + \frac{(1-\lambda_1)\Delta}{64} - \frac{(1-\lambda_1)^3\Delta^3}{16384} + O[\Delta]^3 , \tag{24}$$

Comparing $\eta_E(\lambda_1 = 1/2) = 3/4 + \Delta/128 - \Delta^3/131072 + O[\Delta]^3$ whit (12) it is identical until the linear term in Δ. Finally (24) for Eyring processes ($\lambda_1 = 0$) and Kramers ($\lambda_1 = 1$) [6] have the following forms,

$$\eta_E(\lambda_1 = 0) = \frac{3}{4} + \frac{\Delta}{64} - \frac{\Delta^3}{136384} + O[\Delta]^3 , \tag{25}$$

where $\eta_E(\lambda_1 = 1) = \frac{3}{4}$.

4.1 Concluding Remarks

In this paper an analysis of the thermodynamic performance of an engine model based on a chemical reaction occurs in four states. Expressions for efficiencies were obtained under two optimization criteria; maximum power and maximum ecological function. They are transcendental equations, making a linear approximation to solve algebraic expressions. In order to see the behavior of the efficiencies made developed in the Taylor series, equations (21) and (24) are seen as having a behavior shown by isothermal engines with two states previously analyzed by other authors [6].

References

[1] Díaz-Hernández O., Páez-Hernández R. and Santillán M., Thermodynamics Performance vs. dynamic stability in an enzymatic reaction model, Physica A, **389**, 3476-83.

[2] Houston P. L. 2001 *Chemical Kinetics and Reaction Dynamics* (New York: McGraw-Hill).

[3] McQuarrie D. A. and Simon J.D. 1997 *Physical Chemistry: A Molecular Approach*, (Sausalito :University Science Books).

[4] Lehninger A. L., Nelson D. L. and Cox M. M. 1993 *Principles of Biochemistry* (NewYork :Worth Publisher) .

[5] F.Angulo-Brown, M. Santillán, E. Calleja-Quevedo, Thermodynamic optimality in some biochemical reactions, Nuovo Cimento D 17(1)(1995)87-90.

[6] Van den Broeck C., Kumar N. and Lindenberg K., Efficiency of isothermal molecular machines at maximum power, Phys. Rev. Lett. **108**, 210602.

[7] Howard J. 2001 *Mechanics of Motors Proteins and the Cytoskeleton* (Massachusetts Sinauer Associates, Inc).

[8] Sánchez-Salas N., Chimal-Eguia J. C. and Barranco-Jiménez M. A. Estudio de la eficiencia de un modelo de reacción enzimática *XVII Reunión Nacional Académica de Física y Matemáticas (México D.F., México, 12-16 november 2012)* pp145-147.

Journal of Physics: Conference Series **582** (2015) 012059 doi:10.1088/1742-6596/582/1/012059

Signal differentiation in position tracking control of dc motors

F Beltran-Carbajal[1], A Valderrabano-Gonzalez[2] and J C Rosas-Caro[2]
[1]Universidad Autónoma Metropolitana, Unidad Azcapotzalco, Departamento de Energía, Av. San Pablo No. 180, Col. Reynosa Tamaulipas, C.P. 02200 México, D.F., México.
[2]Universidad Panamericana, Campus Guadalajara, Prol. Calzada Circunvalación Pte. No. 49, Col. Ciudad Granja, C.P. 45010, Zapopan, Jal., México.

E-mail: fbeltran@azc.uam.mx

Abstract. An asymptotic differentiation approach with respect to time is used for on-line estimation of velocity and acceleration signals in controlled dc motors. The attractive feature of this differentiator of signals is that it does not require any system mathematical model, which allows its use in engineering systems that require the signal differentiation for its control, identification, fault detection, among other applications. Moreover, it is shown that the differentiation approach can be applied for output signals showing a chaotic behavior. In addition a differential flatness control scheme with additional integral compensation of the output error is proposed for tracking tasks of position reference trajectories for direct current electric motors using angular position measurements only.

1. Introduction

Control of electric machines is a research subject quite active due to these electric actuators are employed in many practical engineering systems as water pumping systems, electric vehicles, robots and tool machines. Some control schemes for DC electric motors have been reported in the literature (see, e.g., [1-7] and references therein). However, position trajectory tracking controllers for motors based on an efficient motion planning could demand measurements of several state variables increasing the controller implementation costs. Thus, the efficient control problem of electric motors using a minimum number of sensors is challenging, relevant and pertinent, and its solutions admit a wide variety of real applications in the development of engineering products and systems.

In this paper, a differential flatness control scheme with additional integral compensation of the output error is proposed for tracking tasks of position reference trajectories for direct current electric motors using angular position measurements only. The integral compensation is used to improve the robustness property of the output feedback dynamic controller with respect to constant load torque perturbations. An asymptotic differentiation approach with respect to time is proposed for on-line estimation of angular velocity and acceleration. The novel feature of this differentiator of signals is that it does not require any system mathematical model, which allows its use in any engineering system that requires the signal differentiation for its control, identification, fault detection, among other applications. Moreover, the satisfactory performance of the differentiation algorithm is verified for time-varying output signals generated by an uncertain chaotic nonlinear dynamical system of Lorenz. It is widely known that uncertain chaotic systems exhibit complex and unpredictable behavior and hence the state estimation is also an open and challenging research problem. The interested reader about chaotic systems and their control is referred to the works [8-10] and references therein. Some computer simulation results are provided to describe the satisfactory performance of the control scheme proposed in this paper as an alternative solution to the efficient angular position control problem of direct current electric motors.

2. Differential flatness-based control

Consider the mathematical model (1) of a DC permanent magnet motor with gearhead. Here L and R represent the inductance and the resistance of the armature circuit, respectively, k_e is the back electromotive force constant, k_m is the motor torque constant, n is the speed reduction ratio of the gearhead, J_m and J_0 are the inertia moments of the rotors of the motor and the gearhead, respectively, b_m and b_0 are the viscous damping coefficients of the motor and the gearhead, respectively.

$$\begin{aligned} L\frac{d}{dt}i &= -Ri - k_e n\dot{\theta} + u \\ \left(J_0 + n^2 J_m\right)\ddot{\theta} &= -\left(b_0 + n^2 b_m\right)\dot{\theta} + nk_m i - \tau_L \\ y &= \theta \end{aligned} \tag{1}$$

In this system, θ is the angular position of the output shaft of the gearhead, i is the electric current, u is the control voltage applied to the input terminals of the armature circuit, and τ_L denotes the constant load torque.

DC motor with gearhead is a differential flat system, with flat output given by the angular position of the output shaft of the gearhead, $y=\theta$. Hence all the state variables and the control input can be expressed in terms of the flat output and a finite number. System (1) then satisfies the input-output differential equation

$$y^{(3)} + \gamma_2\ddot{y} + \gamma_1\dot{y} = \gamma u - \frac{R}{J_{eq}L}\tau_L \tag{2}$$

with

$$\gamma_1 = \frac{n^2 k_m k_e + Rb_{eq}}{J_{eq}L}, \quad \gamma_2 = \frac{b_{eq}}{J_{eq}} + \frac{R}{L}, \quad \gamma = \frac{nk_m}{J_{eq}L},$$

where J_{eq} and b_{eq} denote the total inertia moment and the equivalent rotational viscous damping coefficient of the system, respectively, both reflected to output shaft of the gearhead.

From equation (2), one obtains the following differential flatness-based controller with integral compensation for tracking tasks of some angular position reference trajectory $y^*(t)$:

$$\begin{aligned} u &= \frac{1}{\gamma}\left(v + \gamma_1 \dot{y} + \gamma_2 \ddot{y}\right) \\ v &= y^{*(3)} - \beta_3 \ddot{e} - \beta_2 \dot{e} - \beta_1 e - \beta_0 \int_0^t e(\tau)d\tau \end{aligned} \quad (3)$$

By replacing the control law (3) into the input-output model (2), and differentiating the resulting expression once with respect to time, it is obtained the closed-loop tracking error dynamics, $e = y - y^*$,

$$e^{(4)} + \beta_3 e^{(3)} + \beta_2 \ddot{e} + \beta_1 \dot{e} + \beta_0 e = 0 \quad (4)$$

Therefore, the asymptotic convergence of the tracking error to zero can then be achieved selecting the design parameters β_i, i = 0, 1, 2, 3, such as the characteristic polynomial associated to the closed-loop tracking error dynamics (4) is a *Hurwitz* polynomial.

Since the controller requires information of the angular acceleration and velocity signals which are assumed to be unavailable due to considerations of cost reduction in the controller implementation, in the next section is applied an approach to differentiate signals with respect to time.

3. Differentiation of signals

In the synthesis of a real-time signal differentiation scheme with respect to time, it is assumed that the position output signal $y(t)$ can be locally approximated by a forth degree Taylor polynomial family as

$$y(t) = p_0 + p_1 t + p_2 t^2 + p_3 t^3 + p_3 t^4, \quad (5)$$

where all the coefficients p_i, i = 0, 1, …, 4, are completely unknown. Then, the position signal can be locally described as

$$\begin{aligned} \dot{y}_1 &= y_2 \\ \dot{y}_2 &= y_3 \\ \dot{y}_3 &= y_4 \\ \dot{y}_4 &= y_5 \\ \dot{y}_5 &= 0 \end{aligned} , \quad (6)$$

where $y_1 = y$, $y_2 = \dot{y}$, $y_3 = \ddot{y}$, $y_4 = y^{(3)}$, $y_5 = y^{(4)}$.

Hence, from (6) we propose the following state observer for asymptotic estimation of the acceleration and velocity signals:

$$\begin{aligned}
\dot{\hat{y}}_1 &= \hat{y}_2 + \beta_4\left(y_1 - \hat{y}_1\right) \\
\dot{\hat{y}}_2 &= \hat{y}_3 + \beta_3\left(y_1 - \hat{y}_1\right) \\
\dot{\hat{y}}_3 &= \hat{y}_4 + \beta_2\left(y_1 - \hat{y}_1\right) \\
\dot{\hat{y}}_4 &= \hat{y}_5 + \beta_1\left(y_1 - \hat{y}_1\right) \\
\dot{\hat{y}}_5 &= \beta_0\left(y_1 - \hat{y}_1\right)
\end{aligned} \quad (7)$$

Then, the estimation error dynamics is governed by

$$\begin{aligned}
\dot{e}_1 &= -\beta_4 e_1 + e_2 \\
\dot{e}_2 &= -\beta_3 e_1 + e_3 \\
\dot{e}_3 &= -\beta_2 e_1 + e_4 \\
\dot{e}_4 &= -\beta_1 e_1 + e_5 \\
\dot{e}_5 &= -\beta_0 e_1
\end{aligned} , \quad (8)$$

where $e_i = y_i - \hat{y}_i$, $i = 1, 2,\ldots,4$.

The characteristic polynomial of the observation error dynamics is then given by

$$P_O(s) = s^5 + \beta_4 s^4 + \beta_3 s^3 + \beta_2 s^2 + \beta_1 s + \beta_0 , \quad (9)$$

which is independent of the coefficients p_i of the Taylor polynomial expansion for $y(t)$. In this work, the following *Hurwitz* polynomial is proposed to select the observer gains:

$$P_{DO}(s) = (s + p_o)\left(s^2 + 2\zeta_o\omega_{no}s + \omega_{no}^2\right)^2 , \quad (10)$$

with p_o, ω_{no}, $\zeta_o > 0$.

The performance of the asymptotic observation scheme (7) was initially tested for the on-line differentiation of the time-varying output signal $y(t)$ generated by the uncertain chaotic system of Lorenz described by

$$\begin{aligned}
\dot{\eta}_1 &= a(\eta_2 - \eta_1) \\
\dot{\eta}_2 &= b\eta_1 - \eta_1\eta_3 - \eta_2 \\
\dot{\eta}_3 &= \eta_1\eta_2 - c\eta_3 \\
y &= \eta_1
\end{aligned} , \quad (11)$$

with $a = 10$, $b = 30$, $c = 1$, and initial conditions $\eta_1(0) = -10$, $\eta_2(0) = -10$, $\eta_3(0) = 20$.

Fig. 1 displays the time-varying output signal of the uncertain chaotic system of Lorenz (11). In Fig. 2 is shown the satisfactory differentiation with respect to time of the output signal. Therefore, the signal differentiation scheme (7) represents an alternative approach to get estimates of some derivatives requested for implementation of the position tracking controller (3).

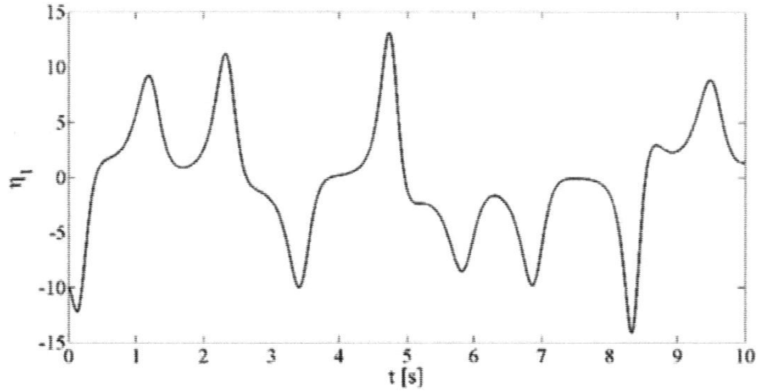

Figure 1. Output signal generated by an uncertain chaotic system of Lorenz.

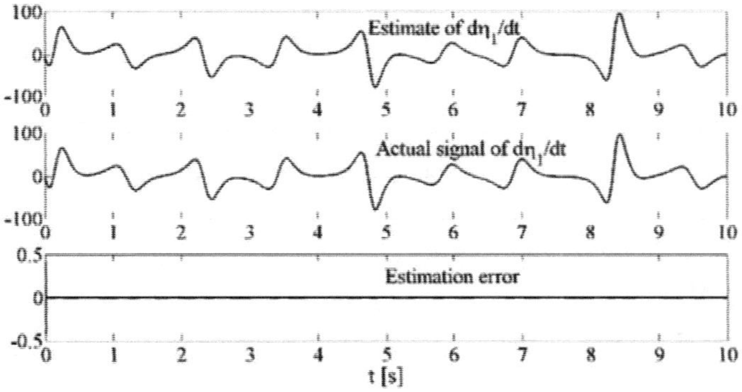

Figure 2. Estimated velocity of the output signal of an uncertain chaotic system of Lorenz.

4. Simulation results

Some computer simulations were performed to verify the performance of the angular position controller (3) using the differentiator of signals (7) for a DC motor with planetary gearhead characterized by the set of parameters: R = 2.5 Ω, L=0.612 H, k_m = 82.2 mNm/A, k_e = 82.3215 mV/rad/s, J_e = 82.3215 mV/rad/s, n = 81. The reference trajectory was specified to smoothly displace the output shaft of the gearhead from the rest position to the nominal operation position of $\pi/4$.

Figs. 3, 4 and 5 describe the closed-loop responses of the output angular position, electric current, and control voltage, when the system is subjected to the load perturbation torque given by $\tau_L = 2\left(1-e^{-5t}\right)$ Nm. The satisfactory tracking of the reference position trajectory can be observed

Journal of Physics: Conference Series **582** (2015) 012059 doi:10.1088/1742-6596/582/1/012059

in Fig. 3 in spite of load perturbation torque. Figs. 6 and 7 show the effective estimation of the velocity and acceleration signals using the differentiation scheme (7).

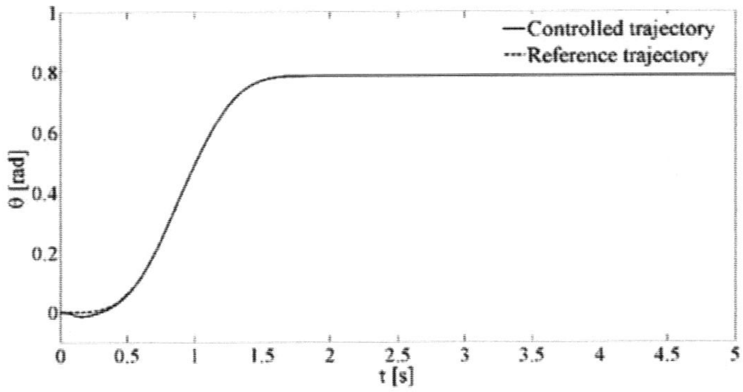

Figure 3. Closed-loop reference position trajectory tracking.

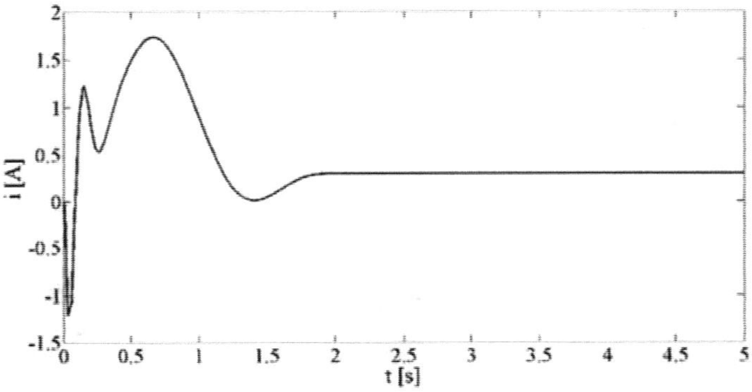

Figure 4. Closed-loop electric current signal.

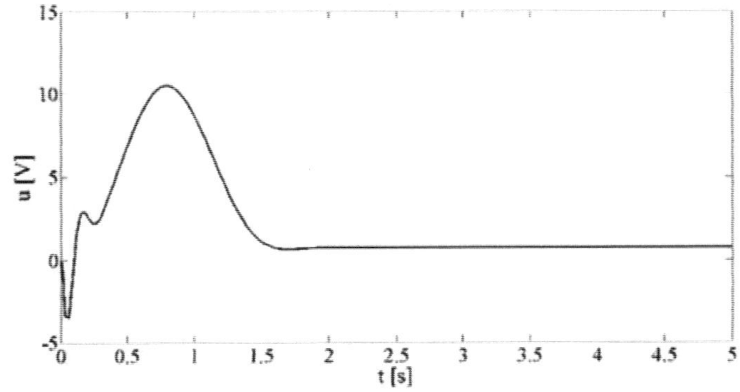

Figure 5. Control voltage applied to motor terminals.

Journal of Physics: Conference Series **582** (2015) 012059 doi:10.1088/1742-6596/582/1/012059

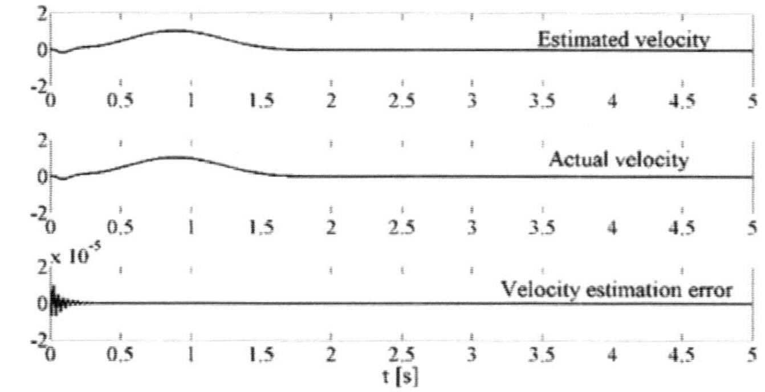

Figure 6. Estimation of the velocity signal.

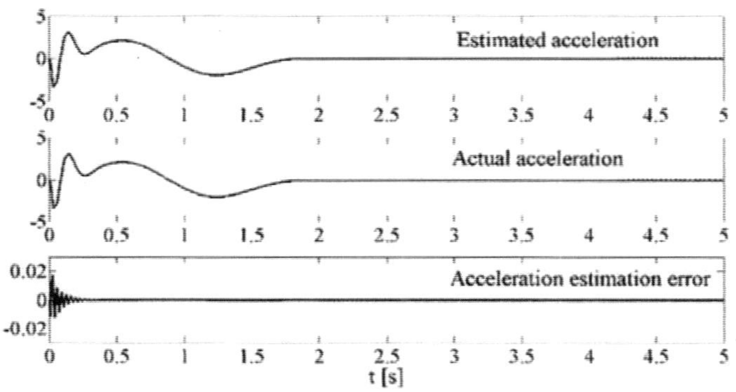

Figure 7. Estimation of the acceleration signal.

5. Conclusions

In this paper, a scheme to get estimates of time derivatives of some signal has been proposed for position tracking control tasks for DC electric motors using measurements of the output variable to be controlled. The main feature of the differentiation approach is that a mathematical model describing the system dynamics is not required. Hence, the real-time signal differentiation algorithm can be easily implemented to other applications of engineering system where unavailable time derivatives are requested. Moreover, the performance of the differentiator was also verified for time-varying output signals generated by an uncertain chaotic nonlinear dynamical system of Lorenz showing satisfactory results. For the synthesis of the signal differentiator, the bounded output signal was locally approximated by a family of forth order Taylor polynomials with unknown coefficients. In addition, the proposed output feedback tracking controller exploits the differential property exhibited by the system. Additional integral compensation of the output error was used in the controller to improve its robustness property against load constant torque perturbations. The computer simulations results manifest a good controller performance using the differentiation of signals. Thence, one can conclude that the tracking control scheme using the presented signal differentiator represents a very good option for position control tasks planned for a DC motor.

References

[1] Alyaqout S F, Papalambros PY and Ulsoy AG 2011 Combined Robust Design and Robust Control of an Electric DC Motor *IEEE/ASME Transactions on Mechatronics* **16** 574-582.

[2] Li Z, Chen J, Zhang G and Gan MG 2011 Adaptive robust control for DC motors with input saturation *IET Control Theory and Applications* **5** 1895–1905

[3] Caruntu CF and Lazar C 2012 Robustly Stabilising Model Predictive Control Design for Networked Control Systems with an Application to Direct Current Motors *IET Control Theory and Applications* **6** 943–952.

[4] Zhou Y, Soh YC and Shen JX 2013 Speed Estimation and Nonmatched Time-varying Parameter Identification for a DC Motor with Hybrid Sliding-mode Observer *IEEE Transactions on Industrial Electronics* **60** 5539-5549.

[5] Rigatos GG, 2012 A Derivative-Free Kalman Filtering Approach to State Estimation-Based Control of Nonlinear Systems *IEEE Transactions on Industrial Electronics* **59** 3987-3997.

[6] Murtaza G and Bhatti AI 2012 Control of DC Motors Using Siding Mode, *Proc. of 9th International Bhurban Conference on Applied Sciences & Technology (IBCAST)* (Islamabad, Pakistan, 9-12 January 2012) pp. 37-42.

[7] Chung C, Lee T, Tien C and Hsu C 2011 Adaptive Wavelet Neural Network Control for DC Motors Via Second-Order Sliding-Mode Approach *Proc. of the 2011 International Conference on Machine Learning and Cybernetics* (Guilin, China, 10-13 July, 2011) pp. 1174-1179.

[8] Zhang X, Liu X Zhu Q 2014 Adaptive Chatter Free Sliding Mode Control for a Class of Uncertain Chaotic Systems *Applied Mathematics and Computation* **232** 431-435.

[9] Senouci A and Boukabou A 2014 Predictive Control and Synchronization of Chaotic and Hyperchaotic Systems Based on a T–S Fuzzy Model *Mathematics and Computers in Simulation* **105** 62-78.

[10] Chen G 1999 *Controlling Chaos and Bifurcations in Engineering Systems* (Boca Raton: CRC Press).

Journal of Physics: Conference Series **582** (2015) 012060 doi:10.1088/1742-6596/582/1/012060

Double vacancy on BN layer: A natural trap for Hydrogen Molecule

J S Arellano

Departamento de Ciencias Básicas, Universidad Autónoma Metropolitana Azcapotzalco. Av. San Pablo 180, Colonia Reynosa Tamaulipas, 02200 México D.F. México.

E-mail: jsap@correo.azc.uam.mx

Abstract. A pair of vacancies, one of boron and other of nitrogen atom at a flat layer becomes a natural trap to capture a hydrogen molecule at the center of the cavity defined by the empty space left by the lack of a nitrogen and a boron atom at the perfect BN layer formed by 16 N atoms and 16 B atoms. The adsorption of the hydrogen molecule is compared with the equivalent graphene layer with a pair of carbon vacancies. The little increase in the BN cell parameter respect to the graphene cell parameter, besides the differences between N, B and C atoms helps to explain the easier adsorption on the defective BN layer.

1. Introduction

Before the great interest on novel materials containing carbon atoms in systems such as fullerenes, nanotubes and other graphitic systems such as the lonely graphene layer, there were many works about the nature of graphite, one of the allotropic forms of carbon. Some years ago we reported the adsorption of the hydrogen molecule above the graphene layer [1]. The main purpose of this work is to study the adsorption of one hydrogen molecule due to a pair of vacancies on the BN layer. A big difference appears when comparison is made for the adsorption of one hydrogen molecule above an equivalent pair of carbon vacancies on a graphene layer with the same amount of atoms as the BN defective layer. With the work of Geim and Novoselov [2] it was pointed out the great importance of the graphene 2D system. Applications now are everywhere and many of them are very innovative ones. In a publication [3] there is a resume about the number of graphene works and graphene patents for the year 2010. The number of works and patents has been increasing the recent years. To show some other application of the graphitic systems it can be pointed out that there are new desalination water systems using the filtering and porous properties of the graphene [4]. On the other side, the possibility to detect and to eliminate or remove defects in graphene has been announced since 2010 [5]. If this can be applied to control the number of double B-N vacancies in the BN layer, the hydrogen adsorption reported here could be controlled in principle.

Journal of Physics: Conference Series **582** (2015) 012060 doi:10.1088/1742-6596/582/1/012060

2. Computer simulations

The results given here were obtained using the free software package Quantum espresso, in particular program PWSCF v.4.3.2 [6]. The package is formed by open source computer code for electronic structure calculations and materials modeling at the nanoscale. As other computer codes, it is based on density functional theory, planes waves and also on pseudopotentials. To simulate the N, B and H atoms, the corresponding pseudopotentials N.pw-mt_fhi.UPF, B.pw-mt_fhi.UPF and H.pw-mt_fhi.UPF were used for all the calculations.

Before [1] it was used a cutoff energy of 40 Ry. Now, while making proofs for the energy convergence for the BN layer, it was noticed the necessity to increase the corresponding parameter, ecutwfc to 70 Ry for the BN system. It was then necessary to calculate for graphene and BN layers with the same values for other quantities refereeing to convergence for example.

3. Results

For the graphene layer it was obtained the value 4.6117 a.u. for the cell parameter. The value obtained for the BN layer is 4.67 a.u., a little bigger than for the layer with only carbon atoms. There is great similitude for both layers because they form hexagonal cavities. The principal difference as is shown in figure 1, boron and nitrogen atoms alternate around any hexagonal cavity.

Those values for the cell parameters were obtained after making tests increasing the cutoff energy from 30 or 40 until 70 Ry. Also several tests were done to conclude that with 26 k points the calculations are well converged and with the required convergence criteria.

For both layers it was designed a 4x4 supercell, with a total of 32 atoms, all carbon atoms for the graphene layer and 16 boron and 16 nitrogen atoms for the BN layer.

a. Results for graphene layer:

To obtain well converged energies for the graphene layer it was enough to use 40 Ry for the cutoff energy. For the lonely graphene layer, the total energy was -365.74482262 Ry.

b. Results for BN layer:

To obtain well converged energies for the BN layer it was used 70 Ry for the cutoff energy. For the lonely BN layer, the total energy was -429.29958813 Ry.

Next results are given for the total energies when a pair of vacancies is created on the graphene layer and on the BN layer.

c. Results for the double vacancy graphene layer:

The total energy is -341.77284704 Ry, when two carbon atoms along the "diameter" of the perfect hexagonal cavity are pulled out. Figure 2 (a) is a picture of the mentioned system. The distance for those C atoms, in the entire system (Figure 1 (a)), was 5.3251 a.u.

d. Results for the double vacancy on the BN layer:

The total energy is -401.71162699 Ry, when two atoms, boron and nitrogen are omitted respect to the perfect BN layer. Figure 2 (b) shows the corresponding vacancies. In the perfect layer (Figure 1 (b)), those atoms were apart one from another 2.66625 a.u. As can be compared this distance is only around one half of the corresponding distance for the lacking carbon atoms at the defective graphene layer. This makes clear a bigger empty space in the BN double vacancy in comparison with the double vacancy for the graphene layer.

e. Adsorption of hydrogen molecule above the double vacancies layers:
Once both defective layers were studied, the next step was to study the adsorption of only one hydrogen molecule above the double vacancies. Figure 3 (a) is a perspective of the final configuration for the hydrogen molecule relaxed above the defective graphene layer. The equilibrium distance between hydrogen atoms is d (H-H) =1.4374 a.u. and its position is 5.43 a.u. above the center of the defective graphene layer. This result agrees with the result given in reference [1].

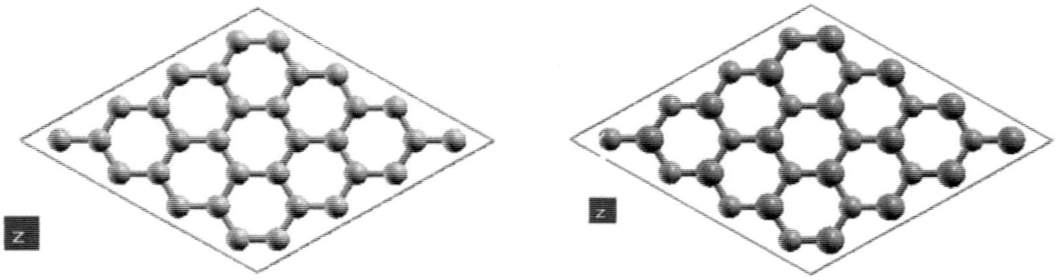

(a) (b)

Figure 1. (a) Graphene layer with 32 carbon atoms. (b) BN layer with 32 atoms. N atoms are smaller than B atoms. The unit cell parameter for both systems is the distance from one atom to the second neighboring atom. The unit cell parameter is 4.6117 a.u. and 4.67 a.u. respectively for graphene and BN layer.

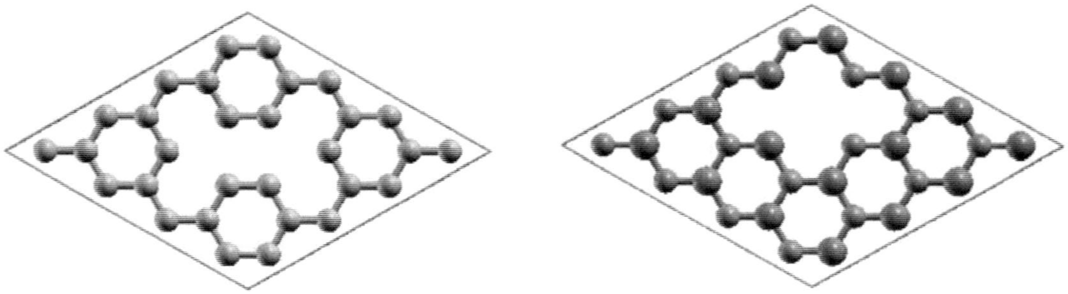

(a) (b)

Figure 2. (a) Graphene layer with 30 carbon atoms. The double vacancy is created at the center of the layer. (b) BN layer with also 30 atoms. The vacancy is created taking out of the perfect BN layer shown on Figure 1 (b), one N atom and one B atom that were initially on a horizontal line.

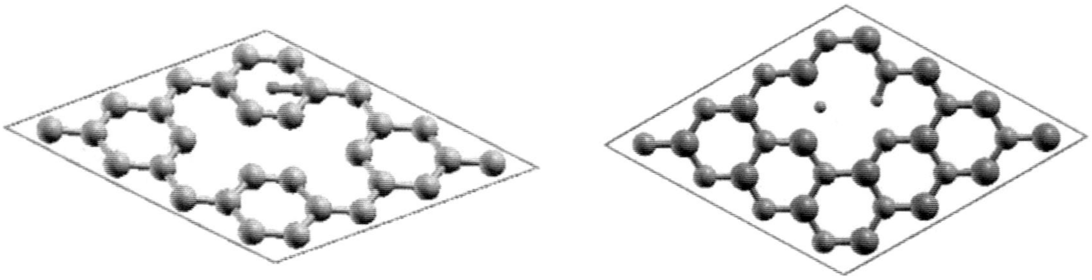

(a) (b)

Figure 3. (a) Hydrogen molecule adsorbed above the double vacancy graphene layer. The molecule is about 5.43 a.u. above the graphene layer. The H-H distance 1.4374 a.u., close value to 1.48 a.u., the calculated value for the lonely molecule. (b) For the BN layer with the double vacancy, the hydrogen molecule is broken and one of the hydrogen atoms makes a bond with the closer N atom.

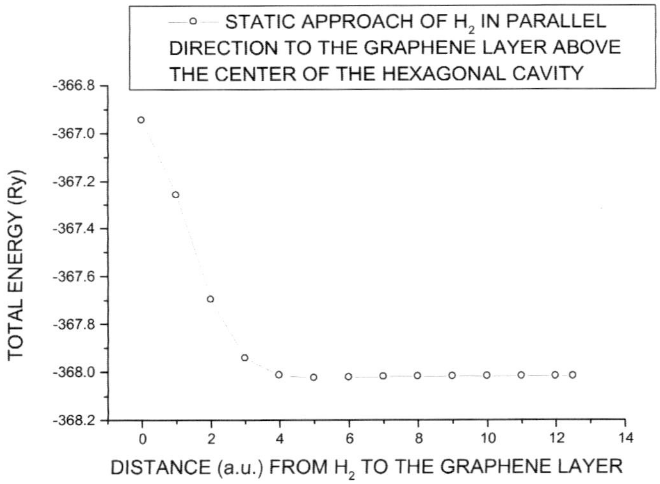

Figure 4. Total energy curve for static calculations with fixed distance H-H equal to 1.48 a.u., the equilibrium interatomic distance obtained for the isolated hydrogen molecule. The maximum separation between the molecule and the perfect graphene layer is z = 12.5 a.u., half of the cell parameter in the perpendicular direction to the layer. The energy difference for z = 12.5 and the minima, around 5.0 a.u. is 0.00639 Ry, close to 0.087 eV.

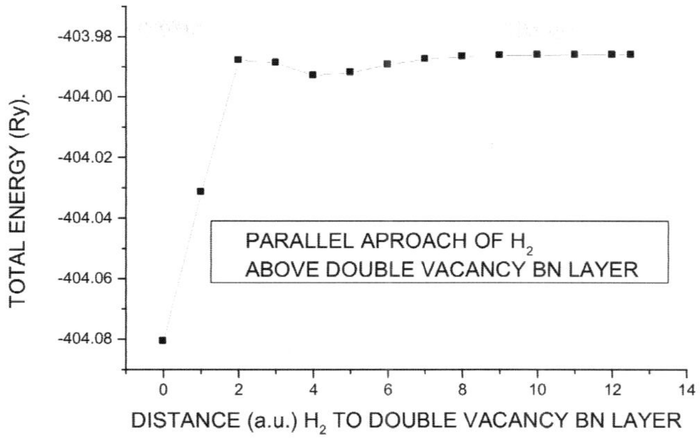

Figure 5. Total energy curve for static calculations with fixed distance H-H equal to 1.48 a.u. The more negative energies for distances less than 2 a.u. means the double vacancy on the BN layer favors the rupture of the hydrogen molecule and at least the bonding of one of the hydrogen atoms with the N atom for example, as shown in figure 3 (b).

In contrast, when the hydrogen molecule is placed above the double vacancy of the BN layer, the most stable configuration is with the hydrogen atoms adsorbed at the same plane of the BN layer and at the middle point of the double vacancy. This was confirmed by a series of static calculations putting the molecule at different distances above the layer, and also making relaxations of the molecule putting it at different distances above the BN layer. All of them finish with the same result: the hydrogen atoms are adsorbed at the center of the double vacancy. This configuration is shown in figure 3 (b). In that panel, the right hydrogen atom makes a bond with an N atom. Their interatomic distance is 2.09 a.u. The left hydrogen is at 2.65 or 2.69 a. u. respect to the boron atom almost above and below this second hydrogen atom, respectively.

4. Discussion

Results show that the hydrogen molecule can be attracted and adsorbed by the BN double vacancy layer at a very short distance, close to 2.0 a.u. This distance is much shorter than the distance close to 5.07 a.u. for the adsorption of the hydrogen molecule by the perfect graphene layer [1] or the distance 5.43 a.u. in the case of the pair of vacancies on the graphene layer. The rupture of the molecule when approaching the layer can be used although there were imperfections in some BN layer synthesized by any method, in such a way that the vacancies can be useful for the purpose of adsorbing hydrogen in a porous material without the presence of any other atomic species. The hydrogen molecule adsorption above the lonely B vacancy or the N vacancy alone also has been studied. The static approach of the hydrogen molecule above one or another of the vacancies for fixed distance H-H = 1.48 a.u. gives very similar results for the total energy curve as a function of the distance from the molecule to de BN defective layer. In both cases the hydrogen molecule is adsorbed at a distance between 5 and 6 a.u. This suggest that carbonaceous planar systems such as the graphene layer with isolated carbon vacancies could be not enough to be able to absorb a good

Journal of Physics: Conference Series **582** (2015) 012060 doi:10.1088/1742-6596/582/1/012060

quantity of molecular hydrogen but perhaps a pair of closer vacancies as shown here for the BN layer could be considered attractive to study experimentally the possibility that hydrogen could be adsorbed in an easier way in such a system.

5. Conclusion

The main conclusion is that a porous system like the BN layer with distributed pairs of B-N vacancies all around the defective material could be a simple and cheaper porous system with a good capacity to adsorb and perhaps to storage an attractive quantity of molecular hydrogen. The proposal is that some experimental groups could study the hydrogen adsorption and desorption hydrogen process to compare with the computer simulations discussed above and compare with other porous systems that are being evaluated as good candidates for hydrogen storage.

References

[1] Arellano J S, Molina L M, Rubio A, and Alonso J A. J. Chem. Phys. 112, 8114, 2000 .
[2] Geim A K and Novoselov K S 2007 *Nature Mater.* **6** 183.
[3] Brumfiel G. Nature 140, V 488, 2012.
[4] http://www.sciencedaily.com/releases/2014/02/140225122504.htm.
[5] http://www.sciencedaily.com/releases/2010/06/100606162131.htm.
[6] P Giannozzi et al. J. Phys. Condens. Matter 21 395502 (2009); URL http://www.quantum-espresso.org".

Journal of Physics: Conference Series **582** (2015) 012061 doi:10.1088/1742-6596/582/1/012061

Image processing applied to measurement of particle size

Fabio Vega[*,1]**,Willian Lasso**[1]**, Cesar Torres**[1]
[1]Laboratorio de Óptica e Informática, Universidad Popular del Cesar, km 4 Via patillal, Valledupar-Colombia

[*]E-mail: fabiovega@unicesar.edu.co

Abstract. Five different types of aggregates have been analyzed, and the size of particles on samples immersed in distilled water as silicon dioxide, titanium dioxide, styrenes and crushed silica particles is made; an attempt at applying the digital image processing (DIP) technique to analyze the particle size, we developed a system of measures microparticles using a microscope, a CCD camera and acquisition software and video processing developed in MATLAB. These studies are combined with laser light using measurements by diffractometry and obtain calibration in the system implemented, in this work we achievement measurement particle size on the order of 4 to 6 micrometers. The study demonstrates that DIP is a fast, convenient, versatile, and accurate technique for particle size analysis; the limitations of implemented setup too will be discussed.

1. Introduction

This paper contains the principal fundaments of a digital measurement model for size particles in order of the micrometers. In this investigation the issue over the measurement of micrometric particles using digital image processing and whose main purpose is to demonstrate that it is possible to measure very small particles optimally, efficiently and easily.

The basic ideas and principles of digital image processing are established for many years, but were not carried out due to lack of computers. With the advent of computers and high-capacity memory, it was natural to begin to develop this field. One of the first places they started making digital processing was at the Jet Propulsion Laboratory in 1959, with the aim of improving the images sent by the rockets. The results obtained in a relatively short time were so impressive that very soon the application of the method is extended to other fields [1].

Finally we can say that the image processing is studied through optical microscope and a simple purchasing system and computer image processing, the main feature of this type of measurement is the ability to obtain results in real time.

2. Procedure

The layout of the paper is as follows: In Section 2, we present the detailed underlying theory of the diffraction method as applied to measurement calibration. This is followed by a brief description of the simulation technique in Section 3. Details of simulation algorithm and simulation results are described respectively, in Sections 4 and 5. Finally some concluding remarks are made in Section 6.

2.1.Diffraction grating

A diffraction grating is a binary repeating set of equally spaced diffracting elements (apertures or obstacles), which produce periodic alterations in the phase and / or amplitude of the light wave studied. The grating can separate a beam of light incident on it in their colors or constituent wavelengths, this phenomenom can also will be measured and obtain the differents wavelengths. We distinguish two types of diffraction gratings: grating reflection and grating transmission . The gratings are constructed by reflection stripes equispaced parallel recording on the polished surface of a metal. The light is reflected in the projections between the marked lines. In gratings for transmission, the parallel lines are written on a glass plate, and the light passes through the transparent spaces between said stripes [2]. In many of college texts the constant (d) is mistakenly referred as the distance of separation between slits, when this constant in reality correspond to the spatial period of the diffraction grating, which is the distance of the slit separation distance over which a new slot (zone begins black grid), so that it can fulfill the equation that relates to the lattice constant, which is the number of lines per millimeter which can be observed in equation (1), a diffraction grating can be seen in Figure 1.

$$C_{gra} = \frac{10^{-3}}{p}. \tag{1}$$

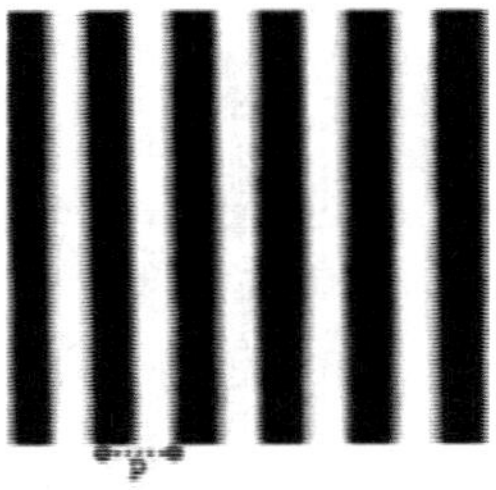

Figure 1. Spatial period of diffraction grating

3. Scheme for experimental measurement of period the grating

The spatial period of the diffraction grating has been experimentally calculated using the setup implemented at the optical bench and is shown in Figure 2.

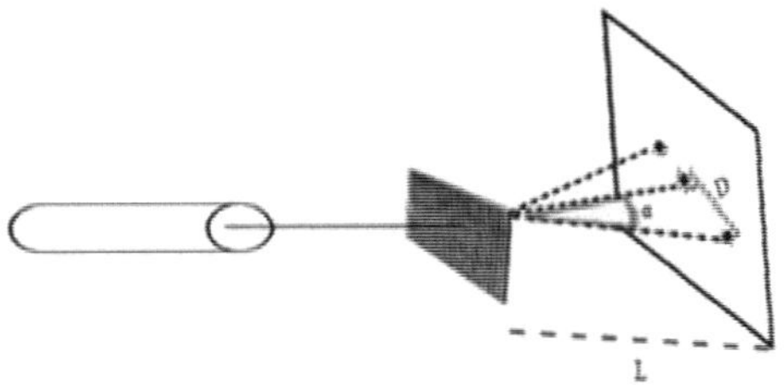

Figure 2. Experimental setup for measurement of spatial period in the grating

The experimental setup consists of a conventional He-Ne laser with a wavelength (λ = 633nm) , a diffraction grating to calculate the spatial period, the theory used for this purpose is the Bragg's Law, this relationship is showed in equation (2) , and consist in an observation plane at a distance L of diffraction grating; in this plane an interference pattern is formed, at correspondence with the

Fourier Theory. The spatial period was referred in terms of the wavelength in accordance to equation (2.1). Then the spatial period has been calculated in the follow form [6].

$$n\lambda = p\sin\alpha, \tag{2}$$

$$p = \frac{\lambda}{\sin\alpha}. \tag{2.1}$$

For obtain the angle α, trigonometric functions has been used in terms of the tangent of the rectangle triangle formed by the distance of the observation plane, to the grating, in other words the distance L and Fourier plane in terms of the separation distance of the maximum intensity of the first-order diffraction (n = 1), taking account the figure 2, and equation (3); then, the angle α is calculated by:

$$\alpha = \tan^{-1}\left(\frac{D}{L}\right). \tag{3}$$

Several measurements obtained, for every cases the distance L was modified and the distance D was experimentally obtained; and then, the angle αwas calculated with the help of equation (3). In Table 1 the data obtained are show.

Table 1. Experimental measuring obtained for spatial grating period.

L (cm)	D (cm)	$\alpha = \tan^{-1}\left(\frac{D}{L}\right)$
108,50	1,45	0,7656583863
130,00	1,65	0,7271766166
150,00	1,85	0,7066121209
175,00	2,25	0,7366194349
200,00	2,55	0,7304816075
220,00	2,80	0,7291796420
230,00	2,90	0,7223867662
Average α		0,7311592249
n=1 λ=633nm α=0,7311592249		$p = 4{,}96051\times10^{-5}$

The standard error of measure has been calculated, using the equation (4) [7], the measurement error obtain was approximately 0.667%. Finally with these data has proceeded to calculate the lattice constant of the reference grating with the aid of Table 1 and Equation 1, which gives a value of 201 lines per millimeter

$$S_{\bar{x}} = \frac{S}{\sqrt{n}}, \tag{4}$$

where S = standard deviation and n = sample size.

4. Software design and graphics interface

A program of acquisition and processing video in real time was been developed using the mathematical tool MATLAB. This software consists of video imported by the card type ENLTV-FM3 [8], and an export of video GeForce 7200 GS-E [9 development]; the captured image using the camera sensor is displayed and digitized [4] in a Independent monitor. Algorithms for different video capture analog sources (CCD cameras, video camera are output RCA Protocol) and USB digital cameras were utilized. The video capture devices (CCD and USB camera) should be placed in the position of the microscope eyepiece, having withdrawn this previously [3], the assembly is as shown in Figure 3.

Figure 3. Experimental setup for acquisition and video processing

Figure 4. Graphical interface designed

4.1 System calibration

The calibrate of the system and the optimal measures using the developed software we proceeded to observe trough of different microscope objectives [5] the diffraction grating used in Section 3; is the reference pattern of measurement using the assembly shown in Figure 3. Then proceeds to measure in pixels the spatial period of the grating which is observed through the microscope, different measures for different images are shown in Figure 5. a mathematical relationship was used for the calculate of the multiplication factor for converting the image size in pixels to its corresponding real value in microns, table 2 shows conversion factors for different microscope objectives.

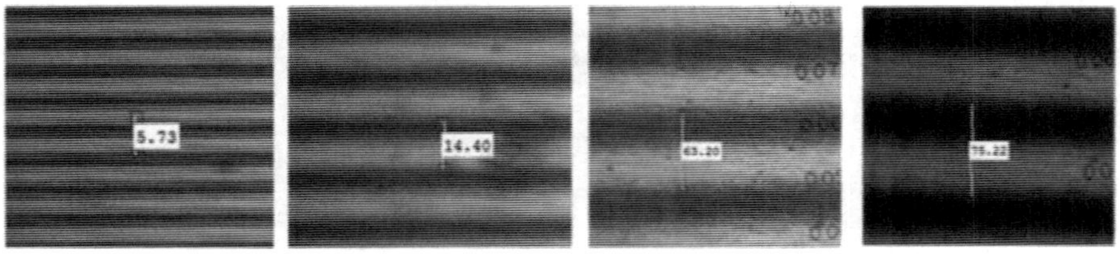

(a) (b) (c) (d)

Figure 5. Reference pattern observed with different microscope objetives a. observed 5X, b. Observed at 10X, 40X Observed c, d. Observed at 100X)

Table 2. Process of converting pixels at distance micrometric

	Objetive 5X	Objetive 10X	Objetive 40X	Objetive 100X
Number of pixels in pattern	5.73	14.4	63.2	75.22
Conversión factor	8.657085515 X 10-6	3.444798611 X 10-6	7.848908228 X 10-7	6.594668171 X 10-7

5. Results

The results obtained finally taking account the measuring of the particle size in pixels [10] and then multiplied by the conversion factor according to the objective pattern that is used to observe, the conversion factors experimentally obtained are showed in Table 2. And finally was developed an algorithm on the computer that allows to select the conversion factor on the menu of the interface graphical and then processing distance between pixels where the user selects the microscope objective that is used to view the sample (see Figure 4.).

For particular application we proceeded to measure the core of a standard optical fiber multimode fabricated by the Newport company , in this process; objectives different of conventional microscope has been used to compared our experimental results with the specification sheet of optical fiber given for the Newport company , which in its specification detailing the extent of 100 μ m for core and 140 μ m for the blindaje . Experimental setup for measures in optical fiber using our system can be showed in Figure 6.

Journal of Physics: Conference Series **582** (2015) 012061 doi:10.1088/1742-6596/582/1/012061

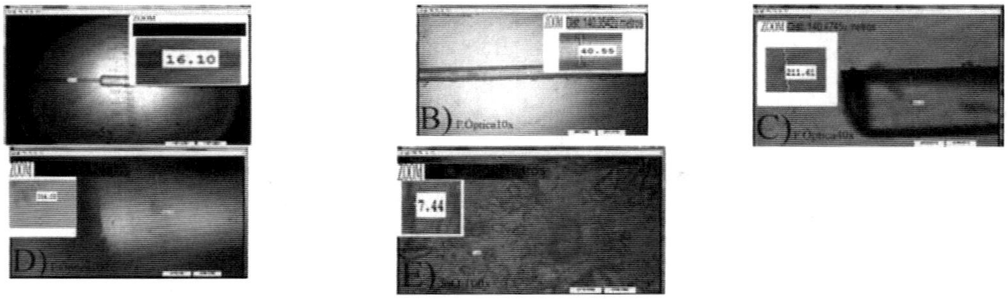

Figure. 6 Images of experimental measurement over geometrical parameters of Optical Fiber multimode of Newport company (a, b, c, d) and Silicon Dioxide in distilled water (e).

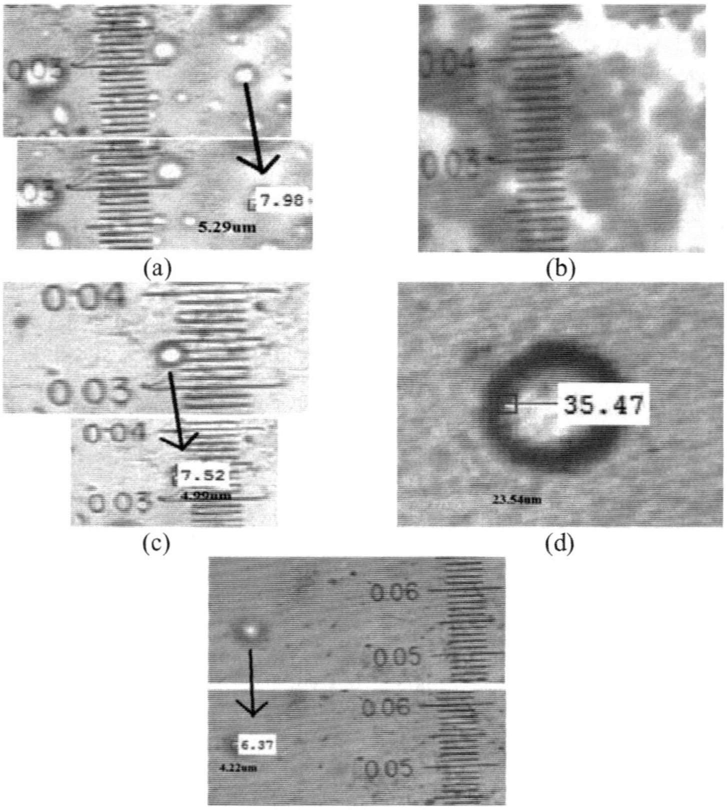

Figure. 7 Observed particles objective 40X: (a).Silicon dioxide more distilled water, measuring 5.29μm in small spherical particle observed. (b) Titanium dioxide more distilled water. (c) Styrene spheres, measuring 4.99μm in small spherical particle observed. (d) Fat globules, measuring 23.54μm in small spherical particle observed. (e) Silica spheres, measuring 6.37μm in small spherical particle observed.

Also, different size particles are measured as fat globules, silicon dioxide, spheres of silica, titanium dioxide, this particles was addition distilled water, these particles has distilled water added to them as a means of immersion. These microparticles are of special interest since it is intended to work in an optical trap using optical fibers and require to know the size of the particles, which partly determines the viability of the trap, the work is under development in the optical e informatics group of popular of Cesar university, Particle measurements can be seen in figure.7.

6. Conclusion

In this paper the design of a hardware and software system that enables the measurement of micrometer particles through digital image processing, based on a optical calibration method (He-Ne laser and diffraction by grating) the physical principle was based experimentally in the concept of the diffraction lattice constant, which is used as yardstick for measurements in microscopic images; in this method the standard error calculated is approximately of 0.667%, and basically is introduced by the manual measurement of researchers. The system captures the image, and then parameterized in pixels and using simple math automated in the software and in accordance with the objective that this under a microscope to observe the sample is multiplied by the conversion factor, and these data finally produced the size of the microparticles observed in the program in real time on the scale of micrometers, the multimode optical fiber of Newport company was used for a particular application, which guarantees through its specifications of design and factory the correct and respective dimensions of the core and blindaje. In this paper using our method is proved that with the use of digital image processing tool of easy form can be obtain micrometric measurements with high precision.

On the basis of experiments with non-biological samples: silicon dioxide, titanium dioxide, styrene and crushed silica spheres, all immersed in distilled water (Figure. 7); It is workable mounting optical traps with silicon dioxide, styrene and crushed silica because the shapes of the particles are approximately spherical and the size is less than or equal (= <5 micron) with respect to those embodied in researches related to the thematic [11].

References

[1] M. Antuono, B. Bouscasse, A. Colagrossi, S. Marrone, A measure of spatial disorder in particle methods, Computer Physics Communications, Available online 16 June 2014, ISSN 0010-4655, http://dx.doi.org/10.1016/j.cpc.2014.06.008.

[2] U. Zafar, C. Hare, A. Hassanpour, M. Ghadiri, Drop test: A new method to measure the particle adhesion force, Powder Technology, Volume 264, September 2014, Pages 236-241, ISSN 0032-5910, http://dx.doi.org/10.1016/j.powtec.2014.04.022.

[3] Hartati, Sri, Agus Harjoko, and Tri Wahyu Supardi. "The Digital Microscope and Its Image Processing Utility." TELKOMNIKA Telecommunication, Computing, Electronics and Control 9.3 (2013): 565-574.

[4] Kılıç, Varlık, and Bülent E. Platin. "Determination of the Position and Orientation of Rigid Bodies by Using Single Camera Images." Proceedings of the World Congress on Engineering. Vol. 1. 2007.

[5] Olivier Hénard, Change of measure in the lookdown particle system, Stochastic Processes and their Applications, Volume 123, Issue 6, June 2013, Pages 2054-2083, ISSN 0304-4149, http://dx.doi.org/10.1016/j.spa.2013.01.015.

[6] Juan M. Vilardy,Jorge Salleg,Ricardo Torres,Cesar Torres Moreno,OBTENCIÓN EXPERIMENTAL Y SIMULACIÓN DIGITAL DE LA LEY DE BRAGG POR MEDIO DE LA ÓPTICA DE FOURIER,REVISTA COLOMBIANA DE FÍSICA, VOL. 38, No. 2, 2006.

[7] Pieter-Jan De Temmerman, Eveline Verleysen, Jeroen Lammertyn, Jan Mast, Semi-automatic size measurement of primary particles in aggregated nanomaterials by transmission electron

microscopy, Powder Technology, Volume 261, July 2014, Pages 191-200, ISSN 0032-5910, http://dx.doi.org/10.1016/j.powtec.2014.04.040.

[8] E. Hamzeloo, M. Massinaei, N. Mehrshad, Estimation of particle size distribution on an industrial conveyor belt using image analysis and neural networks, Powder Technology, Volume 261, July 2014, Pages 185-190, ISSN 0032-5910, http://dx.doi.org/10.1016/j.powtec.2014.04.038.

[9] Teimour Tajdari, Mohd Fua'ad Rahmat, Norhaliza Abdul Wahab, New technique to measure particle size using electrostatic sensor, Journal of Electrostatics, Volume 72, Issue 2, April 2014, Pages 120-128, ISSN 0304-3886, http://dx.doi.org/10.1016/j.elstat.2013.12.009.

[10] Corkidi, Gabriel, Leobardo Serrano-Carreón, and Enrique Galindo. "Análisis digital de imágenes para la caracterización microscópica de parámetros críticos en la producción fermentativa de metabolitos secundarios." Revista Colombiana de Biotecnología 10.1 (2008): 75-93.

[11] Cynthia Jensen-McMullin, Henry P. Lee, and Edward R. Lyons.Demonstration of trapping, motion control, sensing and fluorescence detection of polystyrene beads in a multi-fiber optical trap.Optics Express, Vol. 13, Issue 7, pp. 2634-2642 (2005), http://dx.doi.org/10.1364/OPEX.13.002634.

Journal of Physics: Conference Series **582** (2015) 012062 doi:10.1088/1742-6596/582/1/012062

Crossover scaling evaluation in mixed correlated signals by means of Detrended Fluctuation Analysis

C. R. Martínez-García[1], I. Reyes-Ramírez[2], F. Angulo-Brown[1], L. Guzmán-Vargas[2]

[1]Departamento de Física, Escuela Superior de Física y Matemáticas, Instituto Politécnico Nacional, Edif. No. 9 U.P. Zacatenco, México D. F., 07738, México

[2]Unidad Profesional Interdisciplinaria en Ingeniería y Tecnologías Avanzadas, Instituto Politécnico Nacional, Av. IPN No. 2580, L. Ticomán, México D.F. 07340, México

E-mail: ireyesram@hotmail.com

Abstract. In this work we study the scaling behavior of signals constructed by a class of processes with "intrinsic trends" and correlated noises. We focus our attention in evaluating the appearence of a crossover in the scaling exponents obtained by means of the Detrended Fluctuation Analysis. In particular, we evaluate the conditions of the trend which leads to a crossover where the scaling exponent from small scales (α_s) is smaller or larger than the corresponding exponent from large scales (α_l). We find that decreasing trend leads to $\alpha_s > \alpha_l$, whereas an increasing trend results in $\alpha_s < \alpha_l$. We also find that when one introduces correlated noise into a decreasing or increasing trend map, there is a crossover which separates two regions with exponents related to the trend mainly over short scales for decreasing trends, while over large scales, the scaling exponent resembles the behavior of the increasing trend.

1. Introduction

Many complex real data from different systems are characterized by scaling properties which are often related to the presence of long-range correlations [2,4,13]. This behavior has been observed in systems ranging from physiology to ecology and from seismology to economy[1,8,11,12]. A special case is observed when correlations follow a power-law scaling behavior which permits to characterize the system by a single exponent. In many other cases, we can also find that the scaling is not expressed by one exponent, and two or more exponents are necessary to characterize the fluctuations[1,6,11].These scaling properties also display changes as a function of the time, with transitions between different levels of correlations. The characterization of these transitions together with the presence of a crossover scaling behavior are important to understand the underlying dynamic separated from external trends and artifacts [3,5]. During past decades, several methods have been proposed to evaluate the presence of correlations in complex and irregular signals, ranging from spectral analysis to wavelets. One method which has become the standard for the evaluation of long-range correlations in noisy nonstationary records, is the detrended fluctuation analysis (DFA)[9]. The DFA method has the characteristic of assuming the trend and signal

Journal of Physics: Conference Series **582** (2015) 012062 doi:10.1088/1742-6596/582/1/012062

decomposition to evaluate the fluctuations. The way the mean fluctuations $F(n)$ scales with the box size n, reveals the presence of correlations. In some cases, one find that more than one scaling regime can be defined, which are separated by a particular crossover scale. The appearance of such crossover scale are often related to the influence of a "external trend", affecting the scaling behavior of true long-range correlated signals. However, in many other real or simulated situations, the emergence of a crossover in the scaling, is mainly related to "intrinsic mechanisms" rather than external influences, and thus carrying out important information about the correlations in the system. In this context, an appropriat evaluation of the crossover phenomena together with a detailed inspection of the scaling behavior can lead more reliable results of the presence of correlation in complex signals. In this paper, we analyze the appearance of crossovers in signals constructed with correlated noises and intrinsic trends, which were classified as increasing and decreasing trends. The construction of the intrinsic trend is based on a modified autorregresive process for two cases: exponential process and power law process. In particular, we evaluate the conditions which lead to a crossover where the scaling exponent from small scales (α_s) is less or larger than the corresponding exponent from large scales (α_l). It is found that (i) a correlated noise with a decreasing trend leads to $\alpha_s > \alpha_l$ whereas, (ii) correlated noise with an increasing trend results in $\alpha_s < \alpha_l$. We observe that for case (i), α_s is mainly related to the DFA scaling exponent from the trend alone, while α_l is close to the correlated noise input. In contrast, for case (ii) we find that α_s depends on the correlated noise and α_l is dominated by the scaling exponent of the trend. This paper is organized as follows: in section 1.1 a briefly description of DFA method is made, 1.2 mentions the way the correlated noises were generated; in section 2, we describe the main rule to construct the noises with intrinsic trend for both exponential and power law trends proposed. Finally, conclusions of this work are discussed.

1.1 Detrended Fluctuation Analysis

The DFA method was introduced to quantify long-range correlations in the heartbeat interval time series and DNA sequences [9-11]. The method is briefly described as follows: First, we integrate the original time series to get, $y(k) = \sum_{i=1}^{k} [x(i) - x_{ave}]$, the resulting series is divided into boxes of size n. For each box, a straight line (or a higher order polynomial) is fitted to the points, $y_n(k)$. Next, the line points are subtracted from the integrated series, $y(k)$, in each box. The root mean square fluctuation of the integrated and detrended series is calculated by means of

$$F(n) = \sqrt{\frac{1}{N}\sum_{k=1}^{N}[y(k) - y_n(k)]^2}, \tag{1}$$

this process is taken over several scales (box sizes) to obtain a power law behavior $F(n) \sim n^{\alpha}$, with α an exponent, which reflects self-similar and correlation properties of the signal. It is known that $\alpha = 0.5$ is associated to white noise (non correlated signal), $\alpha = 1$ corresponds to $1/f$ noise and $\alpha = 1.5$ represents a Brownian motion [1].

1.2 Correlated noises

In order to generate fractal signals with power-law correlations, we use the Fourier filtering method [7]. We start with a Gaussian noise with zero mean and unitary standard deviation. The fast Fourier algorithm is used to obtain the transformed signal in the frequency domain. The power spectrum is forced to follow the desired power law form $S(f) \approx 1/f^{\beta}$, with β, the power spectrum exponent with values within the interval $0 < \beta < 2$. Finally, the inverse Fourier transform is applied to send the signal back to the time domain. In this way, the resulting time series is characterized by the desired correlation exponent β, which may take values ranging from uncorrelated random signals, $\beta = 0$, to a random walk process, $\beta = 2$ and long-range correlated noise, $\beta = 1$. There is a linear relationship between β and the DFA exponent α given by $\beta = 2\alpha - 1$ [6], which permits to select one these exponents as a control to generate signals with specific scaling exponent. Here, we will use the exponent α as control exponent to perform our simulations within the range $0.5 < \alpha < 1.5$. From now on, all the noises with correlation exponent α will be denoted as η^{α}.

2. Trends and correlated noise: crossover effect

We study processes with the general map form,

$$x_{t+1} = f(x_t) + \eta_t^{\alpha}, \tag{2}$$

with f a function representing the map and η a stochastic noise with power law correlations characterized by th exponent α. In this case, when the input noise η^{α} is not considered, the map denotes a trend, which in this work are considered to have two behaviors (i) decreasing or (ii) increasing. We remark that our procedure is different to the one reported in [5], where the correlated noise and the trend are linearly combined, allowing to detect the scaling provenient from the noise and from the trend. In our case, the trend is intrinsically mixed with the correlated noise and the superposition rule does not hold [5].

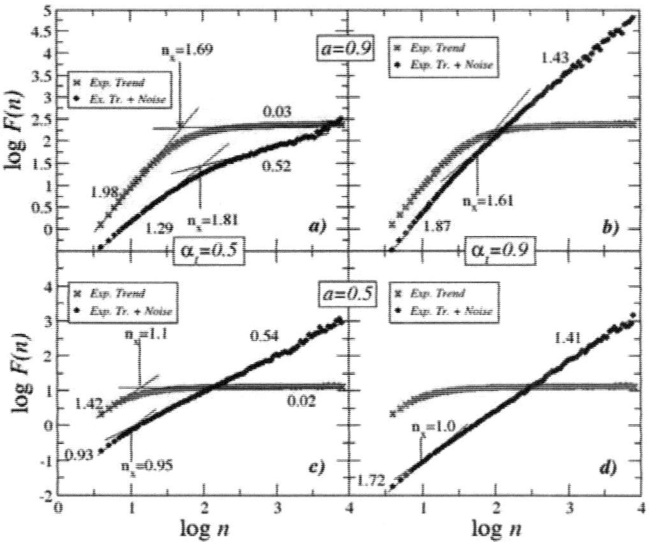

Figure 1: Plots of $F(n)$ *vs.* $logn$ for an exponential trend and for the prosses given by Eq. (3), for different values of the input noise α_i and the control parameter of the trend a.

2.1 Exponential trend

First, we consider the processes of the form,

$$x_{t+1} = ax_t + \eta_t^{\alpha}, \tag{3}$$

with a a parameter within the interval $0 \leq a \leq 1$. In this case, the process is characterized by a negative feedback, so the trend given by the expression $x_{t+1} = ax_t$, is a decaying exponential sequence. First, we analyze the behavior of fluctuation function $F(n)$ for the exponential trend alone. For convenience, we perform the analysis for the values of the parameters $a = 0.9$ and $a = 0.5$. The behavior of $F(n)$ is shown in Fig. 1 a,c. We observe a crossover in $F(n)$ located at $n = n_{\times}$, wich separetes two regimes, over short scales ($n < n_{\times}$) the scaling exponent is $\alpha_s \approx 2$, while for large scales ($n > n_{\times}$), the value is $\alpha_l \approx 0$, indicating a strong transition in the mean fluctuation. This behavior is expected because the fluctuation $[y(k) - y_n(k)] \rightarrow 0$, as the size of the box increases. For smaller values of the parameter a, the crossover position $n = n_{\times}$ moves to the left and the exponent over short scales α_s decreases (Fig. 1*c*).

Next, we explore the scaling behavior of the processes given by Eq. (3), that is, when considering the intrinsic trend map with correlated noise. For convenience, we perform the analysis for the input noise values $\alpha_i = 0.9$ and $\alpha_i = 0.5$. For the case when the input noise is uncorrelated ($\alpha = 0.5$), and $a = 0.9, 0.5$, the fluctuation $F(n)$ exhibits two regimes separated by the crossover point $n_{\times}$; over short-scales ($n < n_{\times}$) the scaling exponent tends to be close to the value observed for the trend (Fig 1*a,c*), indicating that the statistics of the fluctuations $F(n)$ are dominated by the trend. In contrast, over large scales ($n > n_{\times}$), the scaling behavior is quite similar to the one of the input noise α, confirming that the behavior of $F(n)$ is dominated by the noise (Fig. 1*a,c*).

In the case of power-law correlations in the input noise ($\alpha = 0.9$) (Fig. 1*b,d*), the fluctuations $F(n)$ show a weak crossover behavior, where for $n < n_{\times}$, the scaling exponent is close to the one of the trend, while for $n > n_{\times}$ the behavior for $F(n)$ reveals a scaling exponent larger than the input noise ($\alpha > \alpha_i$).

It is noteworthy that the crossover scale n_x remains almost unchanged for the case where the input noise is correlated, whereas for the case when it is uncorrelated ($\alpha = 0.5$), the crossover scale slightly moves to the right (for $a = 0.9$) and to the left (for $a = 0.5$).

2.2 Power law trends

Next, we consider a power-law trend [5] according to,

$$x_{t+1} = x_t + \lambda(x_t^{(1-\lambda)/\lambda)}) + \eta_t^{\alpha}, \tag{4}$$

with λ a constant with values within the interval $-1 < \lambda < 1$. When the stochastic term is not considered, we have an isolated trend; for $\lambda < 0$, the trend decreases whereas for $\lambda > 0$, the trend increases with time. Representative statistics of the fluctuation $F(n)$ for the isolated trends are shown in Fig. 2. We observe that for a decreasing trend ($\lambda = -0.5$), over short scales ($n < n_{\times}$), the scaling exponent is close the value from the corresponding short-term exponent from the trend (see Fig. 2a and 2b), whereas for large scales ($n > n_{\times}$), the exponent resembles the noise α_i. In contrast, for a decreasing trend ($\lambda = 0.5$), the fluctuation $F(n)$ is dominated by the trend in such a way that the input noise $\alpha = 0.5$, a weak crossover is observed (Fig. 2c), and for $\alpha = 1.3$, the scaling exponent is close to two, and no crossover is observed. We also observe that for this case, the scaling behavior of the trend dominates mainly over large scale.

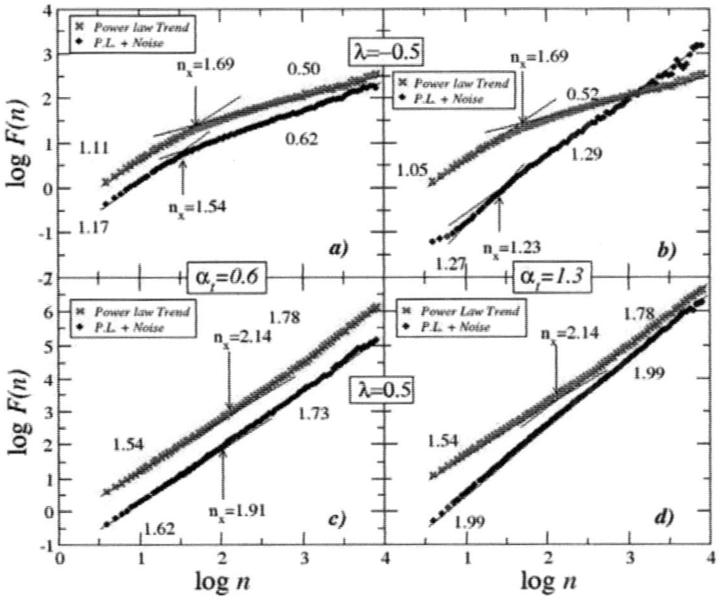

Figure 2: Plots of $F(n)$ *vs.* $logn$ for an exponential trend and for the prosses given by Eq. (4), for different values of the input noise α_i and the control parameter of the trend λ.

3. Conclusions

In this work we have analyzed, by means of the DFA-method, two cases of trend maps. In particular, we found the situation for which a crossover effect is observed. Our results show that a decreasing trend leads to a crossover in the scaling exponents with $\alpha_s > \alpha$, while for increasing trends $\alpha_s < \alpha$. For noisy trend maps, the scaling behavior over short scales is mainly dominated by the trend (increasing or decreasing), whereas over large scales, the fluctuation resemble the scaling of the input noise. All these results for the behavior of $n_{\times}$ have interesting implications to understand the crossovers observed in geoelectric signls assiciated with the seismicity along the Mexican subduction zone [14].

Aknowledgments

We thank to COFAA-IPN, EDI-IPN and CONACyT for partial support.

Journal of Physics: Conference Series **582** (2015) 012062 doi:10.1088/1742-6596/582/1/012062

References

[1] J. L. Del Río A. Muñoz Diosdado, L. Guzmán-Vargas A. Ramírez-Rojas and F. Angulo-Brown. Some crossover behavior in heart interbeat and electroseismic time series. *Fractals*, 13(4):253–263, December 2005.

[2] West B.J. BassingthwaighteJB, LiebovitchLS. *Fractal physiology*. Oxford University Press,, New York, 1994.

[3] Zhi Chen, Plamen Ch. Ivanov, Kun Hu, and H. Eugene Stanley. Effect of nonstationarities on detrended fluctuation analysis. *Phys. Rev. E*, 65:041107, Apr 2002.

[4] A.L. Goldberger. Nonlinear dynamics for clinicians: Chaos theory, fractals and complexity at the bedside. *Lancet*, 347:1312, 1996.

[5] Kun Hu, Plamen Ch. Ivanov, Zhi Chen, Pedro Carpena, and H. Eugene Stanley. Effect of trends on detrended fluctuation analysis. *Phys. Rev. E*, 64:011114, Jun 2001.

[6] N. Iyengar, C. K. Peng, R. Morin, A. L. Goldberger, and L. A. Lipsitz. Age-related alterations in the fractal scaling of cardiac interbeat interval dynamics. *Am J Physiol Regul Integr Comp Physiol*, 271(4):R1078–1084, 1996.

[7] Hernán A. Makse, Shlomo Havlin, Moshe Schwartz, and H. Eugene Stanley. Method for generating long-range correlations for large systems. *Phys. Rev. E*, 53(5):5445–5449, May 1996.

[8] Rosario N. Mantegna and H. Eugene Stanley. *Introduction to Econophysics: Correlations and Complexity in Finance*. Cambridge University Press, 0 edition, 1999.

[9] C. K. Peng, S. V. Buldyrev, S. Havlin, M. Simons, H. E. Stanley, and A. L. Goldberger. Mosaic organization of DNA nucleotides. *Physical Review E*, 49(2), 1994.

[10] C. K. Peng, S. Havlin, H. E. Stanley, and A. L. Goldberger. Long-range anti-correlations and non-gaussian behavior of the heartbeat. *Phys. Rev. Lett.*, 70:1343–1346, 1995.

[11] C. K. Peng, J. Mietus, J.M. Hausdorff, S. Havlin, H. E. Stanley, and A. L. Goldberger. Quantification of scaling exponents and crossover phenomena in nonstationary heartbeat time series, proc. nato dynamical disease conference, edited by l. glass. *Chaos*, 5:82–87, 1995.

[12] Luciano Telesca. Cycles, scaling and crossover phenomenon in length of the day (lod) time series. *Physica A: Statistical Mechanics and its Applications*, 379(2):459 – 464, 2007.

[13] B. J. West. *Fractal physiology and chaos in medicine*. World Scientific, Singapore, 1990.

[14] L Guzmán-Vargas, A Ramírez-Rojas, R Hernández-Pérez, F Angulo-Brown. Correlations and variability in electrical signals related to earthquake activity. *Physica A: Statistical Mechanics and its Applications,* 388(19):4218-4228.

Journal of Physics: Conference Series **582** (2015) 012063 doi:10.1088/1742-6596/582/1/012063

Radial Hilbert Transform in terms of the Fourier Transform applied to Image Encryption

Y Morales, L Díaz and C Torres
Laboratorio de Óptica e Informática, Universidad Popular del Cesar
Valledupar, Cesar - Colombia

E-mail: morales.yaileth@unicesar.edu.co

1. **Abstract.** In the present investigation, a mathematical algorithm under Matlab platform using Radial Hilbert Transform and Random Phase Mask for encrypting digital images is implemented. The algorithm is based on the use of the conventional Fourier transform and two random phase masks, which provide security and robustness to the system implemented. Random phase masks used during encryption and decryption are the keys to improve security and make the system immune to attacks by program generation phase masks.

1. Introduction

The threat of illegal data access has made the data encryption an important subject [1]. Several methods have been proposed [2–8] for optical image encryption including the most popular double random phase encoding (DRPE)[9]. This technique uses two random phase masks and the Fourier planes to encrypt the input image. Optical encryption offers several image parameters, e.g. phase, amplitude, color, spatial frequency, polarization, etc., that can be exploited to perform a robust and highly secure encryption. In this paper, we propose a technique based in the Fourier domain using the properties of Radial Hilbert Transform (RHT). The frequency spectrum of the image is then multiplied with two different random phase function and two Fourier Transform for getting the encrypted images. In the decrypt processing is required a random phase function additional and two fractional order which make this technique safer and more robust compared to the conventional technique of double random phase encoding in the Fourier domain.

2. Methodology

Hilbert Transform in terms of the Fourier

The Hilbert Transform for a function $g(x)$ is defined [10].

$$g_{HT}(x)=\frac{1}{\pi}\int_{-\infty}^{\infty}\frac{g(x)}{x-\tau}d\tau \tag{1}$$

Where $g_{HT}(x)$ Hilbert Transform's signal, $g(x)$ is input signal.

Journal of Physics: Conference Series **582** (2015) 012063 doi:10.1088/1742-6596/582/1/012063

Considering the theory of linear systems such as optical systems can show that the output signal for this type of system can be expressed as the convolution of the input signal with the impulse response.

$$g_{HT} = g(x) * h(x) \tag{2}$$

Assuming that the impulse response $h(x)$ is this form

$$h(x) = \frac{1}{i\pi x} \tag{3}$$

Equation 2 can be written

$$g_{HT} = g(x) * \frac{1}{i\pi x}. \tag{4}$$

As what is sought is to write the Hilbert transform in the terminus Fourier transform proceed to obtain the equation 4

$$F\{g_{HT}\} = F\left\{g(x) * \frac{1}{i\pi x}\right\}. \tag{5}$$

By properties of the TF have

$$F\{g_{HT}\} = F\{g(x)\} * F\left\{\frac{1}{i\pi x}\right\}. \tag{6}$$

Of the table Fourier Transform we have that the TF of impulse response is the sign function, and this correspond to the system transfer function.

$$F\left\{\frac{1}{i\pi x}\right\} = H_P\{\omega_x\} = -i\,\mathrm{sgn}(w_x). \tag{7}$$

Where the function $-i\,\mathrm{sgn}(w_x)$ is defined:

$$-i\,\mathrm{sgn}(w_x) = \begin{Bmatrix} i & if & w_x < 0 \\ 0 & if & w_x = 0 \\ -i & if & w_x > 0 \end{Bmatrix}. \tag{8}$$

This can be expressed in terms of Euler

$$-i\,\mathrm{sgn}(w_x) = e^{iP\theta}. \tag{9}$$

Where

$$e^{iP\theta} = \begin{Bmatrix} i \ \ if \ \ P = \frac{1}{2} \\ \\ -i \ \ if \ \ P = -\frac{1}{2} \end{Bmatrix}, \tag{10}$$

P = 0 indicates a phase mask that does no processing to the image, so it is not taken into consideration.
Therefore, the transfer function can be written

$$H_P\{\omega_x\} = e^{iP\theta}. \tag{11}$$

Then equation 6 can be written

$$F\{g_{HT}\} = F\{g(x)\}.H_P\{\omega_x\}. \tag{12}$$

Now as what we want is to get the output signal g_{HT} by properties of the TF we have:

$$F^{-}\{F\{g_{HT}(x)\}\} = F^{-}\{H_P\{\omega_x\}F\{g(x)\}\}. \tag{13}$$

Substituting 11 in 13 and solving we obtain The P-order Hilbert transform can be written as:

$$g_{HTP}(x) = F^{-}\{e^{iP\theta}F\{g(x)\}\}. \tag{14}$$

Where equation (14) represents the equation of the Hilbert transform of order P in terms of the Fourier Transform and the exponential function represents optically uniform random phase mask and is a technique used for the encrypted image.

(a) (b)

Figure 1. a) Original Image. b) Hilbert Transform.

So far we have found the one-dimensional Hilbert transform for signals, but our case study is two-dimensional signals (images). Then the generalized Hilbert transform in term of the Fourier transform to encrypt an image can be written:

$$g_{HTP}(x, y) = F^{-}\{e^{iP\theta}F(g(x, y))\}, \tag{15}$$

$$g'(x, y) = \{e^{iP\theta}F(g(x, y))\}. \tag{16}$$

Where $g'(x, y)$ is the image encrypted with only one random phase mask. Now for added system security multiplied by another phase mask

$$g_{HTP}(x, y)_{(encripted)} = F^{-}\{g'(x, y)F(e^{im\theta})\}. \quad (17)$$

Where P and m are different from zero, if they are zero we would have the same object image.

Now that we have clear the terms in equation 17, we do a block diagram showing the design of the system image encryption.

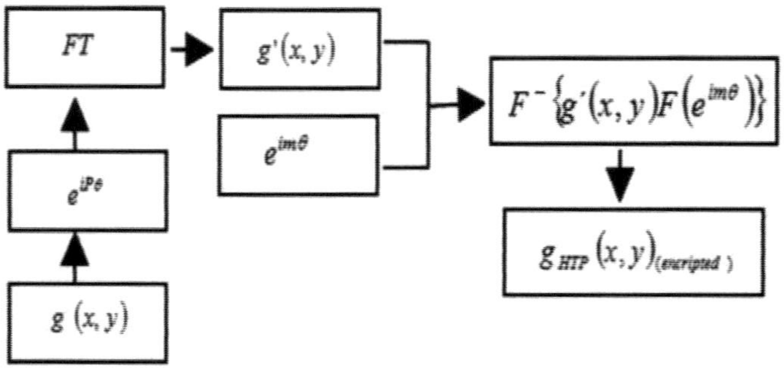

Figure 2. Block Diagram of the system design.

The software takes the original image $g\ (x,y)$ and multiplied by a mask of uniform random phase $e^{iP\theta}$, we find its Fourier transform and get $g'(x, y)$. For added security is again multiplied by another uniform random phase mask $e^{im\theta}$ and then apply the inverse Fourier transform, and finally obtain the encrypted image $g\ (x, y)_{encripted}$.

3. Results

To investigate the quality of encryption, digital simulations are performed on MATLABs 7.6 platform. The input image chosen for encryption is a cameraman of size 256 256 pixels.

Encryption

The input image is multiplied by a random phase mask and he finds the TF. Then for safety multiply it by another random phase mask and finally we calculate the TFI resulting in the encrypted image. Then do the reverse and get the decrypted image.

Decrypted

The image to be encrypted is changed to noise-like patterns and exact recovery of the image is possible only when the correct random phase functions and other key parameters are correctly used during the decryption.

Then to see if the encrypted image is fully recovered do the correlation of the original image with the encrypted image and got this gave us one, ie its maximum value which means that the algorithm manages to get the same image that is sent through the system.

Journal of Physics: Conference Series **582** (2015) 012063 doi:10.1088/1742-6596/582/1/012063

Figure 3. a) Original image. b) Randon phase mask. c) Multiplication phase mask for the original image. d) TF of the multiplication phase mask for the original image. e) Second Randon phase mask. f) Encrypted Image.

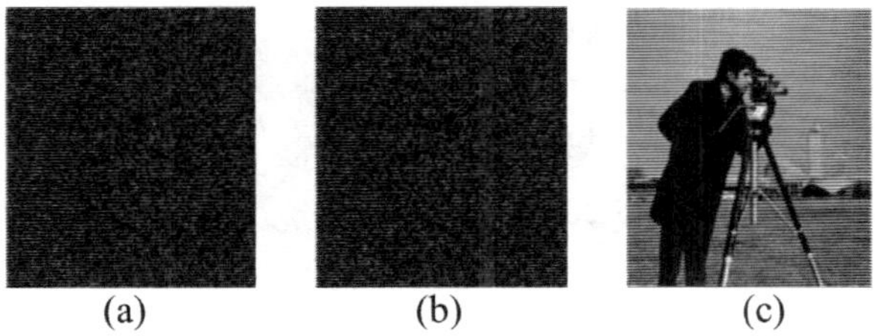

Figure 4. a) Second Randon phase mask. b) First Randon phase mask. c) Decrypted Image.

Figure 5. Correlation of the input image with the decrypted image. a) Original Image. b) Image decrypted with the correct key. c) Correlation peak of the original image with the decrypted image. d) Original image. e) Image decrypted with the incorrect key. f) Correlation peak of the original image with the decrypted image with incorrect key.

Journal of Physics: Conference Series **582** (2015) 012063 doi:10.1088/1742-6596/582/1/012063

Conclusion

Based on conventional Hilbert transform found the Radial Transform Generalized Gilbert order P in terms of the Fourier Transform, which makes possible a visual and computational implementation of a cipher images or data without loss of information.

A double random-phase encoding-based technique for image encryption using the radial Hilbert has been proposed in the terminus of Fourier transform domain. Digital simulations are presented to demonstrate the robustness of the technique against the variation in the key parameters

References

[1] Bracewell RN. The Fourier Transform and Its Application. McGraw-Hill; 1965 (Chapter 12)
[2] Tao R, Xin Y, Wang Y. Double image encryption based on random-phase encoding in fractional Fourier domain. Opt Express 2007;24:16067–79.
[3] Nishchal NK, Joseph J, Singh K. Securing information using fractional Fourier transform in digital holography. Opt Commun 2004;235:253–9.
[4] Hwang H-E, Han P. Fast algorithm of phase mask for image encryption in Fresnel domain. J Opt Soc Am A 2006;23:1870–4.
[5] Javidi B, Nomura T. Polarization encoding for optical security. Opt Eng 2000;39:2439–43.
[6] Nicolas J, Iemmi C, Campos J, Yzuel MJ. Encoding 3D correlation in an optical processor. Opt Commun 2002;209:35–43.
[7] Peng X, Yu LF, Cai LL. Double-lock for image encryption with virtual optical wavelength. Opt Express 2002;10:41–5.
[8] Refregier P, Javidi B. Optical image encryption based on input plane and Fourier plane random encoding. Opt Lett 1995;20:767–9
[9] Ozaktas HM, Zalevsky Z, Kutay MA. The Fractional Fourier Transform with Applications in Optics and Signal Processing. Chichester: Wiley; 2001
[10] MSoo-Chang Pei; Jian-Jiun Ding 2003 The Generalized Radial Hilbert Transform ans its Application to 2D.

Journal of Physics: Conference Series **582** (2015) 012064 doi:10.1088/1742-6596/582/1/012064

Study of the temporal evolution of Whitening Teeth immersed in Peroxide of hidrogen (H2O2) Using Digital Image Processing

L Díaz, Y Morales and C Torres
Laboratorio de Óptica e Informática (LOI), Universidad Popular del Cesar, Valledupar, Cesar, Colombia.

E-mail: leonardodiaz@unicesar.edu.co

Abstract. The esthetic dentistry reference in our society is determined by several factors, including one that produces more dissatisfaction is abnormal tooth color or that does not meet the patient's expectations. For this reason it has been designed and implemented an algorithm in MATLAB that captures, digitizes, pre-processing and analyzed dental imaging by allowing to evaluate the degree of bleaching caused by the use of peroxide of hidrogen. The samples analyzed were human teeth extracted, which were subjected to different concentrations of peroxide of hidrogen and see if they can teeth whitening when using these products, was used different concentrations and intervals of time to analysis or study of the whitening of the teeth with the hydrogen peroxide.

1. Introduction

The teeth have multiple shades and colors depending on certain factors such as age, sex, race; however, the teeth are very vulnerable and sensitive to the effects of toxic chemical contaminants and other drugs, especially during development, can be affected in both the composition of the dental structure common color thereof [1]. Our society is undergoing a process of growing cult of aesthetics dental and for this reason the demand for dental treatment claim that, plus morph functionally rehabilitate the tooth, your appearance will return natural [2].

The aesthetic dental disciplines are responsible for trying to meet these requirements, and highlights including tooth whitening because it is a treatment that allows to achieve satisfactory results for the patient and the dentist within a reasonable time and with a high degree of safety [3] [4]. Today dentistry provides an opportunity to restore the unique characteristics of each patient as each colored natural tooth has many optical features that increase its complexity [5]. To date, consumers worldwide have acquired millions of whitening or dental teams hoping to get a radiant smile. Teeth whitening products come in the form of gel, strips or sleeves dental, and can be used both in the dentist at home. Hydrogen peroxide is the bleaching component of these products. In general, the whitening power of the amount of hydrogen peroxide. Toothpastes and mouthwashes may also contain low concentrations of hydrogen peroxide as a disinfectant to prevent plaque and gum inflammation [6]. This paper proposes a technique to analyze the whitening tooth using digital images processing, however, not is determinated the degree of bleaching because it is the next phase of this research and is treated in another paper our.

This article is developed in four parts: first shown block diagram system design, followed the implementation where is explanation of how was used the materials for development of the same. Then shown results and finally, the conclusions and analysis of the results obtained.

2. Methodology

2.1. Design

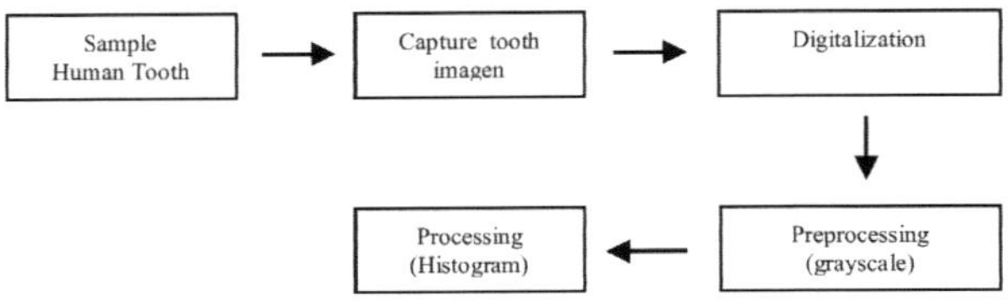

Figure 1. Block Diagram of the system design

The figure 1 show the system implemented, the system input is the image of a human tooth which is digitized, is passed a gray scale and finally applies a histogram which will determine my level of whitening your teeth at the interval of time that has been immersed into hydrogen peroxide.

2.2. Implementation

Was developed an algorithm in MATLAB that captures, digitizes, pre-processing and analyzes dental imaging allowing to obtain a measure whitening by the use of mouthwash. The image Capture is making by a Microfix digital Microscope. The captured image belongs to an upper premolar tooth that has been previously immersed in 20ml of hydrogen peroxide at 12 volumes (3.6%) in different time intervals 0h, 24h, 48h ... 240h. These images are segmented and passed to gray scale and then apply a histogram. Figure (2-3). The Microfix digital Microscope is a digital microscope connects to the USB port of your computer or laptop and allows you to view close ups of objects with up to 200x magnification [7]. Hydrogen peroxide (H2O2) is a colourless liquid with a bitter taste and is highly soluble in water to give an acidic solution. H2O2 is an oxidising agent with a wide number of industrial applications in for example, bleaching or deodorising textiles, wood pulp, hair, fur and foods, in the treatment of water and sewage, as a seed disinfectant and neutralising agent in wine distillation. Low concentrations of H2O2 have been found in rain and surface water, in human and plant tissues, in foods and beverages and in bacteria [8]. Risks commonly reported with tooth whitening include increased tooth sensitivity and mild gingival irritation. The degree of these side effects is directly related to the concentration of the peroxide bleach component, duration of the treatment, and the non-bleach composition of the product used. Tooth sensitivity usually occurs at the time of treatment and can last several days; gingival irritation begins within a day of the treatment and can also last several days. There are additional risks that have been reported from in vitro studies which include tooth erosion, tooth mineral degradation, increased susceptibility to demineralization, and pulpal damage. The ultimate endpoint for tooth whitening is dependent upon the tooth itself, with common wisdom telling us that all treatment regimens will eventually arrive at the same whitening endpoint. This is not exactly true as some very aggressive regimens can damage the tooth through dehydration and demineralization such that the tooth temporarily appears whiter [9].

Histograms are the basis for numerous spatial domain processing techniques. Histogram manipulation can be used effectively for image enhancement and addition to providing useful image statistics. The histogram of a digital image with gray levels in the range [0, L-1] is a discrete function $h(r_k)=n_k$, where r_k is the K the gray level and n_k is the number of pixels in the image having gray level r_k.

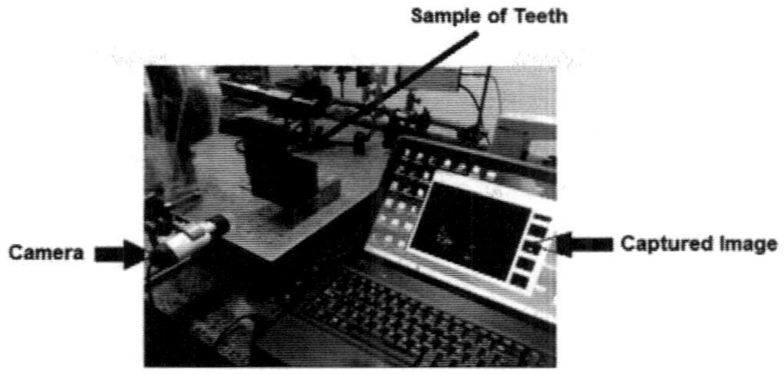

Figure 2. Assembly System

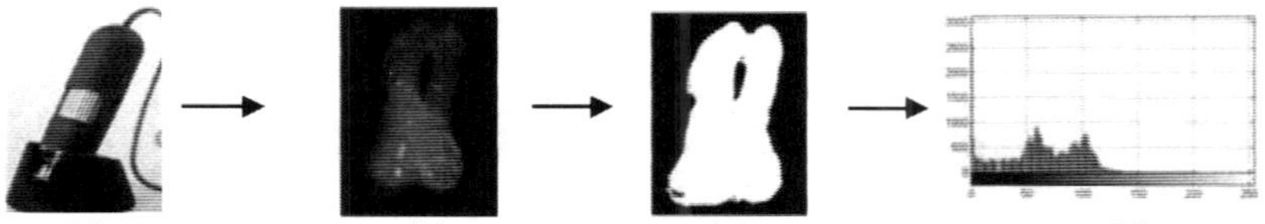

(a) (b) (c) (d)

Figure 3. System implemented. a) Microfix digital Microscope. b) Upper premolar tooth image. c) Segmented image. d) Histogram Imagessembly.

It is common practice to normalize a histogram by dividing each of its values by the total number of pixels in the image, denoted by n. Thus, a normalized histogram is given by $(r_k) = {}^{n_k}/_{n}$, for $k = ,1, ..., L - 1$. Loosely speaking, $p(r_k)$ gives an estimate of the probability of occurrence of gray level r_k. The horizontal axis of each histogram plot corresponds to gray level values, r_k.. The vertical axis corresponds to values of h(r_k)=n_k or $p(r_k) = {}^{n_k}/_{n}$ if the values are normalized. Note that the sum of all components of a normalized histogram is equal to 1 [10]. In Matlab was implemented the instructions imhist(I) that calculates the histogram for the intensity image I and displays a plot of the histogram. The number of bins in the histogram is determined by the image type, if I is a grayscale image, imhist uses a default value of 256 bins or if I is a binary image, imhist uses two bins.

3. Results

The human tooth sample was immersed 20ml of hydrogen peroxide at 12 volumes (3.6%). Intervals were study time 0, 24, 48, 144 and 240 hours and lighting for the capture of the images were the same. Figure (4 - 5).

In figure 4 se showing teeth extracted without submitting to the solution of hydrogen peroxide and is our point of reference for determining the histogram obtained through the degree of whitening.

In the histogram we obtain a graph where the number of pixels displayed per each gray level that appears in the picture. In figure 5 (24 hours). Displacement is not very noticeable, indicating a very poor grade in whitening.

Therefore the sample is left submerged in an interval of 24 hours for a total of 48 hours and then analyzing the behavior. Figure 6. The shift of the histogram at this time was mild and decides to increase the range of Dip tooth in the hydrogen peroxide solution for about 96 hours for a total of 144 hours. In Figure 7, the tooth is shown physically visible changes to its normal color and the histogram obtained.

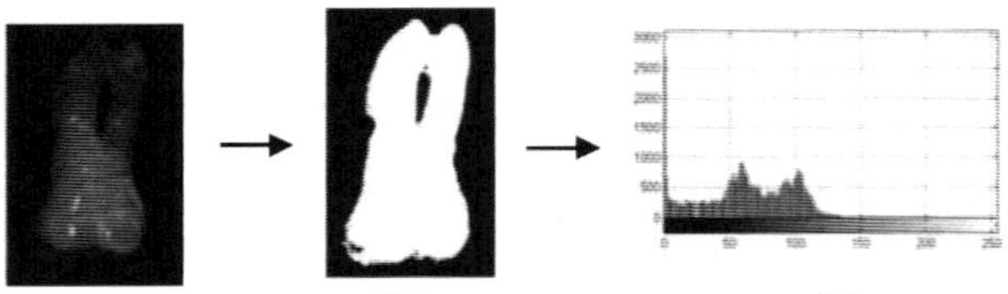

(a) (b) (c)

Figure 4. 0 hours processing image. a) Upper premolar tooth image. b) Segmented image. c) Histogram Image.

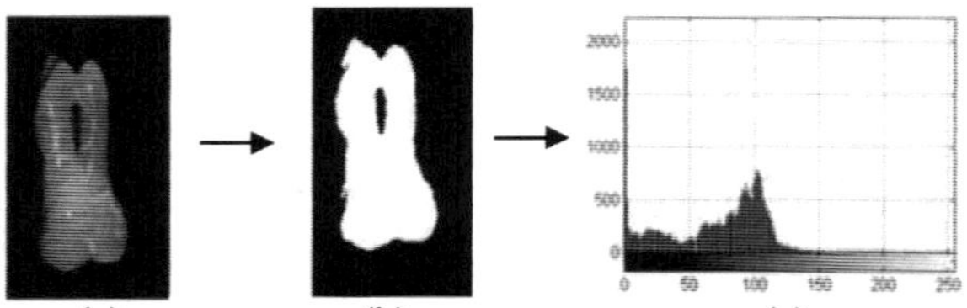

(a) (b) (c)

Figure 5. 24 hours processing image. a) Upper premolar tooth image, b) Segmented image, c) Histogram Image.

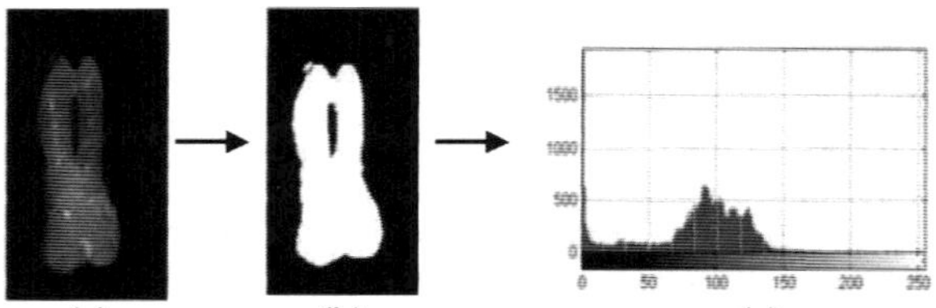

(a) (b) (c)

Figure 6. 48 hours processing image. a) Upper premolar tooth image. b) Segmented image. c) Histogram Image.

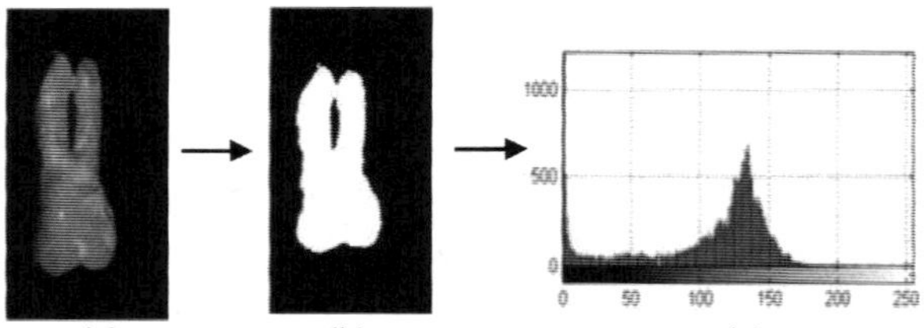

(a) (b) (c)

Figure 7. 144 hours processing image. a) Upper premolar tooth image. b) Segmented image. c) Histogram Image.

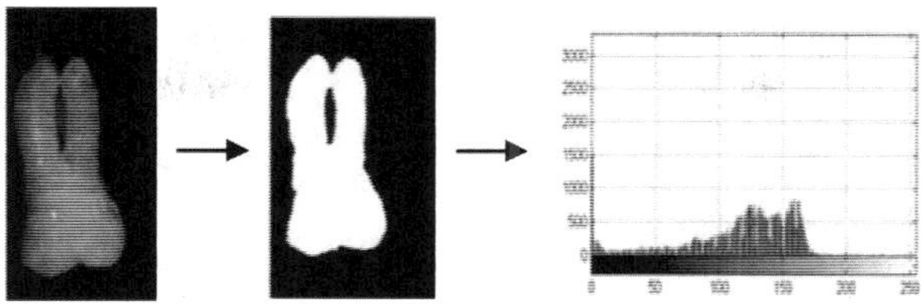

(a) (b) (c)

Figure 8. 240 hours processing image. a) Upper premolar tooth image. b) Segmented image. c) Histogram Image

The variation is noticeable in comparison with the first histogram obtained in Figure 4 also show a degree of whitening although emerging degradation in the tooth enamel.

In Figure 8 the tooth immersed shown in the peroxide solution at an interval of 96 hours to 240 hours total noticing whitening highly found on the tooth and shift the histogram in the. Importantly analyzed in this range there is a large loss of tooth enamel due to the time that the tooth was immersed.

By comparing the images can be seen at a glance that tooth periodically used hydrogen peroxide bleaching has a level which can also check with the running of the white level histogram.

4. Conclusions

Today, hydrogen peroxide is the most popular, easily accessible, and, the most important factor, handy treatment and cheap for home teeth whitening.

Noting the above graphic results, was used a peroxide concentration of (12 volumes at 3.6%). Is observed in the previous figures, the whitening level caused by the differents times at which the extracted human teeth are immersed. The time intervals used in the analysis provide important features in the color change of the tooth as in each of the histograms. Although a degree of whitening as such is not determined in this paper, these characteristics obtained in particular, are important piece is our starting point to determine a level of tooth whitening.

This technique can be a novel method for determining the a level of tooth whitening in accordance to aesthetic the preferred by patient taking into account the supervision of dentist for him to do the analysis optical, and the determinate the diagnosis or changes tooth bleaching of such as the patient tooth glaze. This research will enable future the analysis and study of the differences mouthwashes or other substances that offer dental whitening. Using hydrogen peroxide for teeth whitening might have some unpleasant sensations and burning in the gums but the ultimate result will definitely exceed your expectations. will get clean and snow-white teeth.

Acknowledgments

The authors thank the University Cesar Popular for their support in this research and the dentistry Maria del Pilar Dangond Oñate samples by providing us of human teeth.

5. References

[1] Berasategi A, Alteraciones del Color Dental, Clínica Dental Donnay, pp.1-12, November 2011.

[2] Morley J, "The role of cosmetic dentistry in restoring a youthful appearance", J Am Dent Assoc 1999; 130: 1166-72.

[3] Goldstein RE, In-office bleachig: where we came today, J Am Dent Assoc 1997; vol 128: 11-5.

[4] Lambert DL, Motivación estética y blanqueamiento dental vital, Signature Int 2000, vol 5, pp. 5-10.

[5] Fondriest J, Shade matching in restorative dentistry: the science and strategies, Int J Periodontics Restorative Dent 2003 (5):467-79.
[6] Morris & Chapman, Blanqueadores Dentales y Productos de Higiene Bucal con Peróxido de Hidrógeno, GreenFacts-Facts on Health and the Environment, 2013, Edition, Prentice-Hall, Inc. 2002.
[7] Wondcam.com, All Rights Reserved, Company: Wondcam Innovation Limited 2007~2014.
[8] IARC, Hydrogen peroxide, 1999; 671–689.
[9] Clifton M. Carey, Tooth Whitening: What We Now Know, The Journal of Evidence-Based Dental Practice 2014, pp. 1-19.
[10] Gonzalez Rafael C, Richard E.Woods, Digital Image Processing, -Second Edition, 2002 by Prentice-Hall, Inc.

Journal of Physics: Conference Series **582** (2015) 012065 doi:10.1088/1742-6596/582/1/012065

Fruits and vegetables dehydration

A de Ita, G Flores and F Franco
Área de Ciencias de Materiales, UAM Azcapotzalco, Av. San Pablo 18, México, D.F.

E-mail: add@correo.azc.uam.mx

Abstract. Dehydration diagrams were determined by means of Differential Thermal Analysis, DTA, and Thermo Gravimetric Analysis, TGA, curves of several simultaneous fruits and vegetables, all under the same conditions. The greater mass loss is associated with water containing in the structure of the investigated materials at low temperature. In poblano chile water is lost in a single step. The banana shows a very sharply two stages, while jicama can be observed although with a little difficulty three stages. The major mass loss occurs in the poblano chile and the lower in banana. The velocity and temperature of dehydration vary within a small range for most materials investigated, except for banana and cactus how are very different.

1. Introduction

It is a commonly thought that there are fruits with a high water content and others that don´t. In the case of vegetables in general, we can also make this approximation. To really know if this is so, an experiment was designed to analyze water loss of water from various fruits and vegetables. In the literature no similar data is found [1-6], although there are many studies concerning the drying velocity and water lost with time in many fruits and vegetables. Usually the water lost is associated with low temperatures, that is, normally low than 200 °C. Greater temperatures produce the destruction of the whole structure. In most investigations of dehydration type, the tests are isothermal and mass losses are evaluated with increasing time. The analysis in this paper are isochronous, here the mass loss is measured as the temperature is increased at a fixed heating rate. We don´t know a study of this kind and we think it is very important to characterized the thermal behavior of the chosen materials.

2. Methodology

Different fruits and vegetables were chosen: banana, grape, peach, guava, kiwi, mango, yellow apple, green and chinese melon, poblano chile, jicama and cactus. We determined the weight loss, TGA, curve and the Differential Thermal Analysis, DTA, in a Shimadzu equipment, model DTG-60, located in our Department. Temperature varied from ambient to 300 °C, the scan was performed at a rate of 10 °C per minute under a nitrogen flow rate of 10 cm^3 per minute. The mass used in this work for each material was about one milligram of pulp and always shelled or without skin, such as in the case of cactus, jicama and chile poblano.

3. Results

The results of the material lost can be seen in Figure 1: banana in blue line, is completely different from the rest, first because it presents clearly two stages of water loss and second because their mass loss is lower compared to the other. The first drop ends up to 120 °C, with a mass of 60 %, ie. a loss of 40 %; then a second drop ending at 220 °C as a final mass lost of 40 %, representing a loss of 60 % and then a continuous loss until 300 °C it represents a total mass of 29 % that is, a total lost of 71 % of its initial mass.

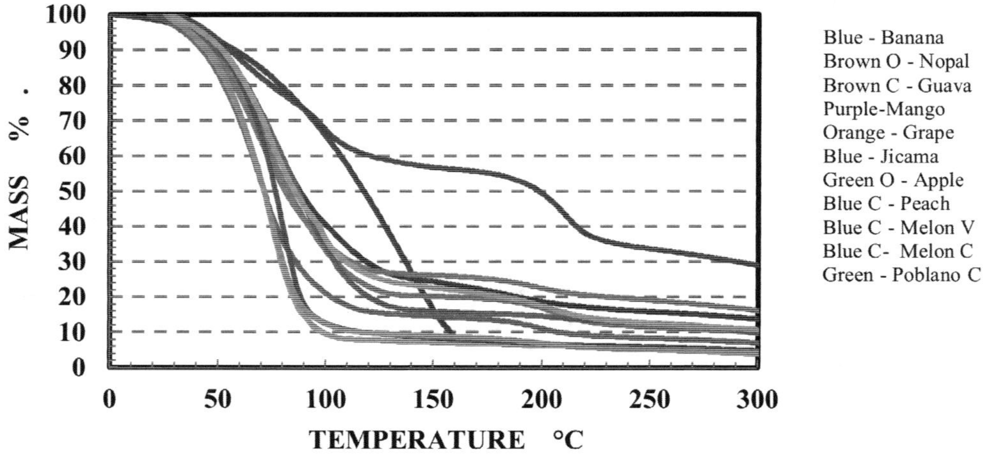

Figure 1. Graph of mass loss of all materials. The sequence of the colors-materials are near the 300 °C temperature.

The second curve is the cactus, in dark brown line, which is also completely different, as it shows a simple fall and that around 160 °C and shows a mass less than 10 %, ie., a greater lost of 90 % of its initial mass and at near 300 °C, perhaps it has only 5 % of its initial mass that is, a lost of 95 % of it´s initial mass.

The rest of the materials are grouped with a more or less similar pattern, with one, two or three small steps and a final and continuous fall between 200 and 300 °C. At this temperature there are the mass of all of them is between 18 and 4 %, which corresponds to a loss of 82 and 96 %. The first case presents guava, in line light brown, and the last value on the graph corresponds to melons, poblano chile and cactus. To observe better the differences we analyze the total graph in small steps, the first until 100 °C, in Figure 2.

The first thing to note is that not all curves begin its loss mass at the same temperature. It shows that the more stables begins at higher temperature like the cactus and the more instable are like the poblano chile.

Second, at 100 °C the materials that have less lost mass is cactus and banana with a mass of 68 %, ie., a lost of 32 % of its original mass. Those who have lost more mass are the poblano chile, green line, melons in blue line and mango, in purple line, with a mass of 10 % ie. a loss of 90 % of the original mass. Follow by peach, in blue color, with a loss of 80 %, that is, with and end mass of 20 %. The rest of materials lost slightly less mass and are grouped between 40 and 36 % by mass, that is, with a loss of 60 to 68 % of the original mass.

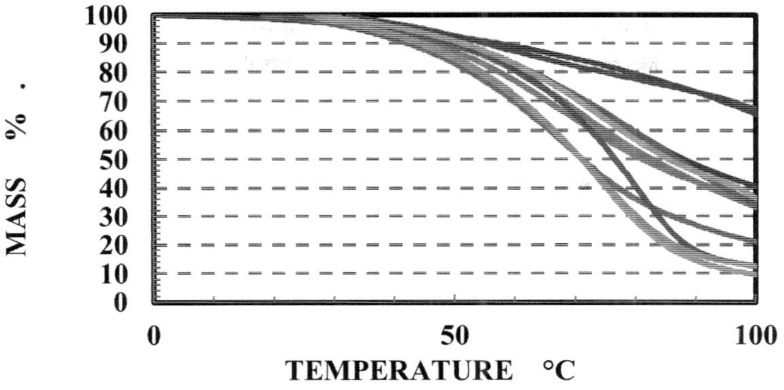

Figure 2. Materials tested until 100 °C.

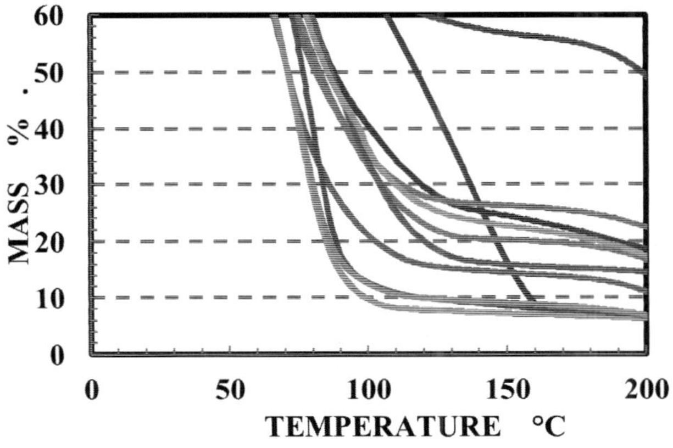

Figure 3. Materials behavior until 200 °C.

For temperature near 200 °C we show in figure 3, the behavior to the materials investigated. Again banana, in blue line, has the lower mass loss about 51 % it represent the 49 % of the original mass. The greatest loss is now 93 % for the poblano chile, in light green line and is about the same with melon, in light blue and nopal in coffee line.

The rest are between 73 and 90 % lost of the original mass, that is, its remains a mass a 27 and 10 % of the initial mass. The first is the guava, in light brown and the second in blue, the peach.

Finally what happens between 200 and 300 °C, is shown in Figure 4. The greatest loss corresponds to the poblano chile, light green, with a little over 96 %, followed by the green melon, blue color, with a little more than 95.8%, followed by Chinese melon, in purple, with 95.5 % Chinese melon, in purple, and the cactus in brown, with 95 %. Below is the peach in strong blue color, with 93 %, followed by the kiwi, light blue, with 92 % then there are almost together jicama and apple 90 %, very close to the grape, orange, with 89.2 % above the handle in purple, with 86.2% and finally guava, coffee, with 83.8 % of the original mass.

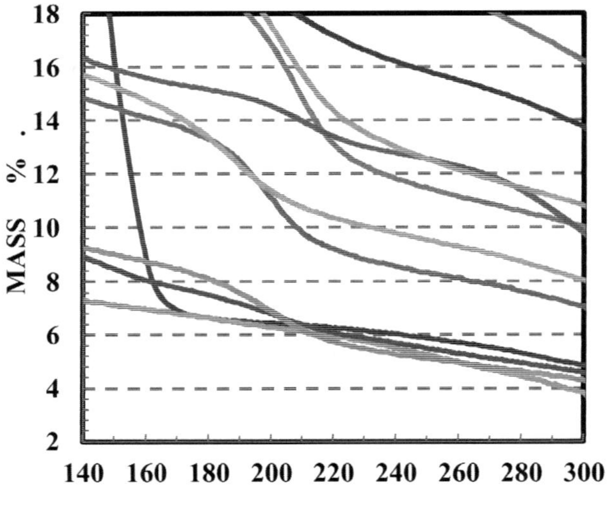

Figure 4. Materials investigated at 300 ° C.

3. Discussion

The values of lost mass obtain are in agreement with the literature values. But we couldn't found in literature is at what temperature this lost happens. And that is the contribution of this work. In general the composition of fruits and vegetables is water 80-90 %, carbohydrates 20-5 %, fiber 2 %, the vitamins, proteins, fat, minerals, scents and pigments less than 0.5 %. Banana is known as the mayor carbohydrate content with 20 %, melon with 5 % and the others 10 %. The carbohydrates are fructose, sucrose and glucose. The fiber is mainly pectin and hemicelluloses. The minerals are potassium, magnesium, iron and calcium. Vitamins: carotenes, vitamin C and B.

The fruits and vegetables lost first water at low temperature, 100-150 °C, then the water on the structure of all the others components 150-250 °C and finely near 300 °C, this structures collapse loosing materials in form of gases like CO or CO_2, and others.

4. Conclusions

The mass lost can be related to the temperature as follows: three types of curves were detected as the loss of water: as the banana with low mass loss (70 %); the second type, as the curve of the poblano chile with a great lost of water (98 %) and finally the third type of curve where the loss is intermediate but large (90 %) but at a slower rate, as are the curves of the other materials investigated.

According to evolution, ie., the number of stages in which meanly water is lost at low temperature, we found that occur in one, two or even three stages during dehydration or mass lost. The poblano chile and cactus have a one stage, while the grape, apple, peach and kiwi have two stages. Finally, jicama is the only one that was very slightly three stages. The mango and melons are not very clearly seen, but are estimated to be two steps. By the way, in the case of melons, green and chinese, it is difficult to find differences between them, only a slight changes by 170 °C is found.

Journal of Physics: Conference Series **582** (2015) 012065 doi:10.1088/1742-6596/582/1/012065

References

[1] Drying of prickly pear cactus cladodes (Opuntia ficus indica) in a forced convection tunnel, R. López, A. de Ita, M. Vaca, *Energy Conversion and Management,* 50 (2009) 2119 – 2126.

[2] Modelado de la cinética de secado del nopal (*opuntia ficus indica*) con energía solar, Raymundo López Callejas, Antonio de Ita, Mabel Vaca Mier, Arturo Lizardi Ramos, Juan Morales Gómez, Araceli Lara Valdivia, 5° Congreso Internacional de Ingeniería Física, UAM - Azcapotzalco, México, D. F., 17-21 de Mayo 2010.

[3] Analysis of the specific energy in microwave drying of pickle pear cactus cladodes, R. Lopez, A. de Ita, M. Vaca, A. Lizardi, 44th Annual Symposium of International Microwave Power Institute´s, Denver, Colorado, USA, Julio 14-16, 2010.

[4] Energy efficiency of drying of prickly cladodes in a forced convention tunnel, R. López, A. de Ita, M. Vaca, Proceedings 3th International Energy, energy and Environment Symposium, Évora, Portugal, 1-5 July 2007.

[5] Kinetics of the Drying of the Poblano Hot Pepper (*Capsicum annuum*), in Forced Convection, R. López, M. Vaca, A. de Ita, A. Lizardi, A. Lara y M. Beltrán, 16th International Drying Symposium (IDS 2008), 9-12 November 2008, Hyderabad, India.

[6] Análisis exergético del secado del nopal y modelado numérico de la cinética del proceso, Raymundo López Callejas, tesis Doctoral, Abril, 2010, UAM-Azcapotzalco, México.